Drug Delivery Technology Development in Canada

Drug Delivery Technology Development in Canada

Special Issue Editors

Kishor M. Wasan
Ildiko Badea

MDPI • Basel • Beijing • Wuhan • Barcelona • Belgrade

MDPI

Special Issue Editors
Kishor M. Wasan
University of Saskatchewan
Canada

Ildiko Badea
University of Saskatchewan
Canada

Editorial Office
MDPI
St. Alban-Anlage 66
4052 Basel, Switzerland

This is a reprint of articles from the Special Issue published online in the open access journal *Pharmaceutics* (ISSN 1999-4923) from 2018 to 2019 (available at: https://www.mdpi.com/journal/pharmaceutics/special_issues/drug_delivery_Canada).

For citation purposes, cite each article independently as indicated on the article page online and as indicated below:

LastName, A.A.; LastName, B.B.; LastName, C.C. Article Title. *Journal Name* **Year**, *Article Number*, Page Range.

ISBN 978-3-03928-004-9 (Pbk)
ISBN 978-3-03928-005-6 (PDF)

Cover image courtesy of pixabay.com.

Contents

About the Special Issue Editors vii

Kishor M. Wasan and Ildiko Badea
Drug Delivery Technology Development in Canada
Reprinted from: *Pharmaceutics* **2019**, *11*, 541, doi:10.3390/pharmaceutics11100541 1

Ellen K. Wasan, Jinying Zhao, Joshua Poteet, Munawar A. Mohammed, Jaweria Syeda, Kevin Soulsbury, Jacqueline Cawthray, Amanda Bunyamin, Chi Zhang, Brian M. Fahlman and Ed S. Krol
Development of a UV-Stabilized Topical Formulation of Nifedipine for the Treatment of Raynaud Phenomenon and Chilblains
Reprinted from: *Pharmaceutics* **2019**, *11*, 594, doi:10.3390/pharmaceutics11110594 5

Babu V. Sajesh, Ngoc H. On, Refaat Omar, Samaa Alrushaid, Brian M. Kopec, Wei-Guang Wang, Han-Dong Sun, Ryan Lillico, Ted M. Lakowski, Teruna J. Siahaan, Neal M. Davies, Pema-Tenzin Puno, Magimairajan Issai Vanan and Donald W. Miller
Validation of Cadherin HAV6 Peptide in the Transient Modulation of the Blood-Brain Barrier for the Treatment of Brain Tumors
Reprinted from: *Pharmaceutics* **2019**, *11*, 481, doi:10.3390/pharmaceutics11090481 23

Waleed Mohammed-Saeid, Abdalla H Karoyo, Ronald E Verrall, Lee D Wilson and Ildiko Badea
Inclusion Complexes of Melphalan with Gemini-Conjugated β-Cyclodextrin: Physicochemical Properties and Chemotherapeutic Efficacy in In-Vitro Tumor Models
Reprinted from: *Pharmaceutics* **2019**, *11*, 427, doi:10.3390/pharmaceutics11090427 39

David Fortin
Drug Delivery Technology to the CNS in the Treatment of Brain Tumors: The Sherbrooke Experience
Reprinted from: *Pharmaceutics* **2019**, *11*, 248, doi:10.3390/pharmaceutics11050248 54

Asmita Poudel, George Gachumi, Kishor M. Wasan, Zafer Dallal Bashi, Anas El-Aneed and Ildiko Badea
Development and Characterization of Liposomal Formulations Containing Phytosterols Extracted from Canola Oil Deodorizer Distillate along with Tocopherols as Food Additives
Reprinted from: *Pharmaceutics* **2019**, *11*, 185, doi:10.3390/pharmaceutics11040185 70

Jiahao Huang, Peter X. Chen, Michael A. Rogers and Shawn D. Wettig
Investigating the Phospholipid Effect on the Bioaccessibility of Rosmarinic Acid-Phospholipid Complex through a Dynamic Gastrointestinal in Vitro Model
Reprinted from: *Pharmaceutics* **2019**, *11*, 156, doi:10.3390/pharmaceutics11040156 86

Kevin J. H. Allen, Rubin Jiao, Mackenzie E. Malo, Connor Frank and Ekaterina Dadachova
Biodistribution of a Radiolabeled Antibody in Mice as an Approach to Evaluating Antibody Pharmacokinetics
Reprinted from: *Pharmaceutics* **2018**, *10*, 262, doi:10.3390/pharmaceutics10040262 104

Hoda Soleymani Abyaneh, Amir Hassan Soleimani, Mohammad Reza Vakili, Rania Soudy, Kamaljit Kaur, Francesco Cuda, Ali Tavassoli and Afsaneh Lavasanifar
Modulation of Hypoxia-Induced Chemoresistance to Polymeric Micellar Cisplatin: The Effect of Ligand Modification of Micellar Carrier Versus Inhibition of the Mediators of Drug Resistance
Reprinted from: *Pharmaceutics* **2018**, *10*, 196, doi:10.3390/pharmaceutics10040196 112

Zaid H. Maayah, Ti Zhang, Marcus Laird Forrest, Samaa Alrushaid, Michael R. Doschak, Neal M. Davies and Ayman O. S. El-Kadi
DOX-Vit D, a Novel Doxorubicin Delivery Approach, Inhibits Human Osteosarcoma Cell Proliferation by Inducing Apoptosis While Inhibiting Akt and mTOR Signaling Pathways
Reprinted from: *Pharmaceutics* **2018**, *10*, 144, doi:10.3390/pharmaceutics10030144 130

Griffin Pauli, Wei-Lun Tang and Shyh-Dar Li
Development and Characterization of the Solvent-Assisted Active Loading Technology (SALT) for Liposomal Loading of Poorly Water-Soluble Compounds
Reprinted from: *Pharmaceutics* **2019**, *11*, 465, doi:10.3390/pharmaceutics11090465 146

Farinaz Ketabat, Meenakshi Pundir, Fatemeh Mohabatpour, Liubov Lobanova, Sotirios Koutsopoulos, Lubomir Hadjiiski, Xiongbiao Chen, Petros Papagerakis and Silvana Papagerakis
Controlled Drug Delivery Systems for Oral Cancer Treatment—Current Status and Future Perspectives
Reprinted from: *Pharmaceutics* **2019**, *11*, 302, doi:10.3390/pharmaceutics11070302 158

Mahdi Roohnikan, Elise Laszlo, Samuel Babity and Davide Brambilla
A Snapshot of Transdermal and Topical Drug Delivery Research in Canada
Reprinted from: *Pharmaceutics* **2019**, *11*, 256, doi:10.3390/pharmaceutics11060256 187

Esen Sokullu, Hoda Soleymani Abyaneh and Marc A. Gauthier
Plant/Bacterial Virus-Based Drug Discovery, Drug Delivery, and Therapeutics
Reprinted from: *Pharmaceutics* **2019**, *11*, 211, doi:10.3390/pharmaceutics11050211 202

Bahman Homayun, Xueting Lin and Hyo-Jick Choi
Challenges and Recent Progress in Oral Drug Delivery Systems for Biopharmaceuticals
Reprinted from: *Pharmaceutics* **2019**, *11*, 129, doi:10.3390/pharmaceutics11030129 240

Courtney van Ballegooie, Alice Man, Mi Win and Donald T. Yapp
Spatially Specific Liposomal Cancer Therapy Triggered by Clinical External Sources of Energy
Reprinted from: *Pharmaceutics* **2019**, *11*, 125, doi:10.3390/pharmaceutics11030125 269

Ada W.Y. Leung, Carolyn Amador, Lin Chuan Wang, Urmi V. Mody and Marcel B. Bally
What Drives Innovation: The Canadian Touch on Liposomal Therapeutics
Reprinted from: *Pharmaceutics* **2019**, *11*, 124, doi:10.3390/pharmaceutics11030124 301

Grace Cuddihy, Ellen K. Wasan, Yunyun Di and Kishor M. Wasan
The Development of Oral Amphotericin B to Treat Systemic Fungal and Parasitic Infections: Has the Myth Been Finally Realized?
Reprinted from: *Pharmaceutics* **2019**, *11*, 99, doi:10.3390/pharmaceutics11030099 327

About the Special Issue Editors

Kishor M. Wasan was Dean of the College of Pharmacy and Nutrition at the University of Saskatchewan from August 2014 until he completed his 5 year term at the end of June 2019. He has published over 550 peer-reviewed articles and abstracts in the area of lipid-based drug delivery and lipoprotein-drug interactions. Dr. Wasan completed his undergraduate degree in Pharmacy at the University of Texas at Austin and his Ph.D. in Cellular and Molecular Pharmacology at MD Anderson, University of Texas Medical Center in Houston, Texas. After completing a postdoctoral fellowship in Cell Biology at the Cleveland Clinic, Dr. Wasan joined the Faculty of Pharmaceutical Sciences at the University of British Columbia in 2014. Dr. Wasan has been the recipient of numerous scientific awards, fellowships, and research chairs, including the American Association of Pharmaceutical Scientists New Investigator Award and the Canadian Institutes of Health Research University-Industry Research Chair, and was named a Fellow of the Canadian Academy of Health Sciences.

Ildiko Badea is a Professor of Pharmacy in the College of Pharmacy and Nutrition at the University of Saskatchewan. Dr. Badea completed her undergraduate degree in Romania and worked as Clinical Pharmacist before obtaining her PhD in pharmaceutical sciences at the University of Saskatchewan. After one year of postdoctoral fellowship at the Vaccine and Infectious Disease Organization, Canada in 2006–2007, she joined the College of Pharmacy and Nutrition at the University of Saskatchewan. Her area of research is drug delivery focusing on lipid-based and solid-core nanoparticle design for biotechnology drugs.

pharmaceutics

MDPI

Editorial

Drug Delivery Technology Development in Canada

Kishor M. Wasan [1,2,*] and Ildiko Badea [1,*]

1   College of Pharmacy and Nutrition, University of Saskatchewan, Saskatoon, SK S7N 2Z4, Canada
2   Faculty of Pharmaceutical Sciences, University of British Columbia, Vancouver, BC V6T 1Z3, Canada
*   Correspondence: kishor.wasan@usask.ca (K.M.W.); ildiko.badea@usask.ca (I.B.)

Received: 10 October 2019; Accepted: 10 October 2019; Published: 17 October 2019

Abstract: Canada has a long and rich history of ground-breaking research in drug delivery within academic institutions, pharmaceutical industry and the biotechnology community. Drug delivery refers to approaches, formulations, technologies, and systems for transporting a pharmaceutical compound in the body as needed to safely achieve its desired therapeutic effect. It may involve rational site-targeting, or facilitating systemic pharmacokinetics; in any case, it is typically concerned with both quantity and duration of the presence of the drug in the body. Drug delivery is often approached through a drug's chemical formulation, medical devices or drug-device combination products. Drug delivery is a concept heavily integrated with dosage form development and selection of route of administration; the latter sometimes even being considered part of the definition. Drug delivery technologies modify drug release profile, absorption, distribution and elimination for the benefit of improving product efficacy and safety, as well as patient convenience and adherence. Over the past 30 years, numerous Canadian-based biotechnology companies have been formed stemming from the inventions conceived and developed within academic institutions. Many have led to the development of important drug delivery products that have enhanced the landscape of drug therapy in the treatment of cancer to infectious diseases. This Special Issue serves to highlight the progress of drug delivery within Canada. We invited articles on all aspects of drug delivery sciences from pre-clinical formulation development to human clinical trials that bring to light the world-class research currently undertaken in Canada for this Special Issue.

Keywords: drug delivery; pharmaceutics; drug development; formulation and dosage form development; translational research; biologicals; small molecules; clinical trials; pharmacokinetics; medical devices; route of administration

---

This special issue in *Pharmaceutics*, entitled *"Drug Delivery Technology Development in Canada"* was put together to highlight the outstanding achievements and international impact of Canadian scientists in the field of drug delivery. For over 30 years Canadian scientists from leading Canadian research-intense academic institutions, pharmaceutical industry and the biotechnology community have played a vital role in the development of and implementation of novel drug delivery technologies that have made an impact on a number of diseases from cancer to infectious diseases.

Drug delivery encompasses a spectrum of approaches, formulations, technologies, and systems for carrying active pharmaceutical ingredients into the body. The main focus is to achieve optimal pharmacokinetic profile, often attained by active targeting. To achieve this goal, the drugs are formulated in chemical drug delivery systems, incorporated in devices or combination of these two strategies [1]. Drug delivery technologies modify drug release profile, absorption, distribution and elimination for the benefit of improving product efficacy and safety, as well as patient convenience and compliance [2].

This Special Issue on Drug Delivery Technologies in Canada highlights the progress of drug delivery research and development within Canada. We invited articles on all aspects of drug delivery sciences from pre-clinical formulation development to human clinical trials that bring to light the world-class research currently undertaken in Canada. In the next paragraphs we summarize the contributions to our special issue.

Babu V.Sajesh et al. [3] discusses the limitations faced by therapeutic agents to reach their target in the brain by crossing the blood-brain barrier by using HAV6, a cadherin binding peptide, the blood-brain barrier was opened transiently, leading to improvement of the delivery of a therapeutic agent in a murine brain tumour model. This proof-of-principle study is a novel avenue for drug delivery to the central nervous system.

David Fortin [4] in his paper entitled "Drug Delivery Technology to the CNS in the Treatment of Brain Tumors: The Sherbrooke Experience" also addresses challenges regarding drug delivery to the central nervous system and reviews strategies encompassing the path of the drug discovery from laboratory explorations to clinical applications.

Waleed Mohammed-Saeid et al. [5], in their article entitled " Inclusion Complexes of Melphalan with Gemini-Conjugated β-Cyclodextrin: Physicochemical Properties and Chemotherapeutic Efficacy in In-Vitro Tumor Models" report on how β-cyclodextrin (βCD) has been widely explored as an excipient for pharmaceuticals and nutraceuticals as it forms host–guest inclusion complexes and enhances the solubility of poorly soluble active agents.

Asmita Poudel et al. [6], in their paper entitled " Development and Characterization of Liposomal Formulations Containing Phytosterols Extracted from Canola Oil Deodorizer Distillate along with Tocopherols as Food Additives investigated formulation strategies for liposomes containing phytosterols obtained from canola oil deodorizer distillate, and tocopherols to overcome the challenges of thermo-sensitivity, lipophilicity and formulation-dependent efficacy of the nutraceuticals. The final aim is the development of functional foods, enriched with phytosterols and tocopherols.

Jiahao Huang and colleagues [7], investigated the effect of phospholipids on a model compound, rosmarinic acid, and established relationship between membrane permeability and bioavailability on a dynamic gastrointestinal in vitro model, providing evidence for the complex interplay of these factors influencing bioaccessibility.

Kevin Allen et al. [8] discuss highly reproducible method of determining its pharmacokinetics of antibodies for further pre-clinical development using 111-indium-labeled antibody in a melanoma tumour model, demonstrating superiority of this strategy compared to mass spectrometry.

Hoda Soleymani Abyaneh et al. [9], in their paper entitled "Modulation of Hypoxia-Induced Chemoresistance to Polymeric Micellar Cisplatin: The Effect of Ligand Modification of Micellar Carrier Versus Inhibition of the Mediators of Drug Resistance" assessed strategies to overcome hypoxia-induced chemoresistance in a triple negative breast cancer cell line. They demonstrated that pharmacological inhibition of hypoxia significantly enhances cytotoxicity ofcisplatin encapsulated in in polymeric micelles.

Zaid H Maayah et al. [10], reported that by chemically conjugating Vit-D to DOX the delivery of DOX into cancer cells increased and chemoresistance associated with DOX was mitigated via inhibition of survival pathways and induction of apoptosis.

Griffin Pauli et al. [11], discuss the advantages of solvent-assisted active loading technology (SALT) for liposomal encapsulation of compounds with low aqueous solubility. This new strategy is characterized by complete encapsulation, high loading efficiency and stable drug retention, leading to improvement of pharmacokinetic and pharmacodynamics parameters of the drugs.

Farinaz Ketabat et al. [12], review treatment options in development for oral squamous cell carcinoma from new delivery systems to chronotherapy, and offer insight into future strategies in the field.

Mahdi Roohnikan et al. [13], showcase research groups interested in the development of state-of-the-art transdermal delivery technologies. Within this short review, they aim to provide a critical overview of the development of these technologies in the Canadian environment.

Esen Sokullu et al. [14], present an overview of applications of plant viruses and phages in drug discovery. Critical assessment of the status of virus-based materials in clinical research are summarized.

The authors provide a critical assessment of challenges and opportunities presented by these highly stable and versatile delivery systems.

Bahman Homayun et al. [15], in their paper entitled "Challenges and Recent Progress in Oral Drug Delivery Systems for Biopharmaceuticals" outlines the advantages of oral drug delivery by reviewing the advantages and disadvantages different administration routes. Additionally mitigation strategies regarding challenges of each route are emphasized.

Courtney Van Ballegooie et al. [16], depict physical strategies aimed towards release of drugs from liposomal formulation at their target site. The mechanism of drug release upon the use of energy sources, including ultrasound, magnetic fields, and external beam radiationis explained.

Ada W.Y. Leung et al. [17], provides a high-level review the most successful Canadian drug delivery systems translated to the clinic, leading to the formation of biotech companies. From the creation of research tools (Lipex Extruder and NanoAssemblr™) todevelopment of pharmaceutical products (Abelcet®, MyoCet®, Marqibo®, Vyxeos®, and Onpattro™) positive impacts on patients' health are numerous. This review highlights the Canadian contribution to the development of these and other important liposomal technologies that have touched patients.

Grace Cuddihy et al. [18], in their paper entitled "The Development of Oral Amphotericin B to Treat Systemic Fungal and Parasitic Infections: Has the Myth Been Finally Realized?" discuss the development of an oral formulation of Amphotericin B to treat systemic fungal and parasitic infections.

Taken together, these articles published in our special issue represents only a fraction of the drug delivery research and development ongoing within Canada but do serve as examples of the outstanding contributions Canadian's have made to the discipline over the past 30 years.

Conflicts of Interest: The authors declare no conflict of interest.

References

1. Delcassian, D.; Patel, A.K.; Cortinas, A.B.; Langer, R. Drug delivery across length scales. *J. Drug Target.* **2019**, *27*, 229–243. [CrossRef] [PubMed]
2. Wen, H.; Jung, H.; Li, X. Drug Delivery Approaches in Addressing Clinical Pharmacology-Related Issues: Opportunities and Challenges. *AAPS J.* **2015**, *17*, 1327–1340. [CrossRef] [PubMed]
3. Sajesh, B.V.; On, N.H.; Omar, R.; Alrushaid, S.; Kopec, B.M.; Wang, W.-G.; Sun, H.-D.; Lillico, R.; Lakowski, T.M.; Siahaan, T.J.; et al. Validation of Cadherin HAV6 Peptide in the Transient Modulation of the Blood-Brain Barrier for the Treatment of Brain Tumors. *Pharmaceutics* **2019**, *11*, 481. [CrossRef] [PubMed]
4. Fortin, D. Drug Delivery Technology to the CNS in the Treatment of Brain Tumors: The Sherbrooke Experience. *Pharmaceutics* **2019**, *11*, 248. [CrossRef] [PubMed]
5. Mohammed-Saeid, W.; Karoyo, A.H.; Verrall, R.E.; Wilson, L.D.; Badea, I. Inclusion Complexes of Melphalan with Gemini-Conjugated β-Cyclodextrin: Physicochemical Properties and Chemotherapeutic Efficacy in In-Vitro Tumor Models. *Pharmaceutics* **2019**, *11*, 427. [CrossRef] [PubMed]
6. Poudel, A.; Gachumi, G.; Wasan, K.M.; Dallal Bashi, Z.; El-Aneed, A.; Badea, I. Development and Characterization of Liposomal Formulations Containing Phytosterols Extracted from Canola Oil Deodorizer Distillate along with Tocopherols as Food Additives. *Pharmaceutics* **2019**, *11*, 185. [CrossRef] [PubMed]
7. Huang, J.; Chen, P.X.; Rogers, M.A.; Wettig, S.D. Investigating the Phospholipid Effect on the Bioaccessibility of Rosmarinic Acid-Phospholipid Complex through a Dynamic Gastrointestinal in Vitro Model. *Pharmaceutics* **2019**, *11*, 156. [CrossRef] [PubMed]
8. Allen, K.J.H.; Jiao, R.; Malo, M.E.; Frank, C.; Dadachova, E. Biodistribution of a Radiolabeled Antibody in Mice as an Approach to Evaluating Antibody Pharmacokinetics. *Pharmaceutics* **2018**, *10*, 262. [CrossRef] [PubMed]

9. Soleymani Abyaneh, H.; Soleimani, A.H.; Vakili, M.R.; Soudy, R.; Kaur, K.; Cuda, F.; Tavassoli, A.; Lavasanifar, A. Modulation of Hypoxia-Induced Chemoresistance to Polymeric Micellar Cisplatin: The Effect of Ligand Modification of Micellar Carrier Versus Inhibition of the Mediators of Drug Resistance. *Pharmaceutics* **2018**, *10*, 196. [CrossRef] [PubMed]
10. Maayah, Z.H.; Zhang, T.; Forrest, M.L.; Alrushaid, S.; Doschak, M.R.; Davies, N.M.; El-Kadi, A.O.S. DOX-Vit D, a Novel Doxorubicin Delivery Approach, Inhibits Human Osteosarcoma Cell Proliferation by Inducing Apoptosis While Inhibiting Akt and mTOR Signaling Pathways. *Pharmaceutics* **2018**, *10*, 144. [CrossRef] [PubMed]
11. Pauli, G.; Tang, W.-L.; Li, S.-D. Development and Characterization of the Solvent-Assisted Active Loading Technology (SALT) for Liposomal Loading of Poorly Water-Soluble Compounds. *Pharmaceutics* **2019**, *11*, 465. [CrossRef] [PubMed]
12. Ketabat, F.; Pundir, M.; Mohabatpour, F.; Lobanova, L.; Koutsopoulos, S.; Hadjiiski, L.; Chen, X.; Papagerakis, P.; Papagerakis, S. Controlled Drug Delivery Systems for Oral Cancer Treatment—Current Status and Future Perspectives. *Pharmaceutics* **2019**, *11*, 302. [CrossRef] [PubMed]
13. Roohnikan, M.; Laszlo, E.; Babity, S.; Brambilla, D. A Snapshot of Transdermal and Topical Drug Delivery Research in Canada. *Pharmaceutics* **2019**, *11*, 256. [CrossRef] [PubMed]
14. Sokullu, E.; Soleymani Abyaneh, H.; Gauthier, M.A. Plant/Bacterial Virus-Based Drug Discovery, Drug Delivery, and Therapeutics. *Pharmaceutics* **2019**, *11*, 211. [CrossRef] [PubMed]
15. Homayun, B.; Lin, X.; Choi, H.-J. Challenges and Recent Progress in Oral Drug Delivery Systems for Biopharmaceuticals. *Pharmaceutics* **2019**, *11*, 129. [CrossRef] [PubMed]
16. Van Ballegooie, C.; Man, A.; Win, M.; Yapp, D.T. Spatially Specific Liposomal Cancer Therapy Triggered by Clinical External Sources of Energy. *Pharmaceutics* **2019**, *11*, 125. [CrossRef] [PubMed]
17. Leung, A.W.Y.; Amador, C.; Wang, L.C.; Mody, U.V.; Bally, M.B. What Drives Innovation: The Canadian Touch on Liposomal Therapeutics. *Pharmaceutics* **2019**, *11*, 124. [CrossRef] [PubMed]
18. Cuddihy, G.; Wasan, E.K.; Di, Y.; Wasan, K.M. The Development of Oral Amphotericin B to Treat Systemic Fungal and Parasitic Infections: Has the Myth Been Finally Realized? *Pharmaceutics* **2019**, *11*, 99. [CrossRef] [PubMed]

pharmaceutics

MDPI

Article

Development of a UV-Stabilized Topical Formulation of Nifedipine for the Treatment of Raynaud Phenomenon and Chilblains

Ellen K. Wasan [1,*], Jinying Zhao [2], Joshua Poteet [1], Munawar A. Mohammed [1], Jaweria Syeda [1], Tatiana Orlowski [1], Kevin Soulsbury [3], Jacqueline Cawthray [1], Amanda Bunyamin [1], Chi Zhang [1], Brian M. Fahlman [1] and Ed S. Krol [1]

[1]   College of Pharmacy and Nutrition, University of Saskatchewan, Saskatoon, SK S7N 5E5, Canada; joshua.poteet@usask.ca (J.P.); munawarali89@gmail.com (M.A.M.); jaweriasyeda@hotmail.com (J.S.); tmo380@mail.usask.ca (T.O.); jacqueline.cawthray@fedorukcentre.ca (J.C.); amb902@mail.usask.ca (A.B.); chz855@mail.usask.ca (C.Z.); Brian.Fahlman@gilead.com (B.M.F.); ed.krol@usask.ca (E.S.K.)
[2]   Faculty of Pharmaceutical Sciences, University of British Columbia, Vancouver, BC V6T 1Z3, Canada; jrzhao0810@gmail.com
[3]   British Columbia Institute of Technology, Burnaby, BC V5G 3H2, Canada; Kevin_Soulsbury@bcit.ca
*   Correspondence: ellen.wasan@usask.ca; Tel.: +1-306-966-3202

Received: 13 August 2019; Accepted: 22 October 2019; Published: 9 November 2019

Abstract: Raynaud's Phenomenon is a vascular affliction resulting in pain and blanching of the skin caused by excessive and prolonged constriction of arterioles, usually due to cold exposure. Nifedipine is a vasodilatory calcium channel antagonist, which is used orally as the first-line pharmacological treatment to reduce the incidence and severity of attacks when other interventions fail to alleviate the condition and there is danger of tissue injury. Oral administration of nifedipine, however, is associated with systemic adverse effects, and thus topical administration with nifedipine locally to the extremities would be advantageous. However, nifedipine is subject to rapid photodegradation, which is problematic for exposed skin such as the hands. The goal of this project was to analyze the photostability of a novel topical nifedipine cream to UVA light. The effect of incorporating the photoprotectants rutin, quercetin, and/or avobenzone (BMDBM) into the nifedipine cream on the stability of nifedipine to UVA light exposure and the appearance of degradation products of nifedipine was determined. Rutin and quercetin are flavonoids with antioxidant activity. Both have the potential to improve the photostability of nifedipine by a number of mechanisms that either quench the intermolecular electron transfer of the singlet excited dihydropyridine to the nitrobenzene group or by preventing photoexcitation of nifedipine. Rutin at either 0.1% or 0.5% (*w/w*) did not improve the stability of nifedipine 2% (*w/w*) in the cream after UVA exposure up to 3 h. Incorporation of quercetin at 0.5% (*w/w*) did improve nifedipine stability from 40% (no quercetin) to 77% (with quercetin) of original drug concentration after 3 h UVA exposure. A combination of BMDBM and quercetin was the most effective photoprotectant for maintaining nifedipine concentration following up to 8 h UVA exposure.

Keywords: nifedipine; emulsion; flavonoids; topical formulation; quercetin; photostabilizers

---

1. Introduction

Raynaud's Phenomenon (RP) is a vascular condition that causes temporary arteriolar vasospasm in cold-exposed hands and feet of affected persons, resulting in numb, ischemic digits. First, there is a characteristic blanching of the skin as circulation is reduced; secondly, the affected area turns a bluish color during resolution of the vasospasm caused by venous blood returning; and thirdly, redness as arteriolar flow resumes. Not only fingers and toes may be affected, but also the tip of the nose, pinnae of the ears, and the nipples. Rewarming is a painful process. For those seriously affected, RP adversely affects quality of life [1]. Thermoregulatory arteriovenous anastomoses, which are enervated by sympathetic nerves, are responsible for the phenomenon, rather than capillaries, which deliver normal circulation. Connective tissue disease, occupational exposure to vibration, and smoking are risk factors, but RP is typically a primary condition, affecting approximately 3–5% of the population or more, depending on climate [2]. Chilblains is a related cold-induced vascular disorder, resulting in papules causing pain and pruritis [3,4]. RP is not always a benign condition; in severe cases associated with scleroderma, rheumatoid arthritis, and other connective tissue diseases, diabetes, or with certain drug exposures, secondary RP can result in tissue damage due to repeated and prolonged ischemia, requiring medical intervention [5]. When adaptive measures to avoid cold exposure are not effective and pharmacological treatment is required to reduce the impact of severe RP, or chilblains, oral calcium channel blockers are the first-line medications, particularly nifedipine, a dihydropyridine compound [6,7]. Alternatives for severe disease include sildenafil and intravenous prostaglandin analogues [8]. Dihydropyridines bind to L-type $Ca_V1.2$ calcium channels [9], and in so doing, effect smooth muscle relaxation including vasodilation of arterioles, the therapeutic target in this case. Other drugs in this pharmacological class include diltiazem, nicardipine, felodipine, amlodipine, and related analogues. Nifedipine has more vascular than cardiac effects [10] and has been demonstrated to have moderate efficacy in the treatment of RP and chilblains [6,11]. Daily oral therapy with nifedipine is not always well-tolerated, however, due to systemic side effects such as dizziness and flushing.

Currently, there is no effective nifedipine topical product marketed for acute RP treatment or prevention of symptoms. Topical application of nifedipine would be advantageous as it would provide a rapid effect on the local tissue while limiting systemic exposure. It is expected that topical nifedipine would be extremely useful for reducing the risk of tissue damage in patients with scleroderma, rheumatoid arthritis, systemic lupus erythematosus, and Sjögen's syndrome, as a part of combination pharmacological therapy for RP and for those who have outdoor occupations with cold exposure. Furthermore, it is anticipated that topical nifedipine, or topical preparations of other calcium channel blockers or vasodilators, will have utility in the future to augment wound healing [12–14] and peripheral vascular insufficiency-related conditions, with a potential role in diabetic ulcer treatment [15–17].

Extemporaneously compounded topical nifedipine has been described, but it has inconsistent efficacy; nifedipine is not stable due to the well-known ultraviolet (UV)-light sensitivity of the drug [18]. Exposure of nifedipine to UVA light (315–400 nm), which accounts for 95% of the UV radiation that reaches the earth's surface, results in the photodegradation of nifedipine to dehydronifedipine, which can undergo further degradation to form dehydronitrosonifedipine [19–23], both of which are inactive compounds (Figure 1). This degradation process is rapid, it is not sensitive to the presence of oxygen, and it is mainly attributed to UVA irradiation [24–26]. One solution to this problem would be to incorporate appropriate photostabilizers; that is, compounds that filter UV energy by absorbing a certain range of high-energy UV wavelengths and releasing the energy at a lower range. We hypothesized that incorporating UV blockers into topical nifedipine formulations would prevent UV-induced decomposition of nifedipine. We describe here a preparation of 2% nifedipine in an oil-in-water emulsion formulation containing photostabilizers that preserves nifedipine from UVA-induced photodegration.

Photostabilization of light-sensitive medications in topical emulsion formulations is not isolated to nifedipine, as a recent analysis of topical products in the United States Pharmacopoeia and the European medicines databases indicated that up to 28% of approved drugs have the recommendation to protect the product from light [27] and the list of new drugs with this recommendation continues to grow [28]. Thus, there is a need for the development of compatible UV blockers for topical formulations. Since topical medications are applied to external body surfaces, they have the potential for significant light exposure. Typically, these preparations are applied as a thin film, which maximizes the surface area of the formulation to UV and visible radiation. In addition to UV or visible light inactivation of topical drug products, other photodegradation products can display toxicities or other unknown effects [29]. Furthermore, light exposure may also influence the physical and technical performance of a topical formulation, such as changes in viscosity, precipitation of components, changes in emulsion droplet size affecting stability, and changes in chemical degradation of materials [27]. Photostabilizers may also serve a role to maintain performance integrity of the topical formulation.

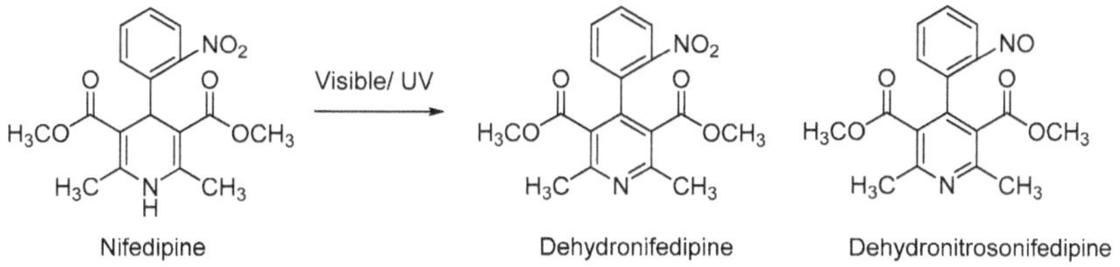

Figure 1. Ultraviolet (UV) radiation-mediated breakdown of nifedipine.

There are several common photostabilizers that could be appropriate for use in a topical nifedipine formulation including butyl methoxydibenzoylmethane, BMDBM, (an approved sunscreen agent also known as avobenzone) [30,31], and octocrylene, an approved photostabilizer sometimes used in combination with BMDBM in sunscreen products [32]. We have recently been exploring the UV blockers rutin and quercetin, polyphenolic compounds that are found to be upregulated by UV stress in a variety of plant sources [33,34], with known antioxidant and UV-protecting properties [35–39]. Both rutin and quercetin can act as photostabilizers via a number of mechanisms, including preventing photooxidation or inhibiting radical formation, both steps involved in the photodegradation of nifedipine. Additionally, these flavonoids and BMDBM (chemical structures are illustrated in Figure 2) are all characterized by regions of broad absorption that overlap with the absorption of nifedipine, and quercetin has been demonstrated to enhance the photostability of BMDBM in vitro, suggesting that both quercetin and rutin may be suitable photostabilizers [40]. All three can then prevent photodegradation of nifedipine through competitive absorption of photons, thus preventing or minimizing the generation of the first excited state of nifedipine.

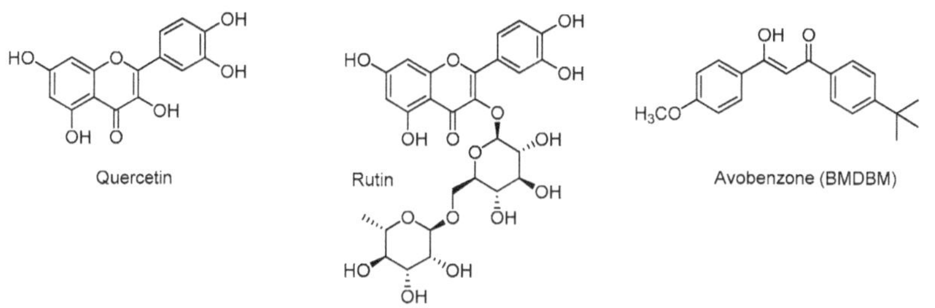

Figure 2. Photostabilizers under investigation in this study.

Extemporaneously compounded topical nifedipine has been observed to undergo UV-induced decomposition during preparation and storage, contributing to the inactivation and inconsistency of these formulations [18,41]. Nifedipine is not water soluble, which presents certain limitations to the pharmacist such as having to use hydrophobic cream bases or to perform relatively complex compounding procedures. The hydrophobic nature of nifedipine, however, makes the use of an oil-in-water (O/W) emulsion an attractive approach. An added theoretical advantage is the solubility of the photostabilizer compounds in the oil phase of the O/W emulsion, where nifedipine is also solubilized and thereby co-localizing protectant and drug, which may be important for optimal photostabilization. It is important to note that some photostabilizers degrade unless used in combination with other UV blockers. BMDBM has been noted to have sensitivity to UVA irradiation, undergoing photoisomerization to the inactive diketone in non-polar solvents. BMDBM decomposes in aqueous solution, but remains stable in polar solvents [42] and in mineral oil or isopropyl myristate [43]. In order to minimize BMDBM degradation under broad spectrum UV light [UVA plus UVB (280–315 nm)], it is usually used in combination with a UVB blocker or a broad spectrum agent such as octocrylene [32]. We hypothesize that the flavonols quercetin and rutin, through antioxidant and UV absorption properties, will stabilize BMDBM and in turn stabilize nifedipine in our formulation [44,45]. In this report, we have compared quercetin + BMDBM vs. octocrylene + BMDBM on maintaining both BMDBM and nifedipine stability to UVA and UVB light.

2. Materials and Methods

2.1. Chemicals

Glyceryl monostearate was purchased from Spectrum Industries (Gardena, CA USA). Stearic acid and glycerin were from BASF (Ludwigshafen, Germany). Liquid paraffin, rutin (>94%), quercetin (>95%), and white petrolatum were bought from Sigma-Aldrich (St. Louis, MO USA), and mixed tocopherols from Lotioncrafter.com(Eastsound, WA USA). Sodium lauryl sulphate was from BioRad(Mississauga, ON Canada). Nifedipine (>98%) was from Alpha Aesar (Ward Hill, MA USA). Butyl methoxydibenzoylmethane (BMDBM) was purchased from Tokyo Chemical Industries(Tokyo, Japan). Diethylene glycol monoethyl ether (Transcutol P®) was a gift from Gattefossé (Saint-Priest, France). Water was purified by reverse osmosis (MilliQ systemFisher Scientific, Ottowa, ON Canada). Analytical references standards of nifedipine, octocrylene and dehydronitrosonifedipine were from Sigma-Aldrich(St. Louis, MO USA) (99% purity).

2.2. Preparation of Topical Nifedipine

Topical nifedipine was prepared as an oil-in-water emulsion using the beaker method [46]. In general, with this method, the water soluble and oil soluble components are separately dissolved and heated, followed by addition of the water phase to the oil phase with continuous mixing for formation of an emulsion, followed by cooling to solidify the cream. In this case, nifedipine was incorporated into the internal oil phase of the emulsion. All excipients in the formula including glyceryl monostearate, stearic acid, liquid paraffin, petrolatum, diethylene glycol monoethyl ether, glycerin, and sodium lauryl sulfate were used within approved US FDA inactive ingredient levels. The photostabilizer BMDBM was used within the US FDA approved usage level [47]. Flavonoids, rutin, and quercetin were included in the formulation to investigate their potential as UV blockers to facilitate photostabilization of nifedipine in the cream.

Where indicated, when quercetin, BMDBM, or rutin were incorporated into the cream, they also went into the oil phase of the emulsion. Work was conducted under yellow light (577–597 nm), which does not cause photodegradation of the compounds of interest. For the oil phase, glyceryl monostearate (6.7% *w/w* of final preparation), stearic acid (9.5% *w/w*), liquid paraffin (9.5% *w/w*), petrolatum (9.5% *w/w*), and Transcutol P (2% *w/w*) were weighed into a 250 mL beaker and warmed in a water bath on a hotplate to 85 °C with stirring until homogeneous, followed by addition of the nifedipine (2% *w/w*). Where indicated, the following additives were included in the oil phase: quercetin (0.5–2% *w/w*), rutin (0.5–2% *w/w*), and/or BMDBM (0.5–2%). For the water phase, Milli-Q purified water (q.s.), glycerin (13.4% *w/w*) and sodium lauryl sulfate (0.95% *w/w*) were warmed in a beaker to 85 °C using a water bath with stirring. The water phase was added slowly to the warmed oil phase with continuous stirring, and within a few minutes, emulsion formation was noted by a visual change to opacity as well as a sudden increase in viscosity. The emulsion in the water bath was removed from heat and stirred continuously at room temperature until reaching 40 °C, followed by homogenization (Virtex23 homogenizer, The Virtex Co., Gardiner, NY USA) for 5 min, then allowed to cool completely at ambient temperature (18–21 °C). Prepared creams were protected from light and stored at 4 °C.

2.3. Light Exposure

Photostability tests were conducted in a manner consistent with ICH photostability testing guidelines [48], although conducted in an academic laboratory. Two F20T12/BL/HO UVA lamps (National Biological Corp., Beachwood, OH, USA) filtered to remove UVC with an intensity of 740–750 μW·cm^{-2} at 365 nm as measured with a UVP UVX-36 sensor (Ultraviolet Products Ltd., Upland, CA, USA) were used for irradiation [35]. This level of flux is roughly equivalent to a bright sunny day in mid-summer. Although the lamps used in these studies conform with ICH guidelines, which focus on product/packaging stability (https://www.ich.org/fileadmin/Public_Web_Site/ICH_Products/Guidelines/Quality/Q1B/Step4/Q1B_Guideline.pdf), the focus of our study was on UV-mediated degradation of a topically applied substance with the result that a flux level previously used to mimic topical stability was chosen. The lamp apparatus was placed inside an enclosure with an access door, to prevent ambient light entering and for worker safety. For samples exposed to "ambient light", these were placed on a laboratory bench where standard fluorescent lighting was used. For UV exposure studies, sample handling was performed under incandescent yellow light to prevent unintentional photodegradation of nifedipine. Nifedipine 20 mg cream samples were spread evenly across the surface of a microscope coverslip to create a thin film. To prevent drying, the samples were covered with Saran Wrap®, a plastic film that was determined to be UVA-transparent (data not shown). After the allotted exposure time, the cream was scraped off the slide for extraction with methanol. Samples that were exposed while in solution and not incorporated in a cream were dissolved in methanol as a 20 mL solution in a 100 mL beaker.

2.4. Extraction of Nifedipine from the Cream

Solvent extractions were performed under yellow light. The sample was warmed in a water bath to 85 °C to melt lipids, followed by addition of 5 mL of methanol, and vortex mixing. The samples were centrifuged at 10,000 rpm for 5 min, and the supernatant retained for analysis by UV spectrophotometry. The extraction efficiency (EE) was 90%, defined as: EE = $(N_{ex}/N_o) \times 100\%$, where N_{ex} is the concentration of nifedipine recovered in the extract and N_o is the concentration of nifedipine in the original sample, based on the measured amount of nifedipine added.

2.5. Spectrophotometric Assay

Nifedipine concentrations were measured in methanol on a UV spectrophotometer (Unico SQ-2800) at 348 nm. The linear range was 5–100 μg/mL ($r^2 > 0.999$). Values reported represent mean ± SD for triplicate measurements.

2.6. Stability Studies

Nifedipine cream was prepared in replicates of 50 g batches and stored at ambient (21 °C) or refrigerated (4 °C) temperatures, protected from light. At the indicated timepoints, triplicate samples of 1 g were removed and extracted as described above, followed by HPLC analysis.

2.7. HPLC Assay

Nifedipine was quantified by reverse-phase HPLC at ambient temperature (23 °C) by an isocratic method on a Waters 2690 instrument equipped with a photodiode array detector (Waters 996, Waters Canada, Mississauga, ON Canada). The column was a C18 5 μm 4.6 × 150 mm (Phenomenex, Torrance, CA USA) and the mobile phase consisted of acetonitrile: Sodium acetate (1 mM, pH 5.3 adjusted with HCl) (70:30 v/v), generating a retention time of 2.9 min for nifedipine (λ = 348 nm; linear range 10–100 μg/mL, $r^2 > 0.99$).

2.8. Emulsion Phase Stability Analysis

A high-end Dispersion Analyzer [LUMiSizer® (LUM Corp., Boulder, CO USA)], which is a multi-sample, temperature-controlled analytical photocentrifuge with dedicated software, was used to predict long-term stability and optimization of nifedipine emulsions by means of creaming rate. This allows for an approximation of the relative stability to phase separation of emulsions that differ in the type or percentage of photoprotectant and for confirming batch-to-batch consistency in stability against phase separation. For each sample type, triplicate samples of 2 mL were loaded into photocentrifuge acrylic cuvettes. The samples were centrifuged at 42 °C × 12 h. Phase separation was detected as an increase in light transmission at the top of the sample, which is interpreted as an "instability index" reflecting the rate of change of light transmission.

2.9. High-Performance Liquid Chromatography–Photodiode Array (HPLC-PDA) For Ketoprofen Sparing

HPLC-PDA analysis was carried out at room temperature using either a Waters 2695 separation module equipped with a Waters 2996 photodiode array detector (Waters, Milford, MA, USA), or an Agilent Series 1200 quaternary pump (G1311A) with online degasser (G1322A), autosampler (G1329A), and photodiode array detector (G1315D) (Agilent Technologies, Mississauga, ON, Canada). Aliquots were injected onto a 250 × 4.6 mm Allsphere ODS-2 column, 5 μm particle size (Alltech, Calgary AB, Canada). Data were processed using Empower software (Waters, Milford, MA, USA) or Chemstation software (Agilent Technologies, Mississauga, ON, Canada). Elution was carried out in gradient mode using two components: A = 1% formic acid in water, B = 1% formic acid in methanol (flow rate 1 mL/min). The gradient for the UVA experiments was as follows: 5 to 15 min linear gradient from 90% A to 10% A; 15 to 19 min, isocratic 10% A; 19 to 22 min, linear gradient from 10% A to 90% A; 22 to 25 min, isocratic 90% A.

2.10. Time Course for UV Irradiation of Rutin and Ketoprofen

For photostability analysis, a 20 mL solution of rutin (50 μM) or ketoprofen (250 μM) in methanol was placed in 50 mL quartz cells fitted with a septa. The quartz cells were exposed to UVA (740 μW·cm^{-2} at 365 nm). For the methanol experiments, aliquots of 100 μL were taken in duplicate at each time point and injected directly on the HPLC. The time-course samples were compared to the 0 h time point to determine the amount of compound remaining by measuring peak area at the λ_{max} for each compound. Experiments were performed on at least 3 separate occasions.

2.11. Ketoprofen Sparing

A 20 mL solution of flavonol (50 μM) or ketoprofen (250 μM) in methanol was placed in 50 mL quartz cells fitted with a septa. The quartz cells were exposed to UVA (740 μW·cm^{-2} at 365 nm). Aliquots of 100 μL were taken in duplicate at each time point and injected directly on the HPLC. The time-course samples were compared to the 0 h time point to determine the amount of ketoprofen and flavonol remaining by measuring peak area at the λ_{max} for ketoprofen and each flavonol. Experiments for each flavonol were performed on at least 3 separate occasions.

2.12. Mass Spectrometry Analysis of Nifedipine and its Photo-degradants

The high-performance liquid chromatography (HPLC) MS/MS system consists of an Agilent series 1200 quaternary pump with an online degasser, auto sampler set to 4 °C, and DAD detector scanning between 190 to 400 nm (Agilent Technologies, Mississauga, ON, Canada) coupled to an AB Sciex API 4000 QTRAP mass spectrometer. Photodegradants of nifedipine were identified following direct infusion of 2.5 ng by observation of the appearance of the protonated and unprotonated product ions for dehydronifedipine [M]$^{1+}$ to [M$^-$ C$_{17}$H$_{16}$N$_2$O$_6$]$^+$ (*m/z* 345 and 344) and for dehydronitrosonifedipine [M]$^{1+}$ to [M$^-$ C$_{17}$H$_{16}$N$_2$O$_5$]$^+$ (*m/z* 329 and 328); peak areas were integrated by Analyst Software v1.6 (SCIEX, Redwood City, CA, USA) [43].

2.13. Statistical Analysis

Descriptive statistics were generated in Microsoft Excel (Office 2016). Comparison of means was analyzed by ANOVA with Tukey's post-hoc test (Astatsa, 2016).

3. Results and Discussion

The goal of this study was to assess the ability of two polyphenolic flavonols, quercetin and its 3'-rutinoside analog rutin, for their ability to attenuate UVA radiation-mediated decomposition of BMDBM and nifedipine in a topical formulation for the treatment of Raynaud Phenomenon. In order to accomplish these goals, we developed oil-in-water (O/W) emulsions containing mixtures of nifedipine, BMDBM, and either quercetin or rutin. We then exposed the nifedipine-containing emulsions to UVA radiation and employed UV spectroscopy, HPLC, and mass spectrometry to assess the ability of quercetin and rutin to act as photostabilizers.

3.1. UVA and Visible Decomposition of Nifedipine

Nifedipine in Methanol

The absorption spectrum of nifedipine (Figure 3) is characterized by strong absorbance at 240 nm and a broad absorption peak near 350 nm. UV spectroscopy clearly shows the change in the absorption spectrum of nifedipine in methanol after exposure to UVA radiation for 2 h with a decrease in absorbance at both 240 and 350 nm and new absorption maxima appearing at 280 and 310 nm. These are consistent with the results of Fasani et al. [49] who observed similar spectral changes for nifedipine in ethanol following UV irradiation and others who have noted rapid degradation of nifedipine in methanol solution exposed to laboratory light [50].

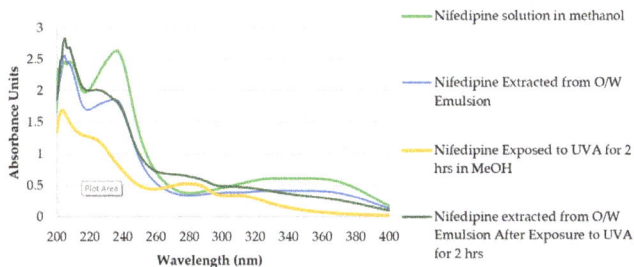

Figure 3. Ultraviolet (UV) absorption spectra of nifedipine as: Nifedipine solution 40 μg/mL in methanol; methanol extract of nifedipine cream (2% (*w/w*) as O/W emulsion); nifedipine 40 μg/mL solution in methanol after 2 h exposure to UVA light at a flux of 750 μW/cm^2; methanol extract of nifedipine cream (2% (*w/w*) as O/W emulsion) that was exposed for 2 h to UVA light at a flux of 750 μW/cm^2 prior to extraction.

3.2. Characterization of Emulsion

We chose to prepare a 2% (*w/w*) oil in water (O/W) emulsion as our topical delivery vehicle because nifedipine is readily incorporated into the internal oil phase, while a non-greasy feel is still achieved. This is advantageous for patient acceptability, as an oily topical preparation is not desirable for use on the hands and feet. In addition to the photostabilizers, the prototype nifedipine formulations contain the approved topical drug penetration enhancer diethylene glycol monoethyl ether (Transcutol HP®). Transcutol HP® is used in these formulations because of its established safety record [51,52], regulatory approval for human use, and ease of incorporation into the emulsions. Nifedipine stability in the cream prepared as 1% or 2% (*w/w*) nifedipine with or without Transcutol HP penetration enhancer (1% or 2% *w/w*) was determined under light-protected conditions at 23 °C and found to be maintained at >95% of original concentration for at least one month (Figure 4).

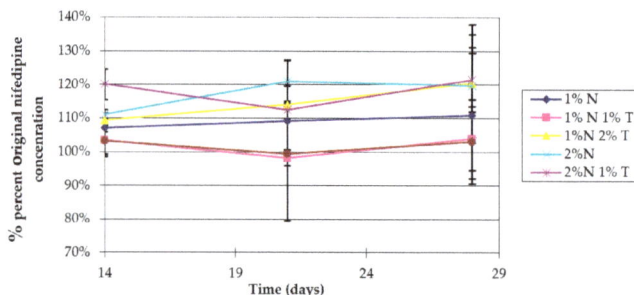

Figure 4. Stability of nifedipine (N) 1% or 2% (*w/w*) cream prepared with or without Transcutol HP (T) (1% or 2% (*w/w*)). The cream was stored protected from light at ambient temperature (23 °C). At 14, 21, and 28 days, nifedipine concentration was measured by UV spectrophotometry and reported as percent of original concentration. Data represent mean ± SD (*n* = 3).

A methanol extract of nifedipine cream (2% (*w/w*) as O/W emulsion) shows a similar spectrum as observed when dissolved in methanol. Upon exposure of a thin film of the O/W emulsion to indoor fluorescent light (ambient laboratory light), nifedipine concentration began to decline at 20 min, and was reduced to 75% of its original concentration by 1 h (Figure 5A). Formation of degradation products was determined by the appearance of a new absorbance maximum at 280 nm, consistent with formation of the aromatic groups in the degradation products and HPLC and mass spectrometric analysis. We compared APCI/ESI (+) mass spectrometry results (discussed below) with literature values to confirm the identity of the decomposition products (dihydronifedipine: m/z 345.10, $[M+H]^+$ and dehydronitrosonifedipine: m/z 329.11, $[M+H]^+$) [53]. Dehydronitrosonifedipine (DHN), the major degradation product, began to appear by 1 h of exposure to visible light. UVA exposure (750 μW/cm^2) of a thin film of the O/W emulsion over 2 h also resulted in loss of nifedipine (Figure 5B), although minimal degradation occurred in the first 45 min of exposure. After 2 h, 72 ± 7.86% of the nifedipine remained, suggesting that the O/W emulsion imparts some, albeit incomplete, photoprotection of nifedipine to UVA radiation.

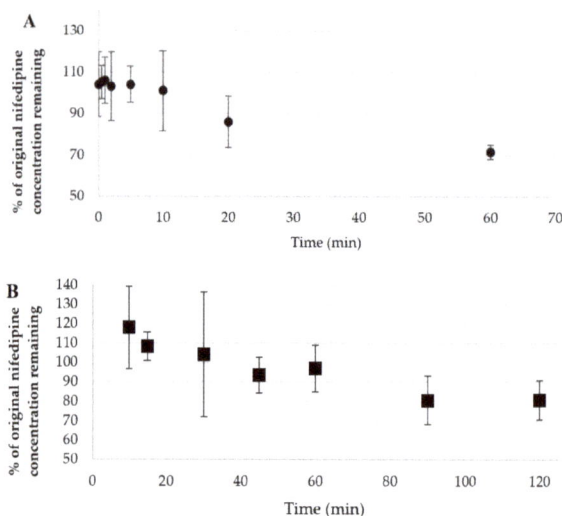

Figure 5. Nifedipine (2% *w/w*) cream exposed as a thin film to (**A**) ambient light over 1 h; (**B**) UVA light over 2 h. Data represent concentration vs. time (mean ± SD (*n* = 3)).

3.3. Characterization of Topical Nifedipine Cream

3.3.1. Influence of Butyl methoxydibenzoylmethane (BMDBM), Quercetin, and Rutin on Nifedipine UV Stability

In an effort to further improve nifedipine stability to light, we assessed the ability of three photostabilizing agents, rutin, BMDBM, and quercetin, either alone or in combination in our O/W emulsions. The photostabilizers were incorporated at varying concentrations up to 3% *w/w* in the emulsion and then treated with UVA radiation. Unfortunately, none of the photostabilizers when used on their own were effective in preventing UVA radiation-mediated decomposition of nifedipine when exposed as a thin film. Incorporation of rutin at concentrations of 0.5–1% (*w/w*) showed insufficient protection in reducing UVA-induced degradation of nifedipine over 3 h (Figure 6). No further studies were done using rutin as an additive.

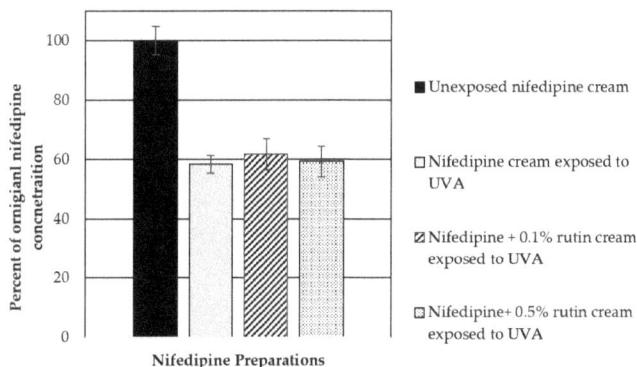

Figure 6. Rutin (0.1% and 0.5% (*w/w*)) was incorporated into the nifedipine cream and evaluated for its ability to decrease nifedipine degradation in the cream due to UVA exposure over 3 h. Data represent mean ± SD (*n* = 3).

Similarly, quercetin or rutin at 0.5% (*w/w*) or BMDBM alone were not effective as single agents in protecting nifedipine from degradation. A combination of quercetin at 0.5% (*w/w*) and BMDBM 3% (*w/w*), however, provided the best protection from UVA radiation-mediated decomposition in terms of the original nifedipine concentration maintained after 8 h of UVA exposure (Figure 7).

Figure 8 illustrates the degradation profile for each combination, demonstrating a two-phase process where the rate of degradation for the first two hours is greater than the rate for the period of 2–8 h. Table 1 lists the first (0–2 h) and second (2–8 h) phase rates, which suggest that the photoprotectants have the most significant effect on reducing the rate of degradation during the first two hours where the rate k is defined as:

$$(\%N_{t1} - \%N_{t2})/(t_2 - t_1)$$

where %N is percentage of original unexposed nifedipine concentration in the indicated O/W cream formulation and t_1 and t_2 represent exposure timepoints. Thus, in Table 1, k_1 represents the rate of degradation in the first 2 h, and k_2 the rate for 2–8 h of continuing exposure.

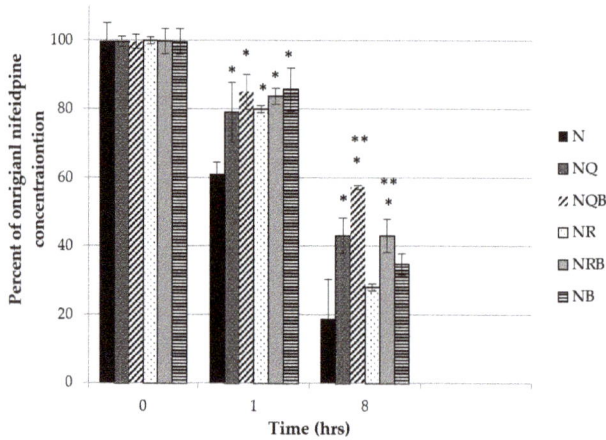

Figure 7. Quercetin or rutin (0.5% (*w*/*w*)) ± avobenzone (BMDBM) (3% (*w*/*w*)) were incorporated into the nifedipine 2% (*w*/*w*) cream as photoprotectant(s). The cream was exposed as a thin film to UVA light. Data represent percentage of original nifedipine concentration vs. exposure time (mean ± SD, *n* = 3). N: Nifedipine only; NQ: Nifedipine with quercetin; NQB: Nifedipine with quercetin and BMDBM; NR: Nifedipine with rutin; NRB: Nifedipine with rutin and BMBDM; NB: Nifedipine with BMDBM. *Significantly different from N (*p* < 0.01); **significantly different from its counterpart without BMDBM (*p* < 0.05).

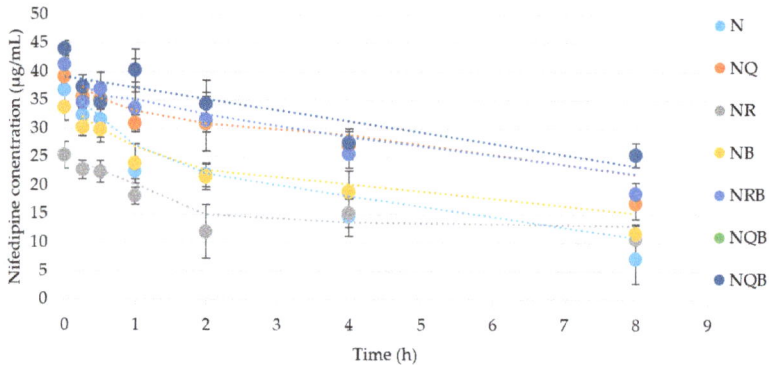

Figure 8. Quercetin or rutin (0.5% (*w*/*w*)) ± BMDBM (3% (*w*/*w*)) were incorporated into the nifedipine 2% (*w*/*w*) cream as photoprotectant(s). The cream was exposed as a thin film to UVA light. Data represent percentage of original nifedipine concentration vs. exposure time (mean ± SD, *n* = 3). N: Nifedipine only; NQ: Nifedipine with quercetin; NQB: Nifedipine with quercetin and BMDBM; NR: Nifedipine with rutin; NRB: Nifedipine with rutin and BMBDM; NB: Nifedipine with BMDBM.

Again, the combination of quercetin with BMDBM is shown to be the most effective at reducing the rate of photodegradation of nifedipine in the emulsion.

Octocrylene (2% *w*/*w*) was also tested as a potential photoprotectant for nifedipine topical emulsion, prepared in the same way as described above, but it did not prevent nifedipine degradation under UVA exposure (Figure 9). Octocrylene (2% *w*/*w*) with BMDBM (3% *w*/*w*) in combination in the nifedipine cream was not physically stable in the O/W emulsion and was not further optimized.

Table 1. Rate of nifedipine degradation vs. time in O/W formulations containing photoprotectants as described in Figure 8. *Significantly different from formulation N ($p < 0.5$) within that phase. #significantly different from formulation NB ($p < 0.5$) within that phase.

FORMU-LATIONS	Phase 1 0–2 h k_1 (%/h)	Phase 2 2–8 h k_2 (%/h)
N	14.3 ± 0.8	2.18 ± 0.7
NQ	8.21 ± 2.6* #	2.32 ± 0.49
NR	8.59 ± 2.87	3.5 ± 0.52
NB	7.86 ± 5.1	4.6 ± 0.46 *
NRB	5.26 ± 1.6 *	2.1 ± 0.26
NQB	3.7 ± 2.14 * #	1.4 ± 0.72*

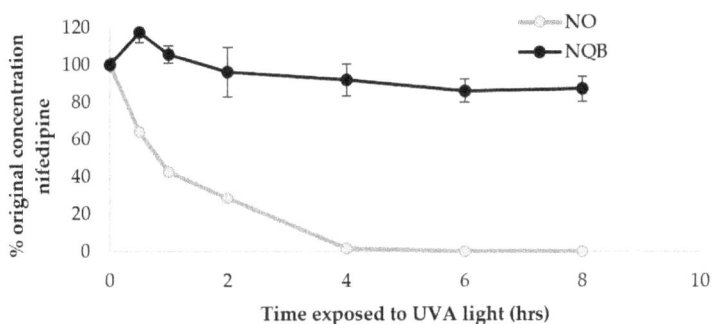

Figure 9. Nifedipine cream was prepared with quercetin 0.5% *(w/w)* + BMDBM 3% *(w/w)* (NQB) or with octylcrylene 2% *(w/w)* (NO) as photoprotectants. The cream was exposed as a thin film to UVA light. Data represent percentage of original nifedipine concentration vs. exposure time (mean $\pm$ SD, $n = 3$).

We used mass spectrometry to determine if the same nifedipine degradation products are formed following UV exposure when one or more of the photostabilizers (rutin, quercetin, BMDBM) is present. Dehydronifedipine and dehydronitrosonifedipine were found in UVA-exposed creams as expected [22], with no alternative degradation pathways identified in the presence of these photostabilizers based on appearance of *m/z* consistent with their expected profiles. A peak for the parent nifedipine (*m/z* of 347 [M]+ and *m/z* of 369.3 [M + Na$^+$]$^+$) can be seen in all panels of Figure 10. The mass spectrum of nifedipine is shown in supplementary Figure S2. The appearance of *m/z* = 329 is consistent with dehydronitrosonifedipine formation and was previously identified as the principal degradation product by HPLC. The mass spectrum of pharmaceutical reference standard dehydronitrosonifedipine is shown in supplementary Figure S3. The nifedipine was not protected from photodegradation by rutin, and the product *m/z* = 329 again made an appearance. Further investigation of the various photoprotectants indicated no unexpected fragment ion formation (data not shown) to indicate any alternate degradation pathways in the presence of rutin, quercetin, or BMDBM. The three panels of Figure 10 shows differential appearance of presumed photodegradation product *m/z* = 329 comparing (A) nifedipine 2% cream containing 0.5% quercetin; (B) nifedipine 2% cream containing 0.5% quercetin that was exposed to UVA × 2 h at 450 μW/cm^2; (C) nifedipine 2% cream containing 0.5% quercetin and BMDBM 3% that was exposed to UVA × 2 h at 450 μW/cm^2. The presence of product ion *m/z* = 329 is consistent with the HPLC results, which analyzed nifedipine and dehydronitrosonifedipine concentration vs. exposure time, as discussed above, although it is acknowledged that secondary MS/MS analysis will be needed to confirm the identity of *m/z* = 329.

The reason behind the inability of two closely related flavonoids to act as photoprotectants was unclear. The decomposition of quercetin in methanol to three major products when exposed to UVA has been described previously [54], however when we examined rutin for UVA stability in methanol, we observed minimal decomposition (Supplementary S1). Furthermore, quercetin has been shown to protect the non-steroidal anti-inflammatory ketoprofen from UV decomposition in vitro [55]. We examined whether rutin could also prevent ketoprofen degradation in an in vitro system. We first confirmed that quercetin could spare UVA-induced ketoprofen degradation by irradiating a methanol solution of ketoprofen with UVA in the presence of quercetin. Consistent with previous reports [55], we observed a loss of quercetin over time and formation of quercetin degradation products, whereas ketoprofen loss was minimal until all of the quercetin had been depleted. Conversely, UVA irradiation of ketoprofen in methanol in the presence of rutin resulted in a more rapid loss of ketoprofen comparable to control exposures and concurrent decomposition of rutin (Figure 11). Since the only structural difference between quercetin and rutin is glycosylation at the 3′−OH of rutin, this suggests to us that 3′−OH substitution confers some UV stability to flavonols. Both quercetin and rutin possess a catechol moiety in the B-ring that undergoes oxidation to an *ortho*-quinone, which in the case of quercetin, appears to lead to photodegradation [54]. This may imply that the decomposition of quercetin is associated with its photoprotective properties; the lack of a photostabilizing effect of rutin on BMDBM is in agreement with these observations.

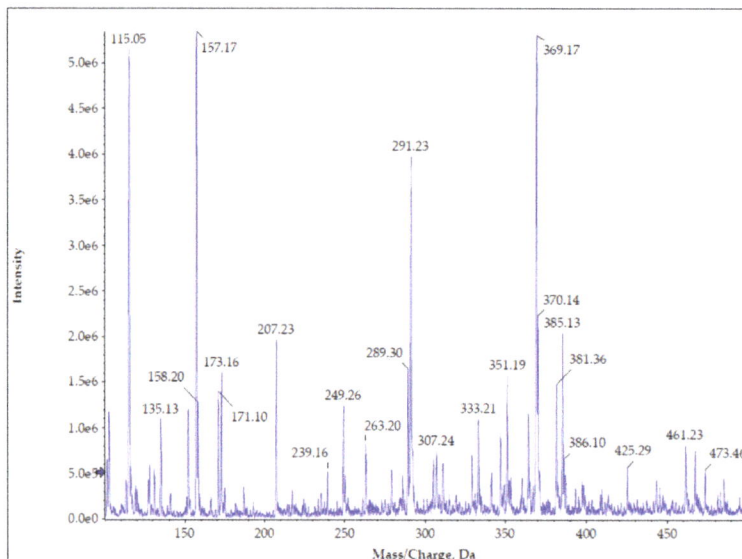

(A)

Figure 10. *Cont.*

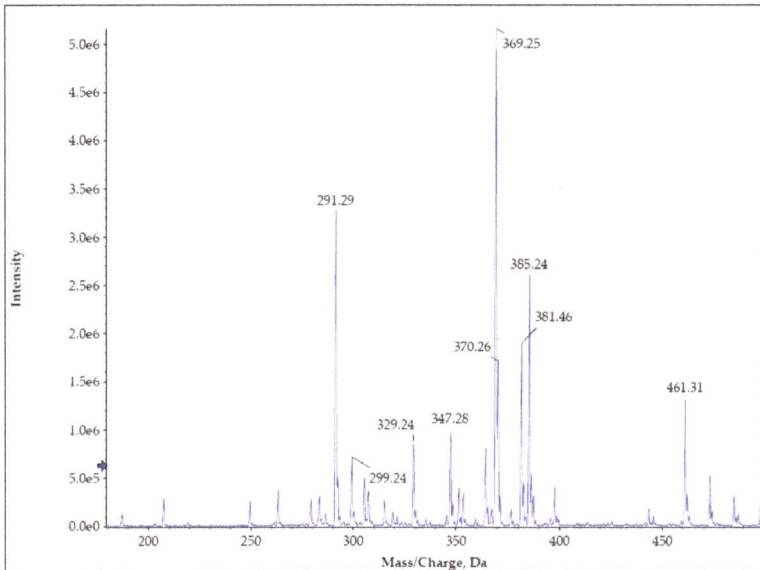

(B)

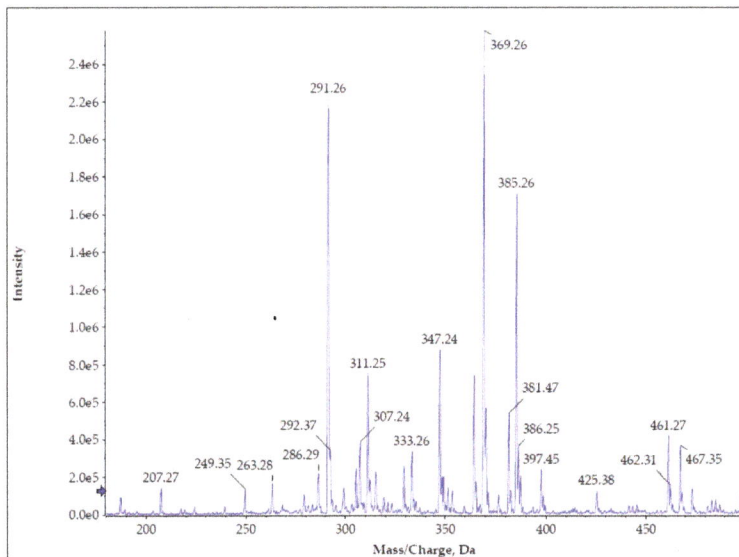

(C)

Figure 10. Mass spectrometry shows differential appearance of photodegradation product comparing (**A**) nifedipine 2% cream containing 0.5% quercetin with no UV exposure; (**B**) nifedipine 2% cream containing 0.5% quercetin that was exposed to UVA × 2 h; (**C**) nifedipine 2% cream containing 0.5% quercetin and 3% BMDBM that was exposed to UVA × 2 h.

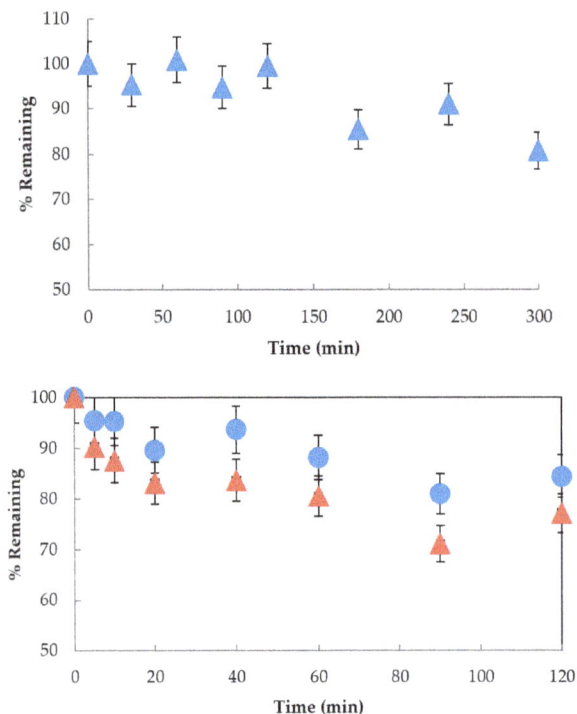

Figure 11. Time-course decomposition for UVA (740 µW·cm^{-2} at 365 nm) exposed compounds in MeOH: (top) (Δ) ketoprofen (250 µM); (bottom) (Δ) ketoprofen (250 µM) in the presence of (●) rutin (50 µM). Data are the mean ± standard deviation of three separate experiments as determined by HPLC-PDA at the λ_{max} for ketoprofen and rutin and are reported as % remaining.

3.3.2. Effect of Photostabilizers on Emulsion Properties

Emulsion composition and viscosity are key features to minimize the potential for phase separation. With our observation that a combination of quercetin and BMDBM can protect nifedipine from UVA photodegradation, we next sought to assess whether the photostabilizers might reduce physical stability of the emulsion by affecting viscosity or emulsion droplet formation. To accomplish this, the creaming rate and extent were assessed in a temperature-controlled photocentrifuge whereby an increased transmission to light indicates phase separation, which is then calculated as an instability index. If an excipient caused a significant change in viscosity, for example, phase separation would occur more quickly and adversely affect emulsion stability on storage. This information can drive the decision to choose between two excipients that are otherwise performing similarly. In our system, the incorporation of quercetin or BMDBM reduced the instability of the nifedipine emulsion (Figure 12 and inset, showing differences between emulsion stability depending on presence of specific photostabilizers), possibly by altering the emulsion viscosity. The improved emulsion stability, together with nifedipine photostabilization, suggest that a topical formulation of nifedipine containing quercetin and BMDBM may be an effective approach for local delivery of nifedipine for RP.

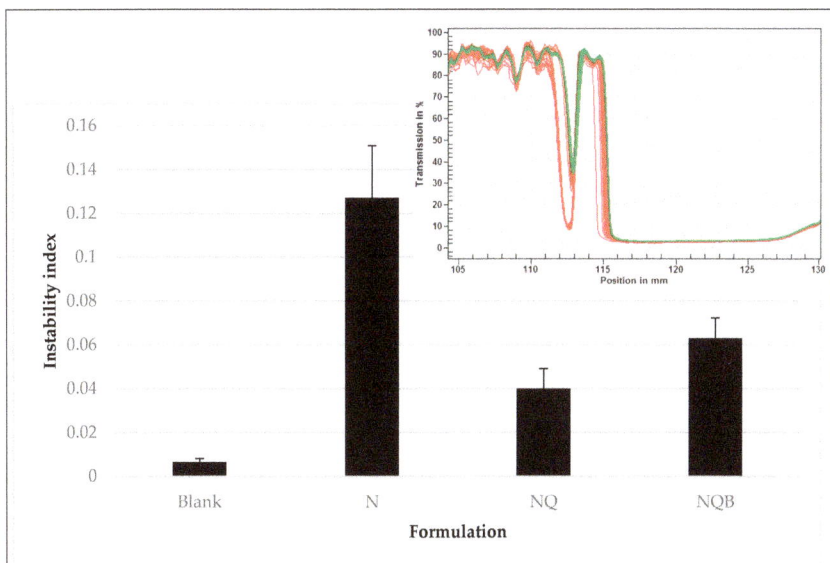

Figure 12. Sedimentation analysis of nifedipine emulsions (N) containing quercetin (NQ) or quercetin plus BMDBM (NQB) show differences in their tendency to exhibit phase separation (at 42 °C over 12 h), as shown by an increase in light transmission at the top of the sample as the oil phase separates to the top. A shift from red to green indicates a change in light transmission between the first and last readings (example shown in inset). This rate of change in light transmission of the sample vs. time is translated into an "instability index".

4. Conclusions

Topical delivery of nifedipine as a treatment for Raynaud's Phenomenon requires a photoprotectant to prevent nifedipine degradation upon exposure to UV radiation. A combination of the flavonoid quercetin and BMDBM in an O/W emulsion was able to protect nifedipine in vitro from UVA radiation-induced decomposition, whereas rutin in combination with BMDBM did not. Although quercetin and rutin share similar scaffolds, only quercetin was able to act as a photostabilizer of BMDBM and nifedipine, the difference in photostabilizing properties would appear to be a function of the unsubstituted 3′−OH in quercetin. Smith et al. [56] suggested that functionalization of the 3′−OH, or absence of a 3′−OH as in luteolin, confers photostability and that the instability of flavonols with an unsubstituted 3′−OH is proposed to be the result of an excited state electron transfer step, although how this would lead to photoprotection is unclear. Rather, it may be that quercetin more readily directs absorption of UV radiation to degradation, thus sparing BMDBM, whereas rutin does not have a comparable path and may form an excited state which then activates BMDBM. These observations provide insight into the development of quercetin or other flavonoids as photoprotectants for nifedipine or other pharmaceutical agents, which possess UV sensitivity. Follow-up studies will include investigation of the effect of this topical nifedipine formulation on vasodilation in vivo.

Supplementary Materials: The following are available online at http://www.mdpi.com/1999-4923/11/11/594/s1, Figure S1: UV absorption spectra of quercetin and rutin before and after UVA exposure. Figure S2. Mass spectrum of nifedipine extracted from the cream (Q1 scan) indicating $m/z = 369$. Figure S3. Mass spectrum of dehydronitrosonifedipine (+MRM) indicating $m/z = 329$.

Author Contributions: Conceptualization, methodology, and formal analysis: E.W. and E.K.; investigation: J.Z., J.P., M.M., J.S., T.O., K.S., J.C., A.B., C.Z., and B.M.F.; resources: E.W. and E.K.; writing—original draft preparation: E.W.; writing—review and editing: E.K., J.Z., and J.C.; supervision: E.W. and E.K.; project administration: E.W.; funding acquisition: E.W.

Funding: This research was funded by British Columbia Institute of Technology: Discovery Parks Applied Research Award; and University of Saskatchewan: VP Research Faculty Recruitment and Retention Fund.

Conflicts of Interest: The authors declare no conflict of interest.

References

1. Garner, R.; Kumari, R.; Lanyon, P.; Doherty, M.; Zhang, W. Prevalence, risk factors and associations of primary Raynaud's phenomenon: Systematic review and meta-analysis of observational studies. *BMJ Open* **2015**, *5*, e006389. [CrossRef] [PubMed]
2. Wigley, F.M.; Flavahan, N.A. Raynaud's Phenomenon. *N. Engl. J. Med.* **2016**, *375*, 556–565. [CrossRef] [PubMed]
3. Bergersen, T.K.; Walloe, L. Acral coldness—Severely reduced blood flow to fingers and toes. In *Handbook of Clinical Neurology*; Elsevier: Amsterdam, The Netherlands, 2018; Volume 157, pp. 677–685. [CrossRef]
4. Joseph, L.; Kim, E.S.H. Non-Atherosclerotic Vascular Disease in Women. *Curr. Treat. Options Cardiovasc. Med.* **2017**, *19*, 78. [CrossRef] [PubMed]
5. Plissonneau Duquene, P.; Pistorius, M.A.; Pottier, P.; Aymard, B.; Planchon, B. Cold climate could be an etiologic factor involved in Raynaud's phenomenon physiopathology. Epidemiological investigation from 954 consultations in general practic. *Int. Angiol. J. Int. Union Angiol.* **2015**, *34*, 467–474.
6. Rirash, F.; Tingey, P.C.; Harding, S.E.; Maxwell, L.J.; Tanjong Ghogomu, E.; Wells, G.A.; Tugwell, P.; Pope, J. Calcium channel blockers for primary and secondary Raynaud's phenomenon. *Cochrane Database Syst. Rev.* **2017**, *12*, Cd000467. [CrossRef]
7. Rustin, M.H.; Newton, J.A.; Smith, N.P.; Dowd, P.M. The treatment of chilblains with nifedipine: The results of a pilot study, a double-blind placebo-controlled randomized study and a long-term open trial. *Br. J. Dermatol.* **1989**, *120*, 267–275. [CrossRef]
8. Belch, J.; Carlizza, A.; Carpentier, P.H.; Constans, J.; Khan, F.; Wautrecht, J.C.; Visona, A.; Heiss, C.; Brodeman, M.; Pecsvarady, Z.; et al. ESVM guidelines—The diagnosis and management of Raynaud's phenomenon. *Vasa* **2017**, *46*, 413–423. [CrossRef]
9. Schaller, D.; Gunduz, M.G.; Zhang, F.X.; Zamponi, G.W.; Wolber, G. Binding mechanism investigations guiding the synthesis of novel condensed 1,4-dihydropyridine derivatives with L-/T-type calcium channel blocking activity. *Eur. J. Med. Chem.* **2018**, *155*, 1–12. [CrossRef]
10. Scholz, H. Pharmacological aspects of calcium channel blockers. *Cardiovasc. Drugs Ther.* **1997**, *10* (Suppl. 3), 869–872. [CrossRef]
11. Gjorup, T.; Kelbaek, H.; Hartling, O.J.; Nielsen, S.L. Controlled double-blind trial of the clinical effect of nifedipine in the treatment of idiopathic Raynaud's phenomenon. *Am. Heart J.* **1986**, *111*, 742–745. [CrossRef]
12. Golfam, F.; Golfam, P.; Khalaj, A.; Sayed Mortaz, S.S. The effect of topical nifedipine in treatment of chronic anal fissure. *Acta Med. Iran.* **2010**, *48*, 295–299. [PubMed]
13. Agrawal, V.; Kaushal, G.; Gupta, R. Randomized controlled pilot trial of nifedipine as oral therapy vs. topical application in the treatment of fissure-in-ano. *Am. J. Surg.* **2013**, *206*, 748–751. [CrossRef] [PubMed]
14. Ashkani-Esfahani, S.; Hosseinabadi, O.K.; Moezzi, P.; Moafpourian, Y.; Kardeh, S.; Rafiee, S.; Fatheazam, R.; Noorafshan, A.; Nadimi, E.; Mehrvarz, S.; et al. Verapamil, a Calcium-Channel Blocker, Improves the Wound Healing Process in Rats with Excisional Full-Thickness Skin Wounds Based on Stereological Parameters. *Adv. Skin Wound Care* **2016**, *29*, 271–274. [CrossRef] [PubMed]
15. Grant, S.M.; Goa, K.L. Iloprost. A review of its pharmacodynamic and pharmacokinetic properties, and therapeutic potential in peripheral vascular disease, myocardial ischaemia and extracorporeal circulation procedures. *Drugs* **1992**, *43*, 889–924. [CrossRef] [PubMed]
16. Hotkar, M.S.; Avachat, A.M.; Bhosale, S.S.; Oswal, Y.M. Preliminary investigation of topical nitroglycerin formulations containing natural wound healing agent in diabetes-induced foot ulcer. *Int. Wound J.* **2015**, *12*, 210–217. [CrossRef] [PubMed]
17. O'Meara, S.; Cullum, N.; Majid, M.; Sheldon, T. Systematic Reviews of Wound Care Management: (3) Antimicrobial Agents for Chronic Wounds; (4) Diabetic Foot Ulceration. *Health Technol. Assess.* **2000**, *4*, 1–237. [CrossRef]
18. McClusky, S.V.; Brunn, G.J. Nifedipine in Compounded Oral and Topical Preparations. *Int. J. Pharm. Compd.* **2013**, *15*, 166–169.

19. Ioele, G.; Gunduz, M.G. A New Generation of Dihydropyridine Calcium Channel Blockers: Photostabilization of Liquid Formulations Using Nonionic Surfactants. *Pharmaceutics* **2019**, *11*, 28. [CrossRef]
20. Hayase, N.; Itagaki, Y.; Ogawa, S.; Akutsu, S.; Inagaki, S.; Abiko, Y. Newly discovered photodegradation products of nifedipine in hospital prescriptions. *J. Pharm. Sci.* **1994**, *83*, 532–538. [CrossRef]
21. Aman, W.; Thoma, K. Particular features of photolabile substances in tablets. *Die Pharm.* **2003**, *58*, 645–650.
22. Görner, H. Nitro group photoreduction of 4-(2-nitrophenyl)- and 4-(3-nitrophenyl)-1,4-dihydropyridines. *Chem. Phys.* **2010**, *373*, 153–158. [CrossRef]
23. Grooff, D.; Francis, F.; De Villiers, M.M.; Ferg, E. Photostability of crystalline versus amorphous nifedipine and nimodipine. *J. Pharm. Sci.* **2013**, *102*, 1883–1894. [CrossRef] [PubMed]
24. Maafi, W.; Maafi, M. Modelling nifedipine photodegradation, photostability and actinometric properties. *Int. J. Pharm.* **2013**, *456*, 153–164. [CrossRef] [PubMed]
25. De Luca, M.; Ioele, G.; Spatari, C.; Ragno, G. Photostabilization studies of antihypertensive 1, 4-dihydropyridines using polymeric containers. *Int. J. Pharm.* **2016**, *505*, 376–382. [CrossRef] [PubMed]
26. Majeed, I.A.; Murray, W.J.; Newton, D.W.; Othman, S.; Al-turk, W.A. Spectrophotometric study of the photodecomposition kinetics of nifedipine. *J. Pharm. Pharmacol.* **1987**, *39*, 1044–1046. [CrossRef] [PubMed]
27. Baertschi, S.W.; Clapham, D.; Foti, C.; Kleinman, M.H.; Kristensen, S.; Reed, R.A.; Templeton, A.C.; Tønnesen, H.H. Implications of In-Use Photostability: Proposed Guidance for Photostability Testing and Labeling to Support the Administration of Photosensitive Pharmaceutical Products, Part 2: Topical Drug Product. *J. Pharm. Sci.* **2015**, *104*, 2688–2701. [CrossRef] [PubMed]
28. Tonnesen, H.H. *Photostability of Drugs and Drug Formulations*; CRC Press: Boca Raton, FL, USA, 2004.
29. Cosa, G. Photodegradation and photosensitization in pharmaceutical products: Assessing drug phototoxicity. *Pure Appl. Chem.* **2004**, *76*, 263. [CrossRef]
30. Kockler, J.; Robertson, S.; Oelgemoller, M.; Davies, M.; Bowden, B.; Brittain, H.G.; Glass, B.D. Butyl methoxy dibenzoylmethane. In *Profiles of Drug Substances, Excipients and Related Methodology*; Harry, G.B., Ed.; Academic Press: New York, NY, USA, 2013; Volume 38, pp. 87–111. [CrossRef]
31. Gange, R.W.; Soparkar, A.; Matzinger, E.; Dromgoole, S.H.; Sefton, J.; DeGryse, R. Efficacy of a sunscreen containing butyl methoxydibenzoylmethane against ultraviolet. A radiation in photosensitized subjects. *J. Am. Acad. Dermatol* **1986**, *15*, 494–499. [CrossRef]
32. Lhiaubet-Vallet, V.; Marin, M.; Jimenez, O.; Gorchs, O.; Trullas, C.; Miranda, M.A. Filter-filter interactions. Photostabilization, triplet quenching and reactivity with singlet oxygen. *Photochem. Photobiol. Sci.* **2010**, *9*, 552–558. [CrossRef]
33. Solovchenko, A.; Schmitz-Eiberger, M. Significance of skin flavonoids for UV-B-protection in apple fruits. *J. Exp. Bot.* **2003**, *54*, 1977–1984. [CrossRef]
34. Wilson, K.E.; Wilson, M.I.; Greenberg, B.M. Identification of the Flavonoid Glycosides that Accumulate in Brassica napus L. cv. Topas Specifically in Response to Ultraviolet B Radiation. *Photochem. Photobiol.* **1998**, *67*, 547–553. [CrossRef]
35. Fahlman, B.M.; Krol, E.S. Inhibition of UVA and UVB Radiation-Induced Lipid Oxidation by Quercetin. *J. Agric. Food Chem.* **2009**, *57*, 5301–5305. [CrossRef] [PubMed]
36. Maini, S.; Fahlman, B.M.; Krol, E.S. Flavonols Protect Against UV Radiation-Induced Thymine Dimer Formation in an Artificial Skin Mimic. *J. Pharm. Pharm. Sci.* **2015**, *18*, 600–615. [CrossRef] [PubMed]
37. Oliveira, C.A.; Peres, D.D.; Graziola, F.; Chacra, N.A.; Araújo, G.L.; Flórido, A.C.; Mota, J.; Rosado, C.; Velasco, M.V.; Rodrigues, L.M.; et al. Cutaneous biocompatible rutin-loaded gelatin-based nanoparticles increase the SPF of the association of UVA and UVB filters. *Eur. J. Pharm. Sci.* **2016**, *81*, 1–9. [CrossRef]
38. Chaiprasongsuk, A.; Onkoksoong, T.; Pluemsamran, T.; Limsaengurai, S.; Panich, U. Photoprotection by dietary phenolics against melanogenesis induced by UVA through Nrf2-dependent antioxidant responses. *Redox Biol.* **2016**, *8*, 79–90. [CrossRef]
39. Choquenet, B.; Couteau, C.; Paparis, E.; Coiffard, L.J. Quercetin and rutin as potential sunscreen agents: Determination of efficacy by an in vitro method. *J. Nat. Prod.* **2008**, *71*, 1117–1118. [CrossRef]
40. Scalia, S.; Mezzena, M. Photostabilization Effect of Quercetin on the UV Filter Combination, Butyl Methoxydibenzoylmethane–Octyl Methoxycinnamate. *Photochem. Photobiol.* **2010**, *86*, 273–278. [CrossRef]
41. Barak, N.; Rice, H.; Kamsler, A. Variability of compounded topical 0.2% nifedipine formulations. *J. Pharm. Pract. Res.* **2019**, *49*, 130–134. [CrossRef]

42. Huong, S.P.; Rocher, E.; Fourneron, J.-D.; Charles, L.; Monnier, V.; Bun, H.; Andrieu, V. Photoreactivity of the sunscreen butylmethoxydibenzoylmethane (DBM) under various experimental conditions. *J. Photochem. Photobiol. A Chem.* **2008**, *196*, 106–112. [CrossRef]

43. Vallejo, J.J.; Mesa, M.; Gallardo, C. Evaluation of the Avobenzone Photostability in Solvents Used in Cosmetic Formulations. *Vitae* **2011**, *18*, 63–71.

44. Afonso, S.; Horita, K.; Sousa e Silva, J.P.; Almeida, I.F.; Amaral, M.H.; Lobao, P.A.; Costa, P.C.; Miranda, M.S.; Esteves da Silva, J.C.; Sousa Lobo, J.M. Photodegradation of avobenzone: Stabilization effect of antioxidants. *J. Photochem. Photobiol. B Biol.* **2014**, *140*, 36–40. [CrossRef]

45. Gaspar, L.R.; Campos, P.M. Photostability and efficacy studies of topical formulations containing UV-filters combination and vitamins A, C and E. *Int. J. Pharm.* **2007**, *343*, 181–189. [CrossRef] [PubMed]

46. Shrewsbury, R.P. *Applied Pharmaceutics in Contemporary Compounding*; Morton Publishing Company: Englewood, CO, USA, 2015.

47. U.S. Food and Drug Administration. Part 352 Sunscreen drug products for over-the-counter human use. In *CFR—Code of Federal Regulations Title 21*; U.S. Food and Drug Administration: Silver Spring, MD, USA, 2018; Volume 5.

48. European Medicines Agency. ICH Topic Q1B Stability testing: Photostability testing of new drug substances and products. In *ICH Harmonised Tripartite Guideline*; European Medicines Agency: London, UK, 1996.

49. Fasani, E.; Dondi, D.; Ricci, A.; Albini, A. Photochemistry of 4-(2-Nitrophenyl)-1,4-Dihydropyridines. Evidence for Electron Transfer and Formation of an Intermediate. *Photochem. Photobiol.* **2006**, *82*, 225–230. [CrossRef] [PubMed]

50. Baranda, A.B.; Alonso, R.M.; Jiménez, R.M.; Weinmann, W. Instability of calcium channel antagonists during sample preparation for LC–MS–MS analysis of serum samples. *Forensic Sci. Int.* **2006**, *156*, 23–34. [CrossRef] [PubMed]

51. Sullivan, D.W., Jr.; Gad, S.C.; Julien, M. A review of the nonclinical safety of Transcutol(R), a highly purified form of diethylene glycol monoethyl ether (DEGEE) used as a pharmaceutical excipient. *Food Chem. Toxicol.* **2014**, *72*, 40–50. [CrossRef] [PubMed]

52. Osborne, D.W. Diethylene glycol monoethyl ether: An emerging solvent in topical dermatology products. *J. Cosmet. Dermatol.* **2011**, *10*, 324–329. [CrossRef] [PubMed]

53. Streel, B.; Zimmer, C.; Sibenaler, R.; Ceccato, A. Simultaneous determination of nifedipine and dehydronifedipine in human plasma by liquid chromatography–tandem mass spectrometry. *J. Chrom. B Biomed. Sci. App.* **1998**, *720*, 119–128. [CrossRef]

54. Fahlman, B.M.; Krol, E.S. UVA and UVB radiation-induced oxidation products of quercetin. *J. Photochem. Photobiol. B Biol.* **2009**, *97*, 123–131. [CrossRef]

55. Nakajima, A.; Tahara, M.; Yoshimura, Y.; Nakazawa, H. Study of Compounds Suppressing Free Radical Generation from UV-Exposed Ketoprofen. *Chem. Pharm. Bull.* **2007**, *55*, 1431–1438. [CrossRef]

56. Smith, G.J.; Thomsen, S.J.; Markham, K.R.; Andary, C.; Cardon, D. The photostabilities of naturally occurring 5-hydroxyflavones, flavonols, their glycosides and their aluminium complexes. *J. Photochem. Photobiol. A Chem.* **2000**, *136*, 87–91. [CrossRef]

pharmaceutics

Article

Validation of Cadherin HAV6 Peptide in the Transient Modulation of the Blood-Brain Barrier for the Treatment of Brain Tumors

Babu V. Sajesh [1,†], Ngoc H. On [2,†], Refaat Omar [2], Samaa Alrushaid [3,4], Brian M. Kopec [5], Wei-Guang Wang [6], Han-Dong Sun [6], Ryan Lillico [3], Ted M. Lakowski [3], Teruna J. Siahaan [5], Neal M. Davies [7], Pema-Tenzin Puno [6,*], Magimairajan Issai Vanan [1,8,*] and Donald W. Miller [2,*]

[1] Research Institute in Oncology and Hematology, University of Manitoba, Winnipeg, MB R3E 0V9, Canada
[2] Department of Pharmacology and Therapeutics, University of Manitoba, Winnipeg, MB R3E 0T6, Canada
[3] College of Pharmacy Pharmaceutical Analysis Laboratory, University of Manitoba, Winnipeg, MB R3E 0V9, Canada
[4] Department of Pharmaceutical Chemistry, Faculty of Pharmacy, Kuwait University, Safat 13110, Kuwait
[5] Department of Pharmaceutical Chemistry, University of Kansas, Kansas, KS 66205, USA
[6] Kunming Institute of Botany, Kunming 650201, Yunnan, China
[7] Pharmacy and Pharmaceutical Sciences, University of Alberta, Alberta, AB T6G 2R3, Canada
[8] Department of Pediatrics and Child Health, University of Manitoba, Winnipeg, MB R3T 2N2, Canada
*   Correspondence: punopematenzin@mail.kib.ac.cn (P.-T.P.); mivanan@cancercare.mb.ca (M.I.V.); donald.miller@umanitoba.ca (D.W.M.); Tel.: +1-204-787-4724 (M.I.V.); +1-204-789-3278 (D.W.M.); Fax: +1-204-786-1095 (M.I.V.)
†   These authors contributed to the work equally.

Received: 21 July 2019; Accepted: 3 September 2019; Published: 17 September 2019

Abstract: The blood-brain barrier (BBB) poses a major obstacle by preventing potential therapeutic agents from reaching their intended brain targets at sufficient concentrations. While transient disruption of the BBB has been used to enhance chemotherapeutic efficacy in treating brain tumors, limitations in terms of magnitude and duration of BBB disruption exist. In the present study, the preliminary safety and efficacy profile of HAV6, a peptide that binds to the external domains of cadherin, to transiently open the BBB and improve the delivery of a therapeutic agent, was evaluated in a murine brain tumor model. Transient opening of the BBB in response to HAV6 peptide administration was quantitatively characterized using both a gadolinium magnetic resonance imaging (MRI) contrast agent and adenanthin (Ade), the intended therapeutic agent. The effects of HAV6 peptide on BBB integrity and the efficacy of concurrent administration of HAV6 peptide and the small molecule inhibitor, Ade, in the growth and progression of an orthotopic medulloblastoma mouse model using human D425 tumor cells was examined. Systemic administration of HAV6 peptide caused transient, reversible disruption of BBB in mice. Increases in BBB permeability produced by HAV6 were rapid in onset and observed in all regions of the brain examined. Concurrent administration of HAV6 peptide with Ade, a BBB impermeable inhibitor of Peroxiredoxin-1, caused reduced tumor growth and increased survival in mice bearing medulloblastoma. The rapid onset and transient nature of the BBB modulation produced with the HAV6 peptide along with its uniform disruption and biocompatibility is well-suited for CNS drug delivery applications, especially in the treatment of brain tumors.

Keywords: blood-brain barrier (BBB); drug delivery; transient modulation; HAV6 cadherin peptide; adenanthin; magnetic resonance imaging (MRI); medulloblastoma

1. Introduction

Chemotherapy and more recently targeted therapy directed at specific tumor targets, are important modalities for treatment of pediatric brain tumors. Numerous chemotherapeutic drugs and small molecule inhibitors that demonstrated significant antitumor activity in preclinical studies have failed in human clinical trials [1]. These disappointing results can partially be explained by the ineffective drug delivery to the CNS [1]. The anatomical features posed by the various specialized physiological barriers (blood-brain barrier-BBB, and blood tumor barrier) together with the various active efflux transporters expressed contribute to drug failure because of the inability to reach the desired target at a therapeutically relevant concentrations [1]. The BBB prevents many drug molecules within the systemic circulation from entering the brain. Even small hydrophobic drugs that could otherwise partition into the plasma membranes and diffuse intracellularly through the vasculature have limited brain penetration due to the presence of multiple efflux transporters [2,3] and drug metabolizing enzymes [4,5] within the brain's endothelial cells. For hydrophilic drugs and macromolecules, the complex tight junctions that form between the brain endothelial cells limits the paracellular diffusion pathway for brain penetration [6]. Indeed, it has been suggested that molecules with hydrodynamic diameters larger than 11 Å or molecular weight of 500 D are too large to pass through the BBB [6,7]. Under pathological conditions, such as brain tumors, the BBB within the tumor area can become leaky, resulting in contrast enhancement on magnetic resonance imaging (MRI) of the tumor [8]. Indeed, the leakage of the MRI contrast media into the tumor site has been cited as evidence for a leaky BBB in brain tumors [9,10]. However, studies in mice have shown that despite the altered BBB integrity within the tumor site at later stages of tumor development, the BBB is still functional and limits solute and drug permeability in and around the tumor [11]. Furthermore, patients with primary brain tumors have clinically significant regions of tumor with an intact BBB and failure to deliver effective therapy to all regions of glioblastoma tumor contributes to treatment failure [12].

The cadherin proteins found within the adherens junction have an important role in the BBB for establishing cell–cell contact and contributing to the tight junction complex. The formation of homodimer complexes of cadherin proteins on adjacent brain microvessel endothelial cells act to physically restrict the paracellular passage of solutes from the blood to the brain extracellular environment [6,7]. Molecularly, the extracellular (EC) domain consists of five tandem repeated units (EC-1 to EC-5). Within the EC1 domain, the highly conserved region of His-Ala-Val (HAV) is crucial for the formation of the cis-dimer formation of cadherins. Synthetic peptides targeting this HAV region sequence of the EC1 domain display concentration-dependent binding to E-cadherin molecules and can prevent homodimer complex formation in brain microvessel endothelial cells [13]. We have previously [14] shown that systemic (intravenous, IV) administration of synthetic E-Cadherin peptide in Balb/c mice caused a reversible disruption of BBB integrity and enhanced the accumulation of permeability markers of various molecular weights ((low molecular weight gadolinium diethylene-triamine-penta-acetate (Gd-DTPA), larger molecular weight infrared Dye (IRDye800CW) and a drug efflux agent, rhodamine 800 (R800)). The magnitude of increase in BBB permeability observed ranged from two-fold to five-fold (depending on size and chemical properties of imaging agent), was rapid, (occurred within 3–6 min following injection of the peptide) and reversible, with complete barrier integrity being restored within 60 to 90 min of injection [14]. Additional studies with both linear and cyclized cadherin peptides demonstrated that the duration and magnitude of BBB opening could be selected based on the stability and binding affinity of the cadherin peptides [14–16]. These studies provided the impetus for use of cadherin peptides to improve brain delivery of therapeutics.

Medulloblastoma (MBL) is the most common malignant brain tumor of childhood and accounts for 10% of all deaths from childhood cancer [17]. Current management consists of surgical resection followed by ionizing radiation (IR) and chemotherapy. Outcome for high-risk patients remains relatively poor, with a five-year event-free-survival of 25–40%. MBL comprises at least four distinct molecular subgroups, with Group-3 having by far the worst prognosis with a 5-year survival of approximately 30% [17]. To date, no specific targeted therapy is available for Group-3 MBLs. Peroxiredoxin-1 (PRDX1) is a multifunctional protein that catalyzes hydrogen peroxide into water and oxygen and thereby prevents free radical mediated oxidative stress [18]. Besides mediating radiation resistance in several cancers, PRDX1 has been suggested to play a role in chemotherapy resistance, cell differentiation, proliferation and apoptosis [18]. We recently validated PRDX1 as a therapeutic target in Group-3 MBL [19]. Ade, a diterpenoid compound isolated from the leaves of *Isodon adenantha* inhibits the peroxidase activity of PRDX 1 and 2 [20]. Adenanthin has shown therapeutic activity in several cancers, like leukemia and liver cancer [20,21], but there is an absence of data pertaining to its effectiveness in treating brain tumors. In the present study, we demonstrated that Ade, when administered alone, did not cross the BBB and was ineffective in treating D425 tumors in a mouse orthostatic medulloblastoma tumor model. However, concurrent use of HAV6 peptide to transiently open the BBB resulted in Ade entry into the brain and significantly prolonged the survival of mice bearing Group-3 MBL tumors. Those tumor bearing mice receiving HAV6 showed no additional adverse responses to the treatment compared to mice receiving placebo. These findings provide proof-of-principle for the use of cadherin peptides in the modulation of BBB permeability and improved treatment of brain tumors.

2. Materials and Methods

2.1. Chemicals and Reagents

Gadolinium diethylene-triamine-penta-acetate (Gd-DTPA) was obtained from Berlex (Lachine, QC, Canada) and used as a contrast agent for MRI monitoring of BBB permeability. Analytical grade Formic acid and HPLC Grade Acetonitrile were purchased from Fisher Scientific. Ultrapure water from a Milli-Q® system (Millipore, Billerica, MA, USA) was used for mobile phase. All other reagents and chemicals were purchased from Sigma Chemical Company (St. Louis, MO, USA).

2.2. HAV6 Peptide Synthesis

The HAV6 peptide (Ac-SHAVSS-NH_2) was synthesized using solid phase Fmoc-chemistry in a Tribute peptide synthesizer (Gyros Protein Technologies, Tucson, AZ), as described previously [13]. After removal from the resin, the peptide was purified using a semi-preparative C18 column in HPLC. The pure fractions were pooled and lyophilized. The purity of the peptide was higher than 98%, as determined by C18 analytical HPLC. The identity of the peptide was confirmed by mass spectrometry.

2.3. Adenanthin Source and Formulation

Adenanthin was isolated from the dried aerial parts of the leaves of Isodon Adenanthus Hara, as described previously [20]. Ade (MW 490.549 g/mol) was reconstituted in 1% DMSO in phosphate buffered saline (PBS) (KCl; 2.66 Mol, KH_2PO_4; 1.47 Mol, NaCl; 137.93 Mol, Na_2HPO_4-$7H_2O$; 8.06 Mol; pH 7.40) at a final concentration of 1 mg/μL to obtain a 10X stock. The solution was diluted to 0.1 mg/μL in physiological saline and was administered to animals so that each animal got a final concentration of 10 mg/kg body weight.

2.4. Animals and Ethics Statement

All experiments described in this study were done at the University of Manitoba and the Research Institute in Oncology and Hematology, as described in animal use protocol 13-051 approved by the Central animal care committee at the University of Manitoba in accordance with the guidelines provided by Canadian council on Animal care on 18 November 2015. All surgical procedures were performed under anesthesia induced and maintained by Isoflurane and every effort to minimize pain, suffering and a reduction in the numbers of animals used were made.

2.5. MRI Imaging of BBB Permeability with HAV6 Peptide

The HAV6 peptide-induced BBB permeability enhancement in Ncr (nu-/nu-) mice was assessed using MRI and Gd-DTPA contrast agent as described previously [11,14]. Briefly, the mice were anesthetized and placed in a 7 Tesla small animal Bruker Biospect MR with a 21 cm bore and 2.5×2.5 cm^2 field of view for spectroscopy. Once secured into MRI, a series of T1-weighted coronal images of the mouse brain were obtained before Gd-DTPA delivery as a contrast agent to acquire background images of the mouse brain. Mice were then administered Gd-DTPA contrast agent (0.4 mmol/kg) along with either HAV6 cadherin peptide (0.01 mmol/kg) or vehicle (PBS) via bolus tail vein injection, and T1-weighted coronal images of the entire brain were obtained at 3 min intervals throughout a 21 min imaging session. After the first imaging session, a second dose of Gd-DTPA was administered and T1-weighted images obtained for an additional 30 min. Quantitative assessment of Gd-DTPA enhancement in various regions of interest in the brain were obtained using Marevisi 7.2 software (Institute for Biodiagnostics, National Research Council, Canada). Changes in Gd-DTPA intensity in the brain as a function of time and treatment were determined using a percent difference analysis of brain slice images obtained as described previously [11,14,22] using the following formula.

(post-Gd-DTPA-T1-weighted image – pre-Gd-DTPA-T1-weighted image) ÷ pre-Gd-DTPA-T1-weighted image) × 100.

2.6. LC-MS/MS Analysis and Conditions

The impact of HAV6 on the BBB permeability of Ade was assessed using the brain to plasma ratio of Ade. For these studies, healthy female Ncr (nu–/nu–) mice were randomly selected to receive bolus injections (1 mL/kg) of either vehicle (Physiological saline), Ade alone (10 mg/kg) or a combination of HAV6 (0.010 mmol/kg) with various concentrations of Ade (5, 10 and 15 mg/kg,) via tail vein injections. The mice were then sacrificed 20 min after the injections, and blood and various tissues, including the brain, were subjected to LC-MS/MS analysis.

2.6.1. Sample Preparation

The extractions of Ade from the plasma and brain tissues were done using acetonitrile and a mixture of acetonitrile with deionized water at a ratio of 2:1 respectively. Frozen plasma samples were thawed followed by the addition of acetonitrile at a ratio of 2:1. For the brain tissue, the samples were weighed, thawed and homogenized with the addition of acetonitrile and water (2:1) at the volume of 5× the weight of the samples. The samples were then vortex mixed continuously for 5 min at a temperature of 4 °C, followed by centrifugation at 4 °C for 10 min at a speed of 10,000 g (for brain) and 1875 g (for plasma). The supernatants were collected and evaporated using a flow of nitrogen gas. Adenanthin in methanol was added to 200 μL blank plasma or blank brain homogenate to achieve concentrations of 0.1, 1, 10 and 100 μg/mL for standardization. Samples and standards were treated with 400 μL cold acetonitrile (−20 °C) to precipitate proteins and were vortexed for 2 min, and centrifuged at 15,000 rpm for 5 min. The supernatants were transferred to new tubes and evaporated to dryness using a Savant SPD1010 SpeedVac Concentrator (Thermo Fisher Scientific, Inc., Asheville, NC, USA) without heat. The dried samples were reconstituted in 400 μL of 50% aqueous methanol with 0.1% formic acid and centrifuged at 15,000 rpm for 5 min. The supernatant was transferred to HPLC vials and 10 mL injected into the LC-MS/MS system.

2.6.2. LC-MS/MS Analysis

A Shimadzu Nexera ultra high performance liquid chromatography system connected to an 8040 triple-quadrupole mass spectrometer (LC-MS/MS) (Shimadzu, Kyoto, Japan) was used to measure Ade. A dual Electrospray ionization (ESI) / atmospheric pressure chemical ionization (APCI) ion source was used in the positive mode multiple reaction monitoring (MRM) for the sodium adduct [M$^+$Na$^+$] of Ade with the transition 513.2 > 453.1 *m/z* using a collision energy of 30 eV. The desolvation line temperature was 250 °C; the heating block temperature was 400 °C; the nebulizing gas flow was 2 L/min; and drying gas flowed at 15 L/min. An XTerra® MS C18 (3.5 μm, 2.1 × 50 mm) (Waters Corporation, Milford, MA, USA) column was used with 0.3 mL/min flow rate at 30 °C. The mobile phases A (50% aqueous methanol with 0.1% formic acid) and B (85% aqueous methanol), and 0.1% formic acid were used in a step gradient starting at 100% A for 1 min, then stepping to 100% B for 3 min and stepping back to 100% A for 2 min before the next injection.

2.7. MBL Xenograft and Animal Experiments

2.7.1. Cell culture and authentication

Early passage Group-3 Medulloblastoma cells (D425-MED) were a kind gift from Dr. Darell Bigner, Duke University, and have been described previously [23]. Cells were cultured at 37 °C in Minimum Essential Medium (Richter's modification; ThermoFisher/Gibco™ Markham, ON, Canada) supplemented with fetal bovine serum (10%); ThermoFisher/Gibco™) in a humidified incubator with 5% CO_2. These cells were transduced by a lentiviral infection to stably express firefly luciferase. The resulting D425-Med-Luc cells were authenticated and validated by morphology and phenotype during recovery from the frozen stock and growth characteristics during culture.

2.7.2. Animal Experiments

An orthotopic Group-3 medulloblastoma mouse model was used to evaluate Ade effectiveness, both alone and in combination with HAV6 peptide for transient BBB disruption. Outbred homozygous nude (NCr nu; Foxn1nu/Foxn1nu, female mice < 4 weeks old) mice with jugular vein access (C20PU-MJV1301; Instech Laboratories, Inc.) and vascular access button (VAB62BS/25; Instech Laboratories, Inc) were obtained from Charles River Laboratories, Saint Constant, QC, Canada. Group 3 MBL (D425-Med-Luc) cells were implanted into the cerebella of these mice as described previously [24]. All mice were serially imaged on an IVIS® spectrum (Perkin Elmer, Waltham, MA), small animal imager, starting on day-7 post tumor cell implantation to confirm the presence and extent of tumor growth using bioluminescence as described previously [25]. Tumor bearing mice were randomly assigned into one of the following treatment groups ($n = 6$): (1) placebo (physiological saline); (2) Ade (10 mg/kg) alone; or (3) HAV6 cadherin peptide (0.01 mmol/kg) with Ade (10 mg/kg). All treatments were given as intravenous injection via the jugular access catheter. Treatments were begun on day-10 post tumor cell implantation. Each treatment cycle consisted of 3 consecutive days of treatment followed by one day of rest, with a maximum of 5 cycles given to each mouse. Mice were assessed daily for tumor progression using humane endpoints, including weight loss (>25% from original weight on day of tumor cell implantation), limb paralysis, locomotion and seizure activity. Tumor progression was monitored by bioluminescence imaging every 3 days starting at day 7-post tumor cell implantation as described in Baumann et al. [25]. Upon reaching humane endpoints tumor-bearing mice were euthanized by cardiac perfusion under iso-flurane anesthesia and the brains were removed and used for histology. Survival data was recorded and Kaplan-Meier curves were generated using Prism6 (V6.0h; GraphPad Software, Inc.; San Diego, CA, USA).

2.8. Immunohistochemistry

When mice bearing tumors presented with moribund phenotype, they were euthanized as described above. Subsequently, brains from each animal were extracted and fixed in 10% phosphate buffered formalin (pH 7.4) for 24 h. They were dehydrated by immersion in graded alcohol solutions for 10 min each (50%, 70%, 80% NS 95%) and for 3 changes in 100% ethanol. Tissue was embedded in paraffin and serial sections (6 μm) of the brain were collected on positively charged glass slides. Paraffin was melted and removed from slides by placing in an incubator at 60 °C for 30 min. Subsequently, slides were immersed in Xylene for 3 min, twice and rehydrated by immersion in a series of graded alcohol solutions for 3 min each (100%, 95%, 80%, 70%, 50%) and were finally washed in deionized water. Immediately, slides were immersed in 1X universal antigen retrieval reagent (ab208572, abcam, Cambridge, MA, USA) and incubated for 20 min at 95 °C in a programmable pressure chamber (Decloaking chamber NxGen, Biocare medical) following the manufacturer's instructions. Slides were stained to detect the expression of PRDX1 (abcam; Anti-PRDX1 antibody ab41906), γ-H2A.X (abcam;

Anti-γ-H2A.X phosphorylated S139 antibody; ab2893), Ki 67 (abcam; Anti Ki67 antibody; ab 15580), 53BP1 (abcam; Anti-53BP1 antibody; ab21083) and cleaved caspase 3 (abcam; Anti-cleaved caspase 3 antibody; ab13847) using an HRP/DAB detection method by using a mouse and rabbit specific HRP/DAB detection IHC kit (abcam; ab80436-EXPOSE mouse and rabbit specific HRP/DAB detection IHC kit), following manufacturers recommendations with minor modifications. Using normal human cerebellar cortex tissue as control, the optimal concentrations of each antibody and the incubation time with DAB substrate was empirically determined to be 1:1000 with an overnight incubation at 4 °C. Subsequently, stained tissue was imaged on a Cytation 5, a cell imaging multi-mode reader (Biotek Instruments Inc, Winooski, VT, USA), and color bright field images were captured on a 16-bit Sony charge coupled device Camera and Gen5 V.3.04 software. An image montage was obtained using a 2.5X objective Meiji, Plan Achromat, working distance (WD) of 6.2 and numerical aperture (NA) of 0.07 to image the entire section. A region of interest was manually imaged using a 20X objective, Plan Flourite, WD 6.6 and NA 0.45. Color images were processed using Gen5 V.3.04. Color images were saved as high-resolution JPEG images and were imported into Adobe Photoshop Creative Suite CS5, Adobe, San Jose, CA, USA to form image panels and additional illustration.

2.9. Cytotoxicity Studies

The D425-Med-luc cells were seeded onto a 96-well plate at a density of 5000 cells per well. Twenty-four hours after seeding, HAV6 (final concentration of 0.01 mM) or phosphate buffered saline (10 μL) was added to the culture media and cells were cultured for 48 h. Viable cells remaining at the end of the experiment were fixed with 4% paraformaldehyde, nuclei were stained with Hoechst 33342 (H3570; Fisher Scientific Inc., St. Loius USA) and imaged on a Cytation 5, a cell imaging multi-mode reader (Biotek Instruments Inc., Winooski, VT, USA) assay as previously described [26].

2.10. Statistical Analyses

Raw data was imported into Prism 6 GraphPad and analysis of variance was performed with post hoc comparison of the means as indicated in figure legends. Survival curve analysis and analysis of variance were performed with post-hoc nonparametric log-rank (Mantel-Cox) tests for significance. Values were represented as mean ± SEM unless otherwise indicated in the figure legends.

3. Results

3.1. HAV6 Peptide Transiently Increases BBB Permeability In-Vivo

In order to test if the novel HAV6 peptide would increase the BBB permeability, we injected mice with HAV6 peptide or vehicle and imaged their brains using MRI and Gd-DTPA contrast. As seen in Figure 1, the BBB integrity was not compromised in vehicle treated mice as shown by similar Gd-DTPA contrast enhancement in representative brain slice images at time 0 (prior to the injection of vehicle) and at 6 min following vehicle injection (Figure 1A,B, respectively). These coronal slice images show minimal Gd-DTPA, within the brain, consistent with its low BBB penetrance under normal conditions (Figure 1A,B). However, there was a significant increase in the signal intensity of the Gd-DTPA contrast agent in the brain (represented by white-gray appearance indicated by blue arrows) following IV injection of HAV6 peptide (0.01 mmol/kg) (Figure 1).

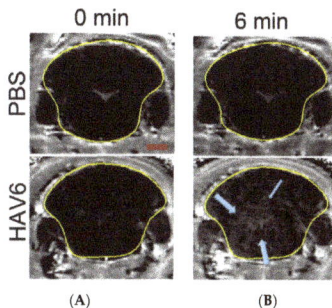

Figure 1. Representative MRI T1-weighted images of mouse brain taken at 0 min (**A**) and 6 min (**B**) following the injection of Gd-DTPA in control (PBS) and HAV6 (0.01 mmol/kg) treated mice. Blue arrows indicate regions of Gd-DTPA enhancement following the injection of the peptide. Scale bar represent 2 mm.

Indeed, quantitative assessment of Gd-DTPA contrast enhancement across the various brain regions indicated an approximately 2–4-fold increase in Gd-DTPA intensity in the HAV6 treatment group compared to control, and this increase was most apparent at 6–9 min following the injection of the HAV6 peptide (Figure 2A–C) and was consistent with our previous findings [14]. Furthermore, while the signal intensity for Gd-DTPA varied as a function of brain region examined, the increases in Gd-contrast enhancement in response to HAV peptide was observed in all regions of the brain examined, as indicated by examining the resulting area under the curve (AUC) from the Gd-DTPA intensity versus time plots taken over a 51 min period following HAV6 administration (Figure 2D). These studies confirmed the rapid onset and relatively brief BBB modulation following the administration of HAV6 cadherin peptide.

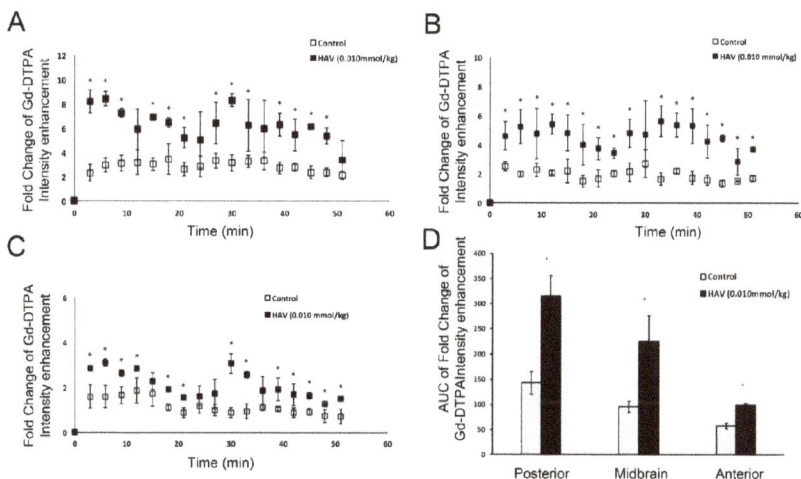

Figure 2. Analysis of pixel intensity for Gd-DTPA from T1-weighted images normalized to the pixel intensity at time 0 of the injection in (**A**) the posterior, (**B**) midbrain and (**C**) anterior regions of the brain. (**D**) Area under the curve for Gd-DTPA obtained from T1-weighted images over the span of 51 min. * p value < 0.05 in comparison to control group. Values represent the mean ± SEM for four mice per treatment group.

3.2. In Silico and In Vivo Determinations of Ade Permeability in the BBB

We employed the BOILED-Egg model (the Brain Or IntestinaL EstimateD permeation predictive model) [27] to predict if Ade would cross the BBB and be available in the brain. The lipophilicity (determined by the n-octanol/water partition coefficient-WLogP) and polarity (determined by the topological polar surface area-tPSA) values for Ade were calculated using the free online chemical property calculation service provided by www.molinspiration.com. The canonical SMILES (Simplified Molecular-Input Line-Entry System) and molecular structure of Ade was determined from a PubChem search (https://pubchem.ncbi.nlm.nih.gov/compound/Ade). Applying this model, we show that Ade has a low probability of crossing the BBB (Figure 3A) [27].

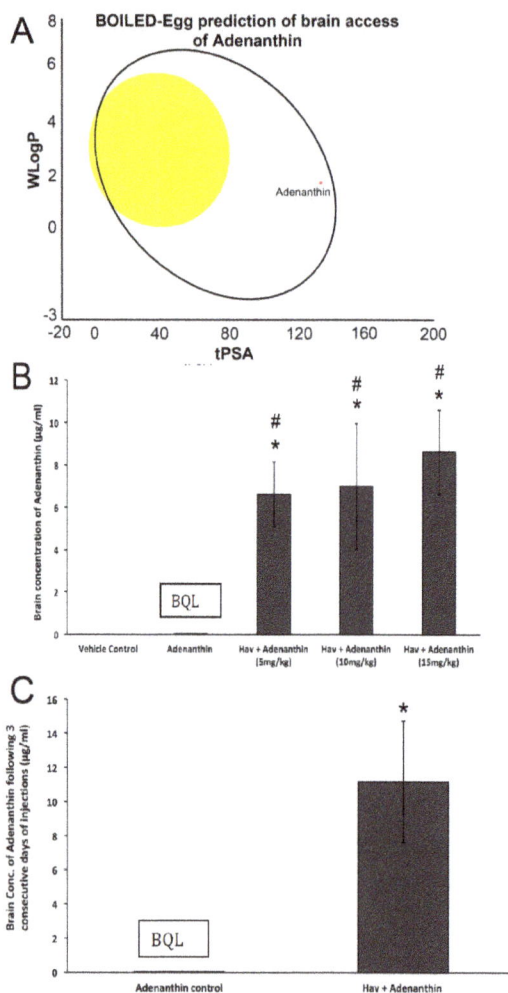

Figure 3. (**A**) Visual representation of in silico prediction of brain access and availability of Ade using the BOILED-Egg model, adapted from Daina, A.; Zoete, V. A BOILED-Egg to predict gastrointestinal absorption and brain penetration of small molecules. *ChemMedChem* **2016**, *11*, 1117–1121. [27]. The tPSA (*x*-axis) and WLogP (*y*-axis) values are plotted and the intersect of those values can predict brain access if it falls within the "yellow yolk". If the intersect lies within the white ellipse, that indicates its gastrointestinal availability. Based on this model, Ade may be available via a gastrointestinal method of delivery but may not access the brain. Quantitative analysis of Ade in the brain using LCMS under control conditions and following treatment with HAV6 after (**B**) a single treatment of different Ade doses and (**C**) three consecutive daily treatments. * p-value < 0.05 in comparison to control group. # p-value < 0.05 compared to Ade group only. Values represent the mean ± SEM for four mice per treatment group.

Consistent with the initial in silico assessment, in vivo analysis in mice revealed that Ade, when administered alone, had limited BBB permeability, with brain concentrations below the quantitative limit (BQL; 0.1 mg/mL) (Figure 3B). However, when Ade was co-administered with HAV6-cadherin peptide, there was a significant and substantial increase in brain concentrations of Ade (Figure 3B). The resulting Ade brain concentrations were statistically similar at all doses of Ade examined (Figure 3B), reaching levels of approximately 10–16 µM. In order to determine the tolerability of a complete cycle of therapy, mice were administered HAV6 cadherin peptide on 3 consecutive days. After three consecutive days of treatment Ade levels in the brain were detectable but below the quantitative limit of analysis (Figure 3C). However when Ade was combined with the HAV6 peptide, brain concentrations of Ade were approximately 22 µM (Figure 3C). Furthermore, analysis of brain samples from mice treated with consecutive injections of either HAV6 or combination of HAV6 and Ade showed no evidence of toxicity (Figure 4A,B), as indicated by absence of astrogliosis and neuro-inflammation (microglia activation) following a complete cycle of HAV6 treatment to enhance BBB delivery of Ade.

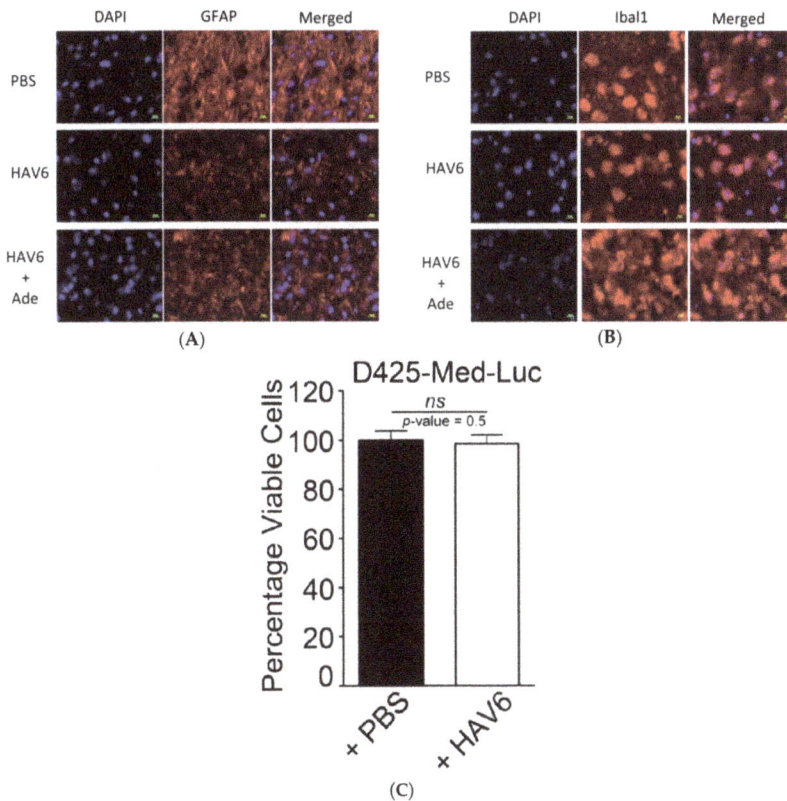

Figure 4. Representative immunofluorescence images of healthy mouse brain to detect (**A**) GFAP and (**B**) Ibal1 following the three consecutive daily treatments of PBS, HAV6 and the combination of HAV6 + Ade. The presence of HAV6 has no impact on the expression level of GFAP and Ibal1 compared to PBS control. Scale bar in bottom represent 10 µm. (**C**) Graphical representation of viability of D425-Med-Luc cells treated with either vehicle control (black bars; PBS) or HAV6 peptide (0.01 mmol; white bars). Values represent mean ± SEM of viable cells from eight monolayers per treatment group. Data are expressed as the percent viable cells compared to D425-Med-Luc cells receiving culture media alone. HAV6 has no negative impact on cell viability; *t*-test indicated *p*-value is 0.496952 and is non-significant (*ns*).

3.3. HAV6 Cadherin Peptide Improves Tumor Response to Ade

To functionally demonstrate that the HAV6 peptide can transiently disrupt the BBB and facilitate delivery of Ade to the brain, and therapeutically target PRDX1 we treated mice bearing Group 3 D425-Med-Luc tumors with either Ade alone or Ade in combination with HAV6 peptide. The mice receiving placebo or Ade alone had a median survival of 20 and 19 days post tumor cell injection respectively (Figure 5C). Furthermore, none of the mice in the placebo or Ade treatment group were able to complete the entire five cycles of treatment or survive past 22 days. In contrast, mice receiving HAV6 peptide and Ade showed significant improvement in tumor response (Figure 5A,B) compared to placebo or Ade alone, with median survival of 30 days post-tumor cell injection (Figure 5C). In addition, half of the mice receiving HAV6 peptide and Ade completed all five cycles of treatment and survived to 45 days post tumor cell injection with no sign of a tumor, as indicated by bioluminescence imaging.

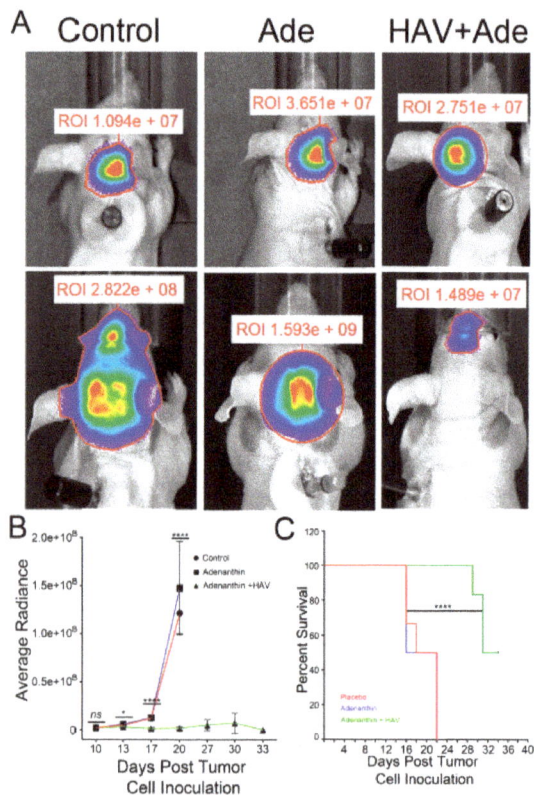

Figure 5. (**A**) Representative bioluminescence image of tumors in mice prior to the treatments (top panels) and following a 5-cycle treatment of control, Ade, or combination of HAV6 and Ade (bottom panel). (**B**) Quantitative analyses of the bioluminescence from tumors as outlined in the regions of interest (ROI) and normalized to background intensity for all tumor mice receiving control, Ade, or combination of HAV6 and Ade. (**C**) Kaplan-Maier survival curve of MBL tumor mice following the five-cycle treatment of control, Ade, or combination of HAV6 and Ade. Each treatment group consisted of six mice.

At the cellular level, tumor cells were positive for PRDX1, an ROS scavenging enzyme that is the molecular target of Ade (Figure 6). Mice receiving placebo or Ade alone showed increased cellular proliferation, compared to the animals that received both HAV6 peptide and Ade (Figure 6). Consistent with the increased tumor responsiveness observed in mice treated with Ade and HAV6 peptide, there was increased intensity of staining observed for γ-H2A.X and 53BP1, surrogate markers of DNA double strand breaks in the tumor cells (Figure 6). Similarly there was increased apoptotic death in tumor cells that received Ade and HAV6. Taken together, the data suggests that HAV6 peptide facilitates the entry of Ade into the brain, allowing Ade to reach therapeutically significant levels that can affect tumor cell death.

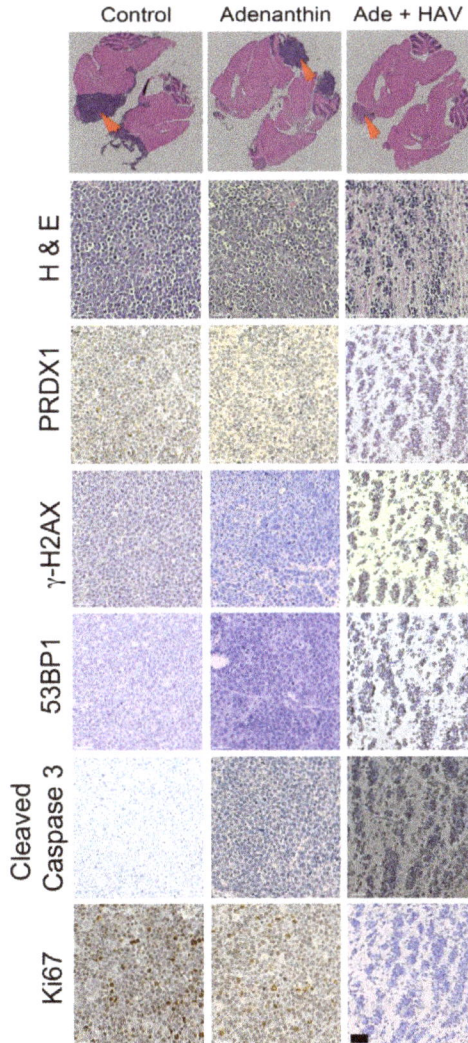

Figure 6. Magnification scans of whole brain sections stained with Hematoxylin and Eosin (H&E) of representative animals that received placebo (control), Ade or a combination of Ade and HAV peptide (Top panel). Arrows indicate those regions that were further examined for various tumor markers (Bottom Panel). White line indicates a scale bar of 100 μm.

4. Discussion

The development of drugs for the treatment of many CNS disorders has long been a difficult process, in part due to the inability of the drugs to reach relevant therapeutic concentration in the brain and the spinal cord. This limited access to the brain and spinal cord for many drugs is a direct result of the protective properties of the BBB and blood cerebral spinal fluid barrier (BCSFB). These barriers are composed of epithelial (BCSFB) or endothelial (BBB) cells with tight junctions and active efflux transporters that together restrict both the paracellular and transcellular passages of solutes into the brain [28,29]. While these barriers protect the brain and spinal cord from exposure to potential neurotoxic agents, under pathological conditions, such as a brain tumor, these barriers represent significant obstacles to therapeutic agents intended for treating these disorders. Several approaches have been used in an effort to circumvent these barriers to improve drug penetration into the brain. For treatment of brain tumors, transient disruption of the tight junctions of the BBB allowing more drug to penetrate have been a successful approach both in preclinical and clinical studies [30–33]. This can be done through the use of high concentration of osmotic agents (i.e., a high concentration of mannitol) or bradykinin analogues [30–33], or more recently, through focused, high-intensity ultrasound [34–36].

Approaches such as osmotic disruption and focused, high-intensity ultrasound, have shown promise in preclinical studies and have been used in the clinic to enhance chemotherapeutic drug delivery to brain tumors. However, the long duration of BBB disruption (up to 8 h or more) can lead to neurotoxicity and inflammation [37,38]. A long recovery time (6–72 h) has also been reported for focused high intensity ultrasound [36,39]. Other approaches, such as use of bradykinin receptor agonists, result in a shortened BBB disruption timeframe compared to either hyperosmotic or ultrasound disruption, but lack clinical effectiveness, due in part to a non-uniform distribution of bradykinin receptors in the brain microvasculature, resulting in a non-uniform distribution of the drugs [14,40]. Finally, none of the current transient BBB disruption approaches provide much control over the magnitude of BBB opening. Clearly the limitations of the existing transient disruption approaches highlight the need for alternative methods of achieving a controlled BBB disruption profile that is able to uniformly enhance BBB permeability to improve drug delivery to the brain.

Synthetic HAV6 peptide disrupts the BBB by inhibiting homodimer interactions between E-cadherin, an essential protein that forms the adherens junctions of the BBB [13,40]. Based on our previous studies, HAV6 peptide-induced changes in BBB permeability were apparent within 3 min with the HAV6 peptide being able to produce similar magnitudes of BBB disruption in all regions of the brain examined [14–16]. Furthermore, these studies also demonstrated that the magnitude of disruption could be modulated with some linear cadherin peptides, such as HAV6, producing BBB openings for small and medium sized permeability markers for short time periods, while other cyclic cadherin peptides had longer durations of action and enhanced BBB penetration of large macromolecules [14–16]. Based on these previous findings, the HAV6 cadherin peptide was selected for the present study, as it demonstrated short durations of BBB opening for small molecules without wholesale opening of the BBB to large macromolecules like albumin.

The initial characterization of HAV6 peptide mediated BBB disruption in nude mice was examined using Gd-DTPA and MRI. Following HAV6 peptide administration, there was an approximately two to four-fold enhancement in Gd-DTPA contrast throughout all areas of the brain examined. Both the magnitude and time frame of BBB disruption mediated by HAV6 peptide in the current study were consistent with our previous findings [14]. The ability to produce a rapid and uniform disruption of BBB with the HAV6 peptide is crucial for eliminating the "sink effect" that is often observed with non-uniform disruption approaches that result in reduced drug concentrations at the tumor site due to the diffusion of drugs from an area of high concentration to an area of low concentration. Furthermore, the Gd-DTPA contrast MRI studies also allowed a quantitative assessment of the duration of BBB modulation produced with the HAV6 peptide and indicated that the time frame for BBB opening was of short duration.

The anticancer properties of Ade were originally reported by Liu et al. [19] in studies examining a series of diterpenoids that formed adducts with PRDX1 and PRDX2. While initially reported as a PRDX selective inhibitor, there is evidence of potential inhibitory activity of Ade in both the thioredoxin and protein disulfide isomerase enzymes that are involved in redox metabolism in cells [41]. Studies in multiple cancer cells indicate Ade has cytotoxic responses linked with PRDX inhibition [19–21,42]. Ongoing work in our laboratory has shown that PRDX1 is overexpressed in Group-3 MBL and may contribute to radiation resistance and poor outcomes [19]. While the effects of Ade in treating primary tumors of the central nervous system have not been reported, preliminary studies with Ade in the D425 Med cells indicated cytotoxicity with a fifty percent effective concentration (EC50) of around 1.0 μM [19]. Based on the available information concerning Ade's effects in tumor cells, and its physico-chemical properties, the compound seemed an ideal choice for examining our drug delivery approach.

Adenanthin had little to no permeability through the BBB when administered alone; however, co-administration of Ade with HAV6 peptide significantly enhanced its accumulation in the brain. The concentration of Ade in the brain following HAV6 administration was similar for all doses of Ade examined. This could be due to the saturation of the influx of Ade into the brain following the disruption with HAV6 peptide. Of importance for the present study is the observation that concurrent administration of Ade (10 mg/kg dose) and HAV6 peptide resulted in brain concentrations of Ade (>10 μM) that would be expected to be in the therapeutic range required for anti-tumor activity. In addition, it should be noted that despite the large increases in Ade accumulation in the brain, there were no overt clinical signs of toxicity (i.e., no weight loss and no changes in locomotive activity in the mice receiving HAV6 peptide). Furthermore, examination of GFAP and Iba1 expression in the brain showed no enhancement in these neuroinflammatory markers as a result of HAV6 either alone or in combination with Ade. This is an important finding, as increased GFAP and Iba1 expression, markers of reactive astrocytosis and activated microglia, respectively, have been reported with both osmotic and focused ultrasound BBB modulation [43,44].

We had three cohorts of mice (placebo, Ade and Ade + HAV6 peptide) to evaluate the in vivo effect of concurrent administration of Ade and HAV6 peptide in an orthotopic Group-3 MBL model. The HAV6 peptide was not included as a cohort, as the peptide had no effect on Group-3 MBL cells (Figure 4C). The average survival of animals receiving both HAV6 and Ade was 30 days post tumor implantation, compared to 20 and 19 days for the Ade and placebo groups respectively. Furthermore, a complete 5-cycle treatment was only possible in the Ade + HAV6 treatment group. Histological studies also confirm the enhanced response of tumor mice to Ade in the presence of HAV6. Together these findings suggest that the improved delivery of Ade to the brain using HAV6 cadherin peptide to transiently open the BBB impacted favorably on treatment outcome.

In summary, we have shown that HAV6 peptide reversibly modifies BBB permeability allowing for effective delivery of Ade for the treatment of brain tumors. The transient modulation of BBB permeability produced by HAV6 cadherin peptide appeared to be well tolerated. While additional studies are required to validate both the effectiveness of adenanthin and our drug delivery approach in treating Group 3 MBL, the current studies suggest that HAV cadherin peptide can be used to enhance the brain delivery and effectiveness of poorly BBB permeable therapeutic agents. The effects of cadherin peptide based BBB modulation in terms of both the duration and magnitude of BBB opening, coupled with its effectiveness in repeated use settings could prove advantageous for treatment of brain tumors.

Author Contributions: Conceptualization, M.I.V. and D.W.M.; Data curation, N.H.O., B.V.S. and R.O.; Formal analysis, N.H.O., B.V.S., S.A., R.L. and T.M.L.; Funding acquisition, T.J.S., M.I.V. and D.W.M.; Investigation, N.H.O., B.V.S., R.O.; Methodology, B.V.S., N.H.O., N.M.D., M.I.V. and D.W.M.; Project administration, M.I.V. and D.W.M.; Resources, N.M.D., T.J.S., W.-G.W., H.-D.S., P.-T.P., M.I.V. and D.W.M.; Supervision, M.I.V. and D.W.M.; Validation, B.V.S., N.H.O., M.I.V. and D.W.M.; Writing–original draft, N.H.O., B.V.S., M.I.V. and D.W.M.; Writing–review & editing, B.V.S., T.M.L., T.J.S., M.I.V. and D.W.M.

Funding: This work was supported by the Health Sciences Center Foundation (HSCF); the Manitoba Medical Sciences Foundation (MMSF); the Children's Hospital foundation of Manitoba (CHFM); and the Cancer Care Manitoba Foundation (CCMF) to M.I.V. M.I.V. is a PJ McKenna St Baldrick's Cancer Research Scholar (Career Development Award), supported by the St Baldrick's Foundation. Additionally, funding came from an R01-NS075374 grant from the National Institute of Neurological Disorders and Stroke (NINDS), the National Institutes of Health (NIH) to T.J.S.; and the CIHR Project Grant to D.W.M. B.M.K. is an NIH pre-doctoral trainee on pharmaceutical aspects of biotechnology (T32-GM008359).

Acknowledgments: We thank Marc Symons, Maria Ruggieri and Spencer Gibson for critically reviewing the manuscript.

Conflicts of Interest: The authors declare no conflict of interest.

References

1. Siegal, T. Which drug or drug delivery system can change clinical practice for brain tumor therapy? *Neuro-Oncol.* **2013**, *15*, 656–669. [CrossRef] [PubMed]
2. McCaffrey, G.; Davis, T.P. Physiology and pathophysiology of the blood-brain barrier: P-glycoprotein and occludin trafficking as therapeutic targets to optimize central nervous system drug delivery. *J. Invest. Med.* **2012**, *60*, 1131–1140. [CrossRef] [PubMed]
3. Mandery, K.; Glaeser, H.; Fromm, M.F. Interaction of innovative small molecule drugs used for cancer therapy with drug transporters. *Br. J. Pharmacol.* **2012**, *165*, 345–362. [CrossRef] [PubMed]
4. Brownlees, J.; Williams, C.H. Peptidases, peptides, and the mammalian blood-brain barrier. *J. Neurochem.* **1993**, *60*, 793–803. [CrossRef] [PubMed]
5. Meyer, R.P.; Gehlhaus, M.; Knoth, R.; Volk, B. Expression and function of cytochrome p450 in brain drug metabolism. *Curr. Drug Metab.* **2007**, *8*, 297–306. [CrossRef]
6. Zheng, K.; Trivedi, M.; Siahaan, T.J. Structure and function of the intercellular junctions: Barrier of paracellular drug delivery. *Curr. Pharm. Des.* **2006**, *12*, 2813–2824. [CrossRef]
7. Van Itallie, C.M.; Anderson, J.M. Measuring size-dependent permeability of the tight junction using PEG profiling. *Methods Mol. Biol.* **2011**, *762*, 1–11.
8. Schneider, G.; Kirchin, M.A.; Pirovano, G.; Colosimo, C.; Ruscalleda, J.; Korves, M.; Salerio, I.; Noce, A.L.; Spinazzi, A. Gadobenate dimeglumine-enhanced magnetic resonance imaging of intracranial metastases: Effect of dose on lesion detection and delineation. *J. Magn. Reson. Imaging* **2001**, *14*, 525–539. [CrossRef]
9. Groothuis, D.R.; Vick, N.A. Brain tumors and the blood-brain barrier. *Trends Neurosci.* **1982**, *5*, 232–235. [CrossRef]
10. Vick, N.A.; Khandekar, J.D.; Bigner, D.D. Chemotherapy of brain tumors. *Arch. Neurol.* **1977**, *34*, 523–526. [CrossRef]
11. On, N.H.; Mitchell, R.; Savant, S.D.; Bachmeier, C.J.; Hatch, G.M.; Miller, D.W. Examination of blood-brain barrier (BBB) integrity in a mouse brain tumor model. *J. Neurooncol.* **2013**, *111*, 133–143. [CrossRef] [PubMed]
12. Sarkaria, J.N.; Hu, L.S.; Parney, I.F.; Pafundi, D.H.; Brinkmann, D.H.; Lacca, N.N.; Giannini, C.; Burns, T.C.; Kizilbash, S.H.; Laramy, J.K.; et al. Is the blood-brain barrier really disrupted in all glioblastomas? A critical assessment of existing clinical data. *Neuro. Oncol.* **2018**, *20*, 184–191. [CrossRef] [PubMed]
13. Lutz, K.L.; Siahaan, T.J. Modulation of the cellular junction protein E- cadherin in bovine brain microvessel endothelial cells by cadherin peptides. *Drug Delivery* **1997**, *10*, 187–193. [CrossRef]
14. On, N.H.; Kiptoo, P.; Siahaan, T.J.; Miller, D.W. Modulation of blood-brain barrier permeability in mice using synthetic E-cadherin peptide. *Mol. Pharm.* **2014**, *11*, 974–981. [CrossRef] [PubMed]
15. Alaofi, A.; On, N.; Kiptoo, P.; Williams, T.D.; Miller, D.W.; Siahaan, T.J. Comparison of linear and cyclic His-Ala-Val peptides in modulating the blood-brain barrier permeability: Impact on delivery of molecules to the brain. *J. Pharm. Sci.* **2016**, *105*, 797–807. [CrossRef]
16. Ulapane, K.R.; On, N.; Kiptoo, P.; Williams, T.D.; Miller, D.W.; Siahaan, T.J. Improving brain delivery of biomolecules via BBB modulation in mouse and rat: Detection using MRI, NIRF, and mass spectrometry. *Nanotheranostics.* **2017**, *1*, 217–231. [CrossRef]
17. Taylor, M.D.; Northcott, P.A.; Korshunov, A.; Remke, M.; Cho, Y.-J.; Clifford, S.C.; Eberhart, C.G.; Parsons, D.W.; Rutkowski, S.; Gajjar, A.; et al. Molecular subgroups of medulloblastoma: The current consensus. *Acta. Neuropathol.* **2012**, *123*, 465–472. [CrossRef]
18. Ding, C.; Fan, X.; Wu, G. Peroxiredoxin 1- an antioxidant enzyme in cancer. *J. Cell. Mol. Med.* **2017**, *21*, 193–202. [CrossRef]
19. Sajesh, B.; On, N.; Refaat, O.; Fediuk, H.; Li, L.; Alrushaid, S.; Kopec, B.M.; Wang, W.G.; Pu, J.; Sun, H.D.; et al. Peroxiredoxin 1 is a therapeutic target in Group-3 medulloblastoma (MBRS-50). *Neuro-Oncol.* **2018**, *20* (Suppl. 2), i139. [CrossRef]
20. Liu, C.-X.; Yin, Q.-Q.; Zhou, H.-C.; Wu, Y.-L.; Pu, J.-X.; Xia, L.; Liu, W.; Huang, X.; Jiang, T.; Wu, M.-X.; et al. Adenanthin targets peroxiredoxin I and II to induce differentiation of leukemic cells. *Nat. Chem. Biol.* **2012**, *8*, 486–493. [CrossRef]
21. Hou, J.-K.; Huang, Y.; He, W.; Wan, Z.-W.; Fan, L.; Liu, M.-H.; Xiao, W.-L.; Sun, H.-D.; Chen, G.-Q. Adenanthin targets peroxiredoxin I/II to kill hepatocellular carcinoma cells. *Cell Death Dis.* **2014**, *5*, e1400. [CrossRef]

22. On, N.H.; Savant, S.; Toews, M.; Miller, D.W. Rapid and reversible enhancement of blood-brain barrier permeability using lysophosphatidic acid. *J. Cereb. Blood Flow Metab.* **2013**, *33*, 1944–1954. [CrossRef]
23. He, X.M.; Wikstrand, C.J.; Friedman, H.S.; Bigner, S.H.; Pleasure, S.; Trojanwski, J.Q. Differentiation characteristics of newly established medulloblastoma cell lines (D384 Med, D425 Med, and D458 Med) and their transplantable xenografts. *Lab. Invest.* **1991**, *64*, 833–843.
24. Huang, X.; Sarangi, A.; Ketova, T.; Litingtung, Y.; Cooper, M.K.; Chiang, C. Intracranial orthotopic allografting of medulloblastoma cells in immunocompromised mice. *J. Vis. Exp.* **2010**, *44*, 2153. [CrossRef]
25. Baumann, B.C.; Dorsey, J.F.; Benci, J.L.; Joh, D.Y.; Kao, G.D. Stereotactic intracranial implantation and in vivo bioluminescent imaging of tumor xenografts in a mouse model system of glioblastoma multiforme. *J. Vis. Exp.* **2012**, *67*, e4089. [CrossRef]
26. Sajesh, B.V.; McManus, K.J. Targeting SOD1 induces synthetic lethal killing in BLM- and CHEK2-deficient colorectal cancer cells. *Oncotarget.* **2015**, *6*, 27907–27922. [CrossRef]
27. Daina, A.; Zoete, V. A BOILED-Egg to predict gastrointestinal absorption and brain penetration of small molecules. *ChemMedChem* **2016**, *11*, 1117–1121. [CrossRef]
28. On, N.H.; Miller, D.W. Transporter-based delivery of anticancer drugs to the brain: Improving brain penetration by minimizing drug efflux at the blood-brain barrier. *Curr. Pharm. Des.* **2014**, *20*, 1499–1509. [CrossRef]
29. Abbott, N.J.; Patabendige, A.A.; Dolman, D.E.; Yusof, S.R.; Begley, D.J. Structure and function of the blood-brain barrier. *Neurobiol. Dis.* **2010**, *37*, 13–25. [CrossRef]
30. Angelov, L.; Doolittle, N.D.; Kraemer, D.F.; Siegal, T.; Barnett, G.H.; Peereboom, D.M.; Stevens, G.; McGregor, J.; Jahnke, K.; Lacy, G.A.; et al. Blood-brain barrier disruption and intra-arterial methotrexate-based therapy for newly diagnosed primary CNS lymphoma: A multi-institutional experience. *J. Clin. Oncol.* **2009**, *27*, 3503–3509. [CrossRef]
31. Boockvar, J.A.; Tsiouris, A.J.; Hofstetter, C.P.; Kovanlikaya, I.; Fralin, S.; Kesavabhotla, K.; Seedial, S.M.; Pannullo, S.C.; Schwartz, T.H.; Stieg, P.; et al. Safety and maximum tolerated dose of superselective intrarterial cerebral infusion of bevacizumab after osmotic blood-brain barrier disruption for recurrent malignant glioma. *J. Neurosurg.* **2011**, *114*, 624–632. [CrossRef]
32. Kroll, R.A.; Pagel, M.A.; Muldoon, L.L.; Roman-Goldstein, S.; Fiamengo, S.A.; Neuwelt, E.A. Improving drug delivery to intracerebral tumor and surrounding brain in a rodent model: A comparison of osmotic versus bradykinin modifications of the blood-brain and/or blood-tumor barriers. *Neurosurgery* **1998**, *43*, 879–886. [CrossRef]
33. Fortin, D. Drug delivery technology to the CNS in the treatment of brain tumors: The Sherbrooke experience. *Pharmaceutics* **2019**, *11*, e248. [CrossRef]
34. Meairs, S. Facilitation of drug transport across the blood-brain barrier with ultrasound and microbubbles. *Pharmaceutics* **2015**, *7*, 275–293. [CrossRef]
35. Meng, Y.; Suppiah, S.; Surendrakumar, S.; Bigioni, L.; Lipsman, N. Low-intensity MR-guided focused ultrasound mediated disruption of the blood-brain barrier for intracranial metastatic diseases. *Front. Oncol.* **2018**, *8*, 338. [CrossRef]
36. Alkins, R.; Burgess, A.; Ganguly, M.; Francia, G.; Kerbel, R.; Wels, W.S. Hynynen, K. Focused ultrasound delivers targeted immune cells to metastatic brain tumors. *Cancer Res.* **2013**, *73*, 1892–1899. [CrossRef]
37. Siegal, T.; Rubinstein, R.; Bokstein, F.; Schwartz, A.; Lossos, A.; Shalom, E.; Chisin, R.; Gomori, J.M. In vivo assessment of the window of barrier opening after osmotic blood-brain barrier disruption in humans. *J. Neurosurg.* **2000**, *92*, 599–605. [CrossRef]
38. Suzuki, M.; Iwasaki, Y.; Yamamoto, T.; Konno, H.; Kudo, H. Sequelae of the osmotic blood-brain barrier opening in rats. *J. Neurosurg.* **1988**, *69*, 421–428. [CrossRef]
39. Shiekov, N.; McDannold, N.; Sharma, S.; Hynynen, K. Effect of focused ultrasound applied with an ultrasound contrast agent on the tight junctional integrity of the brain microvascular endothelium. *Ultrasound Med. Biol.* **2008**, *34*, 1093–1104. [CrossRef]
40. Kiptoo, P.; Sinaga, E.; Calcagno, A.M.; Zhao, H.; Kobayashi, N.; Tambunan, U.S.F.; Siahaan, T.J. Enhancement of drug absorption through the blood-brain barrier and inhibition of intercellular tight junction resealing by E-cadherin peptides. *Mol. Pharm.* **2011**, *8*, 239–249. [CrossRef]

41. Muchowicz, A.; Firczuk, M.; Chlebowska, J.; Nowis, D.; Stachura, J.; Barankiewicz, J.; Trzeciecka, A.; Kłossowski, S.; Ostaszewski, R.; Zagożdżon, R.; et al. Adenanthin targets proteins involved in the regulation of dishulphide bonds. *Biochem. Pharmacol.* **2014**, *89*, 210–216. [CrossRef] [PubMed]

42. Bajor, M.; Zych, A.O.; Graczyk-Jarzynka, A.; Muchowicz, A.; Firczuk, M.; Trzeciak, L.; Gaj, P.; Domagala, A.; Siernicka, M.; Zagozdzon, A.; et al. Targeting peroxiredoxin 1 impairs growth of breast cancer cells and potently sensitises these cells to prooxidant agents. *Br. J. Cancer* **2018**, *119*, 873–884. [CrossRef] [PubMed]

43. Godinho, B.M.D.C.; Henninger, N.; Bouley, J.; Alterman, J.F.; Haraszti, R.A.; Gilbert, J.W.; Sapp, E.; Coles, A.H.; Biscans, A.; Nikan, M.; et al. Transvascular delivery of hydrophobically modified siRNAs: Gene silencing in the rat brain upon disruption of the blood-brain barrier. *Mol. Ther.* **2018**, *26*, 2580–2591. [CrossRef] [PubMed]

44. Alonso, A.; Reinz, E.; Fatar, M.; Hennerici, M.G.; Meairs, S. Clearance of albumin following ultrasound-induced blood-brain barrier opening is mediated by glial but not neuronal cells. *Brain Res.* **2011**, *1411*, 9–16. [CrossRef] [PubMed]

pharmaceutics

MDPI

Article

Inclusion Complexes of Melphalan with Gemini-Conjugated β-Cyclodextrin: Physicochemical Properties and Chemotherapeutic Efficacy in In-Vitro Tumor Models

Waleed Mohammed-Saeid [1,2], Abdalla H Karoyo [3], Ronald E Verrall [3], Lee D Wilson [3] and Ildiko Badea [1,*]

[1] Drug Discovery and Development Research Group, College of Pharmacy and Nutrition, University of Saskatchewan, 107 Wiggins Rd, Saskatoon, SK S7N 5E5, Canada

[2] College of Pharmacy, Taibah University, Medina 42353, Saudi Arabia

[3] Department of Chemistry, University of Saskatchewan, 110 Science Place, Saskatoon, SK S7N 5C9, Canada

[*] Correspondence: ildiko.badea@usask.ca; Tel.: +1-306-966-6349

Received: 12 July 2019; Accepted: 10 August 2019; Published: 22 August 2019

Abstract: β-cyclodextrin (βCD) has been widely explored as an excipient for pharmaceuticals and nutraceuticals as it forms stable host–guest inclusion complexes and enhances the solubility of poorly soluble active agents. To enhance intracellular drug delivery, βCD was chemically conjugated to an 18-carbon chain cationic gemini surfactant which undergoes self-assembly to form nanoscale complexes. The novel gemini surfactant-modified βCD carrier host (hereafter referred to as 18:1βCDg) was designed to combine the solubilization and encapsulation capacity of the βCD macrocycle and the cell-penetrating ability of the gemini surfactant conjugate. Melphalan (Mel), a chemotherapeutic agent for melanoma, was selected as a model for a poorly soluble drug. Characterization of the 18:1βCDg-Mel host–guest complex was carried out using 1D/2D ^{1}H NMR spectroscopy and dynamic light scattering (DLS). The 1D/2D NMR spectral results indicated the formation of stable and well-defined 18:1βCDg-Mel inclusion complexes at the 2:1 host–guest mole ratio; whereas, host–drug interaction was attenuated at greater 18:1βCDg mole ratio due to hydrophobic aggregation that accounts for the reduced Mel solubility. The in vitro evaluations were performed using monolayer, 3D spheroid, and Mel-resistant melanoma cell lines. The 18:1βCDg-Mel complex showed significant enhancement in the chemotherapeutic efficacy of Mel with 2–3-fold decrease in Mel half maximal inhibitory concentration (IC_{50}) values. The findings demonstrate the potential applicability of the 18:1βCDg delivery system as a safe and efficient carrier for a poorly soluble chemotherapeutic in melanoma therapy.

Keywords: cationic gemini surfactant; melphalan; inclusion complex; ROESY NMR spectroscopy; 3D spheroid; drug-resistant melanoma

1. Introduction

Melanoma, the malignant cancer of melanocytes, is the most aggressive form of skin cancer which causes the most skin-cancer related deaths [1]. According to the World Health Organization (WHO), over 132,000 new cases of melanoma are diagnosed annually [2]. In its early stages, melanoma can be treated by surgical incision with high survival rate. In the stage of in-transit metastases, in which the metastases are >2 cm from the primary lesion but within the nodal basin, the response to local and systemic therapeutic options is moderate with 5-year survival of 32.8% [3,4]. However, advanced metastatic melanoma shows limited response to current therapeutic options with very low survival rate of less than 5% over 5 years [5]. Systemic chemotherapy is the first-line option for most patients with metastatic melanoma.

Melphalan (Mel) (Figure 1a) is used regionally as an adjunctive therapy for in-transit metastatic melanoma [6]. The lipophilic nature of Mel requires the use of a co-solvent (e.g., propylene glycol) for parenteral administration. Propylene glycol is known to cause toxicity that includes nephrotoxicity, cardiac arrhythmia, and metabolic acidosis [7]. As a result, the use of Mel in melanoma therapy is limited to isolated limb perfusion/infusion which is an invasive method that requires special medical care [6]. Therefore, attempts to improve the solubility and stability of Mel were conducted by either chemical modification of the molecule or by engineering novel drug delivery systems, such as nano-systems [8–11]. A nanoscale drug delivery system has several advantages, it can: (1) improve solubility of poorly soluble drugs, (2) enhance chemical/biological stability, (3) improve pharmacokinetic profile and biodistribution, (4) increase tumor-specific uptake (passive targeting), (5) minimize drug resistance, (6) achieve drug controlled release, and (7) afford delivery of multiple drug components [12,13]. Our strategy is to create delivery systems that could improve the therapeutic use of Mel addressing both issues of solubility and biological activity at the same time. Cyclodextrins (CDs) form stable host–guest complexes with a variety of organic and inorganic molecules and have been widely employed as versatile carriers for poorly soluble drugs [14]. CDs (Figure 1b,c) are naturally occurring cyclic oligosaccharides consisting of 6 (αCD), 7 (βCD), or 8 (γCD) α-D-glucopyranose units linked by (α-1,4) glycosidic bonds [15]. CDs form a truncated cone with a toroidal structure (Figure 1c) in which the hydroxyl groups of glucopyranose units reside at the narrow (primary) and wide (secondary) tori of the CD annular structure. The CD macrocycle has a hydrophilic outer surface and lipophilic inner cavity that is capable of forming noncovalent inclusion complexes with a large variety of guest molecules [16]. Capitalizing on the ability of the CDs to form host–guest inclusion complexes, such systems have been used in pharmaceutical formulations to increase the apparent water solubility of poorly soluble drugs so as to improve their bioavailability. In addition, CDs and its derivatives serve to provide: (1) enhanced drug stability (thermal, photosensitivity, and chemical), (2) reduced drug mucosal irritation, (3) reduced drug resistance, and (4) controlled release of the drug [17,18]. Several synthetic strategies have been employed to engineer CD-based carriers with enhanced pharmaceutical properties along with reduced systemic toxicity [18]. For example, the introduction of bulky derivatives can limit the formation of intramolecular hydrogen bonding, consequently enhancing the aqueous solubility of CDs while improving their inclusion capacity [14]. Luke et al. reported a 35-fold increase in aqueous solubility for a sulfobutyl-ether-β-cyclodextrin (SEB-β-CD) derivative relative to native βCD [19]. In addition to chemical derivatization, CDs have been chemically conjugated to a variety of functional moieties (i.e., polymers, lipids, and peptides) to create biofunctional and supramolecular complexes [20,21]. For instance, several amphiphilic moieties have been conjugated to CDs to create self-assembling supramolecular structures with improved drug loading capacity and enhanced cellular uptake [22–25].

In the present work, we evaluate a novel βCD-based carrier modified with an unsaturated 18-carbon chain gemini surfactant conjugate, herein referred to as 18:1βCDg (Figure 1d), as a potential advanced drug delivery system for Mel in melanoma therapy. The objectives of the current study are generally two-fold: 1) to synthetically engineer a novel CD-based carrier for Mel with improved therapeutic efficiency and low carrier-cellular toxicity, and 2) to characterize the structure of the host–guest interaction between the carrier and drug. The host–guest complex of the 18:1βCDg-Mel system was investigated using 1D/2D ^{1}H NMR spectroscopy in aqueous solution. NMR results herein show the formation of well-defined carrier–drug inclusion complexes. We previously reported a cationic gemini surfactant-βCD conjugate with 12-carbon chain (12 βCDg) for the delivery of a poorly soluble drugs including curcumin analogs [26–28] and Mel in a melanoma cell line model [29]. The 12 βCDg-Mel complex significantly improved the efficiency of Mel drug and showed no intrinsic toxicity as it did not alter the cellular death triggered by Mel [29]. However, the stability of the formed host–drug inclusion complex and the efficiency of the produced 12 βCDg-Mel system were limited due to the self-inclusion/self-assembly of the terminal alkyl chains of the carrier agent within the βCD cavity [27–29]. Therefore, the newly developed delivery system herein using a cationic gemini surfactant-βCD conjugate with 18-carbon tail (18:1βCDg) is anticipated to overcome these limitations. The in vitro efficiency of the 18:1βCDg-Mel complex using monolayer, 3D spheroid, and Mel-resistant melanoma cell lines demonstrates the potential applicability of the 18:1βCDg delivery system as a safe and efficient carrier for poorly soluble chemotherapeutic in melanoma therapy. This study is anticipated to provide a greater understanding of the structure–function relationship of 18:1βCDg as a carrier agent for poorly soluble drug with optimal therapeutic properties.

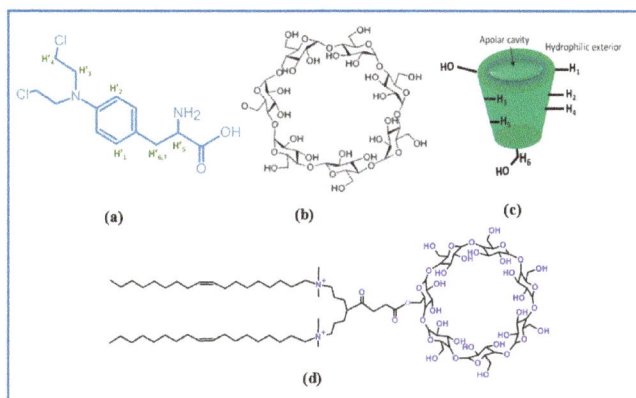

Figure 1. Chemical structures of (**a**) Melphalan (Mel), (**b**) β-Cyclodextrin (βCD) macrocycle, (**c**) βCD toroidal structure showing cavity and external protons, and (**d**) 18:1βCDg host (carrier).

2. Material and Methods

2.1. Preparation of Inclusion Complexes

Melphalan (Mel) and β-Cyclodextrin (βCD) were purchased from Sigma-Aldrich (Oakville, ON, Canada). Synthesis and characterization of 18:1-7NβCD-18:1 gemini surfactant [18:1βCDg] was described elsewhere [26,30]. Figure 1 shows the chemical structures of Mel, βCD, and 18:1βCDg. For physicochemical characterization, the complexes of Mel with βCD or 18:1βCDg were prepared in different carrier-to-drug molar ratios (1:1, 2:1, 3:1, and 5:1). A stock solution of Mel (1 mg/mL) was prepared in acidified ethanol (0.1% HCl). Stock solutions of βCD and 18:1βCDg were prepared in Milli-Q water at 10 mM concentration. An appropriate volume of Mel solution was mixed with βCD and an aqueous 18:1βCDg dispersion to yield the required carrier-to-drug molar ratios. Formulations were frozen at −80 °C for 2 h and transferred to a cascade freeze dryer (Labconco, Kansas City, MO, USA) at −80 °C and 0.03 mBar vacuum and lyophilized for 24 h. The lyophilized formulations were rehydrated in water (or in deuterium oxide for ^{1}H NMR and 1D/2D ROESY experiments) and shaken in the orbital shaker for 1 h at room temperature prior to evaluation.

2.2. Physicochemical Characterization

2.2.1. Size and Zeta Potential Measurements

Eight hundred microliters of rehydrated formulations were transferred into a special cuvette (DTS1061, Malvern Instruments, Worcestershire, UK) for size distribution and zeta potential measurements using a Zetasizer Nano ZS instrument (Malvern Instruments, Worcestershire, UK). Each sample was measured in triplicate and the results are expressed as an average ± standard deviation (SD), where $n = 3$ with a corresponding polydispersity index (PDI) value.

2.2.2. NMR Spectroscopy

1D/2D ^{1}H ROESY NMR spectra in solution were recorded on a 500 MHz 3-channel Bruker Avance spectrometer in D_2O at 298 K. Chemical shifts (δ) are reported in ppm with respect to trimethylsilane (TMS; δ 0.0 ppm) as external standard. ^{1}H 1D spectra were obtained with water solvent suppression, a 2s recycle delay, and a 90° pulse length (10 μs). 2D ROESY NMR spectra were obtained at variable parameters which were optimized as follows: spin-lock time of 350 ms, recycle delay of 3 s with 8 scans and 1k data points. Complexation-induced chemical shift (CIS) values were calculated as $\Delta\delta = \delta_{\text{free}} - \delta_{\text{complex}}$.

2.3. In Vitro Evaluation

Human malignant melanoma (A375) cell line (ATCC®® CRL-1619™, Cedarlane, Burlington, ON, Canada) was cultured in Dulbecco's modified Eagle's medium (DMEM) supplemented with 10% fetal bovine serum and 1% antibiotic and incubated at 37 °C under an atmosphere of 5% CO_2/95% air. For all experiments, culturing conditions and passage numbers were kept constant.

2.3.1. Determination of Half Maximal Inhibitory Concentration (IC_{50}) in Monolayer Melanoma Cell Culture

A375 cell lines were seeded at a density of 1×10^4 cells/well and incubated for 24 h. Cells were treated with serial concentrations of Mel (32 nM to 250 μM), either alone or as 18:1βCDg-Mel complexes at a 2:1 molar ratio, and 18:1βCDg alone (64 nM to 500 μM) in quadruplicate. After 24 h of treatment, cell viability was assessed using MTT (3-(4,5-dimethylthiazolyl-2)-2,5-diphenyltetrazolium bromide) (Sigma-Aldrich, Markham, ON, Canada) assay. The supplemented DMEM containing the treatment was removed from the wells and replaced with 0.5 mg/mL MTT in supplemented media and incubated at 37 °C for 3 h. The supernatant was removed and each well washed with phosphate-buffered saline (PBS). The formed formazan was dissolved in DMSO. Plates were incubated for 10 min at 37 °C. Absorbance was measured at 580 nm using BioTek microplate reader (Bio-Tek Instruments, Winooski, VT, USA). The half maximal inhibitory concentration (IC_{50}) was determined by calculating the fraction of dead cells and plotting the data with a 4-parameter curve generated by GEN5 software from BioTek.

2.3.2. In Vitro Efficiency in Spheroid Melanoma Cell

To evaluate the efficiency of developed formulations in three-dimensional cell culture, melanoma cells (A375) were cultured at a density of 1×10^4 cells/well in 96-well spheroid microplate (Corning Inc., Tewksbury, MA, USA). Cells were incubated at 37 °C under an atmosphere of 5% CO_2/95% air for 48 h prior to treatment with Mel and 18:1βCDg-Mel complexes. After 48 h, cells were treated twice (2nd treatment 24 h after the first) with Mel alone and with 18:1βCDg-Mel complexes at 2:1 molar ratio with final Mel concentrations of 30 and 80 μM in quadruplicate. The CytoTox-ONE Homogeneous Membrane Integrity Assay (Promega Corporation, Madison, WI, USA) was used to determine the lactate dehydrogenase (LDH) activity 12 h after the 2nd treatment. An equal volume of CytoTox-ONE™ Reagent (100 μL) was added to cell culture medium in each well and incubated at 23 °C for 10 min. Fifty microliters of stop solution was added to each well and fluorescence was measured using an excitation wavelength of 560 nm and an emission wavelength of 590 nm by using a BioTek microplate reader. Maximum LDH release was used as a control by adding 2 μL of lysis solution to 4 wells of nontreated cells. % cell toxicity was calculated as follows:

$$\% \text{ Cell Toxicity} = \frac{(\text{Experimental} - \text{Culture Medium Background})}{(\text{Maximum LDH Release} - \text{Culture Medium Background})}$$

2.3.3. In Vitro Efficiency in Mel-Resistant Melanoma Cell Lines

Mel-resistant melanoma cultures were created by using A375 cell line, as described previously [29]. In brief, A375 cells cultured in 25 cm^2 tissue culture flasks and treated with increasing concentrations of Mel from 100 nM to 60 μM over 9 weeks to induce drug resistance. The Mel-resistant cells were seeded at a density of 1.5×10^4 cells/well in 96-well plate and incubated for 24 h. After the incubation period, cells were treated with Mel alone and with 18:1βCDg-Mel complexes at 2:1 molar ratio with final Mel concentrations of 30 and 80 μM in quadruplicate. The MTT assay, as described above, was used to determine % cell toxicity (compared to nontreated cells).

2.4. Statistical Analysis

Statistical analyses were performed using SPSS software (v 24.0). Independent *t*-test and one-way analysis of variance (Bonferroni's post hoc tests) were used. Significant differences were considered at $P < 0.05$ level.

3. Results and Discussion

3.1. Physicochemical Characterization

3.1.1. H NMR and 1D/2D ROESY

Several spectroscopy methods (NMR, circular dichroism, and FT-IR) have been utilized to characterize the structure of host–guest inclusion complexes, along with X-ray diffractometry (XRD) and mass spectrometry (MS) [31,32]. Solution state NMR spectroscopy is a powerful tool for elucidating the molecular level structure and host/guest stoichiometry by analyzing the complexation-induced shifts (CIS). In particular, two-dimensional NMR affords an understanding of the *through-space* dipolar interactions and inclusion geometry of the guest within the βCD cavity [27,33,34]. In 2D ROESY NMR, the nuclear Överhauser effect (NOE also ROE) is employed to elucidate the noncovalent interactions between nuclei that reside in close spatial proximity (~5 Å) [35,36]. Although 2D NMR ROESY cross-correlations are related to NOE, other correlations may arise due to chemical and conformational (rotational) exchanges [37].

In the current work, 1D/2D ^{1}H ROESY NMR was employed to elucidate the structure of the 18:1βCDg and its complexes with Mel. The ^{1}H signals in Figure 2 were assigned according to previous reports [27,38] and the simulated spectra (cf. Figures S1 and S2 Supporting information). Starting with the 18:1βCDg carrier molecule, evidence of the formation of the gemini surfactant-grafted βCD is shown by the substantial broadening of the βCD resonance lines (~3.0–4.5 ppm) relative to the native βCD (Figure 2a,c, respectively), along with the emergence of the gemini surfactant signals at δ ~0.5–1.0 ppm (Figure 2c). Similar line broadening effects were reported previously for grafted βCD-based hosts in D$_2$O [37]. The downfield shifts ($\Delta\delta$ ~0.04–0.10 ppm, Table 1) of the external and framework protons of the 18:1βCDg (i.e., H$_1$, H$_4$, and H$_6$; cf. Figure 1b) indicate an induced conformational change of the βCD macrocycle upon grafting. The internal cavity protons H$_3$ and H$_5$ of the βCD moiety of the 18:1βCDg carrier (Figure 2 highlighted area II) are characterized by upfield shifts (~−0.03 and −0.08 ppm; cf. Table 1) that indicate a possible inter- or intramolecular inclusion of part of the gemini surfactant moiety (cf. Figure 2c), consistent with shielding effects [38]. The alkenic signals of the gemini surfactant ~5.4 ppm (Figure 2 highlighted area I) are characterized by substantial broadening and upfield shift indicating a change of conformation and/or environment upon drug complexation as described above.

Table 1. Complexation-induced chemical shift (CIS) data of the host (βCD and 18:1βCDg) and its complexes with Melphalan. The CIS values (in brackets) were calculated as Δδ = δ$_{complex}$ − δ$_{free}$, where negative and positive values represent shielding (upfield) and deshielding (downfield) effects, respectively. * Chemical shift values are difficult to decipher.

Host:Guest	Ratio	¹H Nuclei					
		H1	H2	H3	H4	H5	H6
βCD	–	4.98	3.56	3.87	3.49	3.77	3.76
βCD:Mel	2:1	4.97 (−0.01)	3.57 (0.01)	3.85 (−0.02)	3.48 (−0.01)	3.70 (−0.07)	3.78 * (0.02)
18:1βCDg	–	5.02 (0.04)	*	3.84 (−0.03)	3.59 * (0.10)	3.69 (−0.08)	3.83 (0.07)
18:1βCDg:Mel	1:1	4.97 (−0.01)	3.57 (0.01)	3.80 (−0.07)	3.50 (0.01)	3.61 (−0.16)	3.77 (0.01)
18:1βCDg:Mel	2:1	4.98 (0.00)	3.57 (0.01)	3.81 (−0.06)	3.52 (0.03)	3.62 (−0.15)	3.79 (0.03)
18:1βCDg:Mel	3:1	4.99 (0.01)	3.57 (0.01)	3.81 (−0.06)	3.52 (0.03)	3.63 (−0.14)	3.79 (0.03)
18:1βCDg:Mel	5:1	5.01 (0.03)	3.60 * (0.04)	3.83 (−0.04)	3.55 * (0.06)	3.67 (−0.10)	3.82 (0.06)

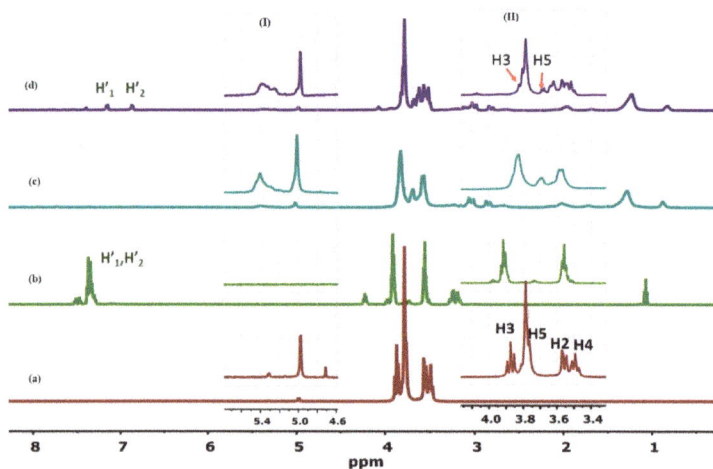

Figure 2. ¹H NMR spectra of (**a**) βCD, (**b**) Mel, (**c**) 18:1βCDg, and (**d**) 18:1βCDg-Mel complex at 2:1 molar ratio obtained in D₂O at 298 K. The insets show enlarged resonance regions for the alkenic signal (H11) of the gemini (**I**) and βCD signals (**II**). Some signals for the drug (H′₁, H′₂) and the interior and exterior protons of βCD are labeled.

In the case of the 18:1βCDg-Mel, various host:guest complexes (1:1, 2:1, 3:1 and 5:1) were studied. The 1:1 and 2:1 18:1βCDg-Mel complexes showed the most affected CIS values as listed in Table 2, indicating that the 18:1βCDg carrier forms 1:1/2:1 stoichiometry with the drug. The 1D ¹H NMR spectrum of the 18:1βCDg-Mel complex at the 2:1 mole ratio is represented in Figure 2d and reveals substantial shielding of the βCD internal protons (~3.7–4.0 ppm) and the gemini surfactant (~0.5–1.0 ppm) resonances. This provides evidence for the inclusion of the Mel within the cavity of the βCD moiety with possible involvement of the gemini surfactant moiety.

Table 2. Physiochemical properties (size and zeta potential) of different melphalan formulations. (–) indicates unmeasurable size or zeta potential. [a] Results are an average of n ≥ 3 ± standard deviation (SD).

Component[a]	Size nm ± SD	PDI	Zeta Potential mV ± SD
Mel (1 mM)	–	–	–
βCD (2 mM)	–	–	–
18:1βCDg (2 mM)	170 ± 17	0.329 ± 0.047	+14 ± 3
βCD-Mel (2:1 mole ratio)	–	–	–
18:1βCDg-Mel (2:1 mole ratio)	160 ± 15	0.430 ± 0.04	+46 ± 2

The 2D ^{1}H ROESY NMR spectra of 18:1βCDg and its complexes with Mel are shown in Figures 3–5. In Figure 3, the spectra of the unbound 18:1βCDg carrier displayed cross-peaks due to typical interactions between the backbone H_1 proton of βCD with the cavity interior (panel a). Furthermore, various cross-peaks are arising from intra-/intermolecular interactions of the 18:1βCDg carrier (Figure 3, panel c). More recently, the self-inclusion of a 12-carbon chain gemini (12 βCDg) within the βCD cavity was reported [27]. In Figure 3, cross-peaks between the βCD cavity and the alkyl chain of the gemini surfactant (highlighted area b) are relatively weak and characterized by positive correlations (green contours in Figure 3) due to possible conformation (rotational) and chemical exchanges [37]. The formation of host–guest inclusion complexes is expected to be thermodynamically more favorable for a saturated long hydrocarbon chain than shorter ones [39]. However, intermolecular aggregation of the gemini surfactant chain is anticipated to be more prominent in the case of 18:1βCDg due to the presence of the unsaturated bond that causes the tail to bend with less flexibility, which can reduce the self-inclusion of the tail region within the βCD cavity.

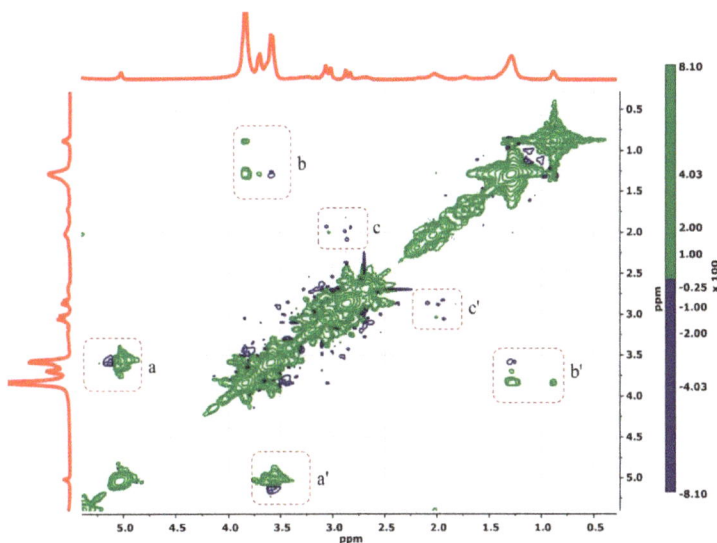

Figure 3. 2D ROESY ^{1}H NMR spectra of unbound 18:1βCDg in D_2O and 295 K. The inter- and intra-molecular cross-peaks are shown as highlighted areas **a–c**, and the bar to the right shows positive and negative correlations.

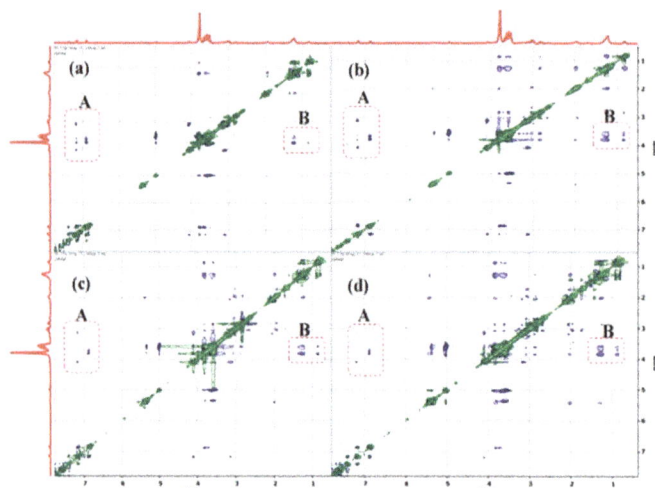

Figure 4. 2D ROESY ^{1}H NMR of the 18:1βCDg-Mel complexes at (**a**) 1:1, (**b**) 2:1, (**c**) 3:1, and (**d**) 5:1 mole ratios. The cross-peaks for the βCD-Mel and βCD-gemini surfactant interactions are shown in panels A and B. (**A**) interaction between Mel and βCD cavity and (**B**) gemini surfactant alkenic region with βCD cavity.

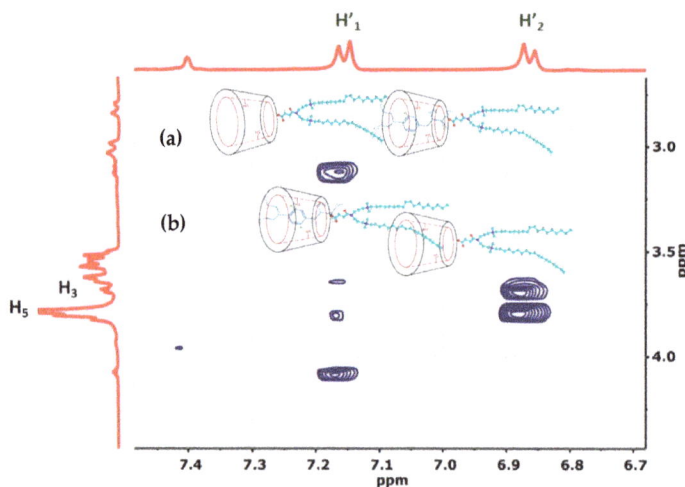

Figure 5. Expanded 2D ROESY ^{1}H NMR of the 18:1βCDg-Mel complex at 2:1 molar ratio in D$_2$O at 295 K. The possible geometries are depicted in the insets.

The relative comparison of the 2D ROESY spectra of the complexes of 18:1βCDg with Mel at various carrier:drug mole ratios (1:1, 2:1, 3:1, and 5:1) were used to further support the effect of stoichiometry on maximum solubility (Figure 4). According to the 2D ROESY spectra in Figure 4, the increased carrier–drug interactions at the 1:1 and 2:1 mole ratios are evidenced by the intense cross-peaks and suggest optimal solubilizing conditions for the drug, more so at the 2:1 mole ratio due to the much greater intensity. It is noteworthy that the interaction between the gemini surfactant and the βCD moiety of the carrier in the presence of the Mel drug as a co-guest illustrates the importance of "cooperativity" in the host–guest inclusion complex. Evidence of cooperative association was reported recently for other βCD-based complexes [37,40]. In the present study, the "cooperative interaction" only occurred when Mel was added to the 18:1βCDg. The foregoing solubility results inferred from the 2D NMR results agree with the CIS data herein (Table 1) and with solubility evaluation results reported previously using mass spectrometry [41]. The ^{1}H NMR CIS data in Table 1 provide evidence of destabilization of the carrier–drug inclusion complex due to hydrophobic effects and conformational motility of the 18:1βCDg at higher carrier loading, as indicated by the downfield shifts of the βCD extracavity (or framework) nuclei (i.e., H_1, H_6, and H_4). The efficiency of 18:1βCDg to increase the aqueous solubility of Mel was evaluated using mass spectrometry, where the highest aqueous solubility was determined at the 2:1 host–guest mole ratio [41]. Increasing the 18:1βCDg-Mel mole ratios to 3:1 and 5:1 caused significant reduction in Mel solubility, consistent with the attenuated βCD-Mel signals at the expense of the βCD-gemini surfactant signals for these systems (cf. Figure 4c,d). The foregoing discussion suggests increased interaction of the gemini surfactant moiety with the βCD cavity at higher host ratio, whereas an alternate geometry involves one 18:1βCDg unit interacting with another 18:1βCDg unit inter-molecularly in a rotaxane-like fashion (cf. Scheme 1). The growing interactions between the gemini surfactant moiety at the alkenic region (~5.4 ppm; Figure 4) with the βCD interior/exterior protons at the higher host ratios indicates a possible intermolecular aggregation of the gemini surfactant around the βCD and further support the rotaxane-like structure depicted in Scheme 1. These proposed structures are further supported by the shielding and deshielding CIS trends of the external and internal βCD resonances, respectively (cf. Table 1). It is noteworthy to mention that the increased gemini surfactant-βCD interaction at the 3:1 and 5:1 mole ratios of 18:1βCDg-Mel correlate with the limited solubility of the drug at these conditions (Scheme 1c).

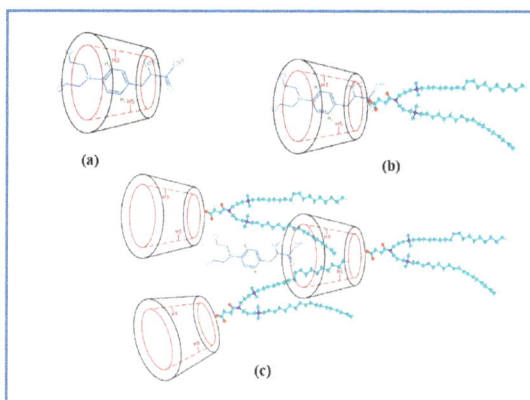

Scheme 1. Schematic presentation of the possible inclusion geometry of the Mel and gemini surfactant; (**a**) 1:1 βCD-Mel, (**b**) 18:1βCDg-Mel at 1:1 molar ratio, and (**c**) 18:1βCD-Mel at 3:1 molar ratio. Note that other possible geometries are possible and also the structures are not drawn to scale.

The 2D NMR results for the 2:1 18:1βCDg-Mel complex are shown as an expanded plot in Figure 5 and reveals NOE correlations due to interactions between the internal cavity of the βCD moiety (H_3, H_5) and the aromatic moiety of the drug at the H'_1 and H'_2 positions (cf. Figures 1a and 5). These results provide unequivocal evidence for the formation of the 18:1βCDg-Mel complexes with well-defined binding and geometry that are consistent with the 1D ^{1}H CIS data. The more intense cross-peaks at H5 compared to H3 in Figure 5 suggests that Mel is directionally encapsulated within the βCD cavity of the 18:1βCDg carrier, as depicted in Figure 5 (see insets). Similar geometry was deduced from the inclusion complex of Mel with pure βCD according to 2D ROESY NMR results of the βCD-Mel complex (cf. Figure S3).

Higuchi and Connors described an analytical approach to study the CD/drug solubility relation known as the phase-solubility method (Figure S4) [42]. This method examines the effect of a solubilizer (CD or ligand) on the drug being solubilized (substrate). The phase-solubility relationship describes only the solubilizing effect of the CD on the drug molecule but not the actual formation of inclusion complexes. Based on the phase-solubility diagram of Higuchi and Connors, the solubility evaluation results reported previously using flow-injection mass spectrometry of 18:1βCDg-Mel [41] and the NMR results discussed herein, we propose that the relationship between the solubility of Mel and the concentration of the host (18:1βCDg) follows either A_N or A-B_S-type model. A_N-type model indicates the formation of a host–drug soluble complex with negatively deviating isotherms, whereas the A-B_S-type indicates the formation of a complex with limited solubility [43]. βCD-drug complexes usually follow a B-type model as a result of the limited aqueous solubility of βCD (1.85 g/100 mL). On the other hand, 18:1βCDg exhibits behavior characteristic of both βCD and cationic gemini surfactant (amorphous molecule). The solubilization effect of 18:1βCDg is combination of formation of host–guest inclusion complex and the association complex formation that involve moieties that lie in the extracavity region of the 18:1βCDg (i.e., the gemini alkyl chain) as shown by the 2D NMR results for the 2:1 18:1βCDg-Mel (Figures 4 and 5).

3.1.2. Size and Zeta Potential

Size of the carrier–drug complex can determine their route of administration and can affect stability, cellular uptake, biodistribution, toxicity, and clearance pathway [44–46]. For instance, optimal endocytosis requires particle size of the nanoparticles to be within the range of 100–200 nm [47]. In the current work, the particle size was measured for the 2:1 18:1βCDg-Mel complex with size of approximately 160 nm (Table 2). Similarly, the 18:1βCDg carrier possesses an average size of ca. 170 nm characteristic of nanoparticles. At identical concentration conditions, βCD-Mel complex and unbound βCD failed to form particles with measurable particle sizes (Table 2). In a previous report, much greater particle size was measured for a 2:1 12 βCDg:Mel complex (ca. 225 nm) in the presence of 0.5% methylcellulose as a suspending agent [29]. In the current work, the reported particle size of 18:1βCDg-Mel complex was measured in an aqueous medium without the need of a suspending agent and no visible precipitation was observed even after 24 h. It is hypothesized that the replacement of saturated 12-carbon tails with unsaturated 18-carbon alkyl tails in the case of 18:1βCDg results in the formation of a stable inclusion complex.

Aside from particle size, the surface charge of a system can affect the physiochemical stability and biological behavior of carrier-drug formulations [48]. In Table 2, the zeta potential (ζ) data reveals much larger positive charge ($\zeta = 46$ mV) for the 18:1βCDg-Mel complex compared to the uncomplexed 18:1βCDg carrier ($\zeta = 14$ mV). This effect can be explained by the difference in the measured pH values between the complex (pH 2.8) and the 18:1βCDg aqueous dispersion (pH 4.7). Low acidic pH can lead to the protonation of melphalan, causing more positive ζ-values [49].

3.2. In Vitro Evaluation

The in vitro activity of 18:1βCDg-Mel complex, optimized at the 2:1 mole ratio, was evaluated in different melanoma cell line models. The in vitro studies were performed to evaluate the ability of the carrier to: (1) enhance the efficiency of Mel in standard monolayer cell lines; (2) enhance the penetration of the Mel in 3D melanoma tissue culture; (3) overcome drug resistance in Mel-resistant cells.

A standard monolayer A375 cell line was used to determine the IC_{50} of Mel (cf. Table 3, Figure S5). The concentration of Mel required to induce cell death in the presence of 18:1βCDg carrier at the 2:1 (18:1βCDg-Mel) molar ratio was significantly lower with IC_{50} values of 27 ± 1 µM compared to Mel-alone with $IC_{50} = 98 \pm 1$ µM ($P < 0.05$). Additionally, the IC_{50} value for the unbound carrier 18:1βCDg was ~89 µM. These results indicate that the 18:1βCDg-Mel complex can significantly improve the efficiency of Mel with low carrier-specific toxicity of the cells. Recently, we reported that βCD-gemini system did not alter the cell death pathway induced by Mel, as shown by apoptosis and cell cycle analyses, indicating that the βCD-modified gemini surfactant bears no intrinsic toxicity to the cells [29].

Table 3. The 50% of inhibitory concentration (IC_{50}) of: [A] Mel-alone, [B] 18:1βCDg alone, and [C] 18:1βCDg-Mel complexes at 2:1 molar ratio. IC_{50} determined in A375 monolayer cell lines by using MTT (3-(4,5-dimethylthiazolyl-2)-2,5-diphenyltetrazolium bromide) assay. IC_{50} was established by calculating the fraction of dead cells and plotting the data with a 4-parameter curve. $N = 3 \pm$ SD.

	Treatment	IC_{50}
[A]	Mel	98 ± 1 μM
[B]	18:1βCDg	89 ± 2 μM
[C]	18:1βCDg-Mel [2:1 molar ratio]	27 ± 1 μM

To mimic the complex three-dimensional (3D) architecture of a solid tumor and to investigate the cell penetration ability of the βCD-modified gemini surfactant and hence the improvement in the drug permeability, a spheroid melanoma cell culture was created. Treatment of spheroid melanoma cell culture with 30 μM 18:1βCDg-Mel complexes at Mel final concentration of 30 μM (the approximate IC_{50} value in monolayer culture) caused 10% cell death, a three-fold increase in cell toxicity compared to Mel alone (cf. Figure 6). These results are based on the differences in IC_{50} values for Mel (98 μM) and 18:1βCDg-Mel (27 μM) in monolayer A375 cell lines (cf. Table 3). Similarly, treatment of spheroid melanoma cell culture at greater Mel final concentration of 80 μM (recapitulating the IC_{50} value of naked Mel in monolayer) showed 24% cell toxicity with 18:1βCDg-Mel complexes, compared to 14% for Mel-alone at the same conditions (Figure 6). While cell toxicity for 18:1βCDg-Mel complexes (at both 30 and 80 μM) are lower than cell death in A375 monolayer cell lines, the results are encouraging for future formulation development upon accounting for the higher complexity of the spheroid cell culture and smaller direct contact area with the drug dispersed in the cell culture medium compared to monolayer cultures. Nevertheless, formulating Mel with 18:1βCDg caused a significant increase in the cytotoxicity of the drug at both concentrations. In our previous work, 12 βCDg did not improve the delivery of Mel in the tumor spheroids of A375 cells as no such difference were found between the IC_{50} values between Mel-alone and the Mel-12 βCDg formulation [29]. Thus, we hypothesize that the 18:1βCDg, with unsaturated 18:1 tail, forms more favorable inclusion complexes with Mel than the 12 βCDg with 12-carbon alkyl tail. This may be attributed to the combined effect of a more prominent inclusion binding along with secondary interactions due to association of the non-included C18:1 gemini chain that resides in the interstitial region of 18:1βCDg, as described in the NMR results (Scheme 1).

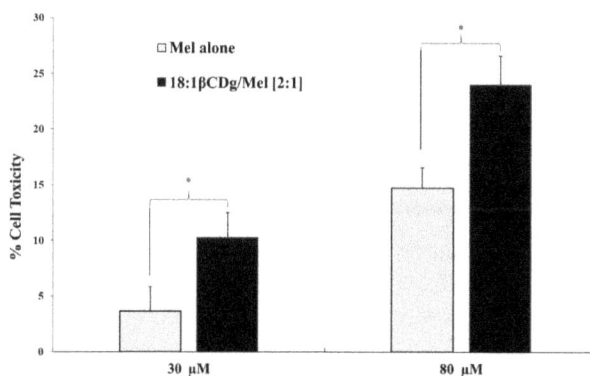

Figure 6. Cytotoxic efficiency of the Mel-alone and 18:1βCDg-Mel complexes (at 2:1 molar ratio) in 3D spheroid A375 melanoma cells. A375 cells were seeded at 1×10^4 cells/well in Corning® spheroid 96-well microplates and treated twice (24 h and 36 h after seeding) with Mel-alone and 18:1βCDg-Mel complexes with final Mel concentrations of 30 μM and 80 μM. Cell toxicity was reported as lactate dehydrogenase (LDH) activity. $N = 3 \pm$ SD * indicates significance ($P < 0.05$).

Advanced drug delivery systems can provide a potential avenue to overcome drug resistance by enhancing the bioavailability of the drug target at the tumor site. This can be achieved by shielding the drug molecule via complex formation to minimize its efflux from the cell [50]. Previously, we reported that formulating Mel with 12 βCDg significantly improved the efficiency of the drug in Mel-resistant A375, whereas Mel alone showed minimum cell death at the same concentration, in line with the anticipated effect [29]. In the current work, we evaluated the capability of newly designed βCD-modified gemini surfactant (18:1βCDg) to enhance the cytotoxic action of melphalan in Mel-resistant cultures (Figure 7). The Mel-resistant A375 cells showed low cell toxicity toward melphalan even at the highest concentration that was evaluated here (500 μM of melphalan). Therefore, we used 30 and 80 μM melphalan based on the known IC_{50} values of melphalan and the corresponding IC_{50} values of 18:1βCDg alone in a standard monolayer cell culture assay (Table 3, Figure S5). Treating the Mel-resistant A375 melanoma cells with Mel-alone at 30 and 80 μM caused low cell death (4% and 27%, respectively). However, after treating the Mel-resistant cells with 18:1βCDg-Mel complexes at final Mel concentration of 30 and 80 μM, a significant recovery of the activity was observed (46 and 76% cell death, respectively), as shown in Figure 7. These results suggest that the 18:1βCDg delivery system was able to overcome the apparent drug resistance and enhance the treatment efficacy.

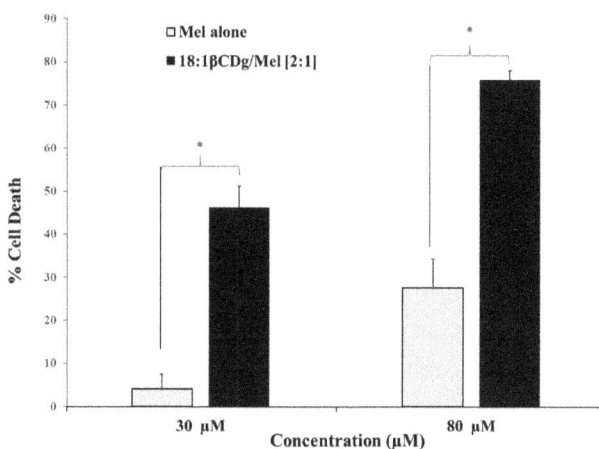

Figure 7. Cytotoxic efficiency of the Mel-alone and 18:1βCDg-Mel complexes (at 2:1 molar ratio) in Mel-resistant A375 melanoma cells. A375 cells were seeded 1.5×10^4 cells/well in 96-well plate and after 24 h treated with Mel-alone and 18:1βCDg-Mel complexes with final Mel concentrations of 30 μM and 80 μM. Cell death was reported as using MTT assay in comparison with nontreated cells. $N = 3 \pm SD$ * indicates significant ($P < 0.05$).

4. Conclusions

In this work, a novel carrier based on derivatization of βCD with an unsaturated 18-carbon chain gemini surfactant conjugate (18:1βCDg) was characterized and its potential as an advanced drug delivery system for Melphalan (Mel) drug in melanoma therapy was evaluated. The 18:1βCDg carrier and its complexes with Mel drug were characterized using 1D/2D NMR spectroscopy, along with the measurement of particle and zeta potential in aqueous solution. The 18:1βCDg carrier improves the solubility of Mel through formation of favorable inclusion complexes at the 2:1 mole ratio, as supported by 1D CIS data and 2D ROESY NMR results. The inclusion of Mel involves a well-defined geometry where the drug is directionally encapsulated within the internal apolar cavity of the βCD carrier system, according to 2D ROESY NMR results. The self-inclusion of the terminal part of the gemini alkyl chain within the βCD cavity cannot be ruled out especially at equimolar carrier/drug ratios. However, these effects are minimized at carrier mole ratios >1:1 due to hydrophobic aggregation of the carrier chains.

The measured particle sizes of the unbound 18:1βCDg carrier (ca. 170 nm) and the 2:1 carrier: drug complex (160 nm) are within nanoparticle size limits (100–200 nm). Thus, the 18:1βCDg carrier affords optimum stability, cellular uptake, biodistribution, toxicity, and clearance pathway of the reported formulation. As well, the in vitro evaluations of the optimized 18:1βCDg-Mel formulation in the presence of various melanoma models (i.e., monolayer, 3D spheroid, and Mel-resistant melanoma cells) resulted in significantly improved cytotoxic efficiency of the Mel in all cases. We are envisioning future studies to elucidate the pathways of cell penetration and of mechanism overcoming drug resistance of the 18:1βCDg-drug complexes. This knowledge will enable us to further optimize the structure that aims to improve efficiency and increase penetration ability into spheroids.

Supplementary Materials: The following are available online at http://www.mdpi.com/1999-4923/11/9/427/s1, Figure S1. Predicted ^{1}H NMR spectra of Melphalan. Spectra was created using nmrdb online tool (www.nmrdb.org). Figure S2. Predicted ^{1}H NMR spectra of 18:1 gemini surfactant. Spectra was created using nmrdb online tool (www.nmrdb.org). Figure S3. 2D ROESY spectrum of βCD-Mel at a 2:1 host-guest mole ratio, showing cross-peaks between βCD internal 1H cavity and Mel nuclei. Figure S4. Phase solubility diagram. Figure S5. Cytotoxic efficiency of Melphalan in human malignant melanoma (A375) cell line. A375 cells were seeded at 1×10^4 cells/well in 96-well plate. Toxicity was reported using MTT Assay in comparison with non-treated cells (100% viability). $N = 3 \pm$ SD.

Author Contributions: Conceptualization, W.M.-S., R.E.V. and I.B.; Methodology, W.M.-S., A.H.K. and L.D.W.; Formal Analysis, W.M.-S., A.H.K. and L.D.W.; Data Curation, W.M.-S. and A.H.K; Writing-Original Draft Preparation, W.M.-S.; Writing-Review & Editing, A.H.K, R.E.V, L.D.W and I.B.; Supervision, R.E.V, L.D.W and I.B Funding Acquisition, L.D.W and I.B.

Funding: This research was funded by The Natural Sciences and Engineering Research Council of Canada grant number 2015-03689. Scholarship W M.-S was provided by Taibah University, Medina 42353, Saudi Arabia.

Conflicts of Interest: The authors declare no conflict of interest.

References

1.   Pilgrim, W.; Hayes, R.; Hanson, D.W.; Zhang, B.; Boudreau, B.; Leonfellner, S. Skin Cancer (Basal Cell Carcinoma, Squamous Cell Carcinoma, and Malignant Melanoma): New Cases, Treatment Practice, and Health Care Costs in New Brunswick, Canada, 2002–2010. *J. Cutan. Med. Surg.* **2014**, *18*, 320–331. [CrossRef] [PubMed]

2.   WHO. Skin Cancers. WHO, 2016. Available online: http://www.who.int/uv/faq/skincancer/en/index1.html (accessed on 26 February 2018).

3.   Edge, S.B.; Compton, C.C. The American Joint Committee on Cancer: The 7th Edition of the AJCC Cancer Staging Manual and the Future of TNM. *Ann. Surg. Oncol.* **2010**, *17*, 1471–1474. [CrossRef] [PubMed]

4.   Read, R.L.; Haydu, L.; Saw, R.P.; Quinn, M.J.; Shannon, K.; Spillane, A.J.; Thompson, J.F. In-transit melanoma metastases: Incidence, prognosis, and the role of lymphadenectomy. *Ann. Surg. Oncol.* **2015**, *22*, 475–481. [CrossRef] [PubMed]

5.   Gershenwald, J.E.; Giacco, G.G.; Lee, J.E. Cutaneous Melanoma. In *60 Years of Survival Outcomes at The University of Texas MD Anderson Cancer Center*; Springer: Berlin/Heidelberg, Germany, 2013; pp. 153–165.

6.   Testori, A.; Verhoef, C.; Kroon, H.M.; Pennacchioli, E.; Faries, M.B.; Eggermont, A.M.; Thompson, J.F. Treatment of melanoma metastases in a limb by isolated limb perfusion and isolated limb infusion. *J. Surg. Oncol.* **2011**, *104*, 397–404. [CrossRef] [PubMed]

7.   Zar, T.; Graeber, C.; Perazella, M.A. Reviews: Recognition, Treatment, and Prevention of Propylene Glycol Toxicity. *Semin. Dial.* **2007**, *20*, 217–219. [CrossRef] [PubMed]

8.   Ajazuddin; Alexander, A.; Amarji, B.; Kanaujia, P. Synthesis, characterization and in vitro studies of pegylated melphalan conjugates. *Drug Dev. Ind. Pharm.* **2013**, *39*, 1053–1062. [CrossRef]

9.   Peyrode, C.; Weber, V.; David, E.; Vidal, A.; Auzeloux, P.; Communal, Y.; Chezal, J.M. Quaternary ammonium-melphalan conjugate for anticancer therapy of chondrosarcoma: In vitro and in vivo preclinical studies. *Investig. New Drugs* **2012**, *30*, 1782–1790. [CrossRef]

10.   Rajpoot, P.; Bali, V.; Pathak, K. Anticancer efficacy, tissue distribution and blood pharmacokinetics of surface modified nanocarrier containing melphalan. *Int. J. Pharm.* **2012**, *426*, 219–230. [CrossRef]

11. Vodovozova, E.L.; Kuznetsova, N.R.; Kadykov, V.A.; Khutsyan, S.S.; Gaenko, G.P.; Molotkovsky, Y.G. Liposomes as nanocarriers of lipid-conjugated antitumor drugs melphalan and methotrexate. *Nanotechnol. Russ.* **2008**, *3*, 228–239. [CrossRef]

12. Cho, K.; Wang, X.; Nie, S.; Chen, Z.; Shin, D.M. Therapeutic Nanoparticles for Drug Delivery in Cancer. *Clin. Cancer Res.* **2008**, *14*, 1310–1316. [CrossRef]

13. Tiwari, G.; Tiwari, R.; Sriwastawa, B.; Bhati, L.; Pandey, S.; Pandey, P.; Bannerjee, S.K. Drug delivery systems: An updated review. *Int. J. Pharm. Investig.* **2012**, *2*, 2–11. [CrossRef] [PubMed]

14. Challa, R.; Ahuja, A.; Ali, J.; Khar, R.K. Cyclodextrins in drug delivery: An updated review. *Aaps Pharmscitech* **2005**, *6*, E329–E357. [CrossRef] [PubMed]

15. Szejtli, J. Introduction and General Overview of Cyclodextrin Chemistry. *Chem. Rev.* **1998**, *98*, 1743–1754. [CrossRef] [PubMed]

16. Loftsson, T.; Jarho, P.; Masson, M.; Järvinen, T. Cyclodextrins in drug delivery. *Expert Opin. Drug Deliv.* **2005**, *2*, 335–351. [CrossRef] [PubMed]

17. Loftsson, T.; Duchene, D. Cyclodextrins and their pharmaceutical applications. *Int. J. Pharm.* **2007**, *329*, 1–11. [CrossRef] [PubMed]

18. Del Valle, E.M. Cyclodextrins and their uses: A review. *Process. Biochem.* **2004**, *39*, 1033–1046. [CrossRef]

19. Luke, D.R.; Tomaszewski, K.; Damle, B.; Schlamm, H.T. Review of the basic and clinical pharmacology of sulfobutylether-β-cyclodextrin (SBECD). *J. Pharm. Sci.* **2010**, *99*, 3291–3301. [CrossRef]

20. Zhang, J.; Ma, P.X. Cyclodextrin-based supramolecular systems for drug delivery: Recent progress and future perspective. *Adv. Drug Deliv. Rev.* **2013**, *65*, 1215–1233. [CrossRef]

21. Chilajwar, S.V.; Pednekar, P.P.; Jadhav, K.R.; Gupta, G.J.; Kadam, V.J. Cyclodextrin-based nanosponges: A propitious platform for enhancing drug delivery. *Expert Opin. Drug Deliv.* **2014**, *11*, 111–120. [CrossRef]

22. Sallas, F.; Darcy, R. Amphiphilic Cyclodextrins—Advances in Synthesis and Supramolecular Chemistry. *Eur. J. Org. Chem.* **2008**, *2008*, 957–969. [CrossRef]

23. Quaglia, F.; Ostacolo, L.; Mazzaglia, A.; Villari, V.; Zaccaria, D.; Sciortino, M.T. The intracellular effects of non-ionic amphiphilic cyclodextrin nanoparticles in the delivery of anticancer drugs. *Biomaterials* **2009**, *30*, 374–382. [CrossRef] [PubMed]

24. Bilensoy, E.; Hincal, A.A. Recent advances and future directions in amphiphilic cyclodextrin nanoparticles. *Expert Opin. Drug Deliv.* **2009**, *6*, 1161–1173. [CrossRef] [PubMed]

25. Cryan, S.; Donohue, R.; Ravoo, B.; Darcy, R.; O'Driscoll, C. Cationic cyclodextrin amphiphiles as gene delivery vectors. *J. Drug Deliv. Sci. Technol.* **2004**, *14*, 57–62. [CrossRef]

26. Michel, D.; Chitanda, J.M.; Balogh, R.; Yang, P.; Singh, J.; Das, U.; El-Aneed, A.; Dimmock, J.; Verrall, R.; Badea, I. Design and evaluation of cyclodextrin-based delivery systems to incorporate poorly soluble curcumin analogs for the treatment of melanoma. *Eur. J. Pharm. Biopharm.* **2012**, *81*, 548–556. [CrossRef] [PubMed]

27. Poorghorban, M.; Karoyo, A.H.; Grochulski, P.; Verrall, R.E.; Wilson, L.D.; Badea, I. A ^{1}H NMR Study of Host/Guest Supramolecular Complexes of a Curcumin Analogue with β-Cyclodextrin and a β-Cyclodextrin-Conjugated Gemini Surfactant. *Mol. Pharm.* **2015**, *12*, 2993–3006. [CrossRef] [PubMed]

28. Poorghorban, M.; Das, U.; AlAidi, O.; Chitanda, J.M.; Michel, D.; Dimmock, J.; Verrall, R.; Grochulski, P.; Badea, I. Characterization of the host–guest complex of a curcumin analog with β-cyclodextrin and β-cyclodextrin–gemini surfactant and evaluation of its anticancer activity. *Int. J. Nanomed.* **2015**, *10*, 503–515.

29. Michel, D.; Mohammed-Saeid, W.; Getson, H.; Roy, C.; Poorghorban, M.; Chitanda, J.M.; Verrall, R.; Badea, I. Evaluation of β-cyclodextrin-modified gemini surfactant-based delivery systems in melanoma models. *Int. J. Nanomed.* **2016**, *11*, 6703–6712. [CrossRef]

30. Donkuru, M.; Chitanda, J.M.; Verrall, R.E.; El-Aneed, A.; El-Aneed, A. Multi-stage tandem mass spectrometric analysis of novel β-cyclodextrin-substituted and novel bis-pyridinium gemini surfactants designed as nanomedical drug delivery agents. *Rapid Commun. Mass Spectrom.* **2014**, *28*, 757–772. [CrossRef] [PubMed]

31. Singh, R.; Bharti, N.; Madan, J.; Hiremath, S. Characterization of cyclodextrin inclusion complexes—A review. *J. Pharm. Sci. Technol.* **2010**, *2*, 171–183.

32. Mura, P. Analytical techniques for characterization of cyclodextrin complexes in aqueous solution: A review. *J. Pharm. Biomed. Anal.* **2014**, *101*, 238–250. [CrossRef]

33. Figueiras, A.; Carvalho, R.A.; Ribeiro, L.; Torres-Labandeira, J.J.; Veiga, F.J. Solid-state characterization and dissolution profiles of the inclusion complexes of omeprazole with native and chemically modified β-cyclodextrin. *Eur. J. Pharm. Biopharm.* **2007**, *67*, 531–539. [CrossRef] [PubMed]

34. De Araujo, D.R.; Tsuneda, S.S.; Cereda, C.M.; Carvalho, F.D.G.; Preté, P.S.; Fernandes, S.A.; De F.A. Braga, A. Development and pharmacological evaluation of ropivacaine-2-hydroxypropyl-β-cyclodextrin inclusion complex. *Eur. J. Pharm. Sci.* **2008**, *33*, 60–71. [CrossRef] [PubMed]

35. Neuhaus, D.; Williamson, M.P. *The nuclear Overhauser effect in structural and conformational analysis*, 2nd ed.; John Wiley & Sons, Inc.: New York, NY, USA, 2000.

36. Berger, S. One Dimensional and Two Dimensional NMR Spectra by Modern Pulse Techniques. *Angew. Chem.* **1992**, *104*, 108–109. [CrossRef]

37. Karoyo, A.H.; Wilson, L.D. Preparation and Characterization of a Polymer-Based "Molecular Accordion". *Langmuir* **2016**, *32*, 3066–3078. [CrossRef] [PubMed]

38. Schneider, H.-J.; Hacket, F.; Rüdiger, V.; Ikeda, H. NMR Studies of Cyclodextrins and Cyclodextrin Complexes. *Chem. Rev.* **1998**, *98*, 1755–1786. [CrossRef]

39. Wilson, L.D.; Siddall, S.R.; Verrall, R.E. A spectral displacement study of the binding constants of cyclodextrin–hydrocarbon and –fluorocarbon surfactant inclusion complexes. *Can. J. Chem.* **1997**, *75*, 927–933. [CrossRef]

40. Bhasikuttan, A.C.; Mohanty, J.; Nau, W.M.; Pal, H. Efficient Fluorescence Enhancement and Cooperative Binding of an Organic Dye in a Supra-biomolecular Host–Protein Assembly. *Angew. Chem.* **2007**, *119*, 4198–4200. [CrossRef]

41. Mohammed-Saeid, W.; Michel, D.; Badea, I.; El-Aneed, A.; Mohammed-Saeid, W.; El-Aneed, A. Rapid and Simple Flow Injection Analysis-Tandem Mass Spectrometric (FIA-MS/MS) Method for the Quantification of Melphalan in Lipid-Based Drug Delivery System. *Rapid Commun. Mass Spectrom.* **2017**, *31*, 1481–1490. [CrossRef] [PubMed]

42. Higuchi, T.; Connors, A. Phase-solubility techniques. *Adv. Anal. Chem. Instrum.* **1965**, *4*, 117–210.

43. Zhengyu, J. *Cyclodextrins: Preparation and Application in Industry*; World Scientific: Singapore, 2018.

44. Gaumet, M.; Vargas, A.; Gurny, R.; Delie, F. Nanoparticles for drug delivery: The need for precision in reporting particle size parameters. *Eur. J. Pharm. Biopharm.* **2008**, *69*, 1–9. [CrossRef] [PubMed]

45. Decuzzi, P.; Godin, B.; Tanaka, T.; Lee, S.-Y.; Chiappini, C.; Liu, X.; Ferrari, M. Size and shape effects in the biodistribution of intravascularly injected particles. *J. Control. Release* **2010**, *141*, 320–327. [CrossRef] [PubMed]

46. Dhand, C.; Prabhakaran, M.P.; Beuerman, R.W.; Lakshminarayanan, R.; Dwivedi, N.; Ramakrishna, S. Role of size of drug delivery carriers for pulmonary and intravenous administration with emphasis on cancer therapeutics and lung-targeted drug delivery. *Rsc Adv.* **2014**, *4*, 32673–32689. [CrossRef]

47. Gratton, S.E.A.; Ropp, P.A.; Pohlhaus, P.D.; Luft, J.C.; Madden, V.J.; Napier, M.E.; DeSimone, J.M. The effect of particle design on cellular internalization pathways. *Proc. Natl. Acad. Sci. USA* **2008**, *105*, 11613–11618. [CrossRef] [PubMed]

48. Fröhlich, E. The role of surface charge in cellular uptake and cytotoxicity of medical nanoparticles. *Int. J. Nanomed.* **2012**, *7*, 5577–5591. [CrossRef] [PubMed]

49. Stout, S.A.; Riley, C.M. The hydrolysis of l-phenylalanine mustard (melphalan). *Int. J. Pharm.* **1985**, *24*, 193–208. [CrossRef]

50. Hu, C.-M.J.; Zhang, L. Nanoparticle-based combination therapy toward overcoming drug resistance in cancer. *Biochem. Pharm.* **2012**, *83*, 1104–1111. [CrossRef] [PubMed]

pharmaceutics

MDPI

Article

Drug Delivery Technology to the CNS in the Treatment of Brain Tumors: The Sherbrooke Experience

David Fortin

Division of Neurosurgery and Neuro-Oncology, Department of surgery, Faculty of Medicine and Health Science, University of Sherbrooke, Sherbrooke, QC J1H-5N4, Canada; David.fortin@usherbrooke.ca;
Tel.: +1-819-346-1110 (ext. 73324)

Received: 8 April 2019; Accepted: 22 May 2019; Published: 27 May 2019

Abstract: Drug delivery to the central nervous system (CNS) remains a challenge in neuro-oncology. Despite decades of research in this field, no consensus has emerged as to the best approach to tackle this physiological limitation. Moreover, the relevance of doing so is still sometimes questioned in the community. In this paper, we present our experience with CNS delivery strategies that have been developed in the laboratory and have made their way to the clinic in a continuum of translational research. Using the intra-arterial (IA) route as an avenue to deliver chemotherapeutics in the treatment of brain tumors, complemented by an osmotic breach of the blood-brain barrier (BBB) in specific situations, we have developed over the years a comprehensive research effort on this specialized topic. Looking at pre-clinical work supporting the rationale for this approach, and presenting results discussing the safety of the strategy, as well as results obtained in the treatment of malignant gliomas and primary CNS lymphomas, this paper intends to comprehensively summarize our work in this field.

Keywords: blood-brain barrier; intra-arterial chemotherapy; malignant gliomas; primary central nervous system lymphomas

1. Introduction

Chemotherapeutic drug trials for brain tumor treatment have been conducted worldwide for many decades, with marginal improvements in patient outcomes. Indeed, the standard of care in the 1st-line management of glioblastoma (grade 4 primary brain tumors) was the addition of temozolomide, an alkylating drug, to radiotherapy, which led to an improvement in survival of 2 months [1]. This regimen is dubbed the "Stupp regimen". Any further attempts to improve on the outcome have produced disappointing results. Interestingly, one of the only reported approaches with seemingly improved outcomes is the addition of a local device emitting low-intensity, intermediate-frequency alternating electric fields (TTF) [2]. As this device is applied directly to the scalp, and its effect does not require a specific delivery paradigm to reach the CNS. Indeed, amongst the factors that can explain a lack of improvement in the care of brain tumor patients, one stands as a major culprit: Impaired delivery to the CNS, related to the presence of the blood-brain barrier (BBB) [3]. Thus, in the presence of a brain tumor, the first barrier to treatment options is just that, a barrier: The BBB.

2. The Blood-Brain Barrier

It has been a long process to recognize the extent to which the BBB really impacts CNS delivery. It is often still debated in some publications, as some authors continue to argue that the presence of contrast enhancement on computed tomography(CT) scans (iodine-based) or on magnetic resonance scans (para-magnetic contrast) remains clear evidence that the integrity of the BBB is altered and access to the

CNS is granted [4,5]. Hence, in that context, these authors claim that the BBB entity does not represent a significant obstacle to therapeutic delivery to the CNS in the presence of pathological lesions, implying that the breach in permeability is sufficient to allow adequate diffusion of therapeutics. This type of all or none argumentation basically translates a lack of knowledge and understanding of BBB alterations and CNS delivery subtleties in the presence of a tumor. Indeed, different pharmacokinetic compartments are defined by the presence of a brain tumor, with a wide variation of the effects on the BBB permeability, and thus, on delivery [6]. This aspect is frequently neglected and under-estimated.

Indeed, looking at modern data on the subject, there is no doubt that the BBB prevents chemotherapy entry to the CNS, even in the presence of a lesion, thereby limiting therapeutic concentration from reaching clinically efficient levels [7,8]. Part of the confusion arises from the fact that within a brain tumor, as well as to the close proximity of the tumor nodule, the BBB is often replaced by a brain-tumor barrier (BTB) which pertains to entirely different pharmacokinetics, displaying a permeability that is classically intermediate between normal BBB and breached BBB. This increase in permeability is a function of the breach in the integrity of the BBB and BTB, and is highly variable, heterogeneous, and dependent on tumor size and type [9–11]. Thus, within any tumor, drug distribution is inherently uneven, with preferential accumulation in the necrotic central core areas [12], whereas drug penetration at the edge of the tumor is classically nonexistent, or marginal at best [13]. As such, although the BBB and the BTB are often partially breached, there remains a significant delivery impediment, and therapeutic levels of drugs are insufficient within the breached areas to mount a clinically significant response [6,14]. As was eloquently exposed by Reichel, despite enormous efforts, achieving an effective concentration profile in the brain remains a significant challenge in CNS drug development [6]. Another factor further complicates the matter: The majority of malignant brain tumors are infiltrative, with tumor cells permeating at a distance from the main tumor nodule, and away from even the most sensitive imaging MR scan sequence (FLAIR) [14,15]. Obviously in these areas, the BBB permeability is unaltered, and tumor cells are shielded by an intact BBB.

3. Alternate Drug Delivery

Different approaches have been designed and tested to circumvent this delivery impediment, bypass the BBB and BTB, and maximize therapeutics delivery to the brain. Indeed, in order to increase the number of therapeutic options available to treat CNS tumors, alternative delivery strategies have to be considered. For a detailed review on the subject covering different strategies, please consult the review by Drapeau et al. [3]. Of all the approaches we have tested in the laboratory, one is currently used in the clinic: The cerebral intra-arterial infusion of chemotherapy (CIAC) with and without BBB permeabilization. This paper will give a thorough description of the efforts carried by our group to successfully deploy this strategy in the treatment of brain tumors. Indeed, we have implemented a continuum of translational research on this topic that will be described in detail in this publication.

The Cerebral Intra-arterial Infusion of Chemotherapy (CIAC) and the Blood Brain Barrier Disruption (BBBD) Adjunct

When one realizes the extensiveness of the vascular network supplying the brain, it becomes obvious that a global delivery strategy is rational and plausible by using this vascular network as a delivery corridor [16]. The importance of this network has already been exposed by Bradbury and colleagues, these authors claiming that the entire network covers an area of 12 m^2/g of cerebral parenchyma [17]. To understand the extensiveness of the cerebral vascularization in a more prosaic statement, let us just consider that the brain receives 20% of the total systemic circulation even though its weight amounts to less than 3% of the total body weight [9].

Interestingly, it is technically easy and actually commonly performed in the clinic to repeatedly access this cerebral vascular network in a patient [18]. Via a simple puncture to access the femoral artery, a catheter is introduced and navigated intra-arterially to reach one of the four major cerebral arteries. Once in position, the chemotherapy is administered via the catheter that is withdrawn at the conclusion of the procedure. The CIAC produces a paradigm of regional chemotherapy distribution within the area deserved by the vessel treated [19].

Through the first pass effect, an increase in the local plasma peak concentration of the drug produces a significantly improved AUC (the concentration of the drug according to the time) [19,20]. This consequently translates in increased local exposure of the target tissue to the therapeutic agent. Interestingly, as our lab has shown, it is also accompanied by a decreased systemic drug distribution, hence reducing systemic toxicities and potential side-effects [21]. Classically, the therapeutic concentration at the tumor cell target is increased by a 3.5–5-fold factor [20]. This procedure is performed under local anesthesia, and typically lasts around 30 min.

The delivery can be further improved by adding as an adjunct to the procedure an osmotic blood-brain barrier disruption (BBBD). This strategy is based on the cerebral intravascular infusion of a hypertonic solution to produce a transient increase in permeabilization of the BBB and BTB, prior to the administration of the chemotherapy. As with CIAC, the parent vessel treated is selected based on the tumor localization in the brain. This approach, which is an adjunct to the CIAC, is physiologically more demanding, requiring general anesthesia, and needs careful preparation, but it does increase significantly delivery across the BBB and BTB [3,9]. It involves the IA infusion of a hyperosmolar solution (usually mannitol) in a flow rate sufficient to allow a complete filling of the vessel. Two parameters are paramount in the ability to mediate a hyperosmolar modification of the barrier: The osmolality of the solution, and the infusion time. Using a solution of 1.6 molal arabinose in pentobarbital-anesthetized rats, Rapoport determined an interval duration of 30 s as the optimal infusion time to produce a BBBD [9]. The same infusion time was applied to the use of mannitol with similar findings in the same animal model [9]. These parameters have made their way to the clinic, albeit the anesthetic agents are now different.

The combination of IA infusion of a molecule with osmotic BBBD has been shown to further increase the effect of the first pass through the brain, increasing maximal peak concentration as well as AUC of the administered molecule [16,22,23]. Sato et al. elegantly presented in vivo data showing that BBBD produces a marked increase in permeability at the edge of the tumor. Interestingly, this area is typically associated with active tumor cells proliferation, whereas the permeability of the BBB and BTB tends to renormalize [13]. Theoretically, the concept of beaching the permeability of the BBB is quite compelling, as it could help evade the "sink effect" by providing higher and more uniform delivery to a whole CNS vascular territory, allowing prolonged tumor cell exposure to higher concentrations of the administered therapeutics [6,16]. This sink effect is triggered by areas of necrosis within the tumor, which tends to attract and concentrate the chemotherapy crossing the CNS, stealing the peripheral areas of the tumor where the drug would be most useful [16]. Obviously, this includes the neoplastic cells at the tumor edges that are often the most proliferative and protected by an intact BBB and/or BTB [14–16,24].

4. Pre-clinical Data

While numerous investigators have studied CIAC and BBBD over the years, we undertook a thorough pre-clinical characterization in the Fischer-F98 model to ascertain, objectivate, and measure the delivery advantage provided by both approaches. We first characterized the F98-Fischer glioma model as a benchmark for our delivery studies. The model was found to be highly predictive and reproducible in term of tumor growth dynamics and animal survival (Figure 1).

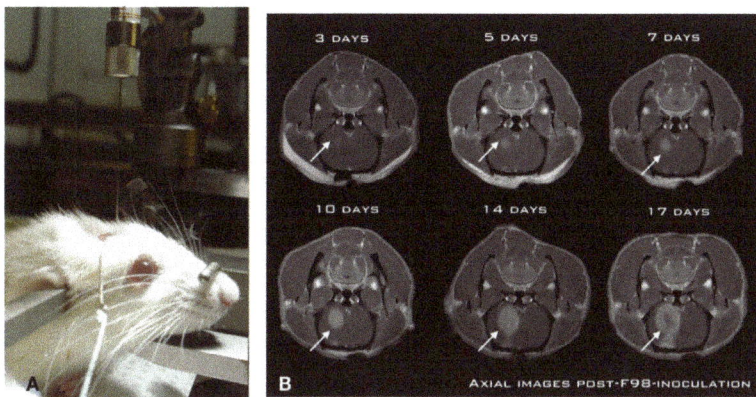

Figure 1. The Fischer-F98 glioma model shows a reproducible and predictable growth pattern. (**A**) The infusion of the cell suspension is accomplished using a slow steady perfusion with a micro-infusion pump. Also, 10,000 cells are infused at a rate of 1 μL/min over 5 min. (**B**) Coronal views of an implanted animal at days 3, 5, 7, 10, 14 and 17 post-implantation. Notice the gradual progression of the gadolinium enhancement on the MR scans in the right hemisphere (arrows) depicting the steady tumor progression. The animal starts to develop faint subtle symptoms (lateralization) at day 14, that culminate at day 26 ± 2 days.

Using a standardized implantation procedure, the tumor-take has systematically been 100%, with a median survival of 26 ± 2 days [25,26]. Figure 1A shows the position of the animal in the stereotactic frame for precise insertion of the needle in the brain of the animal using a precise and standardized coordinate system [25]. This is paramount for reproducibility across experiments. Indeed, a free-hand implantation technique which is frequently employed in the literature is inadequate for these types of studies. Likewise, we found that the use of a micro-infusion pump is essential in minimizing tissue damage and associated inflammatory reaction triggered by the implantation process [25]. The slow (1 μL/min) and steady infusion rate and the low volume (10,000 cells in 5 μL) ensures minimized cerebral tissue disruption and prevents the backflow along the implantation track commonly associated with these models [25,26]. This produces a constant pattern of tumor growth in the right hemisphere of the animal, where the tumor is already noticeable at day 3 post-implantation (Figure 1B), and starts to produce an alteration in consciousness around day 26. Experimental treatments are performed at day 10 post-implantation, when the tumor has reached a significant size (Figure 1B), without altering the neurological functions of the animal [27–30].

Using this model, and based on slight alterations of the methods described by Neuwelt and his team [16], we developed a technique allowing the perfusion of therapeutics via the intra-arterial (IA) route in the carotid of the Fischer rats, while under general anesthesia in an MR gantry. This allowed us to study the dynamics of real-time imaging during the infusion of any selected MR traceable molecule (Figure 2) [22].

In this particular surgical montage, the right external carotid artery has been identified, incised, and cannulated using a PE50 catheter. Once in position, any solution can be perfused in a retrograde fashion via the external carotid artery into the internal carotid artery (Figure 2). When high flow solutions are infused, such as when we perform a BBBD, a clip is secured on the common carotid artery to isolate the system from the heart, and prevent downstream backflow of mannitol to the heart. Once terminated, the clip is removed, the external carotid artery is simply ligated, and the incision is closed.

As an initial experiment, we first characterized the baseline level of CNS entry for 2 paramagnetic compounds, Magnevist (743 Da) and Gadomer (17,000 Da) in tumor-bearing F98-Fischer rats. As expected, the smaller Magnevist displayed a greater than 3-fold baseline penetration in the tumor compared to Gadomer across the BTB, whereas penetration in the BBB around the tumor was negligible, and was no different than in the contralateral hemisphere for both molecules [22,23].

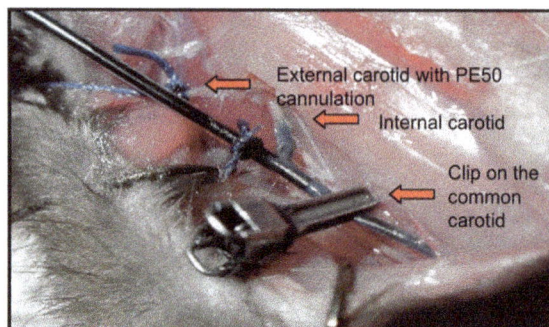

Figure 2. The surgical montage for intraarterial infusion and blood-brain barrier disruption (BBBD) in the Fischer rat. Once the montage is ready, the animal can then be inserted in the magnetic resonance (MR) gantry for real-time imaging. The intraarterial carotid perfusion is accomplished in a retrograde fashion via the external carotid artery. As can be appreciated on this image, a clip is also placed on the common carotid artery to prevent downstream backflow. As soon as the infusion is completed, the clip is removed, and the external carotid artery is sutured.

Next, we studied the concentrations of different platinum drugs when administered via different routes: Intravenous (IV), IA, and IA + BBBD using inductively coupled plasma mass spectrometry (ICP-MS) in the Fischer-F98 rat model. We did so for 5 platinum: Cisplatin, Carboplatin, Oxaliplatin, Lipoplatin and Lipoxal [28]. Figure 3 shows the summary of these experiments. Ten days after the F98 glioma cells implantation, the platinum drugs were administered according to the selected route of administration. Equivalent doses of platinum to those used in humans were established based on the body surface area of the animals [28]. Animals were euthanized 24 h after the drug perfusion, brains were harvested, and cut in sections with a brain matrix [27]. The tumor was separated and divided into cytoplasmic and nuclear compartments using a commercial Nuclear Extract Kit (Active Motif, Carlsbad, CA, USA) for analysis by ICP-MS [28].

Looking specifically at the concentration of platinum reaching the nucleus of the tumor cells, we observed significant differences between the different routes of administration (Figure 3). Comparing IA against IV, an increase in the order of 20-fold was observed for IA Carboplatin, whereas it reached 40-fold for Lipoplatin and 90-fold for Lipoxal! Interestingly, these studies also depicted significant neurotoxicity when experimenting IA infusion of either Oxaliplatin or Cisplatin, hinting at the fact that these 2 drugs were not suitable candidates for IA delivery [27–29]. These increases in the tumor cells nuclei delivery were even more dramatic when a BBBD was added to the IA infusion. Specifically looking at Carboplatin, the IA + BBBD further increased the delivery by a 17-fold factor compared to IA alone, a 320-fold factor compared to the IV infusion [29].

Using the same experimental design, we also assessed the delivery of Temozolomide. Temozolomide is the first-line standard of care in the treatment of primary brain tumors. As the bio-disponibility of the oral formulation is close to 100%, the IV formulation is available but rarely used in the clinic. In the present study, the IV formulation was used to emulate clinical oral administration. Hence, we tested the delivery of IA, IA + BBBD, and IV Temozolomide in the Fischer-F98 glioma model. The animals were once again treated 10 days after implantation. Using liquid chromatography with tandem mass spectrometry (LC-MS/MS), we measured Temozolomide in plasma, CSF and brain at 3 timepoints post-Temozolomide infusion [21]. Compared to IV, we found a fourfold increase in Temozolomide peak concentrations in brain tumor tissues with IA infusion, and a 5-fold increase with BBBD [21] (Tables 1 and 2).

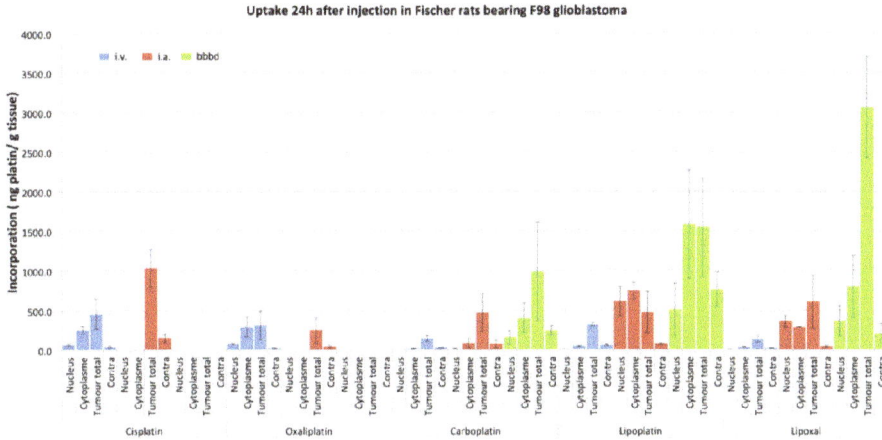

Figure 3. A comparison of the 5 platinum drug accumulations as related to the route of infusion, measured by inductively coupled plasma mass spectrometry (ICP-MS) in the Fischer-F98 rat. As can be observed, the intra-arterial (IA) and IA + BBBD routes were not tested for Oxaliplatin and Cisplatin because of significant toxicity. Results are reported as measurements of platinum (ng pt/g tissue) in the nucleus, cytoplasm, and whole tumor. The magnitude of the increase observed for each platinum agent can be appreciated in relation to the route of delivery. There is a significant increase in platinum delivery (ng platinum/g tissue) with all molecules, except Oxaliplatin.

Table 1. IV administration of Temozolomide (TMZ) (200 mg/m^2).

CNS Compartment Analyzed	$T_{1/2}$ (h)	T_{max} (h)	C_{max}	AUC_{0-t}
Plasma	1.08	0.25	63,581 µg/mL	53,409 h·µg/mL
CSF	0.87	0.25	7,628 µg/mL	6,658 h·µg/mL
Tumor	1.51	0.25	10,582 µg/g	9,521 h·µg/g
Lpsilateral brain	0.66	0.25	10,273 µg/g	8,530 h·µg/g
Contralateral brain	0.83	0.25	9,790 µg/g	8,547 h·µg/g

Table 2. IA administration of TMZ (200 mg/m^2).

CNS Compartment Analyzed	$T_{1/2}$ (h)	T_{max} (h)	C_{max}	AUC_{0-t}
Plasma	n/a	0.5	40,676 µg/mL	38,759 h·µg/mL
CSF	1.89	0.25	8,436 µg/mL	7,681 h·µg/mL
Tumor	0.34	0.25	42,989 µg/g	31,934 h·µg/g
Lpsilateral brain	0.35	0.25	31,056 µg/g	23,930 h·µg/g
Contralateral brain	n/a	0.5	11,714 µg/g	10,130 h·µg/g

Temozolomide (TMZ) pharamacokinetic parameters measured by Liquid chromatography tandem-mass spectrometry (LC-MS/MS) in Fischer-F98 rats treated 10 days after tumor implantation. Parameters are compared between the IV route (Table 1) and the IA route (Table 2).

The increase was not as dramatic using the BBBD as an adjunct with IA of Temozolomide, compared to the platinum compounds. The values of c max according to the route of delivery were as follows: 10.582 (IV), 42.989 (IA), and 50.751 (IA + BBBD), respectively. In this paper, although we could measure a significant increase in Temozolomide delivery as described above, we did not observe a parallel increase in survival of the treated animals. In vitro characterization of the F-98 glioma cell line showed it to be resistant to temozolomide [21]. Hence, it is obvious that delivery is not the only factor at play, as will be discussed later.

These pre-clinical results really highlight the potency of IA and IA + BBBD as an adequate route of delivery to improve the different pharmacokinetic parameters of CNS therapeutic delivery. The pre-clinical research continuum to improve and maximize these procedures continues, as each therapeutic offered by this route first needs to be tested for innocuity in animal models to rule out any major toxicities. Indeed, Taxol, cisplatin, and oxaliplatin were found to be extremely toxic in pre-clinical testing, excluding these drugs as eventual candidates for IA delivery. Moreover, as can be derived from the results obtained with the temozolomide experiments, an increase in delivery is not necessarily associated with an improvement in outcome. Hence, delivery is only one of many aspects of therapeutic success in the treatment of CNS tumors, albeit an important one. We will further discuss this issue in the next section on the clinical applications of these procedures.

5. Clinical Procedures

The access to the arterial system is obviously accomplished differently in humans. The human cerebral arterial system is organized in such a way that there basically are 4 major arteries responsible for the brain irrigation (2 carotids, and 2 vertebral arteries). The vascular anatomy can be variable from one individual to another, and thus the precise anatomy must be determined during the first treatment session by a formal cerebral angiography. If a lesion covers more than one vascular distribution, or if there are multiple lesions, the treatment is delivered by equally splitting the chemotherapy dose in the different distributions (vessels) involved. Parameters such as catheter placement, dilution, and rate of infusion are all standardized. In the human, the arterial system is accessed via a percutaneous transfemoral puncture. Once accessed, the catheter is navigated in the arterial system using radiological imaging (fluoroscopy). As shown in Figure 4A, the catheter has been placed in the left carotid artery, and a contrast infusion shows the distribution of this vessel.

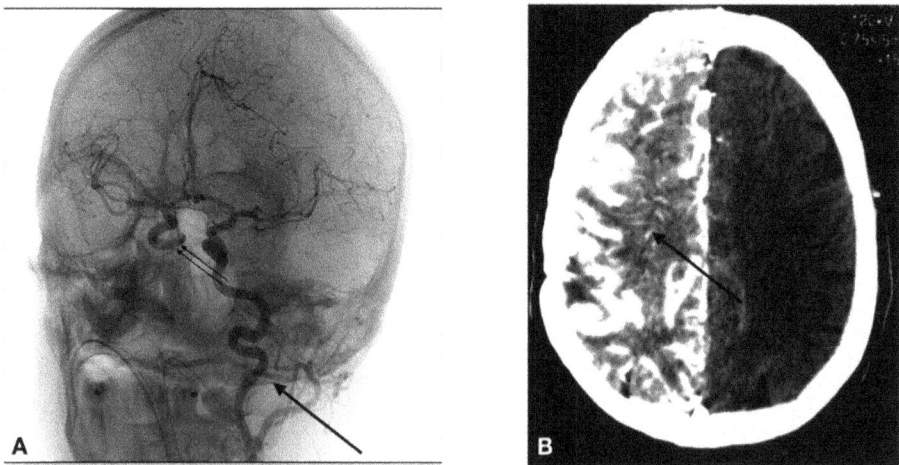

Figure 4. (**A**) Catheter placement in a glioblastoma patient treated for BBBD in the left carotid artery (arrow). An iodine contrast was infused, opacifying the left carotid distribution, as well as the contralateral carotid (double arrow) via the polygon of Willis. (**B**) The image produced by a BBBD of the right carotid artery on a computed tomography (CT) scan in a patient afflicted by a primary central nervous system (CNS) lymphoma after an infusion of iv iodine contrast. As can be appreciated, the whole hemisphere is bathed by the contrast, an evidence by the fact that the BBB is breached.

The technique involves the following steps:

1. Selective catheterization is performed via percutaneous transfemoral puncture of the left internal carotid artery, right internal carotid artery, left vertebral artery or right vertebral artery. The tip

of the catheter is positioned at the C2-C3 vertebral level in the carotid (Figure 4), or at the C6-C7 vertebral level in the vertebral artery.

2. Infusion of the drug IA: When infusing intra-arterial solutions, the concentration of the solution and the rate of infusion are critical factors that need consideration in avoiding neurotoxicity. The phenomenon of streaming defines an inhomogeneous distribution of the administered solution because of poor mixing at the infusion site [9]. Density and viscosity of fluid, lumen diameter of the infused vessels, and velocity of flow are all important determinants to control in order to avoid streaming. In this case, the Caelyx, Melphalan, and Etoposide phosphate are infused at a rate of 0.12 cc/s, whereas the Carboplatin and Methotrexate are infused at a standard rate of 0.2 cc/s.

3. In the case of a BBBD: BBBD procedures require general anesthesia. Hence, after general anesthesia with Propofol, we proceed to a selective catheterization via percutaneous transfemoral puncture of the treated artery. We then determine the individual rate of infusion of Mannitol. We use iodinated contrast injection and fluoroscopy to establish, for each patient, the ideal infusion rate; it is the rate that will fill the entire vessel distribution, without producing significant reflux in the common carotid artery. Once established (usually between 3 and 6 cc/s × 30 s), the patient is prepared for the hemodynamic repercussions of the procedure. Indeed, the osmotic disruption is a physiologically stressful procedure. It can induce focal seizures in 5% of procedures. It can also trigger a vaso-vagal response with bradycardia and hypotension. In order to prevent the occurrence of these adverse effects, the following medications are administered just prior to the disruption: Diazepam 0.2 mg/kg IV (maximum dose = 10 mg), and Atropine IV, titrated to increase heart rate 10–20% from baseline (0.5–1 mg). We then proceed to the BBBD, after which IA infusion of chemotherapy is accomplished. Figure 4B shows the repercussion of BBBD on delivery. In this image, an IV contrast material was infused shortly after the BBBD (within 5 min), showing a diffuse penetration of the contrast compound in the brain parenchyma (arrow).

5.1. CIAC or CIAC + BBBD? A Question of Intensity of Delivery

The question of whether to use CIAC alone or with an adjunctive BBBD really is a question of intensity in the amount of delivery. There is no question that BBBD will increase delivery compared to an isolated CIAC. When studying platinum compounds, this increase has been shown to be variable for each molecule, providing a 2-fold increase for Carboplatin (overall), and up to a 5-fold increase for Lipoxal compared to IA alone (Figure 3). However, the use of BBBD requires general anesthesia, and is significantly heavier for the patient. Hence, its use can be limited by the availability of anesthesia and all it implies (recovery room, etc.). On the other hand, CIAC is easy to perform, and virtually devoid of these limitations. The procedure is cheap, and the only limitation is the access to the angiography suit. Hence, we have traditionally reserved the use of BBBD for patients with potentially curable diseases, such as primary CNS lymphomas (PCNSL). Metastatic brain disease, as well as glial tumors, are typically treated by CIAC. We built most of our clinical studies around a model in which the patient receives a monthly treatment session, typically up to 12 sessions. Only in patients of these 2 groups of pathologies presenting a complete response or near-complete response will we consider using BBBD to consolidate the treatment response in the last 2 cycles of treatment.

5.2. Clinical Data: Safety

Neurotoxicity is a legitimate concern when deploying a strategy that increases CNS delivery of therapeutics. Indeed, transgressing the BBB could result in an increase in neurotoxicity. Thereby each therapeutic used in the clinic has been previously tested in the animal model to screen for compatible drug candidates for CIAC/CIAC + BBBD clinical use. Obviously, this does not entirely preclude the risks of toxicity. However, now looking at the modern series of CIAC/CIAC + BBBD, we can confidentially claim it to be safe, when performed in expert centers.

Doolittle et al. reported on the experience of the BBBD consortium, a multi-site consortium performing CIAC with and without BBBD for malignant brain tumors [30]. These authors concluded that with standardized protocols, CIAC was safe across multiple centers, with a low incidence of catheter-related complications. In their series of 221 patients treated between 1994 and 1997, they observed a sub-intimal tear rate of 5%, whereas the rate of strokes was 1.7%.

We undertook a detailed review of our own experience in terms of complications, going into further details. We analyzed our entire cohort of CIAC patients to brush the best possible picture in terms of innocuity. Between January 2000 and June 2015, a total of 3583 arteriographic procedures for CIAC/CIAC + BBBD were performed on 722 patients in the treatment of brain tumors at CHUS (centre hospitalier universitaire de Sherbrooke, Sherbrooke, Canada). All patients were afflicted by a malignant brain tumor (463 primary brain tumor, 158 metastasis, 101 lymphomas). To our knowledge, this is the largest such series available in the literature [31].

As clinical data have been cumulated prospectively in the context of clinical studies, data were extracted from all hospitalization records for care events related to a CIAC procedure in the treatment of brain tumors (glial tumors, PCNSL, and metastatic tumors). Complications were studied and grouped under 3 different headings: Vascular complications, per-procedural epileptic manifestations, and hematological toxicities. The results are detailed in Table 3.

Table 3. Angiographic, seizure-related and hematologic complications in the series of CIAC/CIAC + BBBD patients treated in Sherbrooke, from 2000–2015. A total of 3583 procedures in 722 patients were accomplished.

Angiographic + Vascular Complications	Number of Event MRI + Angiographic Findings	MRI Findings	Symptomatic Lesions (the Lesion was Accompanied by Clinical Symptoms)	Asymptotic (Lesion Found at MRI or Angiography without Consequent Symptoms)
Dissections	5	1	0	5
Stenosis	9	2	0	9
Occusions	3	2	2	1
Hemorragic lesions	5	5	1	4
Lacunar Strokes	38	38	20	18
Strokes	6	6	4	2
Total of events on 3586 procedures	66 (1.84%)	54 (1.5%)	27 (0.75%)	39 (1.08%)

Focal Seizures (# of Events)	Generalized Seizures (# of Events)	Lymphomas	Metastasis	Glial Tumors	MTX	Carboplatin
65	9	23	4	12	62	12
74 seizure events (2%)				39 patients (5.4%)		

Hematologic Toxicites (per NCIC Toxicity Criteria)	Grade 1	Grade 2	Grade 3	Grade 4	Total
Neutropenia	70 (9.7%)	67 (9.3%)	22 (3.1%)	21 (2.9%)	180 (24.9%)
Thrombocytopenia	43 (5.9%)	37 (5.1%)	35 (4.9%)	21 (2.9%)	136 (18.8%)
Anemia	115 (15.9%)	78 (10.8%)	25 (3.5%)	10 (1.4%)	228 (31.6%)
Total	228 (31.6%)	182 (25.2%)	82 (11.4%)	52 (7.2%)	

5.2.1. Vascular Complications

Overall, a total of 66 vascular angiographic or MRI incidents were uncovered (1.84%). More specifically, 5 asymptomatic dissections were observed, 9 asymptomatic carotid stenosis and 3 occlusions were identified, 2 which were symptomatic. (Table 3).

In terms of cerebral newly described lesions, the MRI identified 5 acute hemorrhagic strokes (1 symptomatic), 38 lacunar strokes (20 symptomatic), and 6 acute ischemic lesions (4 symptomatic). One of these strokes in the posterior fossa was a catastrophic event that led to the patient's death. Overall in this series, the total number of symptomatic vascular complication rate was 27 (0.75%).

5.2.2. Seizure Events

The overall per-procedural seizure incidence was 2% (74 incidents) as can be appreciated in Table 3. Of these, 9 were generalized seizures, whereas 65 were partial seizures. Interestingly, a simple discontinuation of the chemotherapy infusion was sufficient to halt the seizure in all patients, but one.

Most seizure fits (84%) were observed during Methotrexate infusion for primary CNS lymphomas (PCNSL). Only 3 partial seizures were observed in the treatment of glial tumors.

5.2.3. Hematological Complications

Hematological complications were classified according to the National Cancer Institute of Canada (NCIC) toxicity criteria. A total of 11.4% grade 3 and 7.2% grade 4 toxicities were observed.

Hence, from the analysis of this data, we feel justified to conclude that the procedure is safe, and its use is appropriate in this clinical context. This affirmation does imply that the treatments are performed in expert centers, and the therapeutics used with CIAC have been screened with an adequate methodology and are known to be devoid of neurotoxicity. While the technical details of the procedure are beyond the scope of this publication, a few considerations need to be discussed. First, although some might argue that a supra-selective catheter placement might be of interest, we always use a proximal position in the treated vessel (C1-2 for the carotid, C2-C3 for the vertebral). The rationale supporting this has to do with the infiltrative nature of most brain tumors, always lending for a more widespread disease than the MR scans actually reveal. This is true for glial tumors, PCNSL, but also metastasis. Because of that, we see little interest to target for a supra-selective catheterization, especially considering that this approach would likely increase the risks of complications while limiting the actual distribution of chemotherapy to the CNS.

Secondly, each therapeutic comes with its own set of infusion parameters based on the concentration, density, and volume of the infusion solution. This is paramount to minimize the risks of neurotoxicity related to concentration and streaming.

6. Clinical Results

We will focus the discussion on the results obtained for GBM (grade 4 astrocytomas) and PCNSL. As a remainder, all patients with GBM were treated by CIAC, whereas 10% of these receive CIAC + BBBD as a consolidation procedure for the last 2 cycles. PCNSL patients were all treated with CIAC + BBBD.

6.1. Glioblastoma (GBM)

In our series, 319 GBM patients were treated by CIAC. Treatment sessions were performed every 4 weeks, unless hematologic parameters prevented it. Carboplatin is the drug of choice, either alone, or combined with either Melphalan or Etoposide phosphate, depending on the protocol used. All GBM patients were treated at relapse. Seventeen percent were treated at first relapse, 68% at second relapse, 11% at third relapse and 4% at fourth (Figure 5).

The fact that most of our patients were treated at 2nd relapse unfortunately negatively biases our results.

Overall, patients received a median of 4 cycles (1–22 cycles). Progression-free survival was 5 months. The whole series presented an overall median survival of 25 months, and survival from study entry was 8 months for the entire cohort (Figure 6). This is superior to most commonly reported treatments at relapse, which usually produces median survival from treatment initiation of 4–6 months [2,3,7].

Ten patients are still alive, with the longest survival now at 16 years. When looking exclusively at those patients treated at 2nd relapse, median survival jumps at 11 months from the study entry for the entire cohort.

Looking at the best radiological responses obtained according to the RANO criteria, we found the following: 23% of patients have shown progression, 26% have presented a stabilization of their disease, 42% have shown a partial response (Figure 7), and 6% a complete response.

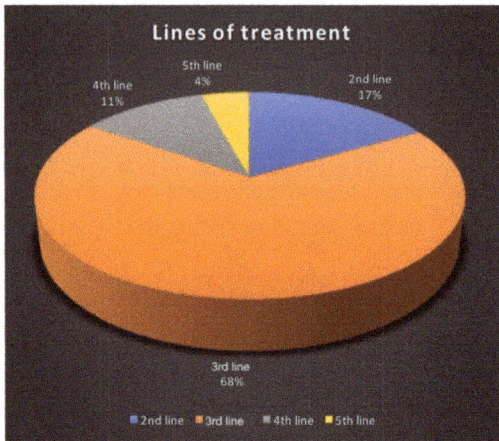

Figure 5. A breakdown of the number of treatment lines to which GBM patients were exposed prior to accrual in our series. As can be appreciated, most patients were exposed to 2 lines of treatment (68%) prior to accrual.

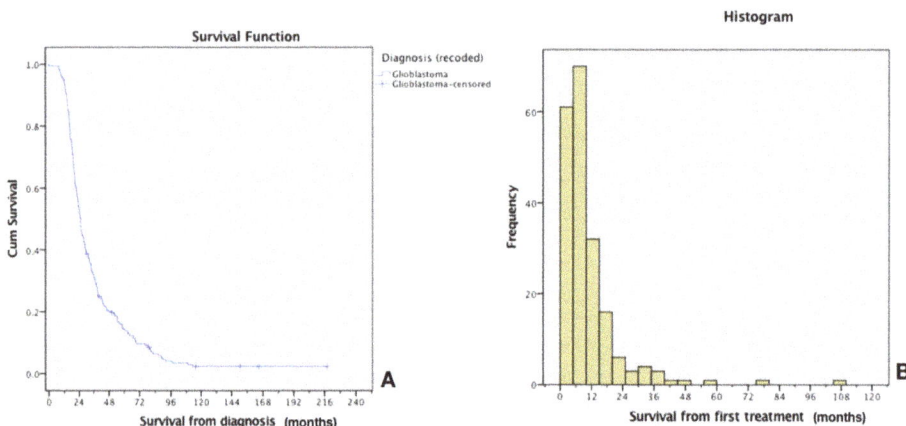

Figure 6. Median survival from diagnosis (**A**) and distribution survival histogram (**B**) of GBM patients exposed to CIAC. Median survival from diagnosis was 25 months, whereas it was 8 months from the study entry. As can be appraised from the distribution histogram, most patients progress and die from their disease in the first 12 months after accrual. This leaves around 25% of patients whose survival is greater than 12 months.

Although these results are encouraging, comparing this data with modern series is complicated by 2 factors. Our data on GBM is plagued by a major weakness: Heterogeneity. Indeed, in 2016, the classification on gliomas changed, and one major overhaul has been the inclusion of molecular markers to better stratify the patients in prognostic classes that strongly determine their evolution [32]. The IDH marker is the stronger determinant of survival, and, nowadays, most modern series stratify patients according to this marker, which we did not. The other source of heterogeneity in this series pertains to the fact that a majority of patients were exposed to multiple treatments prior to accrual (Figure 5). These 2 factors are obviously now considered in the design of our studies. Recruitment and stratification is now refined to eliminate these confounding factors.

Ideally, to avoid this heterogeneity, a randomization process should also be utilized. We tried to launch such a study a few years back, but were faced with difficulty as to what should be the randomized arm. The design of the randomized study is now complete and submitted. It will compare carboplatin/Etoposide phosphate CIAC against oral Lomustine (CCNU) in the control arm, and will constrain accrual at 1st relapse, as well as stratify patients against molecular status. Hopefully, this will allow us to demonstrate the superiority of this approach, once and for all.

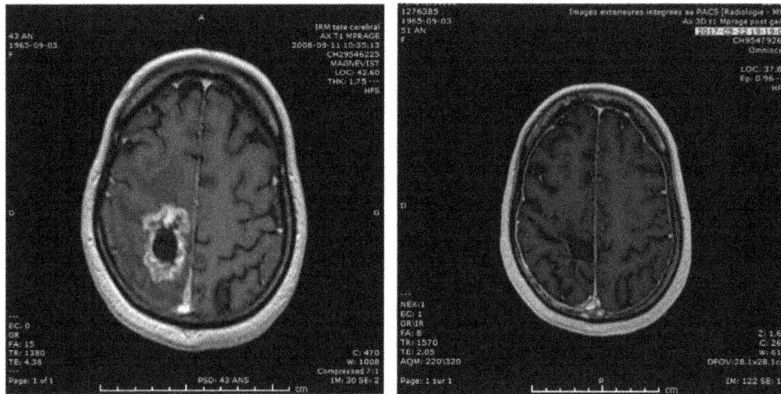

Figure 7. Example of one of our best responder: A 43-year-old female glioblastome (GBM) patient treated with intraarterial Carboplatin/melphalan in 2008, who remains in complete response in 2017. This patient was treated at first relapse, without using any other therapy than the Stupp regimen at the first-line.

6.2. Future Perspectives in the Treatment of Malignant Gliomas

6.2.1. Heterogeneity of Response: The Impossibility to Predict the Best Regimen for Each Patient

Carboplatin is our drug of choice, as it appears to be the most effective, producing responses in 70% of patients for a median PFS of 5 months at relapse. However, we do have an array of agents available for intra-arterial infusion in the clinic, and we continue to expand this list: Carboplatin, Methotrexate, Melphalan, Etoposide phosphate, and more recently, Caelyx. These agents have all been used safely in CIAC for brain tumor treatment by our team. Interestingly, in the case of non-response to Carboplatin, other agents can be used in subsequent cycles of treatment for patients still presenting an adequate functional status. In these cases, we are often confronted with extremely variable results, with some long-term responses (up to 180 months) observed with other agents than Carboplatin, whereas some patients show no response whatsoever to any agents. Indeed, these tumors all appear to have their own distinctive sensitivity profile to chemotherapy agents, and we believe that they should therefore all be approached as a singular disease entity requiring a personalized treatment. Molecular stratification has come a long way in the management of glial tumors [32], but its role is limited in assisting pathological stratification and prognosis. It is not yet used in treatment selection. We propose to combine data from in vitro drug sensitivity testing (DST) and molecular characterization using "The Cancer Genome Atlas" (TCGA) stratification, in addition to a panel of chemoresistance markers, to select the best drug candidates prior to the initiation of CIAC. Hence, in accordance with this scheme in a proposed clinical study at 1st relapse for GBM, all patients will be re-operated prior to the beginning of CIAC. During surgery, a tumor sample will be obtained for the DST, molecular stratification and chemoresistance panel markers, and the treatment will be tailored specifically to each patient.

6.2.2. Radio-Chemotherapy

Another area we have started to explore is the combination of Carboplatin with radiation therapy [33,34]. Indeed, radiotherapy is the most effective single-treatment modality for GBM tumors, but it controls the disease only transiently. A way to improve treatment consists of coupling radiation with a potent radiosensitizer. Carboplatin, a platinum (Pt) drug, is ideally suited for this. Our group has demonstrated that the addition of Carboplatin to ionizing radiation produced significantly more DNA strand breaks [35–39]. In numerous cell lines, combining radiotherapy and Carboplatin was found to increase cell death. In a mouse model, we observed a maximum antitumor effect with Carboplatin administration at 4 or 48 h prior to irradiation. This timing correlated to the highest levels of Pt bound to DNA [35,37,39]. Concurrent Carboplatin and radiation treatment represent a common modality for treating a variety of cancers. Unfortunately, since this class of drug does not readily cross the BBB when administered via the standard IV routes, they are not used to treat GBM. We have just started accrual on a new phase II study in which we administer IA Carboplatin with a re-irradiation protocol in a dose escalation scheme. We feel that this combination has the potential to improve clinical results. We have 6 patients (of a total of 35) recruited, and enrollment is ongoing.

6.3. Primary CNS Lymphomas

PCNSL are a rare and aggressive form of central nervous system tumors. Generally confined to the brain, eyes and/or cerebrospinal fluid compartments, these extra nodal non-Hodgkin large B cell lymphomas typically show no evidence of systemic diffusion [40]. PCNSL is an extremely aggressive disease, with a median survival time of 3 months without treatment [40]. It is a fairly unusual occurrence, accounting for 1% of cases of lymphoma, whereas it represents 4% of primary brain tumors [41,42]. A current trend in the treatment of this disease has been radiation therapy avoidance, as it was shown to be extremely neurotoxic to patients [43]. Over the years, different protocols of IV high-dose Methotrexate have shown encouraging results. Indeed, Da Broi et al. reported the results of 57 patients treated over 12 years with chemotherapy [44]. Overall, they found a median OS of 35.4 months, and a PFS of 15.7 months. Using CIAC + BBBD infusion protocol of high-dose Methotrexate (combined to Etoposide, Cyclophosphamide and/or Procarbazine), Angelov et al. reported a median overall survival of 3.1 years. They also reported the neuropsychological outcome profile in 26 long term survivors from the treatment (median follow up of 12 years), showing the innocuity of this approach [45]. This good quality data shows without a doubt that repeated CIAC + BBBD infusion protocol of high-dose Methotrexate does not impact the neurocognitive functioning of responding patients.

Using CIAC + BBBD Carboplatin (400 mg/m^2) in addition to high dose IA Methotrexate (5 g), we treated 43 newly diagnosed PCNSL patients from 1999–2018. The median age of the cohort was 63, with a mean age of 60 years old. The cohort was comprised of 24 males and 19 females. Overall, remission was induced in 34 patients (79%). The overall median survival was 46.5 months for the entire cohort. Actuarial survival was 88%, 64%, 54%, 39% and 18% at 1, 2, 3, 5 and 10 years. The progression-free survival for the entire cohort was 43.3 months. The actuarial PFS was 83%, 59%, 56%, 30%, and 9% at 1, 2, 3, 5, and 10 years. These are amongst the best results ever published in the treatment of this disease, without the use of radiation therapy. The detailed manuscript presenting these results is in preparation.

7. Conclusions

Intra-arterial chemotherapy is a delivery vehicle allowing the increase of available therapeutics in the treatment of brain cancers. Its initial use many decades ago has been hampered by toxicity, a problem which is no more of concern. Angiographic refinements, combined with intra-arerial infusion of therapeutics carefully selected for this purpose have rendered this approach safe and sound. The addition of an osmotic permeation of the BBB further increases delivery of therapeutics to the CNS. We need to acknowledge the extreme heterogeneity of GBM, and eventually start tailoring treatment to each tumor for individual patients in order to improve the modest results obtained so far. Drug selection is at the core of this process. We also need to keep expanding the pool of agents that can safely be administered via this route. As for the treatment of PCNSL, different refinements are considered to keep improving outcomes in the treatment of this disease. The addition of rituximab, a CD-20 antibody, should be considered as an adjunct to the treatment protocol. In the end, the use of intra-arterial therapeutics infusion combined to osmotic blood-brain barrier permeation answers the need to adequately address an issue that is commonly underestimated: The presence of the BBB, and the complex pharmacokinetic set of compartments it imposes on CNS delivery.

Funding: This research was funded by CIHR, grant # 201950.

Conflicts of Interest: The author declares no conflict of interest.

References

1. Stupp, R.; Mason, W.P.; van de Bent, M.J.; Weller, M.; Fisher, B.; Taphoorn, M.J.B.; Belanger, K.; Brandes, A.A.; Marosi, C.; Bogdahn, U.; et al. Radiotherapy plus concomitant and adjuvant temozolomide for glioblastoma. *N. Engl. J. Med.* **2005**, *352*, 987–996. [CrossRef]
2. Fabian, D.; Guillermo Prieto Eibl, M.P.; Alnahhas, I.; Sebastian, N.; Giglio, P.; Puduvalli, V.; Gonzalez, J.; Palmer, J.D. Treatment of Glioblastoma (GBM) with the addition of Tumor-Treating Fields (TTF): A review. *Cancers* **2019**, *11*, 174. [CrossRef]
3. Drapeau, A.; Fortin, D. Chemotherapy Delivery Strategies to the Central Nervous System: Neither optional nor superfluous. *Curr. Cancer Drug Targets* **2015**, *15*, 752–768. [CrossRef]
4. Fortin, D. The blood-brain barrier: Its influence in the treatment of brain tumors metastases. *Curr. Cancer Drug Targets* **2012**, *12*, 247–259. [CrossRef] [PubMed]
5. Van Den Bent, M.J. The role of chemotherapy in brain metastases. *Eur. J. Cancer* **2003**, *39*, 2114–2120. [CrossRef]
6. Reichel, A. Addressing central nervous system (CNS) penetration in drug discovery: Basics and implications of the evolving new concept. *Chem. Biodivers.* **2009**, *6*, 2030–2049. [CrossRef]
7. Fortin, D.; Morin, P.A.; Belzile, F.; Mathieu, D.; Paré, F.M. Intra-arterial carboplatin as a salvage strategy in the treatment of recurrent glioblastoma multiforme. *J. Neuro-Oncol.* **2014**, *119*, 397–403. [CrossRef] [PubMed]
8. Bellavance, M.-A.; Blanchette, M.; Fortin, D. Recent advances in blood-brain barrier disruption as a CNS delivery strategy. *AAPS J.* **2008**, *10*, 166–177. [CrossRef] [PubMed]
9. Kroll, R.A.; Neuwelt, E.A. Outwitting the blood-brain barrier for therapeutic purposes: osmotic opening and other means. *Neurosurgery* **1998**, *42*, 1083–1099. [CrossRef] [PubMed]
10. Pardridge, W.M. Blood–brain barrier delivery. *Drug Discov. Today* **2007**, *12*, 54–61. [CrossRef]
11. Fortin, D. Altering the properties of the blood-brain barrier: Disruption and permeabilization. *Prog. Drug Res.* **2003**, *61*, 125–154. [PubMed]
12. Tosoni, A.; Ermani, M.; Brandes, A.A. The pathogenesis and treatment of brain metastases: a comprehensive review. *Crit. Rev. Oncol./Hematol.* **2004**, *52*, 199–215. [CrossRef]
13. Sato, S.; Kawase, T.; Harada, S.; Takayama, H.; Suga, S. Effect of hyperosmotic solutions on human brain tumor vasculature. *Acta Neurochir.* **1998**, *140*, 1135–1142. [CrossRef] [PubMed]
14. Silbergeld, D.L.; Chicoine, M.R. Isolation and characterization of human malignant glioma cells from histologically normal brain. *J. Neurosurg.* **1997**, *86*, 525–531. [CrossRef] [PubMed]
15. Chicoine, M.R.; Silbergeld, D.L. Assessment of brain tumor cell motility in vivo and in vitro. *J. Neurosurg.* **1995**, *82*, 615–622. [CrossRef]
16. Kroll, R.A.; Pagel, M.A.; Muldoon, L.L.; Roman-Goldstein, S.; Fiamengo, S.A.; Neuwelt, E.A. Improving drug delivery to intracerebral tumor and surrounding brain in a rodent model: A comparison of osmotic versus bradykinin modification of the blood-brain and/or blood-tumor barriers. *Neurosurgery* **1998**, *43*, 879–886. [CrossRef]
17. Bradbury, M.W.B. Appraisal of the role of endothelial cells and glia in barrier breakdown. In *The Blood-Brain Barrier in Health and Disease*; Suckling, A.J., Rumsby, M.G., Bradbury, M.W.B., Eds.; Ellis Horwood: Chichester, UK, 1986; pp. 128–129.
18. Fortin, D. La barrière hémato-encéphalique: un facteur clé en neuro-oncologie. *Rev. Neurol.* **2004**, *160*, 523–532. [CrossRef]
19. Newton, H.B.; Slivka, M.A.; Volpi, C.; Bourekas, E.C.; Christoforidis, G.A.; Baujan, M.A.; Slone, W.; Chakeres, D.W. Intra-arterial carboplatin and intravenous etoposide for the treatment of metastatic brain tumors. *J. Neuro-Oncol.* **2003**, *61*, 35–44. [CrossRef]

20. Newton, H.B.; Figg, G.M.; Slone, H.W.; Bourekas, E. Incidence of infusion plan alterations after angiography in patients undergoing intra-arterial chemotherapy for brain tumors. *J. Neuro-Oncol.* **2006**, *78*, 157–160. [CrossRef] [PubMed]
21. Drapeau, A.I.; Poirier, M.B.; Madugundu, G.S.; Wagner, J.R.; Fortin, D. Intra-arterial temozolomide, osmotic blood-brain barrier disruption and radiotherapy in a rat F98-glioma model. *Clin. Cancer Drugs* **2017**, *4*, 135–144. [CrossRef]
22. Blanchette, M.; Tremblay, L.; Lepage, M.; Fortin, D. Impact of drug size on brain tumor and brain parenchyma delivery after a blood-brain barrier disruption. *J. Cerebr. Blood Flow Metab.* **2014**, *34*, 820–826. [CrossRef]
23. Blanchette, M.; Pellerin, M.; Tremblay, L.; Lepage, M.; Fortin, D. Real-Time monitoring of gadolinium diethylenetriamine penta-acetic acid during osmotic blood-brain barrier disruption using magnetic resonance imaging in normal wistar rats. *Neurosurgery* **2009**, *65*, 344–351. [CrossRef]
24. Pitz, M.W.; Desai, A.; Grossman, S.A.; Blakeley, J.O. Tissue concentration of systemically administered antineoplastic agents in human brain tumors. *J. Neuro-Oncol.* **2011**, *104*, 629–638. [CrossRef] [PubMed]
25. Mathieu, D.; Lecomte, R.; Tsanaclis, A.M.; Larouche, A.; Fortin, D. Standardization and detailed characterization of the syngeneic Fischer/F98 glioma model. *Can. J. Neurol. Sci.* **2007**, *34*, 296–306. [CrossRef]
26. Blanchard, J.; Mathieu, D.; Patenaude, Y.; Fortin, D. MR-pathological comparison in F98-Fischer glioma model using a human gantry. *Can. J. Neurol. Sci.* **2006**, *33*, 86–91. [CrossRef]
27. Charest, G.; Sanche, L.; Fortin, D.; Mathieu, D.; Paquette, B. Glioblastoma treatment: Bypassing the toxicity of platinum compounds by using liposomal formulation and increasing treatment efficiency with concomitant radiotherapy. *Int. J. Radiat. Oncol. Biol. Phys.* **2012**, *84*, 244–249. [CrossRef]
28. Charest, G.; Sanche, L.; Fortin, D.; Mathieu, D.; Paquette, B. Optimization of the route of platinum drugs administration to optimize the concomitant treatment with radiotherapy for glioblastoma implanted in the Fischer rat brain. *J. Neuro-Oncol.* **2013**, *115*, 365–373. [CrossRef]
29. Charest, G.; Paquette, B.; Fortin, D.; Mathieu, D.; Sanche, L. Concomitant treatment of F98 glioma cells with new liposomal platinum compounds and ionizing radiation. *J. Neuro-Oncol.* **2010**, *97*, 187–193. [CrossRef] [PubMed]
30. Doolittle, N.D.; Miner, M.E.; Hall, W.A.; Siegal, T.; Hanson, E.J.; Osztie, E.; McAllister, L.D.; Bubalo, J.S.; Kraemer, D.F.; Fortin, D.; et al. Safety and efficacy of a multicenter study using intra-arterial chemotherapy in conjunction with osmotic opening of the blood–brain barrier for the treatment of patients with malignant brain tumors. *Cancer* **2000**, *88*, 637–647. [CrossRef]
31. Fortin, D. Safety of intra-arterial chemotherapy in the treatment of brain tumors. In Proceedings of the Society for Neuro-Oncology (SNO), Presentation, Scottsdale, AZ, USA, 17 November 2016.
32. Louis, D.N.; Perry, A.; Reifenberger, G.; Von Deimling, A.; Figarella-Branger, D.; Cavenee, W.K.; Ohgaki, H.; Wiestler, O.D.; Kleihues, P.; Ellison, D.W. The 2016 world health organization classification of tumors of the central nervous system: A summary. *Acta Neuropathol.* **2016**, *131*, 803–820. [CrossRef] [PubMed]
33. Choy, H. *Chemoradiation in Cancer Therapy*; Humana Press: Totowa, NJ, USA, 2003.
34. Mamon, H.J.; Tepper, J.E. Combination chemoradiation therapy: The whole is more than the sum of the parts. *J. Clin. Oncol.* **2014**, *32*, 367–369. [CrossRef]
35. Boudaïffa, B.; Cloutier, P.; Hunting, D.; Huels, M.A.; Sanche, L. Resonant formation of DNA strand breaks by low energy (3 to 20 eV) electrons. *Science* **2000**, *287*, 1658–1660. [PubMed]
36. Rezaee, M.; Hunting, D.J.; Sanche, L. New insights into the mechanism underlying the synergistic action of ionizing radiation with platinum chemotherapeutic drugs: The role of low-energy electrons. *Int. J. Radiat. Oncol. Biol. Phys.* **2013**, *87*, 847–853. [CrossRef] [PubMed]
37. Tippayamontria, T.; Kotb, R.; Paquette, B.; Sanche, L. Efficacy of cisplatin and Lipoplatin™ in combined treatment with radiation of a colorectal tumor in nude mouse. *Anticancer Res.* **2013**, *33*, 3005–3014.
38. Tippayamontri, T.; Kotb, R.; Sanche, L.; Paquette, B. New therapeutic possibilities of combined treatment of radiotherapy with oxaliplatin and its liposomal formulations Lipoxal™ in rectal cancer using nude mouse xenograft. *Anticancer Res.* **2014**, *34*, 5303–5312. [PubMed]
39. Zheng, Y.; Hunting, D.J.; Ayotte, P.; Sanche, L. Role of secondary low-energy electrons in the concomitant chemoradiation therapy of cancer. *Phys. Rev. Lett.* **2008**, *100*, 198101. [CrossRef]
40. Bairey, O.; Siegal, T. The possible role of maintenance treatment for primary central nervous system lymphoma. *Blood Rev.* **2018**, *32*, 378–386. [CrossRef]

41. Campo, E.; Swerdlow, S.H.; Harris, N.L.; Pileri, S.; Stein, H.; Jaffe, E.S. The 2008 WHO classification of lymphoid neoplasms and beyond: Evolving concepts and practical applications. *Blood* **2011**, *117*, 5019–5032. [CrossRef]

42. Darlix, A.; Zouaoui, S.; Rigau, V.; Bessaoud, F.; Figarella-Branger, D.; Mathieu-Daude, H.; Trétarre, B.; Bauchet, F.; Duffau, H.; Taillandier, L.; et al. Epidemiology for primary brain tumors: A nationwide population-based study. *J. Neuro-Oncol.* **2017**, *131*, 525–546. [CrossRef]

43. Doolittle, N.D.; Dósa, E.; Fu, R.; Muldoon, L.L.; Maron, L.M.; Lubow, M.A.; Tyson, R.M.; Lacy, C.A.; Kraemer, D.F.; Butler, R.W. Preservation of cognitive function in primary CNS lymphoma survivors a median of 12 years after enhanced chemotherapy delivery. *J. Clin. Oncol.* **2013**, *31*, 4026–4027. [CrossRef] [PubMed]

44. Da Broi, M.; Jahr, G.; Beiske, K.; Holte, H.; Meling, T.R. Efficacy of the Nordic and the MSKCC chemotherapy protocols on the overall and progression-free survival in intracranial PCNSL. *Blood Cells Mol. Dis.* **2018**, *73*, 25–32. [CrossRef] [PubMed]

45. Angelov, L.; Doolittle, N.D.; Kraemer, D.F.; Siegal, T.; Barnett, G.H.; Peereboom, D.M.; Stevens, G.; McGregor, J.; Jahnke, K.; Lacy, C.A.; et al. Blood-brain barrier disruption and intra-arterial methotrexate-based therapy for newly diagnosed primary CNS lymphoma: A multi-institutional experience. *J. Clin. Oncol.* **2009**, *27*, 3503–3509. [CrossRef] [PubMed]

pharmaceutics

MDPI

Article

Development and Characterization of Liposomal Formulations Containing Phytosterols Extracted from Canola Oil Deodorizer Distillate along with Tocopherols as Food Additives

Asmita Poudel, George Gachumi, Kishor M. Wasan, Zafer Dallal Bashi, Anas El-Aneed * and Ildiko Badea *

Drug Design and Discovery Group, College of Pharmacy and Nutrition, University of Saskatchewan, 107 Wiggins Road, Saskatoon, SK S7N 5E5, Canada; asp170@mail.usask.ca (A.P.); george.gachumi@usask.ca (G.G.); kishor.wasan@usask.ca (K.M.W.); zafer.bashi@usask.ca (Z.D.B.)
* Correspondence: anas.el-aneed@usask.ca (A.E.); ildiko.badea@usask.ca (I.B.); Tel.: +1-306-966-2013 (A.E.); +1-306-966-6349 (I.B.)

Received: 28 February 2019; Accepted: 9 April 2019; Published: 16 April 2019

Abstract: Phytosterols are plant sterols recommended as adjuvant therapy for hypercholesterolemia and tocopherols are well-established anti-oxidants. However, thermo-sensitivity, lipophilicity and formulation-dependent efficacy bring challenges in the development of functional foods, enriched with phytosterols and tocopherols. To address this, we developed liposomes containing brassicasterol, campesterol and β-sitosterol obtained from canola oil deodorizer distillate, along with alpha, gamma and delta tocopherol. Three approaches; thin film hydration-homogenization, thin film hydration-ultrasonication and Mozafari method were used for formulation. Validated liquid chromatographic tandem mass spectrometry (LC-MS/MS) was utilized to determine the entrapment efficiency of bioactives. Stability studies of liposomal formulations were conducted before and after pasteurization using high temperature short time (HTST) technique for a month. Vesicle size after homogenization and ultrasonication (<200 nm) was significantly lower than by Mozafari method (>200 nm). However, zeta potential (−9 to −14 mV) was comparable which was adequate for colloidal stability. Entrapment efficiencies were greater than 89% for all the phytosterols and tocopherols formulated by all three methods. Liposomes with optimum particle size and zeta potential were incorporated in model orange juice, showing adequate stability after pasteurization (72 °C for 15 s) for a month. Liposomes containing phytosterols obtained from canola waste along with tocopherols were developed and successfully applied as a food additive using model orange juice.

Keywords: phytosterols; tocopherols; liposomes; canola oil deodorizer distillate; model orange juice

1. Introduction

Functional foods and nutraceuticals are increasing rapidly due to growing consumer preferences towards natural bioactives rather than synthetic drugs for disease prevention and treatment [1]. Phytosterols and tocopherols are such bioactives (plant metabolites) that have numerous health claims [2,3]. The primary health benefit of phytosterols is to lower low-density lipoprotein (LDL) cholesterol levels in plasma [2,4,5]. Due to this health claim, the National Cholesterol Education Program Adult Treatment Panel III (NCEP ATP III) has recommended phytosterols as adjuvant therapy to statins in hypercholesterolemia [6]. Phytosterols compete with cholesterol for their solubilization in bile salt micelles, hindering the absorption of cholesterol in blood [7,8]. Tocopherols, on the other hand, are free radical scavengers and natural anti-oxidants [9,10]. Due to their anti-oxidant properties, tocopherols are used in the treatment of age related macular degeneration [11], Alzheimer's disease [12], glaucoma [3] and heart diseases [13].

Sources of phytosterols and tocopherols include oil seeds such as canola and sesame, as well as nuts [14–16]. Among these, canola is a major source for edible vegetable oils, and the most abundant oilseed crop in Canada [17]. It is a rich source of four phytosterols, namely beta-sitosterol, campesterol, stigmasterol and brassicasterol, and four tocopherols (alpha, beta, gamma and delta) [16,18]. Canola oil loses some of its valuable components during the refining process [16]. Significant amount of phytosterols and tocopherols are transferred to the waste stream, termed canola oil deodorizer distillate (CODD) [16] which offers an ideal source of these components.

However, formulation of these bioactives in functional foods has always been challenging due to their lipophilicity and light sensitivity [19]. In particular, degradation products of phytosterols, phytosterol oxidation products (POPs), are known to have some negative impact on human health [20,21]. Thus, the selection of suitable formulation approach is crucial during the development of functional food that contain these bioactives. Encapsulation techniques, such as spray drying, fluidized bed coating, microemulsification and liposomal entrapment are emerging in the food industry to address lipophilicity related challenges [22,23]. Unfortunately, most of these techniques have shortcomings such as usage of high temperature (can possibly degrade phytosterols and tocopherols) and the requirement of large quantities of emulsifiers and surfactants, which are deleterious to human health [23–27]. All of these shortcomings can be addressed by employing liposomal formulations that require low heat and low quantities of surfactants or emulsifiers [28].

Phytosterols in both the free and esterified forms have been used in the food industry [2,4]. Solubilization of esterified phytosterols in fat containing foods, like margarine [29,30], salad dressing [31] and yogurt [32] is prevalent in the food industry. However, this approach is not favorable to people who are on low fat diet [33]. To overcome this, various low fat or non-fat food matrices such as low fat milk [34,35], granola bars [36], orange juice [37] and non-fat beverages [38–40] are emerging as food products. However, for these type of food products, lipophilic phytosterols should be well formulated prior to their development into functional food. In addition to the choice of the food matrix, the biological efficacy should also be carefully considered.

Various clinical trials have shown that the efficacy of phytosterols depends on different parameters, such as solubility in the food matrix and the formulation [2,41]. Esterified phytosterols solubilized in fat/oil are driven favorably towards the bile salt micelles in the guts than the crystalline or the insolubilized forms [42,43]. Phytosterols ester containing food products such as milk, spread and yogurt have showed reduction in LDL-cholesterol by 7–12% at daily dose of 1.6–2 g relative to control research participants [44–46]. In contrast, some failed clinical trials are also prevalent [47,48]. For example, Ottestad et al. showed that phytosterol ester in the capsular formulation revealed no significant reduction of LDL cholesterol [47]. Similarly, Denke et al. showed no significance in cholesterol reduction by sitostanol capsule relative to control [48]. Unlike phytosterol capsule-based trials, lecithin-based free phytosterol formulations have shown to impart efficacy as high as 14.3% at a daily dose of only 1.9 g relative to control [39]. In sum, literature reports show that the efficacy of phytosterols depends greatly on the formulation approach, which provides insights regarding the possibility of further enhancing their efficacy by well formulating in suitable delivery systems.

The work of Shin at al. [38] and Spilburg at al. [39] provides a strong evidence that lecithin (phosphatiylcholine) can be effective carrier of phytosterols to increase cholesterol-lowering efficacy. Both of these studies used lecithin micelles to formulate sterol/stanol which have shown promising cholesterol-lowering efficacy [38,39]. Liposomes, which have same building blocks as micelles that is lecithin (i.e., phosphatidylcholine) but different architecture are another formulation strategy in which lecithin can be utilized, thus have potential of further enhancing its cholesterol-lowering efficacy. In addition, liposomes can prevent oxidation of thermo-sensitive bioactives and are biocompatible and biodegradable [49]. Further, co-formulation of tocopherols along with phytosterols can enhance oxidative stability of phytosterols [50].

Thus, in this work, with the aim of enhancing phytosterols' oxidative stability and increasing its efficacy, we formulated phytosterols (obtained from CODD) and commercially available tocopherols into liposomes employing three different approaches, namely thin film hydration homogenization, thin film hydration ultra-sonication and Mozafari method. The liposomal formulation showing the highest entrapment efficiency, adequate size and zeta potential was incorporated into model orange juice (acidified solution). Thus, functional orange juice containing liposomal phytosterols and tocopherols was developed and its stability was assessed.

2. Materials

Chemicals and Reagents

Phytosterols were extracted from CODD obtained from LDM foods (Yorkton, SK, Canada). Briefly, 5 g of CODD was saponified with 1 M potassium hydroxide in 95% ethanol for 1 h at 65 °C after which water was added and the mixture was chilled at 9.5 °C for 1 h. After the crystallization of phytosterols, vacuum filtration was performed and the residue was washed before being dried under high vacuum. Tocopherols, chloroform, ethyl acetate and potassium hydroxide were purchased from Sigma Aldrich (Oakville, ON, Canada), and phosphatidylcholine (PC) was purchased from Avanti Polar Lipids (Alabaster, AL, USA). Purified water was obtained from Millipore (Bedford, MA, USA).

3. Methods

3.1. Formulation of Liposomes

Three different formulation techniques namely thin film hydration homogenization; thin film hydration ultrasonication and Mozafari method were used for formulation in order to evaluate the formulation technique that can produce liposomes with optimum physicochemical properties for both oral delivery and industrial scale up.

3.1.1. Method I. Thin Film Hydration–Homogenization

This method was adopted from Chung et al. [51] with some modifications. In brief, tocopherols (alpha, gamma and delta tocopherol), phytosterols mixture (brassicasterol, campesterol and beta-sitosterol) and PC were dissolved in 5 mL ethyl acetate (food grade) in 0.1:0.9:2, 0.1:0.9:3, 0.1:0.9:4 and 0.1:0.9:5 ratio of tocopherol: phytosterol: PC. Ethylacetate was evaporated using rotary evaporator at pressure of 90 mmHg. The thin lipid film containing bioactives and PC formed on the wall of the flask was lyophilized for 10 h to remove traces of ethylacetate and was hydrated with 20 mL of purified water for 3 h at 55°C with occasional vortexing in the presence of glass beads. The lipid dispersion was homogenized using recirculating high fluid pressure homogenizer (Microfluidics Corporation, Westwood, MA, USA) at 60 psi for 20 min. The prepared liposomes were left overnight at 4 °C prior to size analysis.

3.1.2. Method II: Thin Film Hydration Ultrasonication

This method was adopted from Akbarzadeh et al. [52]. Similar to thin film hydration homogenization; tocopherols (alpha, gamma and delta tocopherol), phytosterols mixture (brassicasterol, campesterol and beta-sitosterol) were dissolved, along with PC in 5 mL ethyl acetate, in 0.1:0.9:2 ratio of tocopherols: phytosterols: PC. Ethylacetate was evaporated using rotary evaporator at pressure of 90 mmHg. Thin lipid film containing bioactives and PC was formed at the bottom of the flask. Lipid film was lyophilized for 10 h to remove traces of ethylacetate and was hydrated with 20 mL of purified water maintained at 55°C. Lipid dispersion was ultrasonicated using bath sonicator (ELMA Corp.,Singen, Germany) for 30 min maintained at 55 °C then was allowed to cool at room temperature. The prepared liposomes were left overnight at 4 °C prior to size analysis.

3.1.3. Method III: Mozafari Method

This method was adopted from Colas et al. [53]. 50 mg of PC was hydrated with 20 mL of purified water for 1 h and was heated to 55 °C. Nine mg of the phytosterol mixture and 1 mg of the tocopherol mixture were heated with 3% *v/v* glycerol at 110 °C and 55 °C temperature, respectively for 15 min on a hot plate stirrer at approximately 1000 RPM (Corning Corporation, Midland, ON, Canada) and then was cooled down to 55 °C. PC dispersion, phytosterols and tocopherols were mixed together with stirring on a hot plate for 30 min at approximately 1000 RPM. The formed liposomes were cooled down to room temperature and kept overnight at 4 °C prior to size analysis.

3.2. Characterization of Particle Size, Size Distribution and Zeta Potential

Particle size and zeta potential measurement of the liposomes were performed using Zeta sizer, Nano ZS instrument, Malvern instruments Ltd. (Worcestershire, England). All measurements were conducted in triplicates at 25 °C and reported as mean ± SD.

3.3. Transmission Electron Microscopy (TEM) Analysis

TEM analysis was performed by negative staining. Briefly, a drop of liposomal sample was placed on copper- formvar coated TEM grid and was allowed to settle on grid surface for 1 min. Excess of the liquid was removed using absorbent tissue. Staining of grid was done using 0.5% phosphotungstic acid for 30 s and excess of stain is removed. Imaging was done using aHT 7700 TEM (Hitachi, Japan) at 80 kV.

3.4. LC-MS/MS Method Development and Validation

LC-MS/MS method was developed and was validated as per International Council for Harmonization of Technical Requirements for Pharmaceuticals for Human Use (ICH) guidance for bioanalytical method validation guideline [54]. Briefly, chromatographic separation of the analytes was carried out on an Agilent Acquity UPLC (Agilent Technologies, Mississauga, ON, Canada) with an Agilent Poroshell C18 column (2.1 mm × 150 mm, 5μm) protected by a guard column (2.1 mm × 4.7 mm, 2.7 μm) of the same packing material. The column temperature was set at 30 °C and the injection volume was 2.5 μL. An isocratic elution consisting of acetonitrile: methanol (99:1 *v/v*) with 0.1% acetic acid was used at a flow rate of 0.8 mL/min. The detection and quantification were performed using an API 6500 QTRAP® quadruple-linear ion trap (QqQ-LIT) mass spectrometer equipped with an atmospheric pressure chemical ionization (APCI) source obtained from AB Sciex(Mississauga, ON, Canada). The instrument was operated in the positive ion mode and tandem mass spectrometric analysis (MS/MS) was employed using the following interface parameters: source temperature 380 °C, curtain gas 30 psi (gas), nebulizer current 2.5 μA, declustering potential 30 V and an ion source gas1 30 psi (gas) [55].

The parameters, selectivity, accuracy, precision, reproducibility, sensitivity, matrix effects, dilution integrity, stability were assessed [55]

3.5. Entrapment Efficiency (EE)

In order to determine entrapment efficiency, free and entrapped bioactives were separated using ultracentrifugation. Ultracentrifuge (Beckman coulter, Inc., Indianapolis, IN, USA) with rotor SW 60Ti was used for ultracentrifugation. Briefly, 5 mL of liposomes was ultracentrifuged at 30, 60, 90 and 120 min at constant RPM 32,000 (G-force of 138000). The sediment at each time were analyzed using a validated LC-MS/MS method to optimize ultracentrifugation parameters. The liposomes (present in sediment) separated by ultracentrifugation were lyophilized using freeze dryer for 24 h. Similar lyophilization process was employed with 5 mL of unseparated liposomes for 24 h. Dried unseparated and separated liposomes were dissolved in 2 mL of chloroform separately. Aliquot of each were spiked with internal standard and diluted with acetonitrile. Samples were injected in LC-MS along with freshly prepared calibration and quality control standards, as described [55]. The entrapment efficiency was calculated by measuring the ratio of entrapped bioactives in the formulation to the total bioactives present in the formulation and was determined using following equation:

$$\%\text{Entrapment efficiency} = \frac{\text{E bioactives}}{\text{T bioactives}} \times 100 \tag{1}$$

where, E bioactives = Entrapped bioactives in liposomes (present in sediment of separated liposomes); T bioactives = Total bioactives in liposomes (present in unseparated liposomes).

3.6. Development of Functional Juice Using Model Orange Juice

In order to preserve particle size during freeze drying, sucrose was added to liposomes as a cryoprotectant by adopting the procedure of Shaikh et al. [56]. Briefly, 5% *w/v* of sucrose was added to liposomes of well-defined size and was vortexed. Lyophilization then was employed for 24 h.

Freeze dried liposomes were re-suspended in model orange juice which is an orange juice mimic at 3.2 pH. A mimic was used instead of real orange juice to enable particle size analysis without the interference of particulate components existing in the orange juice. In fact, the acidified solution is considered a model juice and was prepared by using acetic acid as per the protocol of Marsansco et al. [57]. This protocol can be applied for fruit juice with a pH less than 5.0, such as orange juice and pineapple juice. Finally, liposomes with optimum entrapment efficiency in a dried form were incorporated into the model orange juice by vortexing for 5 min.

3.7. Pasteurization

High temperature short time (HTST) pasteurization technique was employed as described [58]. HTST is a commonly used strategy for the pasteurization of juice [59]. The liposomes containing model orange juice was pasteurized at 72 °C for 15 s. Unpasteurized formulation was used as a control. Both pasteurized and unpasteurized model juice were stored at 4 °C for stability evaluation.

3.8. Chemical Stability Studies

Both pasteurized and non-pasteurized model juice containing liposomal bioactives were analyzed using LC-MS/MS to assess the degradation of bioactive upon exposure to pasteurization temperature. Briefly, 5 mL each of pasteurized and non-pasteurized liposomal model juice were lyophilized. Dried sample were dissolved in chloroform and were diluted with acetonitrile for LC-MS/MS analysis. The LC-MS/MS response was compared to obtain relative quantification data.

3.9. Physical Stability Studies

Physical stability evaluation was conducted at the interval of 7 days for a month. Particle size of pasteurized and non-pasteurized liposomes incorporated both in model orange juice were analyzed.

4. Statistical Analysis

The statistical analysis of the samples was conducted with SPSS statistical software version 24 (SPSS Inc., Chicago, IL, USA) using student *t*-test, and *p*-values < 0.05 were considered statistically significant. All data are reported as means ± standard deviations.

5. Results and Discussions

5.1. Physicochemical Characterization

Size is an important parameter to assess the stability, the biological fate and the efficacy of formulated bioactives [60,61]. Optimization of bioactive to lipid weight ratio (B/L ratio) was performed at ratio of 1:5, 1:4. 1:3, 1:2, 1:1 using liposomes prepared by thin film hydration- homogenization approach. An increase in particle size was observed at high B/L ratio (1:1) as shown in Figure 1. At 1:5 B/L ratio, the particle size was 149.53 nm; however, when B/L ratio increased to 1:1, the particle size increased to 258.31 nm (Figure 1). This observation is reported previously [62,63], in which incremental vesicle size was observed when increasing the cholesterol (a sterol) concentration. While 1:1 ratio is preferable from a commercial point of view (less PC required for formulation), smaller particle size (less than 200 nm) attained at 1:2 ratio, is optimum for liposomal stability. This optimum vesicle size (less than 200 nm) is consistent with several food-based liposomes [64–66]. Thus, 1:2 B/L ratio was selected for follow-up experiments. Same optimum B/L ratio was selected for thin-layer ultrasonication approach. However, in case of the Mozafari method, B/L higher than 1:5 led to the appearance of visible white precipitate. Loading techniques along with the preparation procedures are found to influence drug/lipid ratio of liposomes [67]. In both thin film hydration homogenization and thin film hydration ultrasonication, hydrated bioactives-PC film is subjected to cavitation and shearing forces unlike the Mozafari method where less intense magnetic stirring is used during the loading process. This might have led to differences in the B/L ratio of the mozafari method in comparison with ultrasonication and homogenization methods. In this way, 1:5 B/L as optimum ratio was selected for formulations prepared by the Mozafari method.

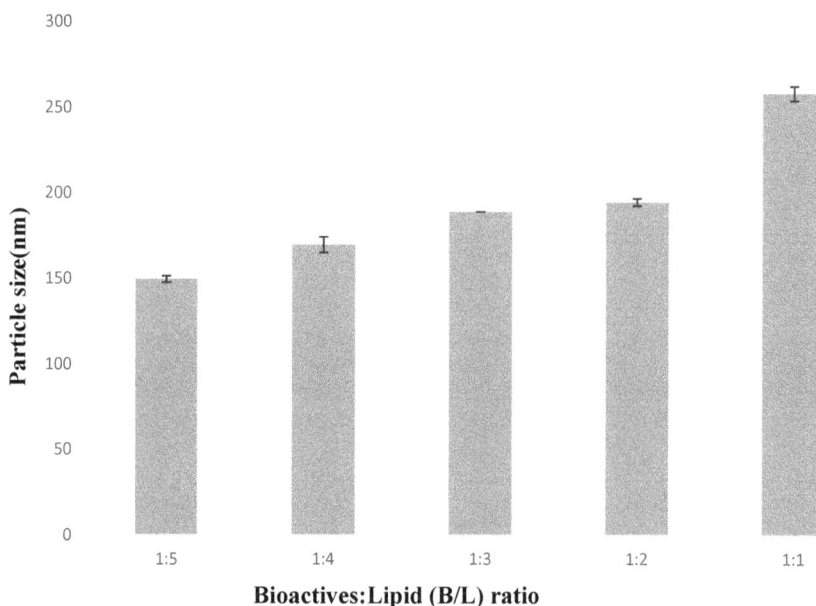

Figure 1. Particle size of liposomes prepared at different B/L ratio by homogenization method expressed as mean ± standard deviation.

The comparison of the vesicle size using the different formulation strategies is presented in Table 1. Thin film hydration homogenization and thin film hydration ultrasonication were comparable, showing sizes at 186.33 ± 4.38 nm and 196.2 ± 16.1 nm, respectively. On the other hand, the size was significantly larger in the case of the Mozafari method (260 ± 22.98 nm). It is possible that the high shear force and cavitation involved in size reduction during the homogenization and ultrasonication methods is the reason for the obtained smaller vesicles [68]. The Mozafari method uses a less intense magnetic stirring [69], probably yielding larger particles. Polydispersibility index (PDI) shown in Table 1 was found to be in the range from 0.29 to 0.37, which shows the desirable narrow size distribution for all formulations.

Table 1. Average particle size (nm), polydispersibility index and zeta potential (mv) of liposomes prepared by thin film hydration homogenization, thin film hydration ultrasonication and Mozafari method expressed as mean ± standard deviation where * represents statistical significant (*p < 0.05) in average particle size of Mozafari method in comparison with homogenization and ultrasonication method.

Formulation Techniques	Average Particle Size (nm)	Polydispersibility Index (PDI)	Zeta Potential (mV)
Thin film hydration Homogenization	186.3 ± 4.4	0.370 ± 0.001	−13.0 ± 5.0
Thin film hydration ultra-sonication	196.2 ± 16.1	0.294 ± 0.084	−14.0 ± 3.4
Mozafari method	260.0 ± 23.0*	0.348 ± 0.087	−9.8 ± 0.3

Zeta potential (surface charge) is another important parameter that determines the stability of liposomal dispersions [70]. All the liposomes, demonstrated similar zeta potential values (Table 1), that is in the range of -9 mV to -14 mV, indicating relatively stable systems [71]. Thus, based on particle size and zeta potential, the developed liposomal formulations are theoretically stable that was confirmed experimentally by conducting the stability studies.

Finally, TEM analysis of liposomes shows spherical shaped particles with a single lipid bilayer (Figure 2), representing the expected morphology of unilamellar liposomal vesicles (ULV) [72,73]. The size of approximately 200 nm is consistent with the size range measured by dynamic light scattering (DLS) (Table 1). Some aggregated particles were observed in ultra-sonication and Mozafari method as shown in Figure 2.

5.2. Entrapment Efficiency (%EE)

The developed LC-MS/MS method (representative chromatogram shown in Figure 3) was able to separate and quantify four phytosterols (brassicasterol, campesterol, stigmasterol and β-sitosterol) and three tocopherols (alpha, gamma and delta). Both ultracentrifugation parameters and entrapment efficiencies were determined by analyzing bioactive using LC-MS/MS. The separation of the liposomes during ultracentrifugation was time-dependent. Relatively low amounts of liposomes sedimented after 30 min (around 80% for all bioactives) of ultracentrifugation, whereas high sedimentation of liposomes was observed at 60, 90 and 120 min. There was no significant difference in sedimentation at 60, 90 and 120 min of ultracentrifugation. This supports the notion that after 60 min of ultracentrifugation at RPM 32,000 (G-force of 138000), a significant amount of liposomes was sedimented, leaving free bioactives in the supernatant.

The optimum entrapment efficiencies of phytosterols and tocopherols into the liposomes obtained by the thin film hydration homogenization, thin film hydration ultra-sonication and Mozafari method is shown in Table 2. The results demonstrate that all three methods resulted in entrapment efficiency > 89% for phytosterols and tocopherols. Table 2 does not show any specific pattern in entrapment efficiency for bioactives. For example, in case of thin film hydration homogenization method, brassicasterol showed the highest entrapment efficiency among all phytosterols; however, in the case of the Mozafari method, brassicasterol has the lowest entrapment efficiency. Similarly, the Mozafari method showed the highest entrapment efficiency for gamma tocopherols among all tocopherols. On the other hand, thin film hydration ultrasonication showed the lowest entrapment efficiency for gamma tocopherol. Thus, no concrete conclusion was obtained regarding entrapment differences between these bioactives. The entrapment efficiency of some of lipophilic compounds were reported to be almost close to 100% [74,75]. However, Table 2 shows EE in the range of 89–97% for various bioactives, evaluated in our work. It is possible that some of the liposomes were too small and failed to sediment during ultracentrifugation. This will lead to decreased EE (the amount of bioactives in the sediment were taken as a basis to calculate EE). Nevertheless, the obtained EE (shown in Table 2) is consistent with entrapment efficiency of nutraceuticals such as vitamin E, resveratrol and retinol specified in the literature [57,65,76]. High entrapment efficiency, that is, greater than 85% is considered economical for industrial application because it eliminates the cost of separating free and entrapped bioactives that will be required in case of low entrapment efficiency.

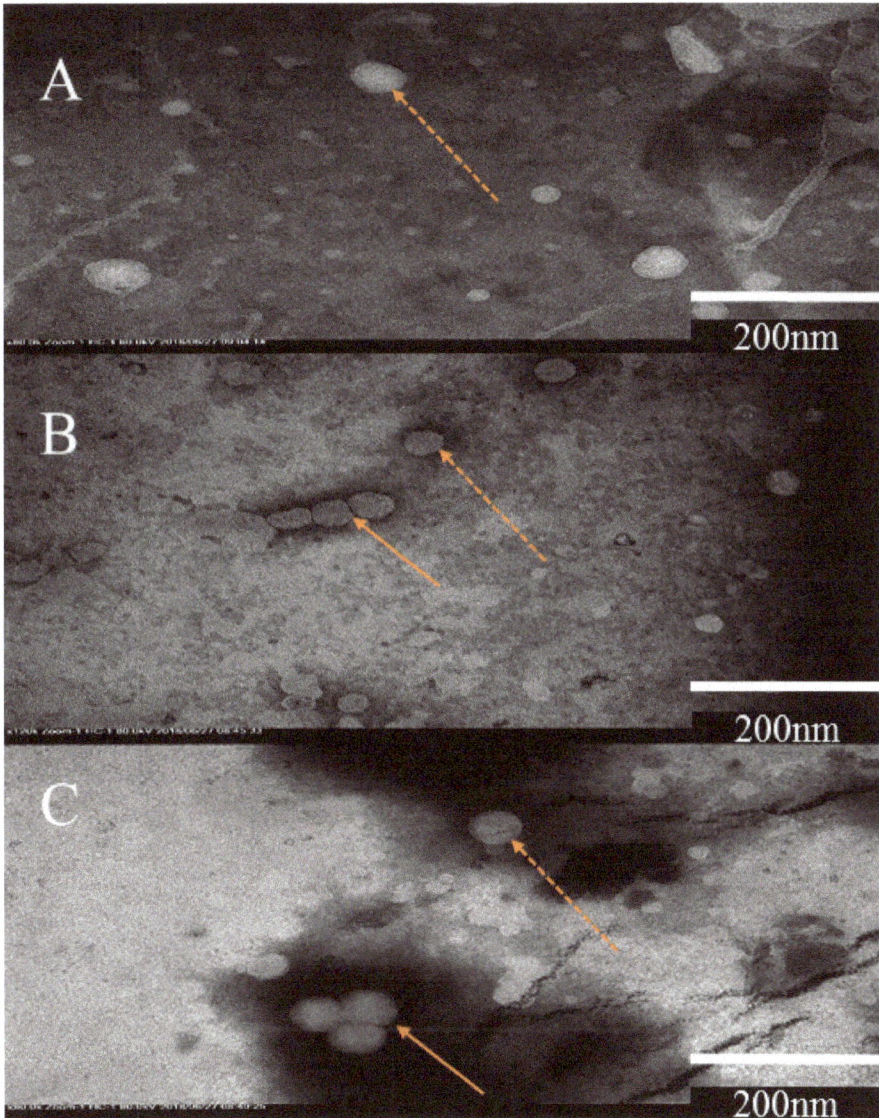

Figure 2. Transmission electron microscopy (TEM) analysis of liposomes prepared by; (**A**) homogenization method, (**B**) Ultrasonication and (**C**) Mozafari method. Sample of unilamellar vesicles are shown with a dotted arrow while aggregates are indicated by solid arrows. Scale bar in the figure A, B and C indicates 200nm, which represents the size of vesicle.

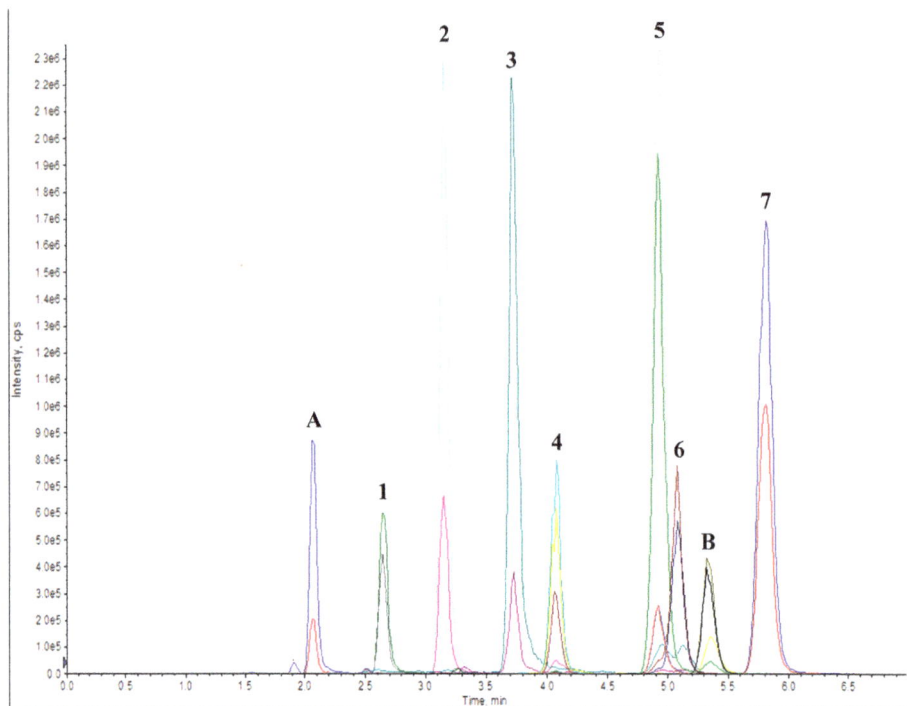

Figure 3. LC-MS/MS chromatogram of tocopherols: 1-δ tocopherol, 2-γ tocopherol, 3-α tocopherol; and phytosterols: 4-brassicasterol, 5-campesterol, 6-stigmasterol and 7-β-sitosterol. A-Rac tocol and B-cholestanol are internal standard.

Table 2. Entrapment efficiency of bioactives (phytosterols and tocopherols) into liposomes prepared by the thin film hydration homogenization, thin film hydration ultra-sonication and Mozafari method expressed as mean ± standard deviation.

Methods	Entrapment Efficiency (EE %)					
	Brassicasterol	Campesterol	β-Sitosterol	Alpha Tocopherol	Gamma Tocopherol	Delta Tocopherol
Thin film hydration-Homogenization	95.9 ± 1.7	94.0 ± 2.2	94.8 ± 3.0	91.6 ± 2.4	90.5 ± 2.9	91.6 ± 3.6
Thin film hydration-Ultrasonication	91.5 ± 2.4	92.3 ± 3.4	90.1 ± 1.9	91.2 ± 2.1	89.8 ± 3.1	90.1 ± 2.3
Mozafari method	89.4 ± 2.8	93.7 ± 6.0	93.1 ± 6.0	92.3 ± 7.5	97.4 ± 1.9	95.3 ± 1.4

5.3. *Effect of Lyophilization on the Physicochemical Properties*

Freeze-drying of liposomes resulted in the increment in particle size, reaching up to 500 nm (Figure 4). Various food compatible cryoprotectants such as sucrose, mannitol and lactose can be used to address this issue [77]. Thus, food compatible sucrose was tested as a cryoprotectant [56]. The addition of sucrose maintained the desired particle size (Figure 4). The lyophilized liposomes were then incorporated into model orange juice. Lyophilization is one of the crucial steps used for stabilization of liposomes [78]. It extends the shelf life of liposomes and can prevent thermosensitive bioactives from degradation [78].

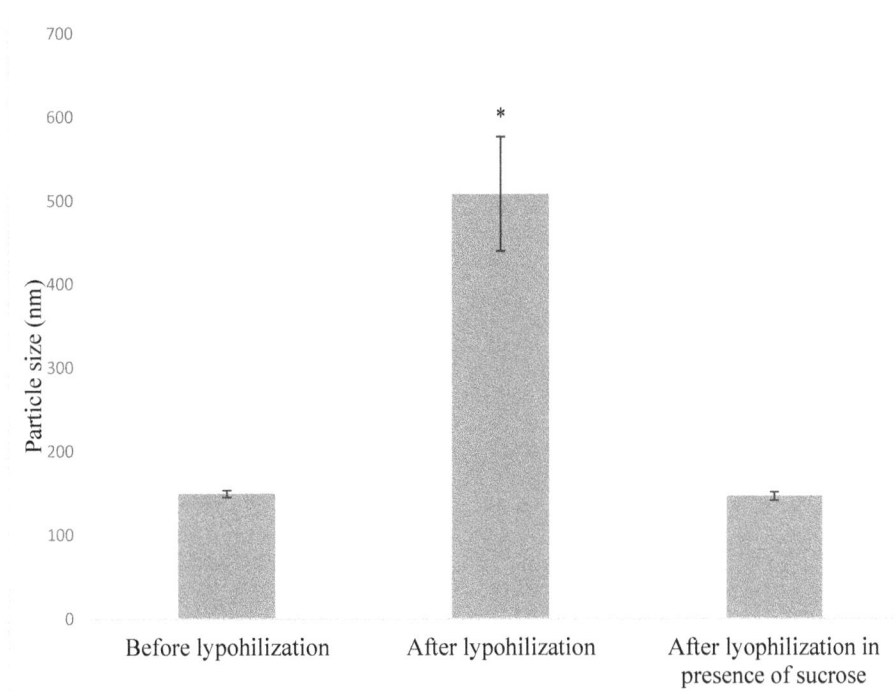

Figure 4. Effect of addition of sucrose as a cryo-protectant on particle size of liposomes before and after lyophilization expressed as mean ± standard deviation where * represents statistical significant (*$p < 0.05$) in particle size after lyophilization in comparison to before lyophilization and after lyophilization in presence of sucrose.

5.4. Chemical Stability Studies

Pasteurization technique did not compromise the stability of bioactives as shown in Table 3. There was no significant change in the LC-MS/MS response for pasteurized and non-pasteurized formulations, ranging from 0.5 to 2.59% (Table 3). This shows that exposure to temperature of 72 °C for short time of 15 s does not degrade bioactives entrapped within liposomes in the model juice

Table 3. Relative change in the concentration (represented by area under curve, AUC) of phytosterols and tocopherol before and after pasteurization.

Bioactives	AUC of Non-Pasteurized Bioactives	AUC of Pasteurized Bioactives	Percentage Relative Change in AUC (%) of Pasteurized and Non-Pasteurized
Brassicasterol	5.63×10^6	5.60×10^6	0.53
Campesterol	2.32×10^7	2.26×10^7	2.59
β-sitosterol	3.40×10^6	3.37×10^6	0.88
α-tocopherol	2.76×10^7	2.72×10^7	1.45
γ-tocopherol	4.94×10^6	4.89×10^6	1.01
δ-tocopherol	4.84×10^6	4.73×10^6	2.27

5.5. Physical Stability Studies

Both pasteurized and non-pasteurized liposomes in model orange juice showed similar trend in particle size (Figure 5). This implies that high temperature in HTST pasteurization process did not compromise the stability of the liposomes. Further, particle size of vesicle did not change significantly during the one-month storage at 4 °C (Figure 5). This shows that liposomal orange juice can be stored in 4 °C for a month with adequate stability. Regarding zeta potential, unlike liposomes in purified water, liposomal model juice was found to have positive zeta potential in the range of 5.6–8.9 mV. Even though this zeta potential value is generally considered an indicator of instability to the colloidal system [71], liposomal model orange juice showed adequate storage stability. Probably, the optimized smaller vesicular size maintained the stability of particles preventing its aggregation.

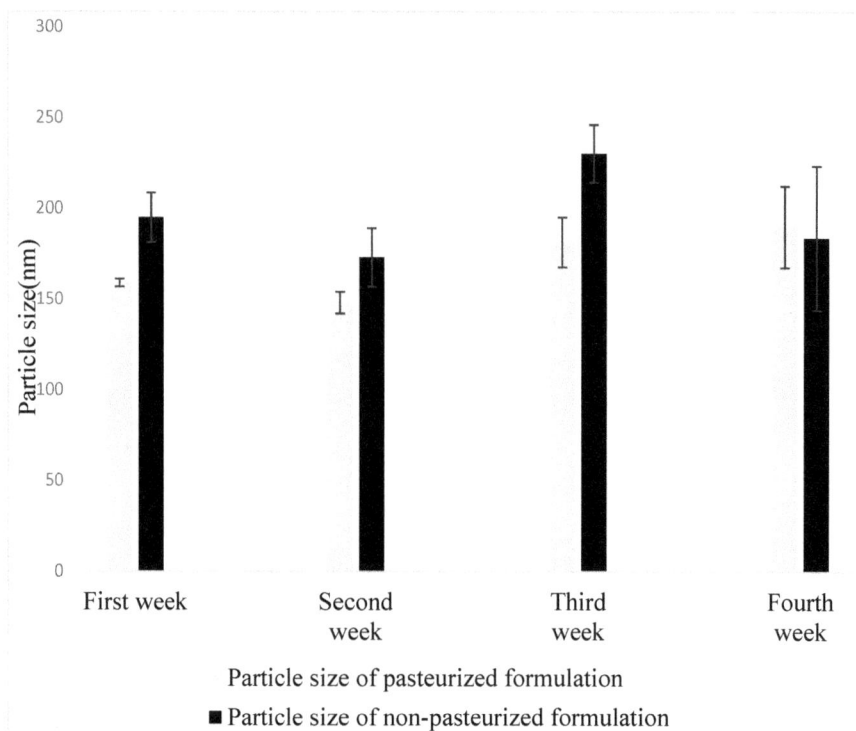

Figure 5. Particle size of pasteurized and non-pasteurized liposomes incorporated in model orange juice during storage period of 1 month at 4 °C expressed as mean ± standard deviation.

6. Conclusions

To address the lipophilicity, heat and light sensitivity challenges, unilamellar liposomes containing phytosterols obtained from CODD and tocopherols were formulated and were applied to develop a functional juice. Three different formulation approaches were employed and were compared for their suitability in formulating phytosterols and tocopherols. All three methods showed optimum physicochemical properties and excellent entrapment efficiencies that were greater than 89%. Mozafari method was found to be simple and quick for formulating liposomes; however, the use of high temperature can possibly degrade thermosensitive bioactives. In addition, its low B/L ratio (not economical for scaling up) makes the Mozafari method less suitable method for phytosterols and tocopherols in comparison to thin film hydration ultrasonication and thin film hydration homogenization method. Both ultrasonication and homogenization seemed to be equally suitable at a laboratory scale. At an industrial scale, however, the homogenization method is more feasible due to the availability of homogenizers of large capacity. Thus, thin film hydration-homogenization seems to be the best method for scaling-up the liposomal formulation containing phytosterols and tocopherols. The pasteurization technique did not affect the chemical stability of tested bioactives.

Moreover, model orange juice containing liposomes maintained an adequate physical stability during a period of one-month storage at 4 °C. In the future, liposomes containing phytosterols will be tested for cholesterol-lowering efficacy by conducting animal and human trials.

Author Contributions: Conceptualization, I.B., A.E., Z.D.B., K.M.W.; Methodology, A.E., I.B., A.P., G.G.; Software, A.E., A.P., G.G.; Validation, A.E., A.P., G.G.; Formal analysis, I.B., A.E., A.P., G.G.; Investigation, A.E., I.B., A.P.; Resources, A.E., Z.D.B.; Data curation, A.P., G.G.; Writing original draft preparation, A.P.; Writing—Review and Editing, A.E., I.B., K.M.W., G.G.; Visualization, A.E., I.B., A.P.; Supervision, A.E., I.B.; Funding administration, A.E.; Funding acquisition, A.E., Z.D.B., I.B.

Funding: This research was funded by Agriculture Development Fund, Ministry of Agriculture, Government of Saskatchewan, grant number 20150216. QTRAP 6500 mass spectrometer was funded by Western Diversification Canada grant. Poudel was provided scholarship funding from Sask Canola.

Acknowledgments: Ellen K Wasan is acknowledged for her great insights to this work. Amal Makhlouf and Deborah Michel is acknowledged for their technical help and training of Poudel. We acknowledge LaRhonda Sobchishin from Western College of Veterinary Medicine Imaging Centre, University of Saskatchewan, for her technical help with transmission electron microscopy.

Conflicts of Interest: The authors declare no conflict of interest.

Abbreviations

LDL	Low-density lipoprotein
CODD	Canola oil deodorizer distillate
POPs	Phytosterol oxidation products
LC-MS/MS	Liquid Chromatography Tandem Mass Spectrometry
PC	Phosphatidylcholine
TEM	Transmission electron Microscopy
HTST	High temperature short time
B/L	Bioactive/lipid
RPM	Revolution per minute
%EE	percentage entrapment efficiency
DLS	Dynamic light scattering
PDI	Polydispersibility Index

References

1. Siró, I.; Kápolna, E.; Kápolna, B.; Lugasi, A. Functional food. Product development, marketing and consumer acceptance—A review. *Appetite* **2008**, *51*, 456–467. [CrossRef]
2. Demonty, I.; Ras, R.; van Der Knaap, H.; Duchateau, G.; Meijer, L.; Zock, P.; Geleijnse, J.; Trautwein, E. Continuous Dose-Response Relationship of the LDL-Cholesterol-Lowering Effect of Phytosterol Intake1,2. *J. Nutr.* **2009**, *139*, 271–284. [CrossRef] [PubMed]
3. Engin, K.N.; Engin, G.; Kucuksahin, H.; Oncu, M.; Guvener, B. Clinical evaluation of the neuroprotective effect of alpha- tocopherol against glaucomatous damage. *Eur. J. Ophthalmol.* **2007**, *17*, 528–533. [CrossRef] [PubMed]
4. Abumweis, S.; Barake, R.; Jones, P. Plant sterols/stanols as cholesterol-lowering agents: A meta-analysis of randomized controlled trials. *Food Nutr. Res.* **2008**, *52*. [CrossRef]
5. Katan, M.B.; Grundy, S.M.; Jones, P.; Law, M.; Miettinen, T.; Paoletti, R. Efficacy and Safety of Plant Stanols and Sterols in the Management of Blood Cholesterol Levels. *Mayo Clin. Proc.* **2003**, *78*, 965–978. [CrossRef]
6. NIH Publication No. 01-3305. *ATP III Guidelines at-a-Glance Quick Desk Reference*; NIH: Bethesda, MD, USA, 2001.
7. Ikeda, I.; Sugano, M. Some aspects of mechanism of inhibition of cholesterol absorption by β-sitosterol. *BBA Biomembr.* **1983**, *732*, 651–658. [CrossRef]
8. Plat, J.; Mensink, R.P. Plant Stanol and Sterol Esters in the Control of Blood Cholesterol Levels: Mechanism and Safety Aspects. *Am. J. Cardiol.* **2005**, *96*, 15–22. [CrossRef] [PubMed]
9. Frei, B. Reactive oxygen species and antioxidant vitamins: Mechanisms of action. *Am. J. Med.* **1994**, *97*, S5–S13. [CrossRef]

10. Niki, E.; Noguchi, N. Dynamics of antioxidant action of vitamin E. *Acc. Chem. Res.* **2004**, *37*, 45–51. [CrossRef] [PubMed]

11. Taylor, H.R.; Tikellis, G.; Robman, L.D.; McCarty, C.A.; McNeil, J.J. Vitamin E supplementation and macular degeneration: randomised controlled trial. (Papers). *Br. Med. J.* **2002**, *325*, 11. [CrossRef]

12. Frank, B.; Gupta, S. A review of antioxidants and Alzheimer's disease. *Ann. Clin. Psychiatry Off. J. Am. Acad. Clin. Psychiatr.* **2005**, *17*, 269–286. [CrossRef]

13. Vivekananthan, D.P.; Penn, M.S.; Sapp, S.K.; Hsu, A.; Topol, E.J. Use of antioxidant vitamins for the prevention of cardiovascular disease: Meta- analysis of randomised trials. *Lancet* **2003**, *361*, 2017–2023. [CrossRef]

14. Ryan, E.; Galvin, K.; O'Connor, T.; Maguire, A.; O'Brien, N. Phytosterol, Squalene, Tocopherol Content and Fatty Acid Profile of Selected Seeds, Grains, and Legumes. *Plant Foods Hum. Nutr.* **2007**, *62*, 85–91. [CrossRef]

15. Maguire, L.S.; O'Sullivan, S.M.; Galvin, K.; O'Connor, T.P.; O'Brien, N.M. Fatty acid profile, tocopherol, squalene and phytosterol content of walnuts, almonds, peanuts, hazelnuts and the macadamia nut. *Int. J. Food Sci. Nutr.* **2004**, *55*, 171–178. [CrossRef]

16. Verleyen, T.; Verhe, R.; Garcia, L.; Dewettinck, K.; Huyghebaert, A.; De Greyt, W. Gas chromatographic characterization of vegetable oil deodorization distillate. *J. Chromatogr. A* **2001**, *921*, 277–285. [CrossRef]

17. Heale, J.B.; Karapapa, V.K. The verticillium threat to canada's major oilseed crop: canola. *Can. J. Plant Pathol.* **1999**, *21*, 1–7. [CrossRef]

18. Schwartz, H.; Ollilainen, V.; Piironen, V.; Lampi, A.-M. Tocopherol, tocotrienol and plant sterol contents of vegetable oils and industrial fats. *J. Food Compos. Anal.* **2008**, *21*, 152–161. [CrossRef]

19. McClements, D.J.; Decker, E.A.; Park, Y.; Weiss, J. Structural Design Principles for Delivery of Bioactive Components in Nutraceuticals and Functional Foods. *Crit. Rev. Food Sci. Nutr.* **2009**, *49*, 577–606. [CrossRef]

20. Alemany, L.; Barbera, R.; Alegría, A.; Laparra, J.M. Plant sterols from foods in inflammation and risk of cardiovascular disease: A real threat? *Food Chem. Toxicol.* **2014**, *69*, 140–149. [CrossRef]

21. Liang, Y.T.; Wong, W.T.; Guan, L.; Tian, X.Y.; Ma, K.Y.; Huang, Y.; Chen, Z.-Y. Effect of phytosterols and their oxidation products on lipoprotein profiles and vascular function in hamster fed a high cholesterol diet. *Atherosclerosis* **2011**, *219*, 124–133. [CrossRef]

22. Nedovic, V.; Kalusevic, A.; Manojlovic, V.; Levic, S.; Bugarski, B. An overview of encapsulation technologies for food applications. *Procedia Food Sci.* **2011**, *1*, 1806–1815. [CrossRef]

23. Desai, K.G.; Jin Park, H. Recent Developments in Microencapsulation of Food Ingredients. *Dry. Technol.* **2005**, *23*, 1361–1394. [CrossRef]

24. Ray, S.; Raychaudhuri, U.; Chakraborty, R. An overview of encapsulation of active compounds used in food products by drying technology. *Food Biosci.* **2016**, *13*, 76–83. [CrossRef]

25. Anandharamakrishnan, C. *Spray Drying Technique for Food Ingredient Encapsulation*; John Wiley & Sons: Hoboken, NJ, USA, 2015.

26. Gibbs, B.F.; Kermasha, S.; Alli, I.; Mulligan, C.N. Encapsulation in the food industry: A review. *Int. J. Food Sci. Nutr.* **1999**, *50*, 213–224.

27. Sagalowicz, L.; Leser, M.E. Delivery systems for liquid food products. *Curr. Opin. Colloid Interface Sci.* **2010**, *15*, 61–72. [CrossRef]

28. Mozafari, M.R. Nanoliposomes: Preparation and analysis. *Methods Mol. Biol. (Clifton, N.J.)* **2010**, *605*, 29–50. [CrossRef]

29. Noakes, M.; Clifton, P.; Ntanios, F.; Shrapnel, W. An increase in dietary carotenoids when consuming plant sterols or stanols is effective in maintaining plasma carotenoid concentrations. *Am. J. Clin. Nutr.* **2002**, *75*, 79–86. [CrossRef]

30. Cleghorn, C.L.; Skeaff, C.M.; Mann, J.; Chisholm, A. Plant sterol-enriched spread enhances the cholesterol-lowering potential of a fat-reduced diet. *Eur. J. Clin. Nutr.* **2003**, *57*, 170–176. [CrossRef]

31. Davidson, M.H.; Maki, K.C.; Umporowicz, D.M.; Ingram, K.A.; Dicklin, M.R.; Schaefer, E.; Lane, R.W.; McNamara, J.R.; Ribaya-Mercado, J.D.; Perrone, G.; et al. Safety and Tolerability of Esterified Phytosterols Administered in Reduced-Fat Spread and Salad Dressing to Healthy Adult Men and Women. *J. Am. Coll. Nutr.* **2001**, *20*, 307–319. [CrossRef]

32. Hyun, Y.J.; Kim, O.Y.; Kang, J.B.; Lee, J.H.; Jang, Y.; Liponkoski, L.; Salo, P. Plant stanol esters in low-fat yogurt reduces total and low-density lipoprotein cholesterol and low-density lipoprotein oxidation in normocholesterolemic and mildly hypercholesterolemic subjects. *Nutr. Res.* **2005**, *25*, 743–753. [CrossRef]

33. Miettinen, T.A.; Puska, P.; Gylling, H.; Vanhanen, H.; Vartiainen, E. Reduction of serum cholesterol with sitostanol-ester margarine in a mildly hypercholesterolemic population. *N. Engl. J. Med.* **1995**, *333*, 1308–1312. [CrossRef] [PubMed]

34. Thomsen, A.B.; Hansen, H.B.; Christiansen, C.; Green, H.; Berger, A. Effect of free plant sterols in low-fat milk on serum lipid profile in hypercholesterolemic subjects. *Eur. J. Clin. Nutr.* **2004**, *58*, 860–870. [CrossRef] [PubMed]

35. Noakes, M.; Clifton, P.; Doornbos, A.; Trautwein, E. Plant sterol ester–enriched milk and yoghurt effectively reduce serum cholesterol in modestly hypercholesterolemic subjects. *Zeitschrift für Ernährungswissenschaft* **2005**, *44*, 214–222. [CrossRef] [PubMed]

36. Yoshida, M.; Vanstone, C.A.; Parsons, W.D.; Zawistowski, J.; Jones, P.J.H. Effect of plant sterols and glucomannan on lipids in individuals with and without type II diabetes. *Eur. J. Clin. Nutr.* **2006**, *60*, 529–537. [CrossRef] [PubMed]

37. Devaraj, S.; Jialal, I.; Vega-Lopez, S. Plant sterol-fortified orange juice effectively lowers cholesterol levels in mildly hypercholesterolemic healthy individuals. *Arteriosc. Thromb. Vasc. Biol.* **2004**, *24*, e25–e28. [CrossRef] [PubMed]

38. Shin, M.-J.; Lee, J.H.; Jang, Y.; Lee-Kim, Y.C.; Park, E.; Kim, K.M.; Chung, B.C.; Chung, N. Micellar Phytosterols Effectively Reduce Cholesterol Absorption at Low Doses. *Ann. Nutr. Metab.* **2005**, *49*, 346–351. [CrossRef]

39. Spilburg, C.A.; Goldberg, A.C.; McGill, J.B.; Stenson, W.F.; Racette, S.B.; Bateman, J.; McPherson, T.B.; Ostlund, R.E. Fat-free foods supplemented with soy stanol-lecithin powder reduce cholesterolabsorption and LDL cholesterol. *J. Am. Diet. Assoc.* **2003**, *103*, 577–581. [CrossRef] [PubMed]

40. Jones, P.J.H.; Vanstone, C.A.; Raeini-Sarjaz, M.; St-Onge, M.-P. Phytosterols in low- and nonfat beverages as part of a controlled diet fail to lower plasma lipid levels. *J. Lipid Res.* **2003**, *44*, 1713–1719. [CrossRef] [PubMed]

41. Zhao, J.; Gershkovich, P.; Wasan, K.M. Evaluation of the effect of plant sterols on the intestinal processing of cholesterol using an in vitro lipolysis model. *Int. J. Pharm.* **2012**, *436*, 707–710. [CrossRef] [PubMed]

42. Ostlund, R.E.; Spilburg, C.A.; Stenson, W.F. Sitostanol administered in lecithin micelles potently reduces cholesterol absorption in humans. *Am. J. Clin. Nutr.* **1999**, *70*, 826–831. [CrossRef] [PubMed]

43. Miettinen, T.A.; Vanhanen, H. Dietary sitostanol related to absorption, synthesis and serum level of cholesterol in different apolipoprotein E phenotypes. *Atherosclerosis* **1994**, *105*, 217–226. [CrossRef]

44. Clifton, P.M.; Noakes, M.; Sullivan, D.; Erichsen, N.; Ross, D.; Annison, G.; Fassoulakis, A.; Cehun, M.; Nestel, P. Cholesterol-lowering effects of plant sterol esters differ in milk, yoghurt, bread and cereal. *Eur. J. Clin. Nutr.* **2004**, *58*, 503–509. [CrossRef]

45. Amir Shaghaghi, M.; Harding, S.V.; Jones, P.J.H. Water dispersible plant sterol formulation shows improved effect on lipid profile compared to plant sterol esters. *J. Funct. Foods* **2014**, *6*, 280–289. [CrossRef]

46. Judd, J.T.; Baer, D.J.; Chen, S.C.; Clevidence, B.A.; Muesing, R.A.; Kramer, M.; Meijer, G.W. Plant sterol esters lower plasma lipids and most carotenoids in mildly hypercholesterolemic adults. *Lipids* **2002**, *37*, 33–42. [CrossRef]

47. Ottestad, I.; Ose, L.; Wennersberg, M.H.; Granlund, L.; Kirkhus, B.; Retterstøl, K. Phytosterol capsules and serum cholesterol in hypercholesterolemia: A randomized controlled trial. *Atherosclerosis* **2013**, *228*, 421–425. [CrossRef] [PubMed]

48. Denke, M.A. Lack of efficacy of low-dose sitostanol therapy as an adjunct to a cholesterol-lowering diet in men with moderate hypercholesterolemia. *Am. J. Clin. Nutr.* **1995**, *61*, 392–396. [CrossRef] [PubMed]

49. Allen, T.M. Liposomal drug formulations. Rationale for development and what we can expect for the future. *Drugs* **1998**, *56*, 747–756. [CrossRef]

50. Kaikkonen, J.; Porkkala-Sarataho, E.; Morrow, J.D.; Roberts, L.J.; Nyyssönen, K.; Salonen, R.; Tuomainen, T.-P.; Ristonmaa, U.; Poulsen, H.E.; Salonen, J.T. Supplementation with vitamin E but not with vitamin C lowers lipid peroxidation in vivo in mildly hypercholesterolemic men. *Free Radic. Res.* **2001**, *35*, 967–978. [CrossRef]

51. Chung, S.K.; Shin, G.H.; Jung, M.K.; Hwang, I.C.; Park, H.J. Factors influencing the physicochemical characteristics of cationic polymer-coated liposomes prepared by high-pressure homogenization. *Colloid. Surf. A Physicochem. Eng. Asp.* **2014**, *454*, 8–15. [CrossRef]

52. Akbarzadeh, A.; Rezaei-Sadabady, R.; Davaran, S.; Joo, S.; Zarghami, N.; Hanifehpour, Y.; Samiei, M.; Kouhi, M.; Nejati-Koshki, K. Liposome: classification, preparation, and applications. *Nanoscale Res. Lett.* **2013**, *8*, 1–9. [CrossRef] [PubMed]

53. Colas, J.-C.; Shi, W.; Rao, V.S.N.M.; Omri, A.; Mozafari, M.R.; Singh, H. Microscopical investigations of nisin-loaded nanoliposomes prepared by Mozafari method and their bacterial targeting. *Micron* **2007**, *38*, 841–847. [CrossRef] [PubMed]

54. Guideline, I.H.T. Validation of analytical procedures: text and methodology Q2 (R1). In Proceedings of the International Conference on Harmonization, Geneva, Switzerland, 10 November 2005; pp. 11–12.

55. Poudel, A.; Gachumi, G.; Bashi, Z.D.; Badea, I.; El-Aneed, A. Lipid based liposomal formulation of phytosterols and tocopherols into functional food. In Proceedings of the Canadian Society for Pharmaceutical Sciences Annual Conference, Toronto, ON, Canada, 22–25 May 2018.

56. Shaikh, J.; Ankola, D.D.; Beniwal, V.; Singh, D.; Kumar, M.N.V.R. Nanoparticle encapsulation improves oral bioavailability of curcumin by at least 9-fold when compared to curcumin administered with piperine as absorption enhancer. *Eur. J. Pharm. Sci.* **2009**, *37*, 223–230. [CrossRef]

57. Marsanasco, M.; Marquez, A.L.; Wagner, J.R.; Alonso, S.D.V.; Chiaramoni, N.S. Liposomes as vehicles for vitamins E and C: An alternative to fortify orange juice and offer vitamin C protection after heat treatment. *Food Res. Int.* **2011**, *44*, 3039–3046. [CrossRef]

58. Nelson, P.E.; Tressler, D.K. *Fruit and Vegetable Juice Processing Technology*; AVI Pub. Co.: Westport, CT, USA, 1980.

59. Charles-Rodríguez, A.V.; Nevárez-Moorillón, G.V.; Zhang, Q.H.; Ortega-Rivas, E. Comparison of Thermal Processing and Pulsed Electric Fields Treatment in Pasteurization of Apple Juice. *Food Bioprod. Process.* **2007**, *85*, 93–97. [CrossRef]

60. Zhao, L.; Temelli, F.; Chen, L. Encapsulation of anthocyanin in liposomes using supercritical carbon dioxide: Effects of anthocyanin and sterol concentrations. *J. Funct. Foods* **2017**, *34*, 159–167. [CrossRef]

61. Reza Mozafari, M.; Johnson, C.; Hatziantoniou, S.; Demetzos, C. Nanoliposomes and Their Applications in Food Nanotechnology. *J. Liposome Res.* **2008**, *18*, 309–327. [CrossRef]

62. López-Pinto, J.M.; González-Rodríguez, M.L.; Rabasco, A.M. Effect of cholesterol and ethanol on dermal delivery from DPPC liposomes. *Int. J. Pharm.* **2005**, *298*, 1–12. [CrossRef]

63. Padamwar, M.N.; Pokharkar, V.B. Development of vitamin loaded topical liposomal formulation using factorial design approach: Drug deposition and stability. *Int. J. Pharm.* **2006**, *320*, 37–44. [CrossRef]

64. Marsanasco, M.; Piotrkowski, B.; Calabró, V.; Alonso, S.; Chiaramoni, N. Bioactive constituents in liposomes incorporated in orange juice as new functional food: thermal stability, rheological and organoleptic properties. *J. Food Sci. Technol.* **2015**, *52*, 7828–7838. [CrossRef] [PubMed]

65. Isailović, B.D.; Kostić, I.T.; Zvonar, A.; Đorđević, V.B.; Gašperlin, M.; Nedović, V.A.; Bugarski, B.M. Resveratrol loaded liposomes produced by different techniques. *Innov. Food Sci. Emerg. Technol.* **2013**, *19*, 181–189. [CrossRef]

66. Cui, H.; Zhao, C.; Lin, L. The specific antibacterial activity of liposome- encapsulated Clove oil and its application in tofu. *Food Control* **2015**, *56*, 128–134. [CrossRef]

67. Chountoulesi, M.; Naziris, N.; Pippa, N.; Demetzos, C. *The Significance of Drug-to-Lipid Ratio to the Development of Optimized Liposomal Formulation*; Taylor & Francis: Abingdon, UK, 2018; Volume 28, pp. 249–258.

68. Taylor, T.; Davidson, P.; Bruce, B.; Weiss, J. Liposomal Nanocapsules in Food Science and Agriculture. *Crit. Rev. Food Sci. Nutr.* **2005**, *45*, 587–605. [CrossRef] [PubMed]

69. Mozafari, M.R. Liposomes: An overview of manufacturing techniques. *Cell. Mol. Biol. Lett.* **2005**, *10*, 711–719.

70. Heurtault, B.; Saulnier, P.; Pech, B.; Proust, J.-E.; Benoit, J.-P. Physico-chemical stability of colloidal lipid particles. *Biomaterials* **2003**, *24*, 4283–4300. [CrossRef]

71. Patel, V.; Agrawal, Y. Nanosuspension: An approach to enhance solubility of drugs. *J. Adv. Pharm. Technol. Res.* **2011**, *2*, 81–87. [CrossRef]

72. Peer, D.; Florentin, A.; Margalit, R. Hyaluronan is a key component in cryoprotection and formulation of targeted unilamellar liposomes. *BBA Biomembr.* **2003**, *1612*, 76–82. [CrossRef]

73. Jung, H.; Coldren, B.; Zasadzinski, J.; Iampietro, D.; Kaler, E. The origins of stability of spontaneous vesicles. *Proc. Natl. Acad. Sci. USA* **2001**, *98*, 1353–1357. [CrossRef]

74. Xuan, T.; Zhang, J.A.; Ahmad, I. HPLC method for determination of SN-38 content and SN-38 entrapment efficiency in a novel liposome-based formulation, LE-SN38. *J. Pharm. Biomed. Anal.* **2006**, *41*, 582–588. [CrossRef]

75. Ugwu, S.; Zhang, A.; Parmar, M.; Miller, B.; Sardone, T.; Peikov, V.; Ahmad, I. Preparation, Characterization, and Stability of Liposome-Based Formulations of Mitoxantrone. *Drug Dev. Ind. Pharm.* **2005**, *31*, 223–229. [CrossRef] [PubMed]

76. Lee, S.-C.; Lee, K.-E.; Kim, J.-J.; Lim, S.-H. The Effect of Cholesterol in the Liposome Bilayer on the Stabilization of Incorporated Retinol. *J. Liposome Res.* **2005**, *15*, 157–166. [CrossRef] [PubMed]

77. Wang, L.; Ma, Y.; Gu, Y.; Liu, Y.; Zhao, J.; Yan, B.; Wang, Y. Cryoprotectant choice and analyses of freeze-drying drug suspension of nanoparticles with functional stabilisers. *J. Microencapsul.* **2018**, *35*, 241–248. [CrossRef] [PubMed]

78. Chen, C.; Han, D.; Cai, C.; Tang, X. An overview of liposome lyophilization and its future potential. *J. Control. Release* **2010**, *142*, 299–311. [CrossRef]

![pharmaceutics logo] *pharmaceutics*

MDPI

Article

Investigating the Phospholipid Effect on the Bioaccessibility of Rosmarinic Acid-Phospholipid Complex through a Dynamic Gastrointestinal in Vitro Model

Jiahao Huang [1], Peter X. Chen [1,2], Michael A. Rogers [2] and Shawn D. Wettig [1,3,]*

1 School of Pharmacy, University of Waterloo, Waterloo, ON N2L3G1, Canada;
   jiahao.huang@uwaterloo.ca (J.H.); pchen@uoguelph.ca (P.X.C.)
2 Department of Food Science, University of Guelph, Guelph, ON N1G2W1, Canada; mroger09@uoguelph.ca
3 Waterloo Institute for Nanotechnology, University of Waterloo, Waterloo, ON N2L3G1, Canada
* Correspondence: wettig@uwaterloo.ca; Tel.: +01-519-888-4567 (ext. 21303)

Received: 30 January 2019; Accepted: 29 March 2019; Published: 2 April 2019

Abstract: Phyto-phospholipid complexes have been developed as a common way of improving the oral bioavailability of poorly absorbable phyto-pharmaceuticals; however, the complexation with phospholipids can induce positive or negative effects on the bioaccessibility of such plant-derived active ingredients in different parts of the gastrointestinal tract (GIT). The purpose of this study was to investigate the effects of phospholipid complexation on the bioaccessibility of a rosmarinic acid-phospholipid complex (RA-PLC) using the TNO dynamic intestinal model-1 (TIM-1). Preparation of RA-PLC was confirmed using X-ray diffraction, Fourier-transform infrared spectroscopy, partition coefficient measurement, and Caco-2 monolayer permeation test. Bioaccessibility parameters in different GIT compartments were investigated. Complexation by phospholipids reduced the bioaccessibility of RA in jejunum compartment, while maintaining the ileum bioaccessibility. The overall bioaccessibility of RA-PLC was lower than the unformulated drug, suggesting that the improved oral absorption from a previous animal study could be considered as a net result of decreased bioaccessibility overwhelmed by enhanced intestinal permeability. This study provides insights into the effects of phospholipid on the bioaccessibility of hydrophilic compounds, and analyzes them based on the relationship between bioaccessibility, membrane permeability, and bioavailability. Additionally, TIM-1 shows promise in the evaluation of dosage forms containing materials with complicated effects on bioaccessibility.

Keywords: phospholipid complex; rosmarinic acid; bioaccessibility; dissolution; TNO gastrointestinal model; gastrointestinal simulator

---

1. Introduction

The use of phospholipid complexation has been commonplace in food and pharmaceutical sciences since its first development in 1989, promising much in the delivery of poorly absorbed plant actives and some synthetic compounds, with less complicated preparation methods compared with many other formulations [1–6]. Many poorly soluble or permeable compounds have been formulated to be more effective, systemically, by complexing with dietary phospholipids through various types of interactions (i.e., hydrogen bonding, van der Waals forces, the hydrophobic effect) [7,8]. Of late, burgeoning interest in this technique arose due to its multi-ability to enhance dissolution of hydrophobic compounds, improve permeability of hydrophilic compounds, reduce gastrointestinal toxicity of non-steroidal anti-inflammatory drugs, and protect unstable phyto-pharmaceuticals [9–12]. These properties are mainly attributed to the amphiphilic and biocompatible nature of phospholipids, which possess a polar and a non-polar moiety in their structures and have considerable dispersing ability in both aqueous and oil media. Phospholipids, especially those containing phosphatidylcholine, have shown the ability to incorporate into cell membranes to replace cellular phospholipids and affect membrane fluidity, facilitating the absorption of co-administrated payloads [13–16].

The rosmarinic acid-phospholipid complex (RA-PLC) described in this work is a PLC formulation designed to enhance the oral bioavailability and certain therapeutic activities by improving the intestinal permeability of hydrophilic RA [10,17]. The complexation reduces the contact between RA and gastrointestinal fluids, increasing the possibility for these hydrophilic compounds to simultaneously cross the membrane barrier through the uptake of phospholipids by intestinal membrane. In a recent study, the permeability coefficient and bioavailability of RA-PLC in rats were determined to be 3.15-fold and 1.25-fold higher than unformulated RA [10].

Systemically viewing the intestinal absorption, bioaccessibility is an important factor apart from membrane permeability. A prerequisite for oral bioavailability is bioaccessibility, defined as the amount of a given compound in a form that can be readily absorbed in gastrointestinal tract (GIT) [18–20]. Phospholipid complexation exhibits two opposing effects on the bioaccessibility according to many studies. On the one side, when RA, or similar poorly absorbable compounds, are complexed with phospholipid, it has been observed that the aqueous solubility and dissolution rate were reduced in both acid media and alkaline phosphate buffer, although the amphiphilic structure of phospholipid was expected to enhance the dissolution [21–23]. Resulting from the low gel to liquid crystal transition temperature (T_m), unsaturated phospholipids are difficult to formulate as fine powders under room temperature as the low Tm values and amorphous nature lead to sticky powders that are difficult to deaggregate [24–26]. Due to the cohesive or agglomerated state of phospholipids, the wetting of these drug-phospholipid complexes can be compromised by the decrease in effective surface area contacting digestion fluids. The poor dispersibility retards the release of payloads into aqueous fluids, thus decreasing the drug amount to be readily bioaccessible. PLC may also positively influence the bioaccessibility of RA through a protective effect in GIT. Many studies have described significant content loss of plant-derived active ingredients under gastrointestinal conditions due to instability [27–30]. RA has been reported to undergo a 0–25% intestinal degradation due to its instability in an alkaline environment [31,32]. Through complexation with phospholipids, unstable active ingredients may become less exposed to digestion fluid due to the described poor dispersibility of the phospholipids, which may lead to a decrease in degradation rate [2,33]. Herein, based on the described opposing effects, attention needs to be paid to the uncertainty in the net effect of phospholipid complexation on RA bioaccessibility. Without such data, it would be difficult to explain whether the improved oral absorption of RA is accredited to the sum effect of increased bioaccessibility plus membrane permeability, or the improved permeability overwhelming decreased bioaccessibility. Therefore, there is demand for a systemic evaluation method on the bioaccessibility of PLC dosage forms in order to assess the net effect of phospholipid. Better understanding on the mechanism of oral absorption is expected to be provided by the analysis of bioaccessibility parameters in different compartments of a dynamic GIT, and in turn guide the formulation process.

The TNO intestinal model (TIM-1) is a dynamic, multi-compartmental digestion system simulating physiological processes in the human upper GIT through the use of biorelevant media and computer-controlled hydrodynamics [34–36]. The physical and biochemical parameters used in the TIM-1 were determined on the basis of extensive in vivo data from both human and animal trials. This system has been widely used in the following aspects: To assess the bioaccessibility of natural extracts and nutritional products, to test the bioequivalence of a variety of drug formulations, to evaluate the gastrointestinal stability of phytochemicals, and to predict the drug–food interactions under different conditions including human fasting and fed states [37–41]. The validation of experimental parameters used for TIM-1 can be found in previous literature and the latest research [37,41–43]. To date, no attempt in the literature is traceable to have used the TIM-1 in the bioaccessibility assessment of PLC formulations, and so in this study, TIM-1 was employed to test the bioaccessibility parameters of RA-PLC and investigate the opposing effects of phospholipid in different GIT environments within a single continuous process. We intended to provide insights into the bioaccessibility change by breaking up the digestion process into individual steps and analyzing the interplay between them. This study also demonstrates a broader application of the TIM-1 system in the assessment of PLC formulations and other phospholipid-based dosage forms loaded with hydrophilic active compounds.

2. Materials and Methods

2.1. Materials

RA (96%, 536954), phospholipid (P3644), 2,2-diphenyl-1-picrylhydrazyl (DPPH, D9132), and (±)-6-Hydroxy-2,5,7,8-tetramethylchromane-2-carboxylic acid (Trolox, 238813) were purchased from Sigma-Aldrich (Oakville, ON, Canada). The phospholipid used in this study contains 55% phosphatidylcholine, 25% phosphatidylethanolamine, and other phospholipids, giving an average molecular weight of 776 g/mol according to the product information sheet. Trace components in the phospholipid including triglycerides and cholesterol are not routinely quantified. ReagentPlus® grade (99.5%) 1-octanol, dimethyl sulfoxide (DMSO), high-performance liquid chromatography (HPLC) grade methanol, and trichloromethane were purchased from Sigma-Aldrich (Oakville, ON, Canada). Cell culture reagents and other chemicals were supplied by Fisher Scientific (Hampton, NH, USA) and used as received. Reagents and enzymes used in the TIM-1 system including lipase, pepsin, amylase, pancreatin, trypsin, hydroxypropylmethylcellulose (HPMC), bile salts, sodium chloride, potassium chloride, calcium chloride di-hydrate, sodium bicarbonate, and hydrochloric acid were purchased from Sigma-Aldrich (Oakville, ON, Canada). Fresh porcine bile was obtained from Conestoga Meats in Breslau, ON. Water used in this study was obtained from a Millipore Milli-Q system (MilliporeSigma, Burlington, MA, USA). The chemical structures of phospholipid and rosmarinic acid are shown in Figure 1.

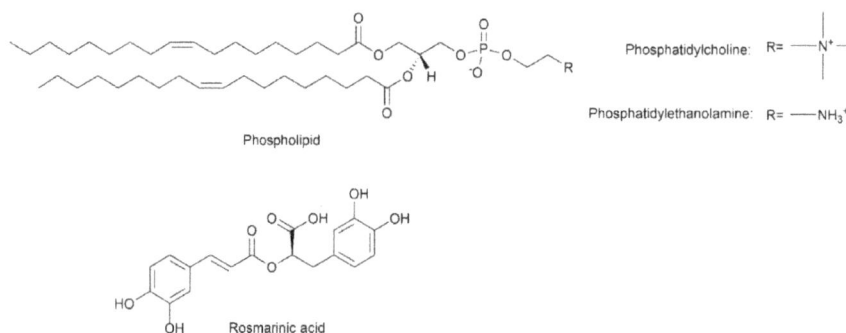

Figure 1. Chemical structures of phospholipid and rosmarinic acid.

2.2. Preparation of the RA-PLC and Physical Mixture of RA and Phospholipid

RA-PLC were prepared by the solvent evaporation method. Unformulated RA and phospholipid with a molar ratio of 1:1.5 were fully dissolved (visually inspected, sonication was used as needed) in anhydrous methanol to give a final solution with a concentration of 2.5 mg RA/mL. The methanol was removed by rotary evaporation at 45 °C. The resulting mixture was solubilized with trichloromethane to obtain a solution of RA-PLC, after which the solution was centrifuged at 5000 rpm for 10 min to remove free RA. The collected supernatant was dried under reduced pressure at room temperature for 24 h to remove the residual solvent. The physical mixture (PM) of RA and phospholipid was prepared by mixing two components at the above molar ratio in a glass mortar.

2.3. Content Determination of RA in RA-PLC

Standard solutions of RA in methanol were prepared and determined spectrophotometrically on a plate reader (SpectraMax M5, Molecular Devices, LLC., San Jose, CA, USA) at 330 nm to give a linear Beer–Lambert calibration curve (0–15 mg/L, $r^2 = 0.998$). A total of 15 mg RA-PLC was fully dissolved in 10 mL methanol and analyzed. The measurements were run in triplicate and the average concentration with standard deviation is reported.

2.4. Powder X-ray Diffraction (PXRD)

To confirm the complexation between RA and phospholipid, the physical state of RA, phospholipid, RA-PLC, and PM were analyzed on a MiniFlex X-ray diffractometer (Rigaku Corporation, Tokyo, Japan). Diffractograms were recorded over a 2θ angle from 5° to 40° at a scanning rate of 2° per minute and a 0.02° step size.

2.5. Fourier-Transform Infrared (IR) Spectroscopy

IR spectra of RA-PLC, unformulated RA, phospholipid, and physical mixture (PM) were recorded in transmission mode using a Perkin-Elmer RX I infrared spectrometer (Perkin-Elmer Inc, Waltham, MA, USA) equipped with diamond ATR attachment, to further validate the formation of RA-PLC. Samples were pressed into thin films without diluent and analyzed within a scan range of 4000 to 450 cm^{-1} with a resolution of 1 cm^{-1}.

2.6. n-Octanol/Water Partition Coefficient (P) Determination

P values of RA-PLC and unformulated RA were determined by the shake flask method at room temperature, to compare the lipophilicity of different samples. Different media including Milli-Q water, hydrochloric acid solution (HCl, pH 1.2), and phosphate buffer solution (PBS, pH 6.8) were pre-saturated with 1-octanol before tests. Different samples equivalent to 5 mg of RA were added to sealed glass containers containing 10 mL of testing medium and stirred on a hotplate at 150 rpm for 24 h. All samples were centrifuged to obtain an aqueous phase. Afterwards, 5 mL of these aqueous samples were added to 5 mL of 1-octanol (pre-saturated with corresponding aqueous media) and stirred at 150 rpm for another 24 h. The liquid mixtures were left to separate for 24 h to obtain aqueous and organic phases. The RA concentration in aqueous phase of each step was determined spectrophotometrically at 330 nm. All measurements were run in triplicate.

The P values of RA in different samples were calculated as following:

$$P = \frac{C_1 - C_2}{C_2},$$

where C_1 represents the RA concentration in the aqueous phase in the first step, and C_2 represents the RA concentration in the aqueous phase after water-octanol phase separation.

2.7. Radical Scavenging Activity Assay

The antioxidant activities of RA and RA-PLC were assessed using an in vitro chemical model system, DPPH, to test the possible change in the bioactivity of RA induced by the co-administration of phospholipid. The antiradical activity was determined spectrophotometrically as described previously [44]. DPPH working solution (350 μM) and Trolox standard solutions (1000, 750, 500, 250, 125, and 62.5 μM) were prepared by dissolving the required amounts of each in anhydrous methanol. Samples equivalent to 5 mg of RA were dissolved in 5 mL of methanol, and 25 μL of this sample solution was mixed with 200 μL of DPPH working solution in a 96-well plate, after which the plate was sealed, and the mixture was allowed to incubate for 6 h at room temperature. Finally, the absorbance was recorded at 517 nm.

The percentage DPPH quenched (%) was determined as following:

$$\text{DPPH quenched (\%)} = \left[1 - \frac{A_{\text{sample}} - A_{\text{blank}}}{A_{\text{control}} - A_{\text{blank}}}\right] \times 100,$$

where A_{sample}, A_{blank}, and A_{control} refer to absorbance values of the sample, methanol, and DPPH methanol solution at 517 nm. DPPH quenched (%) was plotted against the concentrations of Trolox, and the antioxidant capacity of samples were calculated based on the linear calibration curve. The antioxidant activity was expressed as mM Trolox equivalent (TE) per mM RA.

2.8. Cell Culture

The Caco-2 human intestine cell line obtained from Sigma Aldrich Canada was cultured in high-glucose Dulbecco's Modified Eagle's Medium (DMEM) supplemented with 10% fetal bovine serum and 1% penicillin–streptomycin. The cells were incubated under a 5% CO_2 atmosphere at 37 °C and spent media was replaced every 2 days. After reaching 80–90% confluency in a T-75 culture flask, cells were harvested with 0.25% trypsin–ethylenediaminetetraacetic acid (EDTA) solution and sub-cultured to proper culture plates depending on the experiments to be conducted.

2.9. Cell Viability Assay

The cytotoxic effects of unformulated RA and RA-PLC on Caco-2 cells were determined using a 3-(4,5-dimethylthiazol-2-yl)-2,5-diphenyltetrazolium bromide (MTT) assay. RA and RA-PLC were dissolved or dispersed in DMEM and then diluted to different concentrations. Caco-2 cells were sub-cultured in a 96-well plate at a density of 1.0×10^5 cells/cm^2 and allowed to attach. The cells were incubated with sample solutions at RA concentrations of 10, 20, and 50 µg/mL for 4 h, and then the samples were removed and the cells were rinsed with Hank's balanced salt solution (HBSS). The cells were treated with MTT (0.5 mg/mL) for 4 h under 5% CO_2 at 37 °C, after which the medium was aspirated and 100 µL of DMSO was added to each well to dissolve formazan crystals. The plate was shaken for 5 min and the absorbance was recorded at 570 nm with untreated cells as control. The cell viability (%) was calculated as following:

$$\text{Cell viability (\%)} = \frac{A_{\text{sample}} - A_{\text{blank}}}{A_{\text{control}} - A_{\text{blank}}} \times 100,$$

where A_{sample}, A_{blank}, and A_{control} refer to absorbance values of sample, blank media, and untreated cells, respectively.

2.10. Caco-2 Cell Transport Assay

Caco-2 cells were cultured and harvested as described in Section 2.9, followed by seeding in trans-well inserts (Corning Costar Corporation, Tewksbury, MA, USA) in a 6-well plate with a density of 1.0×10^5 cells per cm^2. The media in both upper and lower compartments were replaced every other day and the cells were cultivated over 21 days to achieve a confluent monolayer. RA and RA-PLC were dissolved or dispersed in HBSS buffer to a concentration of 50 µg/mL and applied onto either the apical (AP) side or basolateral (BL) side in an atmosphere of 5% CO_2 at 37 °C, while the recipient compartments were filled with 1 mL blank HBSS. Recipient samples were withdrawn from BL side in AP-BL test, and AP side in BL-AP test at 1, 2, 3, and 4 h after sample treatment, and then a same amount of blank solution was supplied to recipient compartment. RA concentrations of all samples were determined spectrophotometrically on a plate reader at 330 nm based on a linear Beer–Lambert calibration curve with HBSS as blank control. All experiments were run in triplicate. The apparent permeability coefficient (P_{app}) was calculated as following:

$$P_{\text{app}} = \frac{dQ/dt}{A \times C_0},$$

where dQ/dt (μg/s) is the rate of permeation of RA across the monolayer as described by the linear appearance rate of RA in the recipient compartment, A is the surface area of cell monolayer (4.2 cm^2 in this study), and C_0 is the initial RA concentration (49.10 μg/mL) in the donor compartment.

The net efflux ratio was calculated following the equation below:

$$\text{Efflux ratio} = \frac{P_{app}(\text{BL} - \text{AP})}{P_{app}(\text{AP} - \text{BL})},$$

where P_{app} (BL-AP) is the P_{app} value from basolateral side to apical side, and P_{app} (AP-BL) is the P_{app} value from apical side to basolateral side.

2.11. TIM-1 Study

The TNO gastrointestinal model (TIM-1) method has been described in detail [45]. Briefly, the model consists of four compartments that model the stomach, duodenum, jejunum, and ileum, with a fluid volume of 250 mL, 55 mL, 115 mL, and 115 mL, respectively. Each compartment has an inner silicon tubing surrounded by water and encased in a glass exterior. Peristaltic valves determine the transport rate of the digestate between the different compartments. Peristaltic movement is simulated by the squeezing action of the silicon tubes as dictated by the pumping of the surrounding water. The water is set at 37 °C to ensure that all compartments are kept at a physiological temperature. The rate of secretion of digestive juices in each compartment is set in accordance to predetermined physiological data [45]. The pH of the stomach was pre-set in system protocol as follows: 3.0 at 0 min, 2.2 at 10 min, 1.8 at 30 min, and 1.7 from 60min. The pH of duodenum, jejunum, and ileum were kept at 6.3, 6.5, and 7.4, respectively. All parameters are computer-controlled and a protocol for fasted state digestion of a water-soluble compound was selected for this experiment. Transit of the test formulation between compartments is automatically controlled by system sensors that detect the volume of fluids entering the compartments, as well as their solution properties (i.e., viscosity). The residual volume for the stomach compartment was set at 40 mL. For simulation of absorption of potentially available RA, the jejunal and ileal compartments are connected to semi-permeable hollow fiber membrane units (hemodialyzer cut-off of 3–5 kD). Samples equivalent to 20 mg RA were added to 240 mL of water in the stomach compartment containing 10 mL of a gastric enzyme and HPMC/bile mixture solution at pH 2. Gastric enzyme solution consists of 6000 U lipase, 1,440,000 U pepsin and 42,000 U amylase. Each experiment was terminated after 300 min. Jejunal and ileal dialyzable fractions along with the ileal efflux were collected at 15 min intervals for the first hour, at 30 min intervals in the following hour, and at 1 h intervals for the remainder of the run. All samples were run in triplicate. Secretions and enzyme solutions were prepared in accordance with TNO-Triskelion protocols.

2.12. HPLC Analysis

Bioaccessible fractions from TIM-1 digestion were analyzed with an Agilent 1100 series HPLC system equipped with an auto sampler, a degasser, a quaternary pump, a diode-array detector (DAD), ChemStation software, and separated on a Zorbax 300SB-C18 column (150 × 4.60 mm, 5 μm, Agilent Inc., Santa Clara, CA, USA). The binary mobile phase consisted of 0.1% trifluoroacetic acid in water (v/v) (solvent A) and 70% methanol in water (v/v) (solvent B). The solvent gradient was as follows: 0–20 min, 0–100% B; 20–25 min, 100% B; 25.5–30 min, 0% B. Injection volume was 10 μL and the flow rate was constantly kept at 1.0 mL/min for a total run time of 30 min. Peaks were monitored at 330 nm and identified by matching the retention time and UV absorption spectra with the standard RA. Quantification was done using RA standard curve generated from serial dilutions (7.8125–2000 mg/L; $r^2 = 0.99$).

2.13. Statistical Analysis

All samples were run in triplicate and concentrations were expressed as the means ± standard deviation. Data were analyzed using a one factor analysis of variance (ANOVA) with the IBM SPSS Statistics program (IBM, Armonk, NY, USA).

3. Results

3.1. RA Content in RA-PLC

The RA content was determined to be 22.6 ± 1.01% (w/w) in RA-PLC, slightly lower than the weight percent of RA in the starting material blend, which was 23.7%. The result indicated that most of the RA input was complexed to the phospholipids. The complexation efficiency is likely a result of the strong molecular interactions between RA and phospholipids, as discussed below.

3.2. PXRD Pattern

PXRD patterns (Figure 2) were obtained to verify the solid state of the RA-PLC system. The diffraction pattern of unformulated RA showed a high-degree of crystallinity characterized by sharp peaks over the experimental 2θ range, as described in other literature [10]. The phospholipid used in this study showed broad amorphous bands, indicative of its non-crystalline nature. RA-PLC system appeared to be in amorphous state, as revealed by the halo bands in its spectra similar to phospholipid. This amorphous profile suggested that RA was successfully complexed to the phospholipid. In contrast to RA-PLC, the PM diffraction pattern showed a reduction on the intensity of the characteristic peaks of RA, being the sum of RA and phospholipid diffraction patterns. The maintenance of crystalline nature of RA in PM indicated that the molecular interaction between two components was limited. In conclusion, a high-degree of complexation between RA and the phospholipids in RA-PLC validated the successful preparation of a complex structure that differs from the physical mixture of two components.

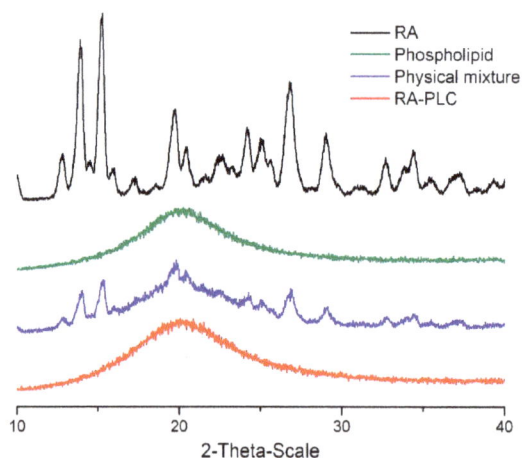

Figure 2. XRD diffractogram of rosmarinic acid (RA), phospholipid, physical mixture, and rosmarinic acid-phospholipid complex (RA-PLC).

3.3. IR Spectroscopy

The IR spectra of RA, phospholipid, RA-PLC, and PM were determined to further examine the interaction between RA and phospholipid. As shown in Figure 3, the spectrum of RA showed characteristic peaks at 3518, 3454, 3396, and 3307 cm^{-1}, assigned to the stretching vibration of the phenolic hydroxyl group in the structure of RA [10,46]. These peaks were also observed in the spectrum of PM, indicative of limited interactions between the phenolic hydroxyl group of RA and phospholipid structure. In contrast, these characteristic peaks were found to disappear in the spectrum of RA-PLC, suggesting that there were strong interactions between the phenolic hydroxyl of RA and the phospholipid induced by complexation. The changes likely result from the formation of hydrogen bonds between the -OH group of RA and the P=O group of the phospholipids.

The spectrum of the phospholipids showed characteristic peaks at 2946, 2829, and 1737 cm^{-1}, assigned to the non-polar saturated long-chain fatty acids of phospholipid. Similarly, these peaks were observed in the spectra of both RA-PLC and the physical mixture of RA and phospholipid, suggesting that the non-polar fatty acids did not interact with RA directly. A possible aggregation behavior of these fatty acid tails is to surround the surface of RA-PLC structure to further improve the lipophilicity of the complex. Strong interactions between RA phenolic group and phosphatidic acid group in phospholipid were considered as an evidence of a high-degree of complexation. IR spectroscopy further supported the PXRD results.

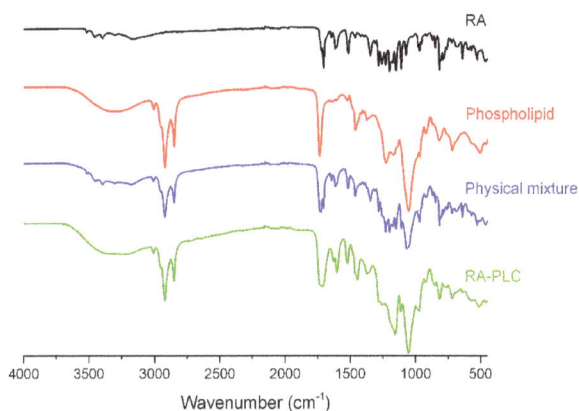

Figure 3. IR spectra of RA, phospholipid, physical mixture, and RA-PLC.

3.4. 1-Octanol/Water Partition Coefficient (P) of RA, RA–PLC, and PM

The P values of RA, RA-PLC, and PM determined in different media are listed in Table 1. Compared to unformulated RA, the *P* values of RA-PLC in water and phosphate buffer solution (PBS, pH 6.8) increased significantly ($p < 0.05$), attributed to the increased partitioning of RA into organic phase due to the improved lipophilicity after complexation with phospholipid. Compared with RA-PLC, PM exhibited a smaller extent of increase in *P* values in water and PBS, respectively, resulting from the slightly increased solubility of RA in 1-octanol phase induced by the in-solution interactions between RA and phospholipids. The *P* value for RA in water showed a 1.07-fold increase after physically mixing with phospholipid, and a 2.47-fold increase ($p < 0.05$) after forming RA-PLC. The difference can be explained by the different extents of interaction between RA and phospholipid as revealed by the PXRD and FTIR results. The partition of RA into the octanol phase relies on its incorporation into the amphiphilic structure of phospholipid, in favor of the RA-phospholipid interactions. Thus, compared with PM, RA-PLC showed a more effective increase in *P* value of due to the stronger interaction. A same trend can be found for PBS, where PM and RA-PLC increased the *P* value by 1.86-fold and 2.43-fold, respectively ($p < 0.05$ for both cases).

Table 1. 1-Octanol/water partition coefficient of RA, physical mixture (PM), and RA–PLC in different aqueous phases ($n = 3$, mean $\pm$ SD).

Sample	Media	Concentration in Original Aqueous Phase (C_1) (μg/mL)	Concentration in Separated Aqueous Phase (C_2) (μg/mL)	Partition Coefficient ($C_1 - C_2$)/C_2
RA	Millie Q water	488 $\pm$ 8	139 $\pm$ 3	2.50 $\pm$ 0.06
	HCl (pH 1.2)	500 $\pm$ 10	4.91 $\pm$ 0.09	100 $\pm$ 4
	PBS (pH 6.8)	483 $\pm$ 3	452 $\pm$ 4	0.068+0.005
PM	Millie Q water	500 $\pm$ 4	136 $\pm$ 4	2.68 $\pm$ 0.09
	HCl (pH 1.2)	350 $\pm$ 8	3.5 $\pm$ 0.3	100 $\pm$ 10
	PBS (pH 6.8)	491 $\pm$ 8	433 $\pm$ 6	0.131 $\pm$ 0.004
RA-PLC	Millie Q water	500 $\pm$ 7	70 $\pm$ 3	6.2 $\pm$ 0.4
	HCl (pH 1.2)	325 $\pm$ 7	3.25 $\pm$ 0.06	99 $\pm$ 3
	PBS (pH 6.8)	493 $\pm$ 6	423 $\pm$ 7	0.165 $\pm$ 0.005

In HCl solution at pH 1.2, both RA-PLC and PM presented similar *P* values to unformulated RA. The concentrations of RA released into original aqueous phase (C_1) were found to be lower than those in pure water and PBS ($p < 0.05$ in all cases). This is likely the result of poor dispersion of the phospholipid in acidic media due to protonation and electrostatic effects. When the pH of the environment is close to the first pK value of a phospholipid, intermolecular acid-anion complexation could occur through strong hydrogen bonding between the protonated phosphatidic acid (P–OH) and deprotonated phosphatidic acid (P–O$^-$) groups [47–49]. The aggregation of phospholipids induced by intermolecular complexation was supposed to reduce the wetting of RA by incorporating a certain portion of free drug molecules and making them less exposed to aqueous media. Thus, the concentrations of RA were observed to decrease in both the first and second aqueous acid phase. As the ratio of RA concentrations in two aqueous layers, the *P* values of RA, PM, and RA-PLC showed similarity in HCl solution.

Both similarities and differences were observed when comparing our data with Yang's results in acidic media [10]. In Yang's study, the amount of RA in the original acidic aqueous phase was significantly decreased when physically mixing or forming complex with phospholipid, which is in agreement with our results, and can be explained by the intermolecular acid-anion complexation of phospholipids in acidic environment. However, Yang's study showed that RA-PLC was more lipophilic than unformulated RA or the physical mixture, which is different from the results in this study. One possible reason for this difference is the composition difference in the phospholipid materials. Differentiating from the phospholipid containing 70–97% phosphatidylcholine used in Yang's study, the one for this study contains 55% phosphatidylcholine and 25% phosphatidylethanolamine. Phosphatidylethanolamine is well characterized by its non-bulky head group with a strong tendency to form intermolecular (N–H to P–O) hydrogen bonds between the amine and phosphate group [50]. Similar to the complexation between protonated and deprotonated phosphatidic acid group, this intermolecular hydrogen bond may further facilitate the phospholipid aggregation in acidic aqueous phase. Alternatively, unlike the preparation of RA-PLC using dissolved phospholipid, the incorporation of RA into phospholipid aggregates in acidic aqueous phase can be seen as an uncontrollable process, which may or may not be accompanied by RA incorporation. When contacting octanol, the RA hosted by phospholipid aggregates can either enter the organic phase with the phospholipid aggregates or release from the aggregates to undergo partitioning individually. The experimentally determined *P* value is a final equilibrium of all these factors, as such could contribute to differences between the two studies. Given the discussed observations, future researchers are suggested to take phospholipid types and their ionization constants into consideration for the formulation work.

The increased lipophilicity of RA in water and alkaline media is expected to provide a faster partition of RA into the lipid cell membranes. Results will be discussed in Section 3.7.

3.5. Radical Scavenging Activity (DPPH) Assay

The major biological effect of RA is demonstrated by its ability to reduce liver damage caused by lipopolysaccharides and D-galactosamine through the scavenging of superoxide molecules [51–53]. As a result, the maintenance of antioxidant activity is expected to preserve the anti-inflammatory bioactivity and other pharmacological effects which are linked to degenerative and chronic diseases caused by oxidative stress [54,55]. The DPPH assay measures the reducing ability of a compound and was used in this study to assess any possible difference in the antioxidant activity between RA and RA-PLC. The TE (Trolox equivalent) values for RA and RA-PLC were calculated based on the equivalent RA content in each sample. The results were calculated to be 3.62 ± 0.08 mM and 3.64 ± 0.09 mM, respectively, suggesting that the antioxidant activity of RA was not compromised by the co-administration of phospholipid. Thus, the possible change in RA biological activity in PLC formulation is considered as the result of bioaccessibility alteration, of which the importance is emphasized in this study.

3.6. Cell Viability

RA and RA-PLC samples equivalent to 10, 20, and 50 µg/mL RA were used to incubate Caco-2 cells to assess the effects of RA and phospholipid on cell viability. It can be seen from Figure 4 that unformulated RA did not induce obvious cell death up to a dose of 50 µg/mL. After complexation with phospholipid, the cell viability was not compromised at these concentrations. Therefore, the Caco-2 cells were treated with an equivalent RA concentration of 50 µg/mL in the following transport assay.

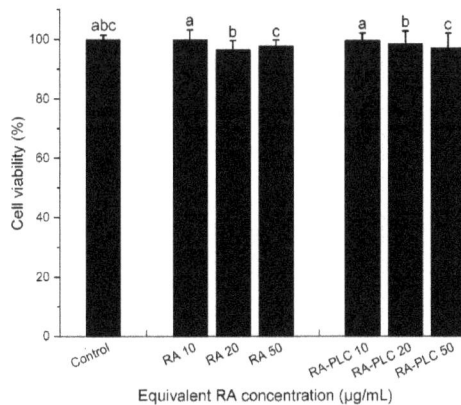

Figure 4. Cell viability (%) of RA and RA-PLC at different concentrations. Shared letters indicate no significant difference in cell viability between compared samples ($n = 3$, mean ± SD).

3.7. Caco-2 Transport Assay

As presented in Figure 5, the membrane permeability of RA was improved from both AP-BL and BL-AP sides ($p < 0.05$). The permeability coefficients were increased by 3.07-fold on AP-BL side and 2.50-fold on BL-AP side. These improvements could be attributed to the enhanced lipophilicity of RA, as revealed by the increased P value in phosphate buffer, and the increased uptake of RA through the simultaneous incorporation of phospholipid into cell membranes. The P value of RA-PLC was 0.17 in phosphate buffer, compared with that of unformulated RA, 0.07. According to Artursson's correlation between drug lipophilicity and apparent permeability coefficients in the Caco-2 model, an increase in drug lipophilicity leads to an increased P_{app}, as it makes the drug partition faster into the lipid cell membranes [56]. Drugs that can be completely absorbed in humans have P_{app} values over 1×10^{-6} cm/s, while those with P_{app} values less than 1×10^{-8} cm/s can only be absorbed to a value of less than 1% [56]. According to the Artursson's correlation, the absorption of unformulated

RA in human should be within the range of 50% to 100%, while RA-PLC is expected to increase the value to approximately 100%. The efflux ratio of RA-PLC was decreased to 0.69, compare with 0.85 of unformulated RA, indicating the uptake of RA through caco-2 monolayer could be effectively improved by RA-PLC. Combined with results of PXRD, FTIR, and o/w partition coefficient tests, the Caco-2 transport assay suggested a successful complexation between RA and phospholipid. RA-PLC was demonstrated to be a complex entity possessing different structure and physiochemical properties from a physical mixture. In this regard, prepared RA-PLC was assessed using the TIM-1 system in the following steps.

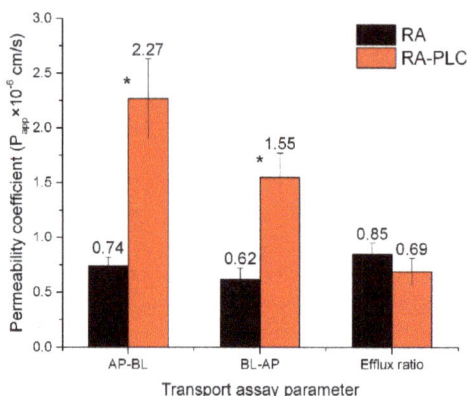

Figure 5. Permeability coefficient of RA and RA-PLC. * $p < 0.05$ between RA and RA-PLC ($n = 3$, mean $\pm$ SD).

3.8. In Vitro Bioaccessibility

3.8.1. Cumulative Bioaccessibility

The bioaccessibility of RA and RA-PLC were studied through the TIM-1 system operating in a water mode, in which the bioaccessible portion was the amount of RA detected in the dialysate. The cumulative contents of bioaccessible RA in jejunum dialysate, ileum dialysate and ileum effluent were determined.

In general, as shown in Figure 6a, the total cumulative amounts of bioaccessible RA in the jejunum compartment increased continuously for both unformulated material and RA-PLC within 5 h. Within each time interval, unformulated RA showed higher bioaccessibility compared to RA-PLC, indicating a retarded release of RA after complexing with phospholipid. At the beginning of jejunum digestion, the slower release from RA-PLC could be explained by a gastric effect from the previous digestion step (i.e., within the stomach compartment). The protonation of the phospholipids in gastric fluid could induce an intermolecular acid-anion complexation as discussed in Section 3.4, resulting in poorly dispersible state for the phospholipid. A certain portion of RA was both complexed with phospholipid molecules and surrounded by the phospholipid aggregate, thus the contact between RA and aqueous media was reduced. Therefore, a smaller amount of RA was released from RA-PLC into aqueous media and transported from stomach to jejunum in a dissolved form that could be detected directly. This observation corresponds well to the findings from the partition coefficient tests, which showed a reduced RA concentration from RA-PLC in acidic media. With the continuous processing of jejunum digestion, the protonation of phospholipids in the gastric environment became less of a determinant in bioaccessibility as the phospholipids could disperse well in alkaline media. The bioaccessibility difference between RA and RA-PLC became smaller gradually, indicative of similar release behaviors for the two samples in alkaline environment. Finally, the maximum bioaccessibility of RA in jejunum compartment within 5 h was determined to be 64.9% of the input amount, slightly higher than the 60.9% of RA-PLC.

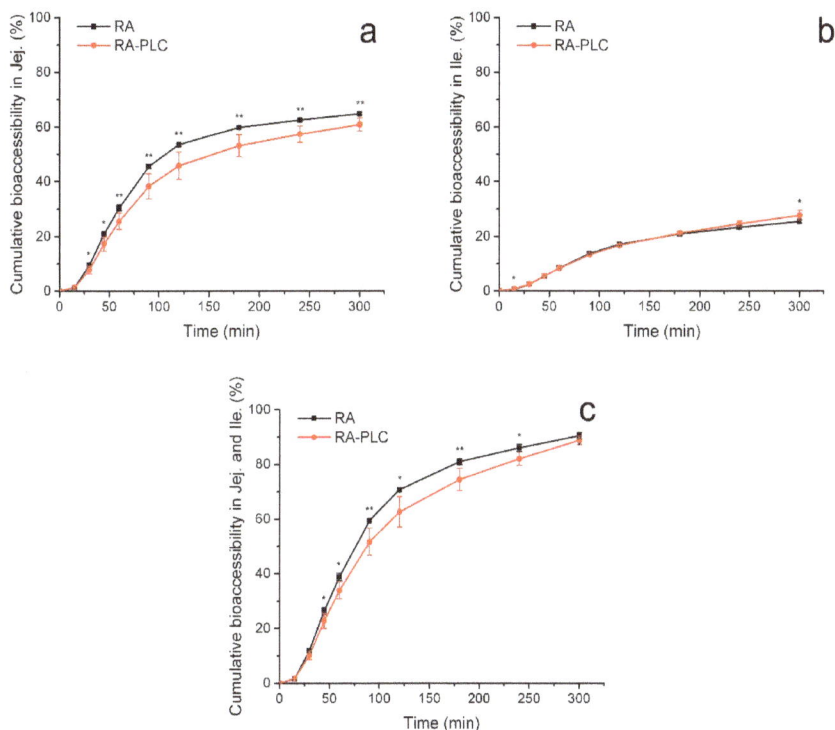

Figure 6. Cumulative bioaccessibility of RA and RA-PLC in jejunum (**a**), ileum (**b**), and the sum (**c**). * $p < 0.2$ and ** $p < 0.1$ between RA and RA-PLC ($n = 3$, mean $\pm$ SD).

In the ileum compartment, as shown in Figure 6b, the digestion of both RA and RA-PLC showed a monotonic increase in bioaccessibility, similar to that observed in the jejunum chamber. Within most time intervals, no significant difference was found between the bioaccessibility of RA and RA-PLC, indicative of similar dissolution behaviors of two samples. An interesting difference between RA and RA-PLC was observed in the profile obtained for the ileum compartment after 4 h, where the cumulative bioaccessibility of RA-PLC became higher than unformulated RA, which suggested the dissolution of RA was enhanced in the presence of phospholipids. In contrast to the bioaccessibility relation between RA and RA-PLC in previous chambers, the release rate of RA from the complex was enhanced by the amphiphilic nature and fine dispersible state of phospholipid in the alkaline intestinal environment, and by virtue of the long exposure time in the intestinal tract. After 5 h digestion, the maximum bioaccessibility of RA-PLC in the ileum compartment was determined to be approximately 28%, slightly higher than the 25% obtained for unformulated RA.

The overall bioaccessibility was defined as the sum of the amount of bioaccessible RA in both the jejunum and ileum dialysates. As shown in Figure 6c, the total bioaccessibility of unformulated RA was higher than RA-PLC for every time interval; however, it should be noticed that the bioaccessibility gap between RA and RA-PLC decreased at longer times, due to the enhancement of bioaccessibility seen for the RA-PLC in the ileum chamber at long times. At the end of the digestion process, the bioaccessibility of unformulated RA was ~90%, essentially equivalent to the ~89% obtained for RA-PLC. A more sustained digestion profile of RA-PLC was shown in Figure 6c.

3.8.2. Non-Cumulative Bioaccessibility

Non-cumulative bioaccessibility of RA and RA-PLC was studied to analyze the digestion behavior of RA in each time interval, further supporting the findings from cumulative bioaccessibility study. Bioaccessibility rate was defined as the amount of RA becoming bioaccessible per minute and was plotted against digestion time.

From Figure 7a, it was observed that the bioaccessibility rate of unformulated RA was higher than for RA-PLC in the jejunum up to 120 min. Within this period, the bioaccessibility rate for unformulated RA rose from 0.08%/min to 0.76%/min and then decreased to a value of 0.26%/min for the 90–120 min timepoint, while the rate of RA-PLC increased from 0.08%/min to the peak of 0.64%/min and then decreased to 0.25% for the 90–120 min timepoint. By virtue of the ready dissolution of unformulated RA in gastric fluid, a larger amount of RA could be transported to the jejunum compartment in a dissolved state to be instantly detected as bioaccessible, showing as a burst release profile for RA in the initial stage of jejunum digestion. In contrast, the lower peak value of bioaccessibility rate of RA-PLC (i.e., a slower rate of increase for bioaccessibility of RA from RA-PLC) could be explained by the poor dispersing behavior of the phospholipids in gastric fluid as discussed above. With less effective wetting process in the stomach, a smaller amount of free RA could be transported to jejunum in a dissolved state to be immediately determined.

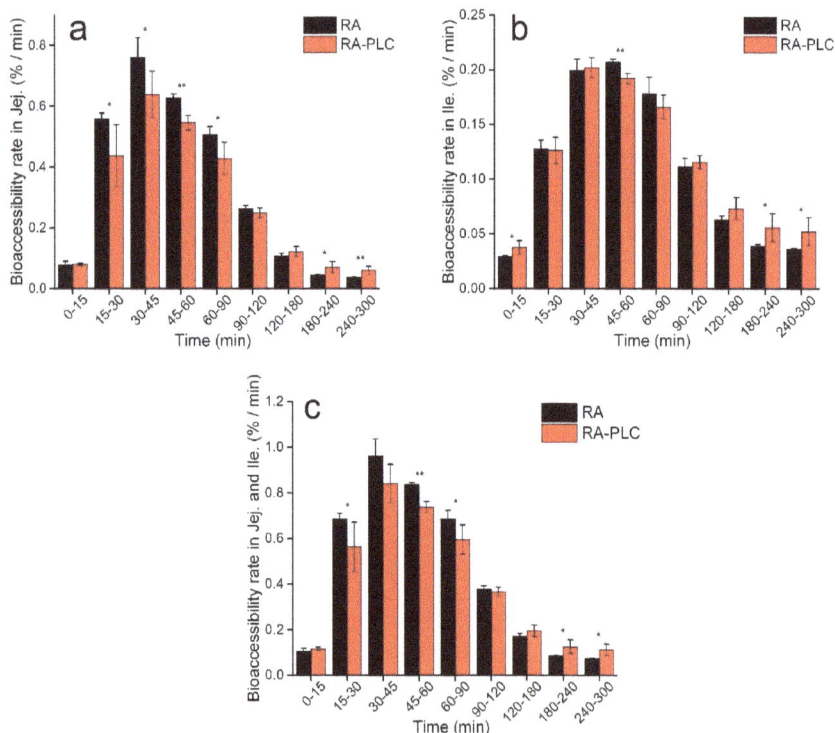

Figure 7. Bioaccessibility rate of RA and RA-PLC in jejunum (**a**), ileum (**b**), and the sum (**c**). * $p < 0.2$ and ** $p < 0.1$ between RA and RA-PLC ($n = 3$, mean ± SD).

At 90 min, the bioaccessibility rate of unformulated RA decreased significantly and the difference in bioaccessibility rate between RA and RA-PLC became non-significant due to two reasons. First, most of the RA had already been dissolved and was bioaccessible, and/or transported to the ileum compartment for the next step of digestion, decreasing the following bioaccessibility rate for RA.

Second, the dissolution of RA-PLC was maintained at a higher level in the jejunum as compared to RA by virtue of the exposure of phospholipid to the alkaline environment (again as described above). After 120 min timepoint, the bioaccessibility rate of RA-PLC became higher than RA. The more sustained bioaccessibility rate profile for RA-PLC from 90 min to 300 min is attributed to the continuous RA release from the reservoir that formed in stomach due to phospholipid complex aggregation. The absolute value of bioaccessibility rate of RA-PLC became larger than that of RA from 120 min, further supporting the discussion.

During ileum digestion, comparable profiles were observed in the bioaccessibility rate of RA and RA-PLC, especially at longer times. According to Figure 7b, the bioaccessibility rate of unformulated RA increased to peak within 60 min and started to decrease significantly thereafter. As a comparison, the rate of RA-PLC increased to maximum in 45 min, presenting a shorter time required to reach maximum. The results suggested that ileum environment is a dissolution-favored environment more for RA-PLC than RA, likely resulting from the amphiphilicity of phospholipid which facilitates the wetting of RA. Compared to unformulated RA, a more sustained bioaccessibility rate profile of RA-PLC was observed similar to that in jejunum compartment. The continuous release of RA from RA-PLC was likely a result of the extra time required for RA to transit from the jejunum into ileum compartment due to the delay in release from the stomach. Another reason could be the continuous separation of RA from the complex which still remained after a long-time digestion.

The total absorption rate was calculated in the same fashion as the total cumulative bioaccessibility, namely by summing the rates of the jejunum and ileum compartments. The total digestion rate is shown to be time-dependent as described above. As shown in Figure 7c, the overall bioaccessibility rate of RA was higher than RA-PLC in the initial stages of digestion, before 90 min, followed by a transition stage where two samples showed comparable rates (90–120 min), after which the digestion of RA-PLC reached an approximate steady state and the bioaccessibility rate became higher than for unformulated RA.

From the profiles of cumulative bioaccessibility, it can be concluded that the bioaccessibility of RA decreased at each time point due to the complexing with phospholipids, or extra time was required for RA-PLC to be bioaccessible. Not only does the analysis on the bioaccessibility rate support the above conclusion, but it also reveals differences that are more detailed, including the starting timepoint for RA-PLC to become more effective than RA, and the quantification of difference in instantaneous performance (i.e., the bioaccessibility rate for a 15-min period) between RA and RA-PLC. These details are expected to benefit the future formulation design of PLCs when modified or sustained release is needed. Besides, with the evaluation on the dispersibility, protonation, and the change in bioaccessibility rate of phospholipid complex in different digestive environments, controlled release of active ingredients to specific GIT regions may be achieved by adjusting phospholipids with different gel to liquid transition temperatures and pK values, as well as drug–phospholipid ratios.

It should be noted that for a BCS class 3 compound with high solubility and low permeability such as RA, bioaccessibility is a necessary but insufficient parameter to determine its bioavailability as the permeation of dissolved drug is the key determinant. Thus, the bioaccessibility data obtained from TIM-1 cannot be quantitatively used to predict drug absorption of this type of compounds. Even if combining with a caco-2 permeation experiment, the prediction of drug absorption may also be inaccurate when there is no quantitative translation model. Nevertheless, TIM-1 can be effectively used to reveal any unfavorable change in drug bioaccessibility induced by excipients, and warns formulation scientists of how much drug loses its potential to permeate into blood circulation. Considering the cost and ethical issues related to animal studies, TIM-1 bioaccessibility study is expected to help select the most worthwhile formulations to move into animal pharmacokinetic studies, in the best effort to improve the success possibility of drug product development.

At this point, it is observed that the RA-PLC dosage form increased the permeability of RA by 3.07-fold and decreased the bioaccessibility by only 1.8%, which is likely to correspond to the increased C_{max}, AUC and the shorten T_{max} in the previous animal study [10]. These results suggest that the bioaccessibility disadvantage of RA-PLC was likely overwhelmed by its permeability enhancement. When formulating phospholipid complexes, it is suggested to conduct a parallel comparison between formulations using bioaccessibility change as indicator before entering animal trial, in order not to overwhelm the permeation enhancement by the decreased bioaccessibility. Several factors were identified as important considerations for future research, including ionization constant, intermolecular complexation of phospholipids, intestinal transition time, and pH changes. Currently, however, there is no way to correlate these factors with pharmacokinetic profiles in a quantitative manner. Here, we do not suppose that the TIM-1 system is able to perfectly mimic the real in vivo situation, instead it is an effective tool to compare formulations in parallel by breaking the digestion process into different stages and testing the complicated effects of excipients like phospholipids.

4. Conclusions

The effect of complexing RA with phospholipids on the RA bioaccessibility was successfully evaluated using the TIM-1 dynamic system where the net effect was shown to be a slight reduction in bioaccessibility for RA-PLC (88.7%) compared to unformulated RA (90.4%). The bioaccessibility profiles of RA-PLC were shown to be dependent on different digestive environments. The complexation with phospholipids decreased the bioaccessibility of RA in the early stage of jejunum digestion, while providing a more sustained digestion profile in the following ileum process. The reduction of jejunum bioaccessibility was considered as a result of poor dispersion of phospholipids in the stomach. As the prerequisite for oral bioavailability, these bioaccessibility profiles are expected to provide rational predictions on the absorption behaviors of tested formulations. In this regard, the improved oral bioavailability of RA in rats from previous research could be considered as a net result of significantly increased intestinal permeability and slightly decreased bioaccessibility.

The insights into the pH-dependent effects of phospholipid materials on the bioaccessibility of a hydrophilic compound acknowledged a potential broader application of the TIM-1 to the characterization of PLC formulations, as well as other types of dosage forms containing components with opposing effects on bioaccessibility. The study on both cumulative and non-cumulative bioaccessibility is expected to benefit the design of controlled-release PLC formulations. The combination of TIM-1 dynamic system and Caco-2 transport assay is expected to provide an alternative approach to better select formulations of low-permeable drug before moving into animal studies.

Author Contributions: Conceptualization, J.H., P.X.C. and S.D.W.; methodology, J.H. and P.X.C.; validation, J.H., P.X.C. and M.A.R.; formal analysis, J.H. and P.X.C.; investigation, J.H. and P.X.C.; resources, M.A.R. and S.D.W.; writing—original draft preparation, J.H.; writing—review and editing, P.X.C., M.A.R. and S.D.W.; supervision, S.D.W.

Funding: This research was funded by the Natural Science and Engineering Research Council of Canada, Grant No. RGPIN-2016-04009.

Acknowledgments: S.D.W. acknowledges the support of the Canadian Foundation of Innovations and the University of Waterloo (UW). M.A.R. acknowledges the generous support of Canada Research Chairs programs. School of Pharmacy and Science Teaching Complex at UW are acknowledged for providing core facilities. Kun Feng (Chemical Engineering, UW) is acknowledged for the PXRD experiments.

Conflicts of Interest: The authors declare no conflict of interest. The funders had no role in the design of the study; in the collection, analyses, or interpretation of data; in the writing of the manuscript, or in the decision to publish the results.

References

1. Bombardelli, E.; Curri, S.; DELLA LOGGIA, R.; Del Negro, P.; Gariboldi, P.; Tubaro, A. Complexes between phospholipids and vegetal derivates of biological interest. *Fitoterapia* **1989**, *60*, 1–9.
2. Khan, J.; Alexander, A.; Ajazuddin; Saraf, S.; Saraf, S. Recent advances and future prospects of phyto-phospholipid complexation technique for improving pharmacokinetic profile of plant actives. *J. Control. Release* **2013**, *168*, 50–60. [CrossRef] [PubMed]

3. Semalty, A.; Semalty, M.; Singh, D.; Rawat, M. Development and characterization of aspirin-phospholipid complex for improved drug delivery. *Int. J. Pharm. Sci. Nanotechnol.* **2010**, *3*, 940–947.

4. Mirza, S.; Miroshnyk, I.; Habib, M.J.; Brausch, J.F.; Hussain, M.D. Enhanced Dissolution and Oral Bioavailability of Piroxicam Formulations: Modulating Effect of Phospholipids. *Pharmaceutics* **2010**, *2*, 339–350. [CrossRef] [PubMed]

5. Jo, K.; Cho, J.M.; Lee, H.; Kim, E.K.; Kim, H.C.; Kim, H.; Lee, J. Enhancement of Aqueous Solubility and Dissolution of Celecoxib through Phosphatidylcholine-Based Dispersion Systems Solidified with Adsorbent Carriers. *Pharmaceutics* **2018**, *11*, 1. [CrossRef]

6. Khazaeinia, T.; Jamali, F. A comparison of gastrointestinal permeability induced by diclofenac-phospholipid complex with diclofenac acid and its sodium salt. *J. Pharm. Pharm. Sci.* **2003**, *6*, 352–359. [PubMed]

7. Peng, Q.; Zhang, Z.-R.; Sun, X.; Zuo, J.; Zhao, D.; Gong, T. Mechanisms of Phospholipid Complex Loaded Nanoparticles Enhancing the Oral Bioavailability. *Mol. Pharm.* **2010**, *7*, 565–575. [CrossRef] [PubMed]

8. Semalty, A.; Semalty, M.; Rawat, M.S.M.; Franceschi, F. Supramolecular phospholipids-polyphenolics interactions: The PHYTOSOME (R) strategy to improve the bioavailability of phytochemicals. *Fitoterapia* **2010**, *81*, 306–314. [CrossRef]

9. Maiti, K.; Mukherjee, K.; Gantait, A.; Saha, B.P.; Mukherjee, P.K. Curcumin-phospholipid complex: Preparation, therapeutic evaluation and pharmacokinetic study in rats. *Int. J. Pharm.* **2007**, *330*, 155–163. [CrossRef]

10. Yang, J.H.; Zhang, L.; Li, J.S.; Chen, L.H.; Zheng, Q.; Chen, T.; Chen, Z.P.; Fu, T.M.; Di, L.Q. Enhanced oral bioavailability and prophylactic effects on oxidative stress and hepatic damage of an oil solution containing a rosmarinic acid-phospholipid complex. *J. Funct. Foods* **2015**, *19*, 63–73. [CrossRef]

11. Lichtenberger, L.M.; Wang, Z.M.; Romero, J.J.; Ulloa, C.; Perez, J.C.; Giraud, M.N.; Barreto, J.C. Non-steroidal anti-inflammatory drugs (NSAIDs) associate with zwitterionic phospholipids: Insight into the mechanism and reversal of NSAID-induced gastrointestinal injury. *Nat. Med.* **1995**, *1*, 154–158. [CrossRef]

12. Belcaro, G.; Cesarone, M.; Dugall, M.; Pellegrini, L.; Ledda, A.; Grossi, M.; Togni, S.; Appendino, G. Product-evaluation registry of Meriva®, a curcumin-phosphatidylcholine complex, for the complementary management of osteoarthritis. *Panminerva Med.* **2010**, *52* (Suppl. 1), 55–62. [PubMed]

13. Kaplan, M.R.; Simoni, R.D. Intracellular transport of phosphatidylcholine to the plasma membrane. *J. Cell Biol.* **1985**, *101*, 441–445. [CrossRef]

14. Singh, A.; Saharan, V.A.; Singh, M.; Bhandari, A. Phytosome: Drug delivery system for polyphenolic phytoconstituents. *Iran. J. Pharm. Sci.* **2011**, *7*, 209–219.

15. Kidd, P.; Head, K. A review of the bioavailability and clinical efficacy of milk thistle phytosome: A silybin-phosphatidylcholine complex (Siliphos^R). *Altern. Med. Rev.* **2005**, *10*, 193–203.

16. Li, Z.Y.; Agellon, L.B.; Allen, T.M.; Umeda, M.; Jewel, L.; Mason, A.; Vance, D.E. The ratio of phosphatidylcholine to phosphatidylethanolamine influences membrane integrity and steatohepatitis. *Cell Metab.* **2006**, *3*, 321–331. [CrossRef]

17. Wirtz-Peitz, F.; Probst, M.; Winkelmann, J. Rosmarinic acid-phospholipide-complex. Google Patents 1982.

18. Holst, B.; Williamson, G. Nutrients and phytochemicals: From bioavailability to bioefficacy beyond antioxidants. *Curr. Opin. Biotechnol.* **2008**, *19*, 73–82. [CrossRef]

19. Chen, J.J.; Zheng, J.K.; Decker, E.A.; McClements, D.J.; Xiao, H. Improving nutraceutical bioavailability using mixed colloidal delivery systems: Lipid nanoparticles increase tangeretin bioaccessibility and absorption from tangeretin-loaded zein nanoparticles. *RSC Adv.* **2015**, *5*, 73892–73900. [CrossRef]

20. Anson, N.M.; van den Berg, R.; Havenaar, R.; Bast, A.; Haenen, G. Bioavailability of ferulic acid is determined by its bioaccessibility. *J. Cereal Sci.* **2009**, *49*, 296–300. [CrossRef]

21. Han, W.-L.; Lu, W.; He, D.-P.; Long, Y.-Q.; Shang, J.-C. Pharmacokinetics and Relative Bioavailability Study of Berberine Hydrochloride Phytosome in Rabbits. *China Pharm.* **2011**, *17*, 1564–1566.

22. Jiang, Q.; Yang, X.; Du, P.; Zhang, H.; Zhang, T. Dual strategies to improve oral bioavailability of oleanolic acid: Enhancing water-solubility, permeability and inhibiting cytochrome P450 isozymes. *Eur. J. Pharm. Biopharm.* **2016**, *99*, 65–72. [CrossRef]

23. Zhang, Z.; Chen, Y.; Deng, J.; Jia, X.; Zhou, J.; Lv, H. Solid dispersion of berberine-phospholipid complex/TPGS 1000/SiO2: Preparation, characterization and in vivo studies. *Int. J. Pharm.* **2014**, *465*, 306–316. [CrossRef]

24. Weers, J.G.; Tarara, T.E.; Dellamary, L.A.; Riess, J.G.; Schutt, E.G. Phospholipid-Based Powders for Drug Delivery. U.S. Patent 7442388B2, 28 October 2008.

25. Van Hoogevest, P.; Wendel, A. The use of natural and synthetic phospholipids as pharmaceutical excipients. *Eur. J. Lipid Sci. Technol.* **2014**, *116*, 1088–1107. [CrossRef] [PubMed]

26. Van Hoogevest, P. Review—An update on the use of oral phospholipid excipients. *Eur. J. Pharm. Sci.* **2017**, *108*, 1–12. [CrossRef]

27. Bermudez-Soto, M.J.; Tomas-Barberan, F.A.; Garcia-Conesa, M.T. Stability of polyphenols in chokeberry (*Aronia melanocarpa*) subjected to in vitro gastric and pancreatic digestion. *Food Chem.* **2007**, *102*, 865–874. [CrossRef]

28. Wu, Z.; Teng, J.; Huang, L.; Xia, N.; Wei, B. Stability, antioxidant activity and in vitro bile acid-binding of green, black and dark tea polyphenols during simulated in vitro gastrointestinal digestion. *RSC Adv.* **2015**, *5*, 92089–92095. [CrossRef]

29. Kamiloglu, S.; Capanoglu, E.; Bilen, F.D.; Gonzales, G.B.; Grootaert, C.; Van de Wiele, T.; Van Camp, J. Bioaccessibility of Polyphenols from Plant-Processing Byproducts of Black Carrot (*Daucus carota* L.). *J. Agric. Food Chem.* **2016**, *64*, 2450–2458. [CrossRef] [PubMed]

30. Siracusa, L.; Kulisic-Bilusic, T.; Politeo, O.; Krause, I.; Dejanovic, B.; Ruberto, G. Phenolic Composition and Antioxidant Activity of Aqueous Infusions from *Capparis spinosa* L. and *Crithmum maritimum* L. before and after Submission to a Two-Step in Vitro Digestion Model. *J. Agric. Food Chem.* **2011**, *59*, 12453–12459. [CrossRef]

31. Costa, P.; Grevenstuk, T.; Rosa da Costa, A.M.; Goncalves, S.; Romano, A. Antioxidant and anti-cholinesterase activities of Lavandula viridis L'Her extracts after in vitro gastrointestinal digestion. *Ind. Crop. Prod.* **2014**, *55*, 83–89. [CrossRef]

32. Gayoso, L.; Claerbout, A.-S.; Isabel Calvo, M.; Yolanda Cavero, R.; Astiasaran, I.; Ansorena, D. Bioaccessibility of rutin, caffeic acid and rosmarinic acid: Influence of the in vitro gastrointestinal digestion models. *J. Funct. Foods* **2016**, *26*, 428–438. [CrossRef]

33. Bhattacharya, S. Phytosomes: The new technology for enhancement of bioavailability of botanicals and nutraceuticals. *Int. J. Health Res.* **2009**, *2*, 225–232. [CrossRef]

34. Blanquet, S.; Zeijdner, E.; Beyssac, E.; Meunier, J.P.; Denis, S.; Havenaar, R.; Alric, M. A dynamic artificial gastrointestinal system for studying the behavior of orally administered drug dosage forms under various physiological conditions. *Pharm. Res.* **2004**, *21*, 585–591. [CrossRef] [PubMed]

35. Lila, M.A.; Ribnicky, D.M.; Rojo, L.E.; Rojas-Silva, P.; Oren, A.; Havenaar, R.; Janle, E.M.; Raskin, I.; Yousef, G.G.; Grace, M.H. Complementary Approaches to Gauge the Bioavailability and Distribution of Ingested Berry Polyphenolics. *J. Agric. Food Chem.* **2012**, *60*, 5763–5771. [CrossRef] [PubMed]

36. Verwei, M.; Arkbage, K.; Havenaar, R.; van den Berg, H.; Witthoft, C.; Schaafsma, G. Folic acid and 5-methyltetrahydrofolate in fortified milk are bioaccessible as determined in a dynamic in vitro gastrointestinal model. *J. Nutr.* **2003**, *133*, 2377–2383. [CrossRef] [PubMed]

37. AlHasawi, F.M.; Fondaco, D.; Ben-Elazar, K.; Ben-Elazar, S.; Fan, Y.Y.; Corradini, M.G.; Ludescher, R.D.; Bolster, D.; Carder, G.; Chu, Y.; et al. In vitro measurements of luminal viscosity and glucose/maltose bioaccessibility for oat bran, instant oats, and steel cut oats. *Food Hydrocoll.* **2017**, *70*, 293–303. [CrossRef]

38. Fondaco, D.; AlHasawi, F.; Lan, Y.; Ben-Elazar, S.; Connolly, K.; Rogers, M.A. Biophysical Aspects of Lipid Digestion in Human Breast Milk and Similac™ Infant Formulas. *Food Biophys.* **2015**, *10*, 282–291. [CrossRef]

39. Barker, R.; Abrahamsson, B.; Kruusmagi, M. Application and Validation of an Advanced Gastrointestinal In Vitro Model for the Evaluation of Drug Product Performance in Pharmaceutical Development. *J. Pharm. Sci.* **2014**, *103*, 3704–3712. [CrossRef]

40. Kong, H.; Wang, M.; Venema, K.; Maathuis, A.; van der Heijden, R.; van der Greef, J.; Xu, G.; Hankemeier, T. Bioconversion of red ginseng saponins in the gastro-intestinal tract in vitro model studied by high-performance liquid chromatography-high resolution Fourier transform ion cyclotron resonance mass spectrometry. *J. Chromatogr. A* **2009**, *1216*, 2195–2203. [CrossRef] [PubMed]

41. Verwei, M.; Minekus, M.; Zeijdner, E.; Schilderink, R.; Havenaar, R. Evaluation of two dynamic in vitro models simulating fasted and fed state conditions in the upper gastrointestinal tract (TIM-1 and tiny-TIM) for investigating the bioaccessibility of pharmaceutical compounds from oral dosage forms. *Int. J. Pharm.* **2016**, *498*, 178–186. [CrossRef]

42. Butler, J.; Hens, B.; Vertzoni, M.; Brouwers, J.; Berben, P.; Dressman, J.; Andreas, C.J.; Schaefer, K.J.; Mann, J.; McAllister, M. In Vitro Models for the Prediction of in Vivo Performance of Oral Dosage Forms: Recent Progress from Partnership through the IMI OrBiTo Collaboration. *Eur. J. Pharm. Biopharm.* **2019**, *136*, 70–83. [CrossRef] [PubMed]

43. Van de Wiele, T.R.; Oomen, A.G.; Wragg, J.; Cave, M.; Minekus, M.; Hack, A.; Cornelis, C.; Rompelberg, C.J.; De Zwart, L.L.; Klinck, B. Comparison of Five in Vitro Digestion Models to in Vivo Experimental Results: Lead Bioaccessibility in the Human Gastrointestinal Tract. *J. Environ. Sci. Health Part A* **2007**, *42*, 1203–1211. [CrossRef] [PubMed]

44. Zhang, B.; Deng, Z.; Ramdath, D.D.; Tang, Y.; Chen, P.X.; Liu, R.; Liu, Q.; Tsao, R. Phenolic profiles of 20 Canadian lentil cultivars and their contribution to antioxidant activity and inhibitory effects on alpha-glucosidase and pancreatic lipase. *Food Chem.* **2015**, *172*, 862–872. [CrossRef]

45. Minekus, M.; Marteau, P.; Havenaar, R.; Huisintveld, J.H.J. Multicompartmental dynamic computer-controlled model simulating the stomach and small intestine. *ATLA* **1995**, *23*, 197–209.

46. Madureira, A.R.; Campos, D.A.; Fonte, P.; Nunes, S.; Reis, F.; Gomes, A.M.; Sarmento, B.; Pintado, M.M. Characterization of solid lipid nanoparticles produced with carnauba wax for rosmarinic acid oral delivery. *RSC Adv.* **2015**, *5*, 22665–22673. [CrossRef]

47. Cevc, G. *Phospholipids Handbook*; CRC Press: Boca Raton, FL, USA, 1993.

48. Boggs, J.M. Lipid intermolecular hydrogen bonding: influence on structural organization and membrane function. *Biochim. Biophys. Acta* **1987**, *906*, 353–404. [CrossRef]

49. Eibl, H. The effect of the proton and of monovalent cations on membrane fluidity. *Membr. Fluidity Biol.* **1983**, *2*, 217.

50. Marsh, D. *Handbook of Lipid Bilayers*; CRC Press: Boca Raton, FL, USA, 2013.

51. Petersen, M.; Simmonds, M.S.J. Molecules of interest—Rosmarinic acid. *Phytochemistry* **2003**, *62*, 121–125. [CrossRef]

52. Frankel, E.N.; Huang, S.W.; Aeschbach, R.; Prior, E. Antioxidant activity of a rosemary extract and its constituents, carnosic acid, carnosol, and rosmarinic acid, in bulk oil and oil-in-water emulsion. *J. Agric. Food Chem.* **1996**, *44*, 131–135. [CrossRef]

53. Erkan, N.; Ayranci, G.; Ayranci, E. Antioxidant activities of rosemary (*Rosmarinus Officinalis* L.) extract, blackseed (*Nigella sativa* L.) essential oil, carnosic acid, rosmarinic acid and sesamol. *Food Chem.* **2008**, *110*, 76–82. [CrossRef]

54. Chen, P.X.; Tang, Y.; Marcone, M.F.; Pauls, P.K.; Zhang, B.; Liu, R.H.; Tsao, R. Characterization of free, conjugated and bound phenolics and lipophilic antioxidants in regular- and non-darkening cranberry beans (*Phaseolus vulgaris* L.). *Food Chem.* **2015**, *185*, 298–308. [CrossRef] [PubMed]

55. Visconti, R.; Grieco, D. New insights on oxidative stress in cancer. *Curr. Opin. Drug Discov. Dev.* **2009**, *12*, 240–245.

56. Artursson, P.; Karlsson, J. Correlation between Oral Drug Absorption in Humans and Apparent Drug Permeability Coefficients in Human Intestinal Epithelial (Caco-2) Cells. *Biochem. Biophys. Res. Commun.* **1991**, *175*, 880–885. [CrossRef]

pharmaceutics

MDPI

Communication

Biodistribution of a Radiolabeled Antibody in Mice as an Approach to Evaluating Antibody Pharmacokinetics

Kevin J. H. Allen [+], Rubin Jiao [+], Mackenzie E. Malo, Connor Frank and Ekaterina Dadachova *

College of Pharmacy and Nutrition, University of Saskatchewan, Saskatoon, SK S7N 5E5, Canada;
kja782@mail.usask.ca (K.J.H.A.); jiaorubin9712@hotmail.com (R.J.); mem510@mail.usask.ca (M.E.M.);
csf876@mail.usask.ca (C.F.)
* Correspondence: ekaterina.dadachova@usask.ca; Tel.: +1-(306)966-5163
[+] The authors contributed equally to this work.

Received: 29 October 2018; Accepted: 1 December 2018; Published: 5 December 2018

Abstract: (1) Background: Monoclonal antibodies are used in the treatment of multiple conditions including cancer, autoimmune disorders, and infectious diseases. One of the initial steps in the selection of an antibody candidate for further pre-clinical development is determining its pharmacokinetics in small animal models. The use of mass spectrometry and other techniques to determine the fate of these antibodies is laborious and expensive. Here we describe a straightforward and highly reproducible methodology for utilizing radiolabeled antibodies for pharmacokinetics studies. (2) Methods: Commercially available bifunctional linker CHXA" and [111]Indium radionuclide were used. A melanin-specific chimeric antibody A1 and an isotype matching irrelevant control A2 were conjugated with the CHXA", and then radiolabeled with [111]In. The biodistribution was performed at 4 and 24 h time points in melanoma tumor-bearing and healthy C57BL/6 female mice. (3) The biodistribution of the melanin-binding antibody showed the significant uptake in the tumor, which increased with time, and very low uptake in healthy melanin-containing tissues such as the retina of the eye and melanized skin. This biodistribution pattern in healthy tissues was very close to that of the isotype matching control antibody. (4) Conclusions: The biodistribution experiment allows us to assess the pharmacokinetics of both antibodies side by side and to make a conclusion regarding the suitability of specific antibodies for further development.

Keywords: pharmacokinetics; antibodies; radiolabeling; biodistribution; mouse models

1. Introduction

The field of immunotherapy is experiencing explosive growth, with new antibodies being approved for clinical use, or being introduced into the research pipeline on a regular basis [1–3]. Monoclonal antibodies find applications in the treatment of multiple conditions, including cancer, autoimmune disorders, and infectious diseases. One of the initial steps in the selection of an antibody candidate for further pre-clinical development is determining its pharmacokinetics (PK) in small animal models. Usually PK studies are performed by administering the antibody candidate to the healthy mice, or a mouse model of a relevant disease, followed by harvesting organs and tissues at pre-determined time points. These samples are then digested and subjected to various downstream analytical techniques, such as mass spectrometry and immune-PCR (Polymerase Chain Reaction), in order to test for the presence of the candidate antibody [4,5]. These techniques are laborious and expensive and require access to state-of-the-art equipment, such as MALDI (Matrix Assisted Laser Desorption Ionization) mass spectrometers, as well as highly trained personnel for interpretation of the results. An alternative technique is to attach a radiolabel to the antibody of interest before administering it to mice, and then to follow its fate in vivo by measuring the amount of radioactivity present in the mouse organs and tissues at the pre-determined time points. Here we describe a straightforward and highly reproducible method for radiolabeling antibodies using commercially available linker and radionuclide, and performing biodistribution in a murine melanoma model.

2. Materials and Methods

2.1. Reagents, Antibodies, Radionuclides, and Cell Lines

The antibody to melanin, Ab1, was produced in our laboratories and human IgG isotype control Ab, referred to as Ab2 in the text, was purchased (Creative Diagnostics, Shirley, NY, USA). ^{111}Indium was purchased as ^{111}InCl$_3$ from Nordion (Vancouver, BC, Canada). Bifunctional CHXA" ligand was purchased from Macrocyclics (Plano, TX, USA). Murine melanoma cell line B16-F10 was purchased from ATCC (Manassas, VA, USA).

2.2. Metal-Free Buffer Preparation

Stock buffers must be prepared as metal-free solutions in order to ensure contaminating metals do not interfere with downstream radiolabeling steps. All buffers were prepared as concentrated stocks with distilled/deionized H$_2$O (ddH$_2$O), using components purchased from Fisher Scientific (Ottawa, ON, Canada). Stock buffers were run through a Chelex cation exchange resin column to scavenge contaminating free metal ions.

The Chelex column was prepared by placing a glass wool plug in a glass chromatography column. The wool plug was rinsed with concentration HCl, followed by water until the eluate was a neutral pH. A slurry of Chelex-100, Na+ form, 200–400 mesh (BioRad, Hercules, CA, USA) was prepared in ddH$_2$O and poured into the column in order to have approximately 5 cm of packed resin. The Chelex column was washed with ddH$_2$O until the eluate returned to a neutral pH.

Conjugation buffer stock was prepared as 0.5 M Carbonate/Bicarbonate (0.02 M/0.48 M), 1.5 M NaCl solution at pH 8.6–8.7 in ddH$_2$O. A prepared Chelex column was equilibrated with 100 mL of 10× stock buffer and the eluate was discarded. The remaining 10× stock buffer was run through the column and collected as a metal-free 10× stock. Conjugation buffer was prepared by diluting the 10× conjugation buffer by 10 with ddH$_2$O and adding EDTA to 5 mM.

Ammonium acetate buffer was prepared as a concentrated stock (5 M, pH 7.5) and run through a Chelex column to scavenge free metal ions. Prior to completing Ab labeling it was diluted with ddH$_2$O and used at 0.15 M concentration.

2.3. Radiolabeling of Antibody-CHXA" Conjugate with ^{111}Indium (^{111}In)

To conjugate the bifunctional chelator CHXA" to the antibody, the Ab must first be exchanged out of the storage buffer and into conjugation buffer. This was achieved by loading the Ab onto a 0.5 mL 30K molecular weight cutoff Amicon microconcentrator (Millipore, Burlington, MA, USA) with conjugation buffer and then centrifuging at 4 °C following the microconcentrator manufacturer's recommended conditions. Conjugation buffer was added and centrifugation was repeated at least 10 times to ensure complete exchange of the Ab storage buffer to conjugation buffer. A 5-fold molar excess of CHXA" (2 mg/mL in conjugation buffer, prepared immediately before use) was added to the antibody solution. The reaction mixture was then incubated at 37 °C for 1.5 h. Upon completion of the reaction the Ab-CHXA" conjugate was then exchanged into the 0.15 M ammonium acetate buffer at 4 °C, the same as described for the exchange into the conjugation buffer. Protein concentration was determined by Bradford assay (BioRad, Hercules, CA, USA) prior to labeling the Ab.

The radiolabeling of the antibody-CHXA" conjugate with ^{111}In was performed to achieve a specific activity of approximately 5 µCi/µg of the antibody. The amount of the radiolabeled antibody to be administered to a mouse was approximately 6 µg, therefore, the amount of radioactivity was 30 µCi. ^{111}In chloride was diluted with 0.15 M ammonium acetate buffer and added to a microcentrifuge tube (MCT) containing the Ab-CHXA" conjugate in the 0.15 M ammonium acetate buffer, a minimum volume was desired, with a typical reaction volume being ~30 μL. The reaction mixture was incubated for 60 min at 37 °C. The reaction was quenched by the addition of 3 μL of 0.05 M EDTA solution to bind any free [111]In.

The percentage of radiolabeling (radiolabeling yield) was measured by instant thin layer chromatography (iTLC) by developing 10 cm silica gel strips (Agilent Technologies, Santa Clara, CA, USA) in 0.15 M ammonium acetate buffer. In this system the radiolabeled antibodies stay at the point of application while free ^{111}In, in the form of EDTA complexes, move with the solvent front. The strips were cut in half and each half was counted on a 2470 Wizard2 Gamma counter (Perkin Elmer, Waltham, MA, USA) that was calibrated for the ^{111}In emission spectrum and only emissions in this range were considered in the counts per minute (CPM). The percentage of radiolabeling was calculated by dividing the CPM at the bottom of the strip (labeled antibody) by the sum of the CPM at the bottom and the top of the strip (total amount of radioactivity) and multiplying the result by 100.

2.4. B16-F10 Melanoma Tumor Model

All animal studies were approved by the Animal Research Ethics Board of the University of Saskatchewan (Animal Use Permit #20180006, approved 1 February 2017). For the biodistribution experiment, 6 week-old C57BL/6 female mice obtained from Charles River Laboratories (Wilmington, MA, USA) were injected subcutaneously with 5×10^6 B16-F10 murine melanoma cells in Matrigel (Corning, Corning, NY, USA) into the right flank, or were not given tumor cells. The radioactivity was administered on Day 8 post tumor cells injection, when the tumors reached 0.7–1.0 cm in diameter.

2.5. Biodistribution of Antibody-CHXA" Conjugate in Tumor-Bearing Mice

For the biodistribution of melanin binding antibody (Ab1) B16-F10 tumor-bearing C57BL/6 mice were injected intravenously (IV) via the tail vein with 30 μCi antibody-CHXA"—[111]In in 100 μL saline. The same activity of the radiolabeled non-specific antibody (Ab2) was injected into healthy C57BL/6 mice. At the predetermined time points of 4 and 24 h groups of 4–5 mice were humanely sacrificed, their tumors and major organs removed, blotted from blood, weighed and counted for radioactivity in a gamma counter. The standard was prepared by diluting 10 μL (1/10 of the injected dose) of the respective radiolabeled antibody with 2 mL 0.15 M ammonium acetate buffer and counted in a gamma counter at the same time as the organ/tissues were counted. Percentage of injected dose per gram (ID/g) organ/tissue was calculated by dividing the CPM in an organ by its weight in grams and by the CPM of the standard, followed by multiplying the resulting number by 10.

3. Results

3.1. Radiolabeling of the Antibodies

Two IgG isotype antibodies, Ab1 to melanin and irrelevant isotype matching control Ab2, were conjugated to the bifunctional chelating agent CHXA" to enable subsequent radiolabeling with [111]In. The concentration of the conjugated antibodies was 16 mg/mL as per Bradford assay which constitutes the 80% recovery of the antibodies after buffer exchange and subsequent purification on the Amicon microconcentrators. The iTLC analysis of the ^{111}In-labeled antibody conjugates revealed a radiolabeling yield of >92%, which allowed for the use of the radiolabeled antibody conjugates in the biodistribution experiment without further purification.

3.2. Biodistribution of the Radiolabeled Antibodies in Melanoma Tumor Bearing Mice

The raw data and the calculated % ID/g for a 4 h time point for the ^{111}In-Ab1 is shown in Table 1. The complete set of data for both ^{111}In-Ab1 and ^{111}In-Ab2 and 4 and 24 h time points is presented in Supplementary Table S1 in the form of an Excel file. The columns in Table 1 include the following: rack number (for record keeping purposes), tube number, Ab name, organ/tissue, empty tube weight into which that organ was subsequently placed, weight of the tube with the organ, calculated weight of the organ, the CPM in the organ obtained by counting it on the gamma counter, and finally, the % ID/g which is calculated using a formula described in the Methods section, in footnotes to Table 1 and incorporated into the Excel file in Supplementary Table S1.

To take into consideration the radioactive decay of ^{111}In during the experiment, the standards (in our case 1/10 of the injected dose) were counted simultaneously with the organs. Each standard was counted two times, at the beginning and end, and the mean of these two measurements was used in the calculations of the % ID/g. The mean of the two readings allow us to account for the minor decay of ^{111}In that occurs during counting of multiple samples in a gamma counter. Table 2 shows the values in CPM for the ^{111}In-Ab1 and ^{111}In-Ab2 standards for 4 and 24 h time points.

The process of biodistribution is shown in Supplementary Video S1. The biodistribution was then used to construct biodistribution plots which are presented in Figure 1. The organs and tissues are shown on the *X* axis while the % ID/g for the respective organ/tissues is shown on the *Y* axis. The biodistribution patterns for both antibodies are typical for IgG biodistribution in mice. The clearance from the blood has begun by 24 h, although antibodies are still in circulation at this time point. The antibodies have started to clear from blood rich organs such as heart and lungs. The organs where the antibodies are processed such as spleen, liver and kidneys show very similar uptake for both antibodies, which starts to decrease by 24 h. No penetration of the antibodies in the brain and into the melanized tissues such as eyes is observed. There is very insignificant uptake into stomach, small and large intestines, femur and muscle. The melanin binding antibody Ab1 accumulates in the tumor in B16F10 tumor bearing mice and its uptake increases with time, indicating that it is specific to melanin present in the tumors.

Table 1. Example of the raw data and the calculated uptake of the ^{111}In-Ab1 into the organs/tissues at 4 h post administration to mice.

Rack #	Tube #	Antibody	Time Point, h	Organ	Tube Weight, g	Tube + Organ, g	Organ Weight, g	Organ CPM *	Formula **
7	1	Ab1	4	Blood	2.6051	2.6621	0.057	1,045,501.78	51.1224844
7	2	Ab1	4	Tumor	2.849	2.9611	0.1121	61,454.48	1.5279529
7	3	Ab1	4	Spleen	2.7385	2.8142	0.0757	52,319.71	1.92633351
7	4	Ab1	4	Kidneys	2.8507	3.1053	0.2546	538,036.52	5.89000124
7	5	Ab1	4	Liver	2.7795	3.2932	0.5137	392,858.5	2.13151611
7	6	Ab1	4	Brain	2.7548	3.0212	0.2664	66,697.57	0.69781102
7	7	Ab1	4	Lungs	2.7711	2.9214	0.1503	832,976.76	15.4467079
7	8	Ab1	4	Stomach	2.8467	3.0603	0.2136	271,074.71	3.53711992
7	9	Ab1	4	Small Ints	2.7852	3.1185	0.3333	462,766.45	3.86980055
7	10	Ab1	4	Large Ints	2.759	3.1762	0.4172	111,965.48	0.74800048
7	11	Ab1	4	Thigh Muscle	2.7371	2.8462	0.1091	9180.22	0.23452566
7	12	Ab1	4	Femur	2.7765	2.9106	0.1341	145,455.67	3.02317932
7	13	Ab1	4	Eye	2.6675	2.6907	0.0232	23,482.75	2.82112935
7	14	Ab1	4	Tail	2.8006	2.843	0.0424	104,294.99	6.85582338
7	15	Ab1	4	Heart	2.731	2.8187	0.0877	267,839.25	8.51209882

#—number; * CPM—counts per minute. ** Formula—Percentage of injected dose per gram (ID/g) organ/tissue was calculated by dividing the CPM (counts per minute) in an organ by its weight in grams and by the CPM of the standard, followed by multiplying the resulting number by 10.

Table 2. CPM for the ^{111}In-Ab1 and ^{111}In-Ab2 standards for 4 and 24 h time points.

Standard	Time Point	Reading 1	Reading 2	Average
Ab1	4	3,619,311	3,556,450	3,587,880.5
Ab2	4	3,391,460	3,328,188	3,359,824
Ab1	24	3,371,690	3,238,291	3,304,990.5
Ab2	24	3,148,796	3,022,212	3,085,504

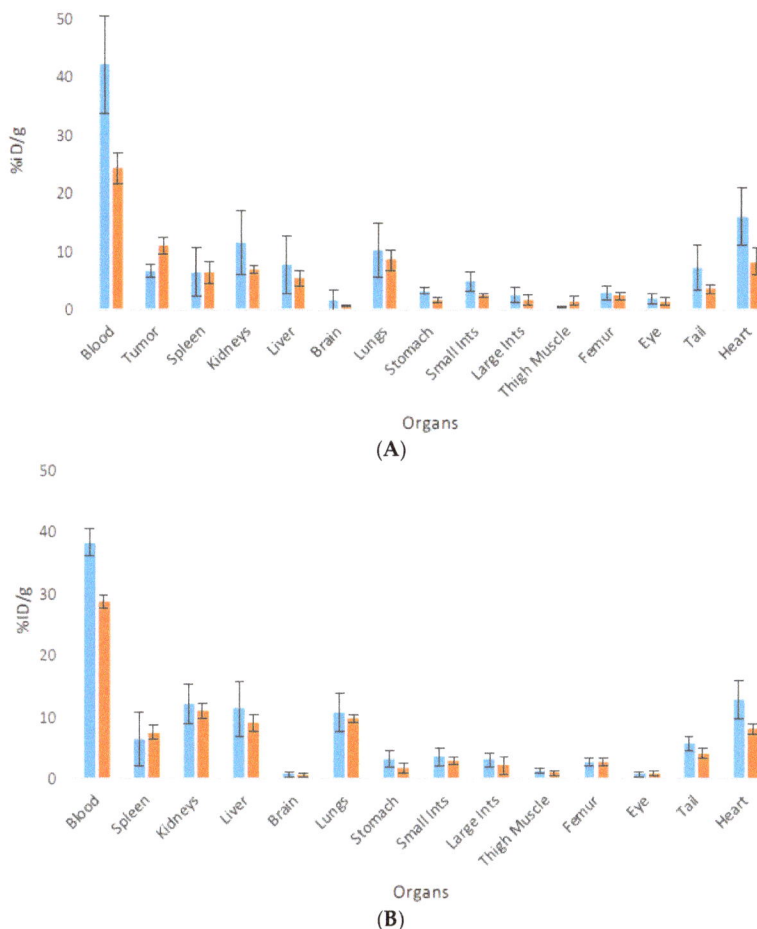

Figure 1. Biodistribution of ^{111}In-labeled antibodies Ab1 to melanin and isotype matching irrelevant control Ab2 in C57BL6 mice at 4 and 24 h after IV administration. (**A**) Ab1 antibody in B16F10 melanoma bearing mice; (**B**) Ab2 in healthy mice. Blue and orange bars represent 4 and 24 h time points, respectively. The error bars represent the SEM of sample groups of 4–5 mice.

4. Discussion

With the increasing pace of drug discovery, in particular antibody-based drugs, faster and more cost-effective methods to assess PK parameters are required. Utilizing commercially available linkers and radioisotopes, we are able to improve upon traditional pharmacokinetic studies by reducing time and cost investments while increasing reproducibility. Here we describe the radiolabeling and biodistribution of two IgG isotype antibodies in tumor bearing versus healthy mice. Ab1 is specific for the pigment melanin, while Ab2 is an isotype matching control. The radiolabeling procedure, which involves using commercially available bifunctional chelating agent CHXA" and radionuclide ^{111}In, is straightforward and does not require post-labeling purification. Mice used in the biodistribution could be either healthy if only PK data is being collected, or could be a model of a disease of interest. For this study we used murine melanoma tumor bearing mice for the melanin-binding antibody and healthy mice for isotype matching controls.

In this study, we collected the data at two time points of 4 and 24 h. These time points were chosen to obtain initial evidence of specific targeting. While these time points do not constitute a complete PK study, they do provide sufficient evidence that this method for assessing PK parameters is sensitive enough to detect the changes in targeting and accumulation over time. Depending on the nature of the research, much earlier or much later time points could be collected for more extensive longitudinal studies assessing adsorption, distribution, metabolism, and excretion. For example, when the ^{111}In radiolabel is used for labeling antibodies the pharmacokinetics in mice could be followed for up to 7 days, as the physical half-life is 2.8 days. If much later time points are desirable, longer lived radionuclides such [177]Lutetium, which also forms stable complexes with CHXA" bifunctional chelating agent [6] and has a 6.7-day physical half-life, could be used in place of [111]In. Collection of data from at least 4–5 time points would enable the modeling of blood clearance of the antibodies using biphasic models, as well as allowing radiation dosimetry calculations if an antibody-based imaging or radioimmunotherapeutic agent is being developed. This method is sensitive as we were detecting ^{111}In at the 150–1200 nCi level, which correlates to 75–600 ng amounts of Ab, based on the 2 µCi/µg specific activity. The use of metabolic cages during the biodistribution experiments would enable collection of urine and feces, which could be counted in a gamma counter in the same way as the organs, and would provide a wealth of information on the metabolic fate of the antibody. Finally, if non-invasive imaging equipment such as microSPECT (micro Single Photon Emission Computed Tomography) or microPET (micro Positron Emission Tomography) is available [7] and is equipped with the software allowing quantification of the radiolabeled antibody uptake in the organs/tissue, one group of 4–5 mice per antibody could be followed longitudinally without the need to sacrifice animals at every time point, greatly reducing the number of animals used in the experiment.

5. Conclusions

In conclusion, here we describe a relatively simple and efficient way to study the pharmacokinetics of antibodies by radiolabeling them with commercially available radionuclides and performing biodistribution, either in healthy mice or in disease models. We, as well as other groups have successfully used this technique for preclinical development of antibodies for therapy or for imaging of cancer, infection, and neurodegenerative diseases [8–20].

Supplementary Materials: The following are available online at http://www.mdpi.com/1999-4923/10/4/262/s1, Table S1: Raw data for biodistribution; Video S1: Evaluating Pharmacokinetics of an Antibody by Radiolabeling and Performing Biodistribution in Mice.

Author Contributions: Conceptualization, E.D.; Methodology, K.J.H.A., M.E.M., R.J., C.F.; Investigation, K.J.H.A., M.E.M., R.J., C.F.; Writing-Original Draft Preparation, E.D.; Writing-Review & Editing, K.J.H.A., M.E.M., R.J.; Funding Acquisition, E.D.

Funding: This research was funded by NCI SBIR Contract HHSN261201600017C.

Acknowledgments: We thank Fedoruk Center for Nuclear Innovation at the University of Saskatchewan for use of its resources.

Conflicts of Interest: The authors declare no conflict of interest. E. Dadachova is a co-inventor on the patent describing radioimmunotherapy of melanoma.

References

1. Hung, S.Y.; Fu, W.M. Drug candidates in clinical trials for Alzheimer's disease. *J. Biomed. Sci.* **2017**, *24*, 47. [CrossRef] [PubMed]
2. Grierson, P.; Lim, K.H.; Amin, M. Immunotherapy in gastrointestinal cancers. *J. Gastrointest. Oncol.* **2017**, *8*, 474–484. [CrossRef] [PubMed]
3. Gleason, C.; Catamero, D.D. Immunotherapy in Multiple Myeloma. *Semin. Oncol. Nurs.* **2017**, *33*, 292–298. [CrossRef] [PubMed]
4. Cong, Y.; Zhang, Z.; Zhang, S.; Hu, L.; Gu, J. Quantitative MS analysis of therapeutic mAbs and their glycosylation for pharmacokinetics study. *PROTEOMICS–Clin. Appl.* **2016**, *10*, 303–314. [CrossRef] [PubMed]

5. Shadid, M.; Bowlin, S.; Bolleddula, J. Catabolism of antibody drug conjugates and characterization methods. *Bioorg. Med. Chem.* **2017**, *25*, 2933–2945. [CrossRef] [PubMed]

6. Pandey, U.; Gamre, N.; Lohar, S.P.; Dash, A. A systematic study on the utility of CHX-A"-DTPA-NCS and NOTA-NCS as bifunctional chelators for ^{177}Lu radiopharmaceuticals. *Appl. Radiat. Isot.* **2017**, *127*, 1–6. [CrossRef] [PubMed]

7. Nayak, T.K.; Bernardo, M.; Milenic, D.E.; Choyke, P.L.; Brechbiel, M.W. Orthotopic pleural mesothelioma in mice: SPECT/CT and MR imaging with HER1- and HER2-targeted radiolabeled antibodies. *Radiology* **2013**, *267*, 173–182. [CrossRef] [PubMed]

8. Dadachova, E.; Revskaya, E.; Sesay, M.A.; Damania, H.; Boucher, R.; Sellers, R.S.; Howell, R.C.; Burns, L.; Thornton, G.B.; Natarajan, A.; et al. Pre-clinical evaluation and efficacy studies of a melanin-binding IgM antibody labeled with 188Re against experimental human metastatic melanoma in nude mice. *Cancer Biol. Ther.* **2008**, *7*, 1116–1127. [CrossRef] [PubMed]

9. Dadachova, E.; Nakouzi, A.; Bryan, R.; Casadevall, A. Ionizing radiation delivered by specific antibody is therapeutic against a fungal infection. *Proc. Natl. Acad. Sci. USA* **2003**, *100*, 10942–10947. [CrossRef] [PubMed]

10. Phaeton, R.; Jiang, Z.; Revskaya, E.; Fisher, D.R.; Goldberg, G.L.; Dadachova, E. Beta emitters Rhenium-188 and Lutetium-177 are equally effective in radioimmunotherapy of HPV-positive experimental cervical cancer. *Cancer Med.* **2016**, *5*, 9–16. [CrossRef] [PubMed]

11. Regino, C.A.; Wong, K.J.; Milenic, D.E.; Holmes, E.H.; Garmestani, K.; Choyke, P.L.; Brechbiel, M.W. Preclinical evaluation of a monoclonal antibody (3C6) specific for prostate-specific membrane antigen. *Curr. Radiopharm.* **2009**, *2*, 9–17. [CrossRef] [PubMed]

12. Milenic, D.E.; Wong, K.J.; Baidoo, K.E.; Ray, G.L.; Garmestani, K.; Williams, M.; Brechbiel, M.W. Cetuximab: Preclinical evaluation of a monoclonal antibody targeting EGFR for radioimmunodiagnostic and radioimmunotherapeutic applications. *Cancer Biother. Radiopharm.* **2008**, *23*, 619–632. [CrossRef] [PubMed]

13. Nedrow, J.R.; Josefsson, A.; Park, S.; Bäck, T.; Hobbs, R.F.; Brayton, C.; Bruchertseifer, F.; Morgenstern, A.; Sgouros, G. Pharmacokinetics, microscale distribution, and dosimetry of alpha-emitter-labeled anti-PD-L1 antibodies in an immune competent transgenic breast cancer model. *EJNMMI Res.* **2017**, *7*, 57. [CrossRef] [PubMed]

14. Nedrow, J.R.; Josefsson, A.; Park, S.; Ranka, S.; Roy, S.; Sgouros, G. Imaging of programmed death ligand-1 (PD-L1): Impact of protein concentration on distribution of anti-PD-L1 SPECT agent in an immunocompetent melanoma murine model. *J. Nucl. Med.* **2017**, *58*, 1560–1566. [CrossRef] [PubMed]

15. James, M.L.; Hoehne, A.; Mayer, A.T.; Lechtenberg, K.; Moreno, M.; Gowrishankar, G.; Ilovich, O.; Natarajan, A.; Johnson, E.M.; Nguyen, J.; et al. Imaging B cells in a mouse model of multiple sclerosis using ^{64}Cu-Rituximab-PET. *J. Nucl. Med.* **2017**, *58*, 1845. [CrossRef] [PubMed]

16. Adams, C.J.; Wilson, J.J.; Boros, E. Multifunctional desferrichrome analogues as versatile ^{89}Zr(IV) chelators for immunoPET probe development. *Mol. Pharm.* **2017**, *14*, 2831–2842. [CrossRef] [PubMed]

17. Cook, B.E.; Adumeau, P.; Membreno, R.; Carnazza, K.E.; Brand, C.; Reiner, T.; Agnew, B.J.; Lewis, J.S.; Zeglis, B.M. Pretargeted PET imaging using a site-specifically labeled immunoconjugate. *Bioconjug. Chem.* **2016**, *27*, 1789–1795. [CrossRef] [PubMed]

18. Cutler, C.S.; Hennkens, H.M.; Sisay, N.; Huclier-Markai, S.; Jurisson, S.S. Radiometals for combined imaging and therapy. *Chem. Rev.* **2012**, *113*, 858–883. [CrossRef] [PubMed]

19. Kikuchi, M.; Clump, D.A.; Srivastava, R.M.; Sun, L.; Zeng, D.; Diaz-Perez, J.A.; Anderson, C.J.; Edwards, W.B.; Ferris, R.L. Preclinical immunoPET/CT imaging using Zr-89-labeled anti-PD-L1 monoclonal antibody for assessing radiation-induced PD-L1 upregulation in head and neck cancer and melanoma. *Oncoimmunology* **2017**, *6*, e1329071. [CrossRef] [PubMed]

20. Deri, M.A.; Zeglis, B.M.; Francesconi, L.C.; Lewis, J.S. PET imaging with ^{89}Zr: From radiochemistry to the clinic. *Nucl. Med. Biol.* **2013**, *40*, 3–14. [CrossRef] [PubMed]

pharmaceutics

MDPI

Article

Modulation of Hypoxia-Induced Chemoresistance to Polymeric Micellar Cisplatin: The Effect of Ligand Modification of Micellar Carrier Versus Inhibition of the Mediators of Drug Resistance

Hoda Soleymani Abyaneh [1], Amir Hassan Soleimani [1], Mohammad Reza Vakili [1],
Rania Soudy [1,2], Kamaljit Kaur [1,3], Francesco Cuda [4], Ali Tavassoli [4] and Afsaneh Lavasanifar [1,5,*]

[1] Faculty of Pharmacy and Pharmaceutical Sciences, University of Alberta, Edmonton, AB T6G 2E1, Canada;
     hoda1@ualberta.ca (H.S.A.); asoleimani212@gmail.com (A.H.S.); vakili@ualberta.ca (M.R.V.);
     soudy@ualberta.ca (R.S.); kkaur@chapman.edu (K.K.)
[2] Faculty of Pharmacy, Cairo University, Kasr El-Aini, Cairo 11562, Egypt
[3] School of Pharmacy, Chapman University, Irvine, CA 92618, USA
[4] School of Chemistry, University of Southampton, Southampton SO17 1BJ, UK;
     francesco_cuda@yahoo.co.uk (F.C.); A.Tavassoli@soton.ac.uk (A.T.)
[5] Department of Chemical & Materials Engineering, Faculty of Engineering, University of Alberta,
     Edmonton, AB T6G 1H9, Canada
*   Correspondence: afsaneh@ualberta.ca; Tel.: +1-780-492-2742; Fax: +1-780-492-1217

Received: 13 September 2018; Accepted: 19 October 2018; Published: 21 October 2018

Abstract: Hypoxia can induce chemoresistance, which is a significant clinical obstacle in cancer therapy. Here, we assessed development of hypoxia-induced chemoresistance (HICR) against free versus polymeric cisplatin micelles in a triple negative breast cancer cell line, MDA-MB-231. We then explored two strategies for the modulation of HICR against cisplatin micelles: a) the development of actively targeted micelles; and b) combination therapy with modulators of HICR in MDA-MB-231 cells. Actively targeted cisplatin micelles were prepared through surface modification of acetal-poly(ethylene oxide)-poly(α-carboxyl-ε-caprolactone) (acetal-PEO-PCCL) micelles with epidermal growth factor receptor (EGFR)-targeting peptide, GE11 (YHWYGYTPQNVI). Our results showed that hypoxia induced resistance against free and cisplatin micelles in MDA-MB-231 cells. A significant increase in micellar cisplatin uptake was observed in MDA-MB-231 cells that overexpress EGFR, following surface modification of micelles with GE11. This did not lead to increased cytotoxicity of micellar cisplatin, however. On the other hand, the addition of pharmacological inhibitors of key molecules involved in HICR in MDA-MB-231 cells, i.e., inhibitors of hypoxia inducing factor-1 (HIF-1) and signal transducer and activator of transcription 3 (STAT3), substantially enhanced the cytotoxicity of free and cisplatin micelles. The results indicated the potential benefit of combination therapy with HIF-1 and STAT3 inhibitors in overcoming HICR to free or micellar cisplatin.

Keywords: hypoxia-induced chemoresistance; cisplatin; polymeric micelle; EGFR-targeted therapy; STAT3; HIF-1; GE11 peptide; pharmacological Inhibitors of HIF-1 and STAT3; combination therapy

1. Introduction

Hypoxia is a common feature of solid tumors. Hypoxic areas in tumors are defined as regions with lower oxygen (O_2) levels than physiological oxygen concentrations [1]. Hypoxia arises when the need for oxygen exceeds its supply. Hypoxia-induced chemoresistance (HICR) has been observed in a number of human cancers, including triple negative breast cancer (TNBC) [2–5], the most deadly and therapy-resistant type of breast cancer [6]. The development of HICR in TNBC is a significant clinical obstacle against effective cancer therapy. This necessitates the development of new strategies that can prevent or overcome HICR in TNBC.

Cisplatin is a platinum drug used as part of a standard chemotherapy regimen in TNBC patients [7,8]. However, its use in cancer patients leads to the emergence of severe side effects including renal damage, deafness, and peripheral neuropathy. The use of nanocarriers has been extensively studied in recent decades as a means to attenuate the toxic side effects of anticancer agents. Several different nanocarriers of cisplatin have been reported in the literature, from which a few have found their way to clinical trials [9]. The developed nano-formulations have been mostly effective in lowering the side effects of cisplatin in preclinical models. However, they failed to potentiate the anticancer effects of drug [10,11].

Reports on the effect of nano-delivery of anticancer drugs on HICR are limited. This is of particular interest, as delivery of anticancer drugs by their nano-formulations may, in fact, restrict drug access and movement to hypoxic regions of the tumor where resistant cancer cells may be present. To circumvent the problem of nano-formulation penetration to hypoxic regions of solid tumors, incorporation of nanoparticles in non-malignant cells that display inherent hypoxia-targeting abilities, such as monocytes, macrophages or even neural stem cells [12,13], or modification of the surface of nanoparticles by tumor-penetrating peptides has been tried by different research groups [14]. These strategies sought to enhance the penetration of nano-drug delivery systems at the solid tumor at a tissue level.

This study aimed to study the effect of cisplatin delivery by a stealth nano-formulation on HICR at a cellular level and explore strategies to circumvent HICR. For this purpose, we used a previously reported polymeric micellar formulation for cisplatin developed by our group [15] and assessed the cytotoxicity of cisplatin as part of this formulation versus free drug in normoxic versus hypoxic MDA-MB-231 cells. We then explored two strategies for the modulation of hypoxia-induced cisplatin resistance in the same cell line; first, by active targeting of cisplatin micelles to MDA-MB-231 cells, to enhance intracellular cisplatin levels, and second, by combining cisplatin or its nano-formulations with modulators of HICR in MDA-MB-231 cells.

For the purpose of active drug targeting, we chose modification of polymeric micellar cisplatin with peptide ligands against epidermal growth factor receptor (EGFR), since coexistence of hypoxia and high levels of EGFR expression is a known feature of TNBC [16,17]. We postulated the high expression of EGFR on the hypoxic TNBC cells can be exploited to achieve enhanced delivery of cancer therapeutics to the cells [17,18].

Targeted nanocarriers can significantly improve drug performance by delivering a high payload of drug to cancer cells [19–22]. Previous studies have provided support for the use of EGFR monoclonal antibody for the development of ligand guided nanocarriers for the purpose of tumor imaging or targeted drug delivery [23–27]. The high molecular weight of the full-length antibody, however, may compromise the penetration of antibody modified nanocarriers into tumor tissue, particularly in hypoxic tumor regions. It may also enhance the chance of nanocarrier removal by the reticuloendothelial system following opsonization in the blood circulation [28]. Furthermore, the high cost of the full-length antibody and limitations on the use of organic solvents for its conjugation to the surface of nanocarriers prohibits the wide use of antibodies as ligands for tumor-targeted nanocarriers. In this context, use of an EGFR-specific peptide, GE11, may be a better option. GE11 is a 12-residue peptide (YHWYGYTPQNVI) which was originally developed using phage display technique [29]. GE11 peptide shows a lower affinity for EGFR than its natural ligand, EGF, however, it provides the advantage of lower mitogenic activity. Overall, due to its high EGFR affinity, minimum immunogenicity, and relatively cheap method of synthesis and scale-up, GE11 has been widely conjugated to a variety of nanocarriers, including liposomes, polymeric micelles, as well as gold and gelatin nanoparticles [28,29].

In this study, we developed GE11 modified polymeric micellar complexes of cisplatin and assessed the success of this approach in enhancing cellular delivery of cisplatin and overcoming HICR to cisplatin in a TNBC associated cell line, i.e., MDA-MB-231 cells. We then investigated the effect of adding pharmacological inhibitors of hypoxia inducing factor-1 (HIF-1) and signal transducer and activator of transcription 3 (STAT3), as the key modulators of HICR in this cell line [4,30], on anticancer activity of polymeric micellar formulations of cisplatin versus free drug.

2. Materials and Methods

2.1. Materials

Cisplatin (cis-diamminedichloroplatinum(II) (CDDP) (purity 99%), #H878, was purchased from, AK Scientific Inc., Union City, CA, USA. Methoxy-PEO 5000 (MePEO), sodium cyanoborohydride, and Stattic were obtained from Sigma, St Louis, MO, USA. Stannous octoate was dried and purified using anhydrous magnesium sulfate, dry toluene, and vacuum distillation [31]. α-Benzylcarboxylate-ε-caprolactone (BCL) was synthesized by Alberta Research Chemicals Inc. (ARCI, Edmonton, AB, Canada) based on methods published previously by our group [32]. All other chemicals and reagents used were of analytical grade.

2.1.1. Synthesis of Block Copolymers with Functionalized Poly(ethylene oxide) (PEO)

Acetal-poly(ethylene oxide) (acetal-PEO) was synthesized based on the previously published method by Nagasaki et al. [33]. Briefly, potassium naphthalene, the catalyst, was freshly prepared before the synthesis of acetal-PEO. To prepare the catalyst, 1.65 g (12.9 mmol) naphthalene and 0.575 g (14.7 mmol) potassium were added into 50 mL anhydrous tetrahydrofuran (THF). The reaction was protected under dry argon gas and kept running for 24 h, until a dark green color was obtained. To prepare acetal-PEO, 0.3 mL (2 mmol) 3,3-diethoxy propanol, the initiator, was first added into 40 mL anhydrous THF in a three-neck round bottom flask. The flask was purged with dry argon gas and maintained under an argon atmosphere. The catalyst solution (7 mL, ~2 mmol) was added dropwise into the reaction solution to activate the initiator. After 10 min of stirring, the flask was transferred into an ice water bath. Ethylene oxide (11.4 mL, 228 mmol) was added into the reaction solution. After 48 h, the reaction was quenched by acidified ethanol (2 mL). Acetal-PEO was recovered by precipitation in ethyl ether. The product was purified by dissolution in THF and precipitation in ethyl ether, and vacuum dried for further use.

Synthesis of acetal-PEO-poly-(ε-caprolactone) (acetal-PEO-PCL) and acetal-PEO-poly(α-benzyl carboxylate-ε-caprolactone) (acetal-PEO-PBCL) block copolymers has been described in our previous publications in detail [34]. Acetal-PEO-PBCL was first prepared through ring opening bulk polymerization of BCL with acetal-PEO as an initiator. This was followed by hydrogen reduction of PBCL block catalyzed by Pd/charcoal. Briefly, 600 mg (0.1 mmol) acetal-PEO was reacted with 500 mg (2 mmol) BCL under vacuum at 145 °C for 6 h using stannous octoate as catalyst. Then the benzyl groups on acetal-PEO-PBCL were removed through hydrogen reduction in anhydrous THF catalyzed by Pd/charcoal. The produced acetal-PEO-PCCL was recovered and purified by precipitation in hexane.

2.1.2. Synthesis of GE11 Peptide and GE11 Conjugation to Poly(ethylene oxide)-poly(α-carboxyl-ε-caprolactone) (PEO-PCCL) Block Copolymers

The GE11 peptide (NH_2-YHWYGYTPQNVI-COOH) (Figure S1a) was synthesized chemically using standard Fmoc solid phase peptide synthesis as described previously by the laboratory of Kaur et al. [35]. Briefly, the first amino acid, isoleucine, was coupled to a 2-chlorotrityl resin (0.1 mM) (NovaBiochem, San Diego, CA, USA) at 5-fold excess using the N,N diisopropyl ethylamine (DIPEA) at room temperature for 5 h. Further amino acids were added automatically using an automated peptide synthesizer (Tribute, Protein Technology, Inc., Tucson, AZ, USA). The completed peptide was ultimately released from the resin with a mixture of 90% trifluoroacetic acid (TFA), 9% dichloromethane, and 1% triisopropylsilane (~10 mL) for 90 min at room temperature. The cleaved peptide combined with TFA was then concentrated, washed with diethyl ether, dissolved in water and purified. Purification was done using C18 semi-preparative (1 cm × 25 cm, 5 μm) reverse-phase high-pressure liquid chromatography (HPLC) (Varian Prostar, MD, USA) with a gradient of acetonitrile−H$_2$O (10−70% containing 0.05% TFA, 2 mL/min, 45 min run time). The peptide solution was freeze-dried to give pure peptide as a white powder. Analytical (0.46 cm × 25 cm, 5 μm) HPLC revealed a purity of 97% at 220 nm with retention time (Rt) = 13 min, and the MALDITOF (Voyager spectrometer, Applied Biosystems, Foster City, CA, USA) mass analysis showed [M + H]$^+$ for the peptide as 1541.6 (calculated 1540.7) (Figure S1b).

The GE11 peptide was conjugated to the micellar surface through a reaction with the functional acetal groups on the micellar shell [34]. First, acetal-PEO$_{6000}$-PCCL$_{3000}$ (with 87% reduction of PBCL to PCCL) was assembled into polymeric micelles. Briefly, a diblock copolymer of acetal-PEO$_{6000}$-PCCL$_{3000}$ (20 mg) was dissolved in 1 mL acetone and added dropwise to 4 mL water while stirring. The solution was stirred for 24 h under a fume hood to remove acetone by evaporation. On the following day, the aqueous solution of polymeric micelles was acidified to pH 2 with 0.5 M HCl and stirred for 1 h at room temperature to produce aldehyde modified polymeric micelles. The resulting solution was then neutralized with 0.5 M NaOH. The osmolarity of the micellar solution was adjusted by addition of an appropriate volume of concentrated 10X P phosphate-buffered saline (PBS). An aqueous solution of the peptide (1.95 mg peptide in 500 μL of 1% dimethyl sulfoxide (DMSO)) (1:2 peptide to polymer, mole:mole ratio) was added and incubated with the aldehyde bearing micelles at room temperature for 2 h under moderate stirring. Subsequently, sodium cyanoborohydride (NaBH$_3$CN) (1 mg) was added to the polymer to reduce the Schiff's base. After 48 h of reaction, the unreacted peptide and reducing agent were removed by extensive dialysis using Spectrapor, MWCO 3500 (Spectrum Laboratories, Inc., Rancho Dominguez, CA, USA) against distilled water (24 h). The conjugation efficiency of the peptide to polymeric micelle was assessed by gradient reversed phase HPLC method measuring unreacted peptide concentration. A μ Bondpack (Waters Corp., Milford, MA, USA) C18 analytical column (10 μm 3.9 × 300 mm) was used. Gradient elution was performed at a flow rate of 1 mL/min using a Varian Prostar 210 HPLC System. Detection was performed at 214 nm using a Varian 335 detector (Varian Inc., Mulgrave, Australia). The mobile phase consisted of 0.1% TFA in H$_2$O (solution A) and acetonitrile (solution B). The mobile phase was programmed as follows: (1) 100% A for 1 min, (2) linear gradient from 100% A to 60% A in 20 min, (3) linear gradient from 60% A to 0% A in 4 min, (4) 0% A for 2 min, (5) linear gradient from 0% A to 100% A in 4 min, and (6) 100% A for 5 min. The concentration of unreacted peptide was calculated based on a calibration curve for the peak height of known concentrations of GE11 peptide in aqueous solution of 1% DMSO. The amount of conjugated peptide was calculated by subtracting the amount of unreacted peptide from the initial peptide added to the reaction. The peptide conjugated polymer was then freeze-dried until further use. ^{1}H NMR was performed on a Bruker, ASENDTM 600 MHz spectrometer (Billerica, MA, USA) to confirm the conjugation of the GE11 peptide on PEO-PCCL block polymer. Samples (GE11 peptide, acetal-PEO-PCCL, and GE11-PEO-PCCL) were dissolved in deuterated DMSO at a concentration range of 3–5 mg/mL and ^{1}H NMR spectra were generated.

2.1.3. Preparation of Plain and GE11 Cisplatin Micelles

GE11-PEO-PCCL or PEO-PCCL block copolymers were assembled into cisplatin polymeric micelles as reported previously with slight modification [15]. Briefly, either of the diblock copolymer (20 mg) was mixed with 4 mL aqueous solution of cisplatin (20 mg) and sodium bicarbonate (4–5 mg). The mixture was stirred for 24 h at room temperature. We couldn't dissolve the whole amount of cisplatin in such volume, but high cisplatin levels were identified to be required to enhance complexation with the polymer and increase its loading levels. After micelle preparation, initially using centrifugation, the undissolved portion of cisplatin was separated from the micellar solution.

In the next step, the unbound cisplatin was removed by ultrafiltration (3600× *g* for 40 min) using Centricon® plus centrifugal filter units (MWCO 3 KDa, Millipore, Billerica, MA, USA) and micelles were re-suspend in 4 mL doubly distilled water. The final concertation of cisplatin was determined using ion coupled plasma mass spectrometer (ICP-MS).

2.1.4. Measurement of the Size and Zeta Potential of Plain and GE11 Cisplatin Micelles

The average hydrodynamic diameter and size distribution of the GE11 cisplatin micelles were estimated and compared to plain cisplatin micelles by dynamic light scattering (DLS) using Malvern Zetasizer (Nano ZEN3600, Malvern, UK). The zeta potential of polymeric micelles was also estimated using the same equipment.

2.1.5. Measurement of the Critical Micellar Concentration (CMC) of Plain and GE11 Cisplatin Micelles

The CMC of the GE11 cisplatin micelles were estimated and compared to plain cisplatin micelles by DLS [36] using Malvern Zetasizer (Nano ZEN3600, Malvern, UK). For this purpose, plain and GE11 cisplatin micelles having polymer concentrations ranging from 1000 to 3 µg/mL were prepared. Briefly, from a stock solution of 1000 µg/mL micellar solution, different concentrations of micelles were prepared by serial dilution. The lowest prepared concentration was 3 µg/mL. The intensity of scattered light for each of concentrations was measured at a scattering angle of 173° at 25 °C. The average intensity of scattered light from three measurements was plotted against polymer concentration. The intersection of the two linear graphs in the sigmoidal curve, i.e., the onset of a rise in the intensity of scattered light, was defined as the CMC value.

2.1.6. Measurement of Cisplatin Encapsulation

The Pt(II) content in the GE11 cisplatin micelles was determined by ion coupled plasma mass spectrometer (ICP-MS, Agilent Technologies, Tokyo, Japan). The ICP operated at a radiofrequency power of 1550 W, and the flow rate of argon carrier gas was 0.9–1.0 L/min. Pt(II) was monitored at m/z 195. A standard curve in the Pt(II) concentration range of 100, 50, 20, 10, and 1 ppb was generated using atomic absorption standard. Appropriate dilutions of the test samples were prepared in 1% nitric acid (HNO_3). Data were acquired and processed by ICP-MS ChemStation (Agilent Technologies, Santa Clara, CA, USA). The encapsulation efficiency (EE) and drug loading (DL) were calculated using the following equations:

$$EE\ (\%) = \frac{the\ amount\ of\ encapsulated\ cisplatin\ (mg)}{the\ total\ feeding\ amount\ of\ cisplatin\ (mg)} \times 100$$

$$DL\ (\%) = \frac{the\ amount\ of\ encapsulated\ cisplatin\ (mg)}{the\ total\ amount\ of\ polymer\ (mg)} \times 100$$

2.1.7. In Vitro Release Studies

The release of free cisplatin and its micellar formulations (plain and GE11 cisplatin micelles) was measured using equilibrium dialysis method in PBS (pH = 7) and acetate buffer saline (pH = 5). Briefly, free cisplatin, plain or GE11 cisplatin micelles (4 mL) containing 30 µg/mL cisplatin were placed into a dialysis bag (Spectrapor, MWCO 3500) in a beaker containing 500 mL PBS or acetate buffer saline. The release study was performed at 37 °C in a Julabo SW 22 shaking water bath (Seelbach, Germany). At selected time intervals, 100 µL samples were withdrawn from the inside of dialysis bag and replaced with fresh medium for ICP-MS analysis. The percent cumulative amount of cisplatin released was calculated and plotted as a function of time. The release profiles of plain and GE11 cisplatin micelles were compared using the similarity factor, f_2, and the profiles were considered significantly different if $f_2 < 50$ [37]. The similarity factor, f_2, was calculated using the following equation [38].

$$f_2 = 50 \times \log\left(\left[1 + \left(\frac{1}{n}\right)\sum_{j=1}^{n}|Rj - Tj|^2\right]^{-0.5} \times 100\right) \tag{1}$$

where n is the sampling number, R_j and T_j are the percents released of the reference and test formulations at each time point j.

2.1.8. Cell Culture

MDA-MB-231 cells was obtained from ATCC (Manassas, VA, USA) and maintained in RPMI 1640 medium supplemented with 10% fetal bovine serum (Invitrogen, Karlsruhe, Germany), 100 units/mL penicillin, and 100 µg/mL streptomycin in a humidified incubator under 95% air and 5% CO_2 at 37 °C. For hypoxic conditions, cells were cultured in a CO_2 incubator maintained at 94% N_2, 5% CO_2, and 1% O_2.

2.1.9. Flow Cytometric Detection of Apoptosis using Annexin V-FITC and Propidium Iodide

Annexin V-FITC (Fluorescein IsoThioCyanate) and propidium iodide (PI) from BD Biosciences (FITC Annexin V Apoptosis Detection Kit I, #556547, BD Pharmingen™) was used to measure apoptotic cells by flow cytometry according to the manufacturer's instructions. Briefly, both floating and adherent cells were harvested; adherent cells were collected by adding a warm solution of 10 mM ethylenediaminetetraacetic acid (EDTA) in PBS. The cells were centrifuged at 500 g for 5 min, washed with ice-cold 1 PBS twice and re-suspended in 400 µL binding buffer containing 5 µL Annexin V-FITC and 5 µL PI for 15 min at room temperature in the dark. Fluorescence was induced on a Beckman Coulter Cytomics Quanta SC MPL flow cytometer (10,000 events per sample). Spectral compensation was performed using Cell Lab Quanta analysis software (Cell Lab Quanta™ SC MPL, Beckman Coulter, Mississauga, ON, Canada). The number of viable and apoptotic cells were quantified by events in the quadrants. The results were expressed as the percentage of apoptotic cells at the early stage (PI negative and Annexin V positive, lower right quadrant), apoptotic cells at the late stage (PI positive and Annexin V positive, upper right quadrant), necrotic cells (PI positive and Annexin V negative, upper left quadrant) and viable cells (PI negative and Annexin V negative, lower left quadrant).

2.1.10. MTT Assay

MDA-MB-231 cells (9×10^3 cells/well) were seeded in 96-well plates overnight, and on the following day, they were exposed to increasing concentration of cisplatin (free drug/cisplatin micelles) and then incubated for 48 h under hypoxia or normoxia. For combination therapy, MDA-MB-231 cells (9×10^3 cells/well) were seeded in 96-well plates overnight. On the following day, cells were treated with Tat-tagged form of cyclic peptide inhibitor of HIF-1 (*cyclo*-CLLFVY) named P1 (50 µM per well) [39] or Stattic (2 µM per well), or combination of P1 and Stattic for 4 h under normoxia, to give a final concentration of 2 µM and 50 µM per well, respectively. After 4 h incubation, cells were treated with cisplatin (50 µM) (as a free drug, plain cisplatin micelles or GE11 cisplatin micelles) and then incubated for additional 48 h under hypoxic or normoxic conditions. Cellular viability was assessed by the (3-(4,5-dimethylthiazol-2-yl)-2,5-diphenyltetrazolim bromide (MTT) assay. Briefly, MTT solution (5 mg/mL) was added to incubated cells for 4 h at 37 °C. Then the medium was replaced by DMSO to dissolve the crystals. Optical density was measured spectrophotometrically using a plate reader (Synergy H1 Hybrid Reader, Biotek, Winooski, VT, USA) at 570 nm. The cellular activity ratio was represented relative to control (untreated group, cells with media only).

2.1.11. Western Blot

To measure the expression level of different proteins, MDA-MB-231 cells (2×10^5 cells/well) were seeded in 6-well plates overnight. After treatment, cells were washed with cold 1X PBS and lysed using radioimmunoprecipitation assay buffer (RIPA lysis buffer) that was supplemented with 0.1 mM phenylmethylsulfonyl fluoride (PMSF) (Sigma-Aldrich), a protease Inhibitor Cocktail Set III, Animal-Free–Calbiochem (#535140, Millipore), and a phosphatase Inhibitor Cocktail Set II (#524625, Millipore). The lysate was then incubated on ice for 30 min, which was followed by centrifugation at $17,000 \times g$ for 20 min to remove genomic DNA. Protein quantification was made by the bicinchoninic acid (BCA) protein assay kit (Pierce, Rockford, IL, USA), and equal amounts of protein (35–40 µg) were loaded in 4–15% Tris-Glycine gradient gel (#456-1084, Biorad, Pleasanton, CA, USA). After gel electrophoresis, proteins were transferred to a nitrocellulose membrane. Membranes were probed with antibodies against phospho-STAT3 (Tyr705) (pSTAT3) (#9131, Cell Signaling Technologies, Danvers, MA, USA), Total-STAT3 (T-STAT3) (#8768s, Cell Signaling Technologies), EGFR (#2232, Cell Signaling Technologies), and glyceraldehyde 3-phosphate dehydrogenase (GAPDH) (#sc-47724, Santa Cruz Biotechnologies). Proteins were then detected using peroxidase-conjugated anti-mouse IgG (#7076, Cell Signaling Technologies) or anti-rabbit IgG (#7074, Cell Signaling Technologies) and visualized by enhanced chemiluminescence (Pierce ECL Western Blotting Substrate, #32106, Thermo Scientific, Rockford, IL, USA). Representative results of three independent Western blot analyses are shown in the Figures 4 and 5, Figure S4 and Figure S5.

2.1.12. Cisplatin Cellular Uptake

Cellular uptake of cisplatin was quantified by ICP-MS (Agilent Technologies, Tokyo, Japan). MDA-MB-231 cells (65×10^4 cells/flask) were seeded in 25 cm^2 flasks overnight. Cells were exposed to free cisplatin or its micellar formulations (plain and GE11 cisplatin micelles) (166 µM) for 24 h under normoxic and hypoxic conditions. On the following day, the medium was aspirated, cells were rinsed with cold PBS, detached using trypsin-EDTA, aliquoted in duplicate in 1.5 mL micro-centrifuge tubes and pelleted by centrifugation at $500 \times g$ for 5 min. One of each duplicate cell pellet was digested with 20% (v/v) HNO$_3$ overnight at 60 °C and analyzed for Pt(II) content by ICP-MS. The other duplicate was lysed using RIPA lysis buffer that was supplemented with 0.1 mM phenylmethylsulfonyl fluoride (PMSF) (Sigma-Aldrich, St. Louis, MO, USA), a protease Inhibitor Cocktail Set III, Animal-Free–Calbiochem (#535140, Millipore), and a phosphatase Inhibitor Cocktail Set II (#524625, Millipore) and quantified for protein content using the BCA protein assay kit (Pierce, Rockford, IL, USA). The cell uptake was expressed as cisplatin/cell protein (µg)/(µg).

2.1.13. Statistical Analysis

The statistical analysis was performed by Graphpad Prism (version 5.00, Graphpad Software Inc., La Jolla, CA, USA). Statistical analysis was performed either using unpaired Student's *t* test or one-way ANOVA (analysis of variance) with Tukey post-test analysis. Statistical significance is denoted by ($p < 0.05$). All graphs represent the average of at least three independent experiments with triplicates unless mentioned otherwise in the text, or graphs. Results were represented as the mean ± standard deviation (SD).

3. Results

3.1. Successful Synthesis of GE11 Conjugated Poly(ethylene oxide)-poly(α-carboxyl-ε-caprolactone)(PEO-PCCL) Block Copolymer and Its Self-assembly

GE11 showed a conjugation efficiency reaching 70% of an added peptide as quantified by reversed phase HPLC to the acetal-PEO-PCCL micellar surface (Figure S2). The molar conjugation percent for GE11 conjugated PEO-PCCL block polymer was ~35%. In other words, for 100 mole block copolymers, there is around 35 mole conjugated peptide.

The peptide conjugation was also confirmed by ^{1}H NMR (Figure S3). Signals from 6.5 to 8.5 ppm in ^{1}H NMR spectra of GE11 peptide correspond to aromatic protons in its structure ($n = 20$). To calculate the molar conjugation percent for GE11-conjugated PEO-PCCL block polymer, first, the summation of integration of peaks from 6.5 to 8.5 ppm in acetal-PEO-PCCL spectra was subtracted from the summation of integration of peaks from 6.5 to 8.5 ppm in GE11-PEO-PCCL spectra. The obtained value corresponded to the integration of aromatic hydrogens of GE11 peptide ($n = 20$). The degree of conjugation of GE11 was then determined by calculating the peak intensity ratio of methylene protons of PCCL segment ($OCH_2CH_2CH_2CH_2$ ((COOH)CO): $\delta = 4.1$ ppm) and the value calculated for the integration of aromatic hydrogens of GE11 peptide. The calculated degree of conjugation was 0.45 which corresponds to 45 mole peptide per 100 mole block copolymers (Figure S3). The calculated peptide conjugation degree by ^{1}H NMR (0.45 mol/mol) was slightly higher than what was calculated by HPLC (0.35 mol/mol).

As summarized in Table 1, both plain and GE11 cisplatin micelles showed similar average diameters around 80 nm with a low polydispersity index. Critical micellar concentration (CMC), cisplatin encapsulation efficiency and drug loading of both micelles were also comparable. Overall, both micellar formulations showed similar characteristics, and it appeared as though the presence of the GE11 peptide did not alter different micellar properties ($p > 0.05$, Student's t test).

Table 1. Characteristics of Cisplatin and GE11 Cisplatin Micelles ($n = 3$).

Micelle [a]	Average Diameter ± SD (nm) [b]	PDI ± SD [c]	Zeta potential ± SD (mV)	CMC ± SD (μg/mL) [d]	EE ± SD (%) [e]	DL ± SD (%) [f]	Drug/polymer ± SD (mol/mol)
Cisplatin plain micelle	84.4 ± 2.6	0.263 ± 0.11	−13.3 ± 1.2	65.1 ± 5.5	12.4 ± 0.99	12.0 ± 1.41	3.93 ± 0.31
GE11 cisplatin micelle	84.1 ± 3.2	0.235 ± 0.18	−13.6 ± 0.95	70.5 ± 7.2	13.0 ± 2.95	15.5 ± 3.53	4.01 ± 0.93

[a] Plain and GE11 cisplatin micelles consist of PEO_{6000}-$PCCL_{3000}$ block copolymers. The number shown in the subscript indicates average number molecular weight of each block determined by ^{1}H NMR spectroscopy. [b] Z average measured by dynamic light scattering (DLS). [c] Average polydispersity index (PDI) of micellar size distribution measured by DLS. [d] Measured from the onset of rise in the intensity values of scattered light as a function of concentration of micelles by DLS. [e] Encapsulation efficiency (%) = $\frac{\text{the amount of encapsulated cisplatin (mg)}}{\text{the total feeding amount of cisplatin (mg)}} \times 100$.
[f] Drug loading (%) = $\frac{\text{the amount of encapsulated cisplatin (mg)}}{\text{the total amount of polymer (mg)}} \times 100$.

The in-vitro release of cisplatin from both micelles was investigated in phosphate buffered saline (PBS) (pH = 7.4), and acetate buffered saline (pH = 5.0) using a dialysis method. Both micellar formulations showed burst release at the early time points (<1 h). The cumulative drug release appeared to be significantly slower at the later time points (>2 h) as compared to free drug for both micelles, however (Student's t test, $p < 0.05$). As shown in Figure 1a, micellar formulations showed ~60% release of cisplatin within 30 min in PBS as compared to ~80% release of free drug in the same media. After 48 h, in PBS, ~85% of the incorporated drug was released from micelles to media, compared to complete 100% release for free drug. Similar results were obtained for micellar formulations of cisplatin in acetate buffered saline (pH = 5.0) (Figure 1b) where a significant reduced drug release was achieved at later time points (>2 h) (Student's t test, $p < 0.05$) for micellar cisplatin compared to free drug. No difference was observed between the release profiles of plain versus GE11 cisplatin micelles in either media ($f_2 > 50$). Of note, in acetate buffered saline (pH = 5) complete release of drug after 48 h was seen, whereas only 80% drug was released in PBS at the same time point (pH = 7.4) (Figure 1b), although the overall profile of drug release did not show a significant difference between the two pHs ($f2 > 50$).

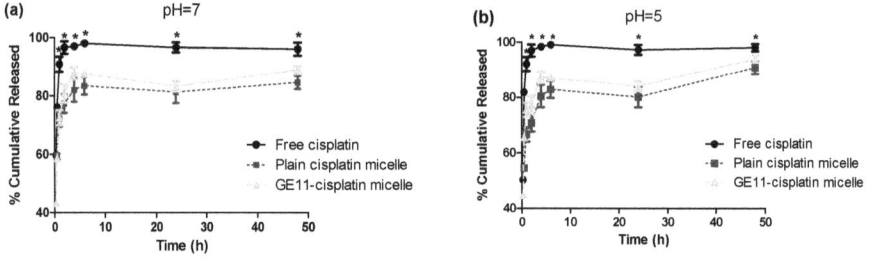

Figure 1. Percent cumulative release profile of cisplatin from plain and GE11 cisplatin micelles at different pHs in (a) PBS (pH = 7.4) and (b) acetate buffer saline (pH = 5.0). (*) denotes where the cumulative release of free drug appeared to be significantly different from plain and GE11 cisplatin micelles at the related time points ($p < 0.05$, Student's t test). Data are represented as mean $\pm$ SD ($n = 3$).

3.2. Hypoxia Induces Chemoresistance to Free Cisplatin in MDA-MB-231 Cells

The MDA-MB-231 cells incubated under hypoxic conditions were shown to be less sensitive to cytotoxic effects of cisplatin as measured by MTT assay. Cells cultured under hypoxia had a significantly higher number of colonies surviving cisplatin treatment than cells grown under normoxic conditions, as well (Figure 2a). Moreover, apoptosis induced by cisplatin was significantly reduced under hypoxia, as evidenced by a significant decrease in the proportion of late apoptotic cells measured by FITC Annexin V/propidium iodide (PI) assay (Figure 2b). The viable proportion of cells also showed a significant increase under hypoxic conditions compared to normoxic ones, when treated with cisplatin (Figure 2b). Interestingly, the proportion of necrotic cells following treatment with cisplatin under hypoxic conditions increased compared to normoxia, pointing to a change in the predominant mode of cisplatin-induced cell death under hypoxic conditions. This phenomenon was also previously reported for prostate carcinoma [40]. Similar to observation on free cisplatin (Figure 3a), we found less sensitivity towards polymeric micellar cisplatin in MDA-MB-231 cells cultured under hypoxic conditions by MTT assay (Figure 3b,c). There was also no significant difference between cell responses to micellar versus free cisplatin irrespective of the oxygen content for cell culture (one-way ANOVA with Tukey post-test, $p > 0.05$).

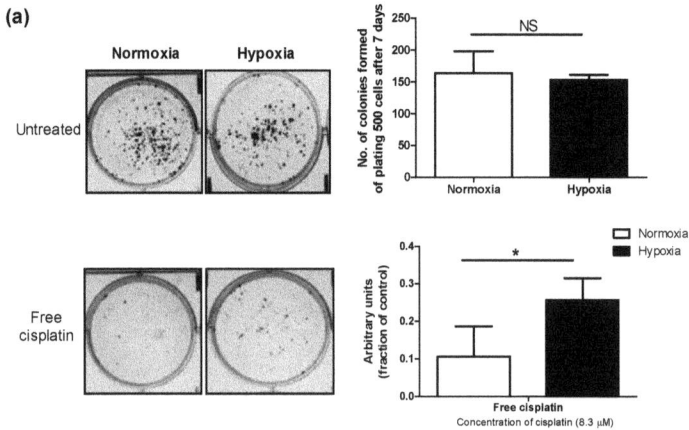

Figure 2. *Cont.*

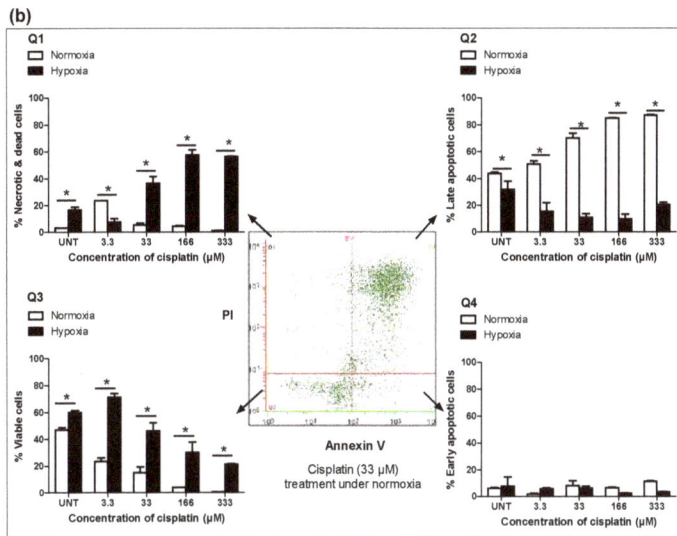

Figure 2. Hypoxia confers chemoresistance to free cisplatin in MDA-MB-231 cells. (**a**) Colony formation ability was assessed for cells treated with cisplatin (8.3 µM) after 24 h incubation under normoxic or hypoxic conditions. Cells were then re-plated at a density of 500 cells/well in duplicate in six-well plates under normoxia. The number of colonies formed from 500 cells after 7 days was graphed. (**b**) Cisplatin induced-apoptosis under normoxic and hypoxic conditions was measured by flow cytometric analysis of Fluorescein IsoThioCyanate (FITC) Annexin V staining in a buffer containing propidium iodide. MDA-MB 231 cells were left untreated or treated with increasing concentrations of cisplatin for 48 h. Flow cytometry analysis showed different populations of (Q1) necrotic or already dead cells (PI positive), (Q2) cells in end-stage apoptosis (FITC Annexin V and PI positive), (Q3) viable cells (FITC Annexin V and PI negative), and (Q4) cells in early stage of apoptosis (FITC Annexin V positive and PI negative). 2D plot is representative of cells treated with cisplatin (33 µM) under normoxia for 48 h. Data are represented as mean $\pm$ SD (n = 3). (*) denotes a significant difference between hypoxic and normoxic groups at each individual concentration (Student's t test, $p < 0.05$).

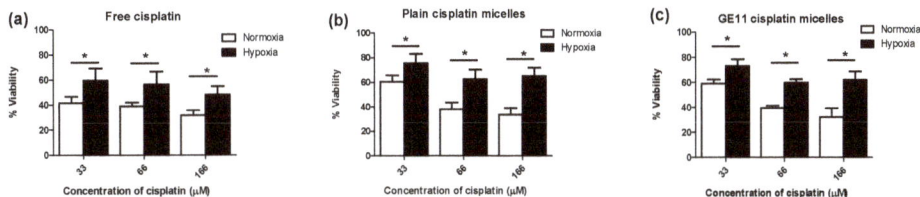

Figure 3. Hypoxia confers chemoresistance to free and micellar formulations of cisplatin in MDA-MB-231 cells. Viability of MDA-MB-231 cells was measured by MTT assay for cells treated with increasing concentrations of (**a**) free cisplatin; (**b**) plain cisplatin micelles and (**c**) GE11 cisplatin micelles under hypoxic or normoxic conditions for 48 h. (*) denotes a significant difference between groups at each individual concentration (Student's t test, $p < 0.05$).

3.3. Modification of Cisplatin Micelles with GE11 Peptide Enhances the Cellular Uptake of Cisplatin, but Does Not Affect its Cytotoxicity in MDA-MB-231 Cells

Modification of polymeric micelles with EGFR targeting peptide, GE11, was not able to increase the cytotoxicity of incorporated cisplatin towards MDA-MB-231 cells under normoxic or hypoxic conditions (Figure 3c versus 3b).

This was despite high levels of EGFR expression by MDA-MB-231 cells under normoxic or hypoxic conditions (Figure 4a) [17], that led to significantly higher cisplatin uptake following modification of polymeric micelles with GE11 peptide compared to plain micelles under both conditions (Figure 4b).

As shown in Figure 4a, the levels of EGFR expression under hypoxia were time-dependent. At the 24-h time point, there was no significant difference between the expression of EGFR under hypoxia and normoxia. However, the expression of EGFR was reduced under hypoxia for longer incubation times. Thus, for the purpose of comparability, we chose the 24-h time point for performing the cell uptake studies. When we treated the cells with the low concentration of cisplatin, due to multiple steps of digestion and washing for a sample preparation, we could not quantify the (low) amount of intracellular cisplatin. To compensate for the limitation of the measurement method, we chose to treat the cells with a high concentration of cisplatin (166 µM). As shown in Figure 4b, intracellular levels of cisplatin were found to be significantly reduced under hypoxia for the free drug ($\sim$1.6-fold decrease) as well as plain cisplatin micelles ($\sim$1.4-fold decrease). Nonetheless, the presence of the GE11 peptide on micelles appeared to compensate for the hypoxia-mediated reduction in cellular cisplatin levels. GE11 modified micelles showed similar intracellular drug levels under both hypoxic and normoxic conditions.

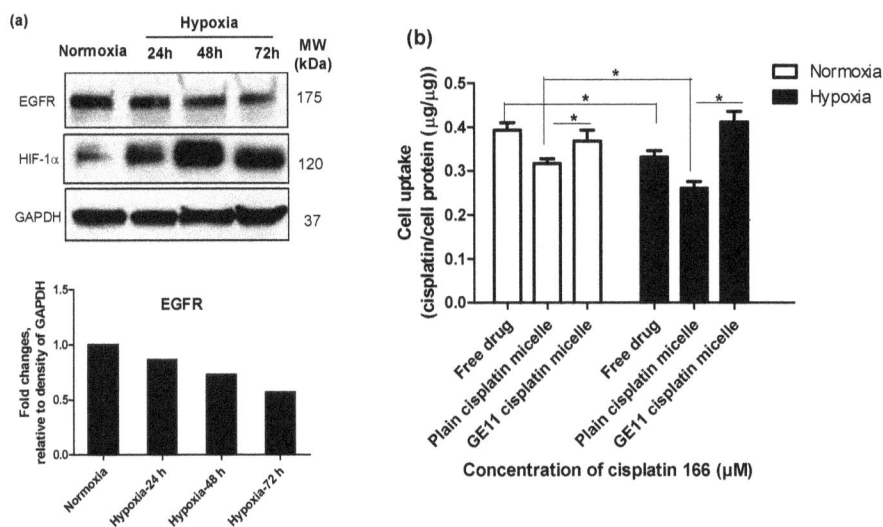

Figure 4. Modification of cisplatin micelles with GE11 peptide enhances the cellular uptake of cisplatin in MDA-MB-231 cells. (**a**) High levels of epidermal growth factor receptor (EGFR) expression under normoxia and hypoxia in MDA-MB-231 cells; (**b**) The GE11-peptide decoration of cisplatin micelles enhanced cellular uptake of cisplatin under hypoxia in MDA-MB-231 cells and bridged the gap of its cellular uptake under hypoxia and normoxia. Cisplatin content was measured by ion coupled plasma mass spectrometer (ICP-MS) after 24 h treatment of cells with cisplatin (166 µM) under hypoxia or normoxia. (*) denotes a significant difference between compared groups (Student's t test, $p < 0.05$).

3.4. Co-Treatment with Pharmacological Inhibitors of HIF-1 and STAT3 Potentiates the Anticancer Activity of Free Cisplatin, as well as Its Micellar Formulations in Hypoxic MDA-MB-231 Cells

We have previously reported on the role of STAT3 up-regulation under hypoxia as a mediator of hypoxia-induced resistance to cisplatin in MDA-MB-231 cells [30,41]. Here, we assessed the chemo-sensitizing effect of pharmacological inhibitors of STAT3 and/or HIF-1 (a known mediator of hypoxia- induced chemoresistance) in MDA-MB-231 cells treated with different formulations of cisplatin. For this purpose, a cyclic peptide that inhibits the assembly and function of the HIF-1 transcription factor (*cyclo*-CLLFVY) [39], as well as a known inhibitor of STAT3, i.e., Stattic, [42] were used in combination with free, plain and GE11 cisplatin micelles and the cytotoxicity of cisplatin against MDA-MB-231 cells was measured using MTT assay under normoxic and hypoxic conditions. The tat-tagged form of *cyclo*-CLLFVY (named P1), a cyclic peptide which has shown to prevent the dimerization of HIF-1α/HIF-1β complex by binding to HIF-1α and subsequently inhibit HIF-1 mediated hypoxia response [39,43], was used as a HIF-1 inhibitor. Stattic, a small molecule shown to inhibit the dimerization and activation of STAT3 mainly through prevention of its phosphorylation [42], was used as a STAT3 inhibitor. As shown in Figure 5a and Figure S4, successful inhibition of phosphorylation of STAT3 was achieved in MDA-MB-231 cells using Stattic (2 μM). Both P1 and Stattic showed minimal non-specific cytotoxicity at their respective effective dose for the inhibition of HIF-1 and STAT3 (50 and 2 μM, respectively) as shown by MTT assay (Figure 5b).

Under normoxic conditions, co-treatment of MDA-MB-231 cells with P1 and cisplatin formulations showed a trend towards potentiating the anticancer effect of cisplatin, plain cisplatin micelles or GE11 cisplatin micelles, as measured by MTT assay, although the difference was not statistically significant (Figure 5c–e, white bars). However, under hypoxic conditions, combination of P1 with free cisplatin (Figure 5c, black bars), plain cisplatin micelles (Figure 5d, black bars) and GE11 cisplatin micelles (Figure 5e, black bars) significantly enhanced the cytotoxicity of cisplatin as part of each formulation (one-way ANOVA with Tukey post-test, $p < 0.05$). This was not observed when different formulations of cisplatin were combined with the inhibitor of STAT3, Stattic, under normoxic conditions (Figure 5c–e, white bars). Under hypoxia, a combination of Stattic with free cisplatin (Figure 5c, black bars), plain cisplatin micelles (Figure 5d, black bars) and GE11 cisplatin micelles (Figure 5e, black bars) showed a trend towards increasing the cytotoxicity of cisplatin, although the results were not statistically significant.

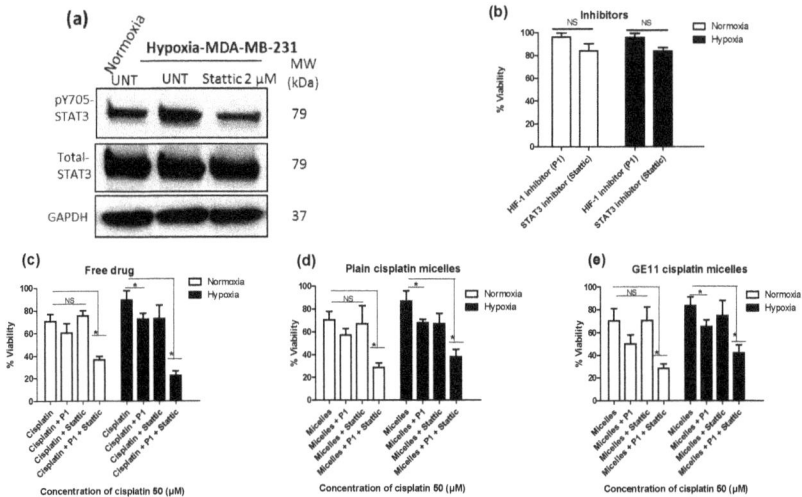

Figure 5. Dual pharmacological inhibition of signal transducer and activator of transcription 3 (STAT3) and hypoxia inducing factor-1 (HIF-1) in combination with free cisplatin or its micellar formulations successfully reversed hypoxia-induced chemoresistance. (**a**) Lower expression of pSTAT3 in MDA-MB-231 cells after treatment with STAT3 inhibitor (Stattic). Phosphorylation of STAT3 Tyr705 was analyzed by Western blot. (**b–d**) Viability of MDA-MB-231 cells was measured by MTT assay for cells which first pre-incubated with the HIF-1 inhibitor (named P1) (50 μM), the STAT3 inhibitor (Stattic) (2 μM) or both under normoxia for 4 h and then incubated under hypoxia for additional 48 h (b) in the absence of cisplatin or in the presence of 50 μM (c) free drug; (d) plain cisplatin micelles and (**e**) GE11 cisplatin micelles. (*) denotes a significant difference between compared groups (one-way ANOVA with Tukey post-test, $p < 0.05$).

Simultaneous co-treatment of cells with inhibitors of HIF-1 and STAT3 enhanced the cytotoxic effects of free cisplatin (Figure 5c), plain cisplatin micelles (Figure 5d) and GE11 cisplatin micelles (Figure 5e) in both normoxic and hypoxic MDA-MB-231 cells (one-way ANOVA with Tukey post-test, $p < 0.05$). Simultaneous co-treatment of cells with inhibitors of HIF-1 and STAT3 was the most effective approach in the reversal of HICR for free cisplatin as well its nano-formulations in this study. Co-treatment of cells with inhibitors of HIF-1 and STAT3 increased cisplatin toxicity under both conditions for all different drug formulations; the enhancement was more noticeable under hypoxia. For example, there was a $\sim$3.51-fold increase of cytotoxicity under hypoxia versus $\sim$1.93 under normoxia for free drug, $\sim$2.42-fold increase of cytotoxicity under hypoxia versus $\sim$2.27 under normoxia for plain cisplatin micelles, and $\sim$2.46-fold increase of cytotoxicity under hypoxia versus $\sim$1.97 under normoxia for GE11 cisplatin micelles. All the comparisons were made to cells treated with the related formulation in the absence of the inhibitors. Overall, free drug showed the highest cytotoxicity enhancement ($\sim$3.51-fold increase) after co-treatment of cells with the inhibitor of HIF-1 and STAT3, following by GE11 cisplatin micelles ($\sim$2.46-fold increase) and plain cisplatin micelles ($\sim$2.42-fold increase) under hypoxia; whereas the cytotoxicity enhancement was similar for all the formulations under normoxia ($\sim$2.05-fold increase).

In all of the above experiments, irrespective of the oxygen pressure under which the cells were cultured, no significant difference between the cytotoxicity of free drug and micellar formulations was observed. In addition, GE11 cisplatin micelles showed a similar profile of toxicity as compared to plain cisplatin micelles in MDA-MB-231 cells.

4. Discussion

Hypoxia is widely known to be associated with chemoresistance in different types of solid tumors [2,3] including TNBC [4,5]. The increase in the expression of HIF-1α has long been associated with the development of chemoresistance [4]. More recently, an increase in the activation of STAT3 following hypoxia is shown to be at least partly responsible for mediating chemoresistance in the human ovarian cancer (i.e., A270 cells) and TNBC (i.e., MDA-MB-231 cells) [3,30]. Overexpression of epidermal growth factor receptor (EGFR) and tumor hypoxia have also been shown to correlate with worse outcomes in several types of cancers including breast cancer [16,44]. This coexistence of hypoxia and EGFR may implicate a survival advantage of hypoxic cells that also express EGFR.

Design and development of nanotechnology products as a means to enhance the therapeutic index of anticancer drugs has been explored intensely in the past few decades. In successful cases, nano-formulations of anticancer drugs have shown to extend their blood circulation time leading to improved accumulation of the drug in solid tumors mostly by the enhanced permeability and retention (EPR) effect and/or decrease drug exposure and toxicity to normal tissues [9,45]. However, owing to their nanoscopic size and slow drug release, nano-formulations of the anticancer drugs are also speculated to provide limited access of the drug to hypoxic cancer cells that are located in areas of tumor away from blood vessels [46].

The objective of this study was first to evaluate the cytotoxic behavior of a newly developed nano-formulations of cisplatin on hypoxia-induced chemoresistant (HICR) MDA-MB-231 cells, a TNBC cell line; and secondly to explore feasible approaches for overcoming HICR against free and nano-formulations of cisplatin. For the latter purpose, we developed EGFR-targeted cisplatin micelles using an EGFR-ligand, GE11 peptide and investigated the effect of combination therapy with inhibitors of HIF-1 and STAT3 as the key mediators of HICR in this cell line.

Similar to plain cisplatin micelles, the GE11 cisplatin micelles showed slightly accelerated drug release in an acidic environment (Figure 1b). The ability of acid-triggered release of these micelles is important; tumors, particularly those with hypoxic regions, usually have lower extracellular pH than that of normal tissues [47]. In this case, the acid-triggered release of cisplatin from the micellar formulations may encourage drug release from the nano-formulation in the hypoxic tumor. This may compensate for the restrictions faced by these nano-formulations for penetrating the tumor core. It is of note that this feature, i.e., the acid-triggered release of cisplatin from its micellar formulations, was not investigated in our cell cytotoxicity studies as the pH of the cell culture media was maintained at 7.4.

In concert with our previous study [30], we found that hypoxia significantly induced resistance against free cisplatin in MDA-MB-231 cells (Figures 2 and 3a). MDA-MB-231 cells also showed similar resistance against nano-formulations of cisplatin under hypoxia (Figure 3b,c). GE11 cisplatin micelles showed increased intra-cellular levels of cisplatin compared to free and plain cisplatin micelles under hypoxia, reaching intracellular cisplatin levels similar to that of the free drug under normoxic conditions (Figure 4b). EGF receptors (EGFR) are internalizing receptors [48] with GE11 peptide as their ligand [29]. Thus, surface modification of cisplatin micelles with the GE11 peptide in this study was made to compensate for lower uptake of cisplatin under hypoxic conditions in MDA-MB-231 cells expressing high levels of EGFR. The upregulation of ATP-binding cassette (ABC) drug transporters, particularly ABCC2 and ABCC6, has been shown as one of the possible mechanisms responsible for active efflux of free cisplatin under hypoxia in MDA-MB-231 cells, which subsequently contributes to a decrease in the cellular levels of free cisplatin in this cell line [30]. Thus, the GE11 modified micelles have a potential to bypass ABC-transporter mediated drug efflux.

In spite of enhanced intracellular cisplatin levels by the GE11 modified micelles, cytotoxicity of cisplatin as part of GE11 micelles was not improved over plain cisplatin micelles, and both formulations showed similar cytotoxicity in MDA-MB-231 cells under normoxic and hypoxic conditions. In addition, irrespective of the oxygen pressure under which the cells were cultured, no significant difference between the cytotoxicity of free drug and micellar formulations was observed. We are speculating two reasons for this observation: (a) release of cisplatin from micelles at > 24 h incubation of micelles with the cells have contributed to the similar profile of toxicity between free drug and micellar formulations; (b) the increased level of cell uptake for cisplatin by GE11 cisplatin micelles is not enough to pass the threshold required for bypassing the mechanisms of cisplatin resistance. More thorough studies are required to elucidate the reason, which will be the subject of our future investigations.

The master regulator of cellular adaptation under hypoxia is believed to be the hypoxia-inducible factor (HIF) protein [47,49]. HIF is a heterodimeric transcription factor comprised of an oxygen-regulated unit, HIF-1α, as well as a constitutively expressed beta unit, HIF-1β. In the presence of oxygen, once HIF-1α is produced; it will be hydroxylated, ubiquitinated, and degraded. In contrast, in the absence of oxygen, HIF-1α is stabilized, dimerizes with HIF-1β, and the HIF heterodimers translocate to the nucleus where it activates the transcription of various downstream targets, many of which are known to be involved in cancer aggressiveness and chemoresistance [47,49,50]. The HIF-1 inhibitor used in our study (P1) specifically inhibits HIF-1α dimerization with HIF-1β and subsequently inhibits HIF-1 transcription factor activity [39]. Our results showed that the combination of P1 with cisplatin significantly enhanced cisplatin toxicity under hypoxia; however, it was not effective under normoxic conditions. Co-treatment of cells with the HIF-1 inhibitor resulted in similarly enhanced cytotoxicity for all cisplatin formulations under hypoxia as compared to cells treated with the related formulation in the absence of inhibitor. The effectiveness of combination therapy under hypoxia is likely due to the higher expression of HIF-1α under hypoxia as compared to normoxic conditions.

When combining a STAT3 inhibitor (Stattic) with cisplatin treatment, we did not observe any significant changes in the profile of cytotoxicity of cisplatin in MDA-MB-231 cells. This was unexpected given the results of previous studies supporting a role for activation of STAT3 in conferring HICR to cisplatin [3,30], and the effectiveness of STAT3 siRNA in sensitization of MDA-MB-231 cells to cisplatin. The exact reason behind the discrepancy is not clear and warrants further investigation. The efficiency of inhibition of STAT3 activation may be varied following siRNA transfection versus Stattic treatment at applied doses. The difference in the expression levels of pSTAT3 and its downstream targets involved in aggressiveness and chemoresistance of cancer cells (e.g., c-Myc [51]) following STAT3 siRNA versus Stattic treatment is speculated to have a role in this observation (Figure S5).

It should be noted that some cancer cells, including MDA-MB-231 cells, constitutively express high levels of HIF-1α [52] and pSTAT3 [53] under normal oxygen conditions, although their levels of expression are significantly lower as compared to hypoxia. The inhibition of HIF-1 or STAT3, alone, was not effective in enhancing the cytotoxic effect of cisplatin in our study under normoxic conditions. However, when HIF-1 and STAT3 both were inhibited, the cytotoxic effects of cisplatin increased under both normoxic and hypoxic conditions.

Previous studies have also provided support for the combined targeting of HIF-1 and STAT3 under hypoxia for enhancing anti-tumor activity. For instance, administration of a series of dual inhibitors of HIF-1α and STAT3 (in the absence of anticancer agent) resulted in significant anti-proliferative activity across a panel of various cancer cell lines [54]. Furthermore, it has been shown that the combination of other pharmacological inhibitors of HIF-1α and STAT3 enhanced prostate tumor growth suppression [55]. The results of our study, however, provide proof-of-principle for the use of HIF-1 and STAT3 inhibitors (individually or in combination) for sensitization of resistant cells under hypoxia to cisplatin and its micellar formulations in TNBC. Our present efforts for combination therapy are made by separate addition of inhibitors and cisplatin micelles in-vitro. Addition of inhibitors to the cell culture medium was done a few hours before the treatment with cisplatin or its micellar formulations. This approach was expected to provide enough time for the inhibitors to execute their inhibitory effect and sensitize the cells to the treatment. However, moving forward to in-vivo studies, delivery of both inhibitors and cisplatin within the same or separate micellar formulation can be explored. The use of nanodelivery systems for drugs is expected to provide a control over the time and extent of their delivery in solid tumors, maximizing their benefit

5. Conclusions

In summary, we have shown that the modification of cisplatin micelles with EGFR ligand (i.e., GE11 peptide) compensated for the hypoxia-mediated reduced cisplatin uptake, although this approach was not successful in increasing the levels of cytotoxicity of the drug. Importantly, our findings suggest that the potency of conventional (i.e., cisplatin) and nano-formulations (i.e., cisplatin micelles) can be enhanced under hypoxia once inhibitors of major cellular and molecular players of hypoxia-induced chemoresistance (i.e., HIF-1 and STAT3) were used in combination. To conclude, we have provided evidence to support that the rational therapeutic drug combination of sensitizing drugs with other therapies should be used to overcome drug resistance.

Supplementary Materials: The following are available online at http://www.mdpi.com/1999-4923/10/4/196/s1, **Figure S1.** (a) GE11 peptide structure and (b) MALDI-MS spectrum of GE11 peptide, **Figure S2.** Monitoring GE11 peptide conjugation to acetal-PEO-PCCL polymers using HPLC with UV detection at 214 nm, **Figure S3.** The conjugation of GE11 peptide onto acetal-PEO-PCCL was confirmed by 1H-NMR spectra of block copolymer before and after peptide conjugation. Samples (3–5 mg/mL) of acetal-PEO-PCCL and GE11-PEO-PCC were prepared in DMSO for 1H NMR analysis, **Figure S4.** Bar graph illustrating quantification of the Western blot densitometry analysis for Figure 5A. Densitometry data are expressed as fold changes compared to untreated normoxic group, normalized to GAPDH band intensity, **Figure S5.** The chosen experimental model for inhibition of STAT3 activation, STAT3 siRNA transfection versus pharmacological inhibition by Stattic, may result in different levels of expression of STAT3 and/or its downstream targets such as c-Myc protein. The differential levels of protein expression in MDA-MB-231 cells after transfection with STAT3 siRNA or treatment with Stattic (2μM) for 24 to 48 hours under hypoxia is depicted by Western blot.

Author Contributions: Conceptualization, H.S.A. and A.L.; Data curation, H.S.A. and A.H.S.; Formal analysis, H.S.A.; Funding acquisition, A.L.; Investigation, H.S.A. and A.H.S.; Methodology, H.S.A.; Project administration, H.S.A. and A.L.; Resources, M.R.V., R.S., K.K., F.C., A.T. and A.L.; Supervision, A.L.; Validation, H.S.A.; Writing—original draft, H.S.A.; Writing—review and editing, H.S.A., A.H.S., K.K., A.T. and A.L.

Funding: This research was funded by Canadian Institute of Health Research (CIHR, grant number 137153) and HSA was funded by Alberta Cancer Foundation (ACF) and Women and Children Health Research Institute (WCHRI).

Acknowledgments: The authors would like to thank X. Chris Le, and Xiufen Lu, Department of Laboratory Medicine and Pathology for their technical assistance with ICP-MS.

Conflicts of Interest: AL is the vice president and chief scientific officer of Meros Polymers Inc. The polymers used for the preparation of cisplatin formulation here are licensed to Meros Polymers Inc from the University of Alberta. The fund and the company had no role in the design of the study; in the collection, analyses, or interpretation of data; in the writing of the manuscript, and in the decision to publish the result.

References

1. Tredan, O.; Galmarini, C.M.; Patel, K.; Tannock, I.F. Drug resistance and the solid tumor microenvironment. *J. Natl. Cancer Inst.* **2007**, *99*, 1441–1454. [CrossRef] [PubMed]
2. Mamede, A.C.; Abrantes, A.M.; Pedrosa, L.; Casalta-Lopes, J.E.; Pires, A.S.; Teixo, R.J.; Goncalves, A.C.; Sarmento-Ribeiro, A.B.; Maia, C.J.; Botelho, M.F. Beyond the limits of oxygen: Effects of hypoxia in a hormone-independent prostate cancer cell line. *ISRN Oncol.* **2013**, *2013*, 918207. [CrossRef] [PubMed]
3. Selvendiran, K.; Bratasz, A.; Kuppusamy, M.L.; Tazi, M.F.; Rivera, B.K.; Kuppusamy, P. Hypoxia induces chemoresistance in ovarian cancer cells by activation of signal transducer and activator of transcription 3. *Int. J. Cancer* **2009**, *125*, 2198–2204. [CrossRef] [PubMed]
4. Sullivan, R.; Pare, G.C.; Frederiksen, L.J.; Semenza, G.L.; Graham, C.H. Hypoxia-induced resistance to anticancer drugs is associated with decreased senescence and requires hypoxia-inducible factor-1 activity. *Mol. Cancer Ther.* **2008**, *7*, 1961–1973. [CrossRef] [PubMed]
5. Notte, A.; Ninane, N.; Arnould, T.; Michiels, C. Hypoxia counteracts taxol-induced apoptosis in MDA-MB-231 breast cancer cells: Role of autophagy and JNK activation. *Cell Death Dis.* **2013**, *4*, e638. [CrossRef] [PubMed]
6. Foulkes, W.D.; Smith, I.E.; Reis-Filho, J.S. Triple-negative breast cancer. *N. Engl. J. Med.* **2010**, *363*, 1938–1948. [CrossRef] [PubMed]
7. O'Reilly, E.A.; Gubbins, L.; Sharma, S.; Tully, R.; Guang, M.H.; Weiner-Gorzel, K.; McCaffrey, J.; Harrison, M.; Furlong, F.; Kell, M.; et al. The fate of chemoresistance in triple negative breast cancer (TNBC). *BBA Clin.* **2015**, *3*, 257–275. [CrossRef] [PubMed]
8. Park, S.R.; Chen, A. Poly(Adenosine diphosphate-ribose) polymerase inhibitors in cancer treatment. *Hematol. Oncol. Clin. N. Am.* **2012**, *26*, 649–670. [CrossRef] [PubMed]
9. Hang, Z.; Cooper, M.A.; Ziora, Z.M. Platinum-based anticancer drugs encapsulated liposome and polymeric micelle formulation in clinical trials. *Biochem. Compd.* **2016**, *4*, 2. [CrossRef]
10. Matsumura, Y.; Kataoka, K. Preclinical and clinical studies of anticancer agent-incorporating polymer micelles. *Cancer Sci.* **2009**, *100*, 572–579. [CrossRef] [PubMed]
11. Zamboni, W.C.; Gervais, A.C.; Egorin, M.J.; Schellens, J.H.; Zuhowski, E.G.; Pluim, D.; Joseph, E.; Hamburger, D.R.; Working, P.K.; Colbern, G.; et al. Systemic and tumor disposition of platinum after administration of cisplatin or STEALTH liposomal-cisplatin formulations (SPI-077 and SPI-077 B103) in a preclinical tumor model of melanoma. *Cancer Chemother. Pharmacol.* **2004**, *53*, 329–336. [CrossRef] [PubMed]
12. Choi, M.-R.; Stanton-Maxey, K.J.; Stanley, J.K.; Levin, C.S.; Bardhan, R.; Akin, D.; Badve, S.; Sturgis, J.; Robinson, J.P.; Bashir, R.; et al. A cellular Trojan horse for delivery of therapeutic nanoparticles into tumors. *Nano Lett.* **2007**, *7*, 3759–3765. [CrossRef] [PubMed]
13. Mooney, R.; Weng, Y.; Garcia, E.; Bhojane, S.; Smith-Powell, L.; Kim, S.U.; Annala, A.J.; Aboody, K.S.; Berlin, J.M. Conjugation of pH-responsive nanoparticles to neural stem cells improves intratumoral therapy. *J. Controll. Release* **2014**, *191*, 82–89. [CrossRef] [PubMed]
14. Zhu, W.; Dong, Z.; Fu, T.; Liu, J.; Chen, Q.; Li, Y.; Zhu, R.; Xu, L.; Liu, Z. Modulation of hypoxia in solid tumor microenvironment with MnO_2 nanoparticles to enhance photodynamic therapy. *Adv. Funct. Mater.* **2016**, *26*, 5490–5498. [CrossRef]
15. Shahin, M.; Safaei-Nikouei, N.; Lavasanifar, A. Polymeric micelles for pH-responsive delivery of cisplatin. *J. Drug Target.* **2014**, *22*, 629–637. [CrossRef] [PubMed]
16. Franovic, A.; Gunaratnam, L.; Smith, K.; Robert, I.; Patten, D.; Lee, S. Translational up-regulation of the EGFR by tumor hypoxia provides a nonmutational explanation for its overexpression in human cancer. *Proc. Natl. Acad. Sci. USA* **2007**, *104*, 13092–13097. [CrossRef] [PubMed]
17. Milane, L.; Duan, Z.; Amiji, M. Development of EGFR-targeted polymer blend nanocarriers for combination paclitaxel/lonidamine delivery to treat multi-drug resistance in human breast and ovarian tumor cells. *Mol. Pharm.* **2011**, *8*, 185–203. [CrossRef] [PubMed]
18. Rojo, F.; Albanell, J.; Rovira, A.; Corominas, J.M.; Manzarbeitia, F. Targeted therapies in breast cancer. *Semin. Diagn. Pathol.* **2008**, *25*, 245–261. [CrossRef] [PubMed]

19. Nie, S.; Xing, Y.; Kim, G.J.; Simons, J.W. Nanotechnology applications in cancer. *Annu. Rev. Biomed. Eng.* **2007**, *9*, 257–288. [CrossRef] [PubMed]
20. Davis, M.E.; Chen, Z.G.; Shin, D.M. Nanoparticle therapeutics: An emerging treatment modality for cancer. *Nat. Rev. Drug Discov.* **2008**, *7*, 771–782. [CrossRef] [PubMed]
21. Wang, M.D.; Shin, D.M.; Simons, J.W.; Nie, S. Nanotechnology for targeted cancer therapy. *Expert Rev. Anticancer Ther.* **2007**, *7*, 833–837. [CrossRef] [PubMed]
22. Nishiyama, N.; Okazaki, S.; Cabral, H.; Miyamoto, M.; Kato, Y.; Sugiyama, Y.; Nishio, K.; Matsumura, Y.; Kataoka, K. Novel cisplatin-incorporated polymeric micelles can eradicate solid tumors in mice. *Cancer Res.* **2003**, *63*, 8977–8983. [PubMed]
23. Nida, D.L.; Rahman, M.S.; Carlson, K.D.; Richards-Kortum, R.; Follen, M. Fluorescent nanocrystals for use in early cervical cancer detection. *Gynecol. Oncol.* **2005**, *99*, S89–S94. [CrossRef] [PubMed]
24. Yang, J.; Eom, K.; Lim, E.K.; Park, J.; Kang, Y.; Yoon, D.S.; Na, S.; Koh, E.K.; Suh, J.S.; Huh, Y.M.; et al. In situ detection of live cancer cells by using bioprobes based on Au nanoparticles. *Langmuir* **2008**, *24*, 12112–12115. [CrossRef] [PubMed]
25. Melancon, M.P.; Lu, W.; Yang, Z.; Zhang, R.; Cheng, Z.; Elliot, A.M.; Stafford, J.; Olson, T.; Zhang, J.Z.; Li, C. In vitro and in vivo targeting of hollow gold nanoshells directed at epidermal growth factor receptor for photothermal ablation therapy. *Mol. Cancer Ther.* **2008**, *7*, 1730–1739. [CrossRef] [PubMed]
26. Patra, C.R.; Bhattacharya, R.; Wang, E.; Katarya, A.; Lau, J.S.; Dutta, S.; Muders, M.; Wang, S.; Buhrow, S.A.; Safgren, S.L.; et al. Targeted delivery of gemcitabine to pancreatic adenocarcinoma using cetuximab as a targeting agent. *Cancer Res.* **2008**, *68*, 1970–1978. [CrossRef] [PubMed]
27. Wu, G.; Barth, R.F.; Yang, W.; Kawabata, S.; Zhang, L.; Green-Church, K. Targeted delivery of methotrexate to epidermal growth factor receptor-positive brain tumors by means of cetuximab (IMC-C225) dendrimer bioconjugates. *Mol. Cancer Ther.* **2006**, *5*, 52–59. [CrossRef] [PubMed]
28. Master, A.M.; Sen Gupta, A. EGF receptor-targeted nanocarriers for enhanced cancer treatment. *Nanomedicine* **2012**, *7*, 1895–1906. [CrossRef] [PubMed]
29. Li, Z.; Zhao, R.; Wu, X.; Sun, Y.; Yao, M.; Li, J.; Xu, Y.; Gu, J. Identification and characterization of a novel peptide ligand of epidermal growth factor receptor for targeted delivery of therapeutics. *FASEB J.* **2005**, *19*, 1978–1985. [CrossRef] [PubMed]
30. Soleymani Abyaneh, H.; Gupta, N.; Radziwon-Balicka, A.; Jurasz, P.; Seubert, J.; Lai, R.; Lavasanifar, A. STAT3 but not HIF-1alpha is important in mediating hypoxia-induced chemoresistance in MDA-MB-231, a triple negative breast cancer cell line. *Cancers* **2017**, *9*, 137. [CrossRef] [PubMed]
31. Storey, R.F.; Sherman, J.W. Kinetics and mechanism of the stannous octoate-catalyzed bulk polymerization of ε-Caprolactone. *Macromolecules* **2002**, *35*, 1504–1512. [CrossRef]
32. Mahmud, A.; Xiong, X.-B.; Lavasanifar, A. Novel self-associating poly(ethylene oxide)-block-poly(ε-caprolactone) block copolymers with functional side groups on the polyester block for drug delivery. *Macromolecules* **2006**, *39*, 9419–9428. [CrossRef]
33. Nagasaki, Y.; Kutsuna, T.; Iijima, M.; Kato, M.; Kataoka, K.; Kitano, S.; Kadoma, Y. Formyl-ended heterobifunctional poly(ethylene oxide): Synthesis of poly(ethylene oxide) with a formyl group at one end and a hydroxyl group at the other end. *Bioconjugate Chem.* **1995**, *6*, 231–233. [CrossRef]
34. Xiong, X.B.; Mahmud, A.; Uludag, H.; Lavasanifar, A. Conjugation of arginine-glycine-aspartic acid peptides to poly(ethylene oxide)-b-poly(epsilon-caprolactone) micelles for enhanced intracellular drug delivery to metastatic tumor cells. *Biomacromolecules* **2007**, *8*, 874–884. [CrossRef] [PubMed]
35. Soudy, R.; Gill, A.; Sprules, T.; Lavasanifar, A.; Kaur, K. Proteolytically stable cancer targeting peptides with high affinity for breast cancer cells. *J. Med. Chem.* **2011**, *54*, 7523–7534. [CrossRef] [PubMed]
36. Topel, Ö.; Çakır, B.A.; Budama, L.; Hoda, N. Determination of critical micelle concentration of polybutadiene-block-poly(ethyleneoxide) diblock copolymer by fluorescence spectroscopy and dynamic light scattering. *J. Mol. Liq.* **2013**, *177*, 40–43. [CrossRef]
37. Shahin, M.; Lavasanifar, A. Novel self-associating poly(ethylene oxide)-b-poly(epsilon-caprolactone) based drug conjugates and nano-containers for paclitaxel delivery. *Int. J. Pharm.* **2010**, *389*, 213–222. [CrossRef] [PubMed]
38. Costa, P.; Sousa Lobo, J.M. Modeling and comparison of dissolution profiles. *Eur. J. Pharm. Sci.* **2001**, *13*, 123–133. [CrossRef]

39. Miranda, E.; Nordgren, I.K.; Male, A.L.; Lawrence, C.E.; Hoakwie, F.; Cuda, F.; Court, W.; Fox, K.R.; Townsend, P.A.; Packham, G.K.; et al. A cyclic peptide inhibitor of HIF-1 heterodimerization that inhibits hypoxia signaling in cancer cells. *J. Am. Chem. Soc.* **2013**, *135*, 10418–10425. [CrossRef] [PubMed]

40. Thews, O.; Gassner, B.; Kelleher, D.K.; Schwerdt, G.; Gekle, M. Impact of hypoxic and acidic extracellular conditions on cytotoxicity of chemotherapeutic drugs. *Adv. Exp. Med. Biol.* **2007**, *599*, 155–161. [PubMed]

41. Soleymani Abyaneh, H.; Gupta, N.; Alshareef, A.; Gopal, K.; Lavasanifar, A.; Lai, R. Hypoxia Induces the Acquisition of Cancer Stem-like Phenotype Via Upregulation and Activation of Signal Transducer and Activator of Transcription-3 (STAT3) in MDA-MB-231, a Triple Negative Breast Cancer Cell Line. *Cancer Microenviron.* **2018**. [CrossRef] [PubMed]

42. Schust, J.; Sperl, B.; Hollis, A.; Mayer, T.U.; Berg, T. Stattic: A small-molecule inhibitor of STAT3 activation and dimerization. *Chem. Biol.* **2006**, *13*, 1235–1242. [CrossRef] [PubMed]

43. Mistry, I.N.; Tavassoli, A. Reprogramming the transcriptional response to hypoxia with a chromosomally encoded cyclic peptide HIF-1 inhibitor. *ACS Synth. Biol.* **2017**, *6*, 518–527. [CrossRef] [PubMed]

44. Hoogsteen, I.J.; Marres, H.A.; van den Hoogen, F.J.; Rijken, P.F.; Lok, J.; Bussink, J.; Kaanders, J.H. Expression of EGFR under tumor hypoxia: Identification of a subpopulation of tumor cells responsible for aggressiveness and treatment resistance. *Int. J. Radiat. Oncol. Biol. Phys.* **2012**, *84*, 807–814. [CrossRef] [PubMed]

45. Anselmo, A.C.; Mitragotri, S. Nanoparticles in the clinic. *Bioeng. Transl. Med.* **2016**, *1*, 10–29. [CrossRef] [PubMed]

46. Aldea, M.; Florian, I.A.; Kacso, G.; Craciun, L.; Boca, S.; Soritau, O.; Florian, I.S. Nanoparticles for targeting intratumoral hypoxia: exploiting a potential weakness of glioblastoma. *Pharm. Res.* **2016**, *33*, 2059–2077. [CrossRef] [PubMed]

47. Brahimi-Horn, C.; Pouyssegur, J. The role of the hypoxia-inducible factor in tumor metabolism growth and invasion. *Bull. Cancer* **2006**, *93*, 10073–10080.

48. Ono, M.; Kuwano, M. Molecular mechanisms of epidermal growth factor receptor (EGFR) activation and response to gefitinib and other EGFR-targeting drugs. *Clin. Cancer Res.* **2006**, *12*, 7242–7251. [CrossRef] [PubMed]

49. Semenza, G.L. Defining the role of hypoxia-inducible factor 1 in cancer biology and therapeutics. *Oncogene* **2010**, *29*, 625–634. [CrossRef] [PubMed]

50. Rohwer, N.; Cramer, T. Hypoxia-mediated drug resistance: Novel insights on the functional interaction of HIFs and cell death pathways. *Drug Resist. Update* **2011**, *14*, 191–201. [CrossRef] [PubMed]

51. Klauber-DeMore, N.; Schulte, B.A.; Wang, G.Y. Targeting MYC for triple-negative breast cancer treatment. *Oncoscience* **2018**, *5*, 120–121. [CrossRef] [PubMed]

52. Robey, I.F.; Lien, A.D.; Welsh, S.J.; Baggett, B.K.; Gillies, R.J. Hypoxia-inducible factor-1alpha and the glycolytic phenotype in tumors. *Neoplasia* **2005**, *7*, 324–330. [CrossRef] [PubMed]

53. Berishaj, M.; Gao, S.P.; Ahmed, S.; Leslie, K.; Al-Ahmadie, H.; Gerald, W.L.; Bornmann, W.; Bromberg, J.F. Stat3 is tyrosine-phosphorylated through the interleukin-6/glycoprotein 130/Janus kinase pathway in breast cancer. *Breast Cancer Res.* **2007**, *9*, R32. [CrossRef] [PubMed]

54. Godse, P.; Kumar, P.; Yewalkar, N.; Deore, V.; Lohar, M.; Mundada, R.; Padgaonkar, A.; Manohar, S.; Joshi, A.; Bhatia, D.; et al. Discovery of P3971 an orally efficacious novel anticancer agent targeting HIF-1alpha and STAT3 pathways. *Anticancer Agents Med. Chem.* **2013**, *13*, 1460–1466. [CrossRef] [PubMed]

55. Reddy, K.R.; Guan, Y.; Qin, G.; Zhou, Z.; Jing, N. Combined treatment targeting HIF-1alpha and Stat3 is a potent strategy for prostate cancer therapy. *Prostate* **2011**, *71*, 1796–1809. [CrossRef] [PubMed]

pharmaceutics

MDPI

Article

DOX-Vit D, a Novel Doxorubicin Delivery Approach, Inhibits Human Osteosarcoma Cell Proliferation by Inducing Apoptosis While Inhibiting Akt and mTOR Signaling Pathways

Zaid H. Maayah [1,2], Ti Zhang [3], Marcus Laird Forrest [3], Samaa Alrushaid [4], Michael R. Doschak [1], Neal M. Davies [1] and Ayman O. S. El-Kadi [1,*]

[1] Faculty of Pharmacy and Pharmaceutical Sciences, University of Alberta, Edmonton, AB T6G 2E1, Canada; almaayah@ualberta.ca (Z.H.M.); mdoschak@ualberta.ca (M.R.D.); ndavies@ualberta.ca (N.M.D.)
[2] Cardiovascular Research Centre, Department of Pediatrics and Medicine, Mazankowski Alberta Heart Institute, Faculty of Medicine and Dentistry, University of Alberta, Edmonton, AB T6G 2E1, Canada
[3] Department of Pharmaceutical Chemistry, School of Pharmacy, University of Kansas, Lawrence, KS 66047, USA; tzh217@gmail.com (T.Z.); lforrest@ku.edu (M.L.F.)
[4] Department of Pharmaceutical Chemistry, Faculty of Pharmacy, Kuwait University, Safat 13110, Kuwait; samaa.alrushaid@hsc.edu.kw
* Correspondence: aelkadi@ualberta.ca; Tel.: +1-780-492-3071; Fax: +1-780-492-1217

Received: 3 August 2018; Accepted: 31 August 2018; Published: 4 September 2018

Abstract: Doxorubicin (DOX) is a very potent and effective anticancer agent. However, the effectiveness of DOX in osteosarcoma is usually limited by the acquired drug resistance. Recently, Vitamin D (Vit-D) was shown to suppress the growth of many human cancer cells. Taken together, we synthesized DOX-Vit D by conjugating Vit-D to DOX in order to increase the delivery of DOX into cancer cells and mitigate the chemoresistance associated with DOX. For this purpose, MG63 cells were treated with 10 μM DOX or DOX-Vit D for 24 h. Thereafter, MTT, real-time PCR and western blot analysis were used to determine cell proliferation, genes and proteins expression, respectively. Our results showed that DOX-Vit D, but not DOX, significantly elicited an apoptotic signal in MG63 cells as evidenced by induction of death receptor, Caspase-3 and BCLxs genes. Mechanistically, the DOX-Vit D-induced apoptogens were credited to the activation of p-JNK and p-p38 signaling pathway and the inhibition of proliferative proteins, p-Akt and p-mTOR. Our findings propose that DOX-Vit D suppressed the growth of MG63 cells by inducing apoptosis while inhibiting cell survival and proliferative signaling pathways. DOX-Vit D may serve as a novel drug delivery approach to potentiate the delivery of DOX into cancer cells.

Keywords: doxorubicin; MG63; Vitamin D; DOX-Vit D

1. Introduction

Osteosarcoma (OS) is one of the most widespread and lethal forms of childhood primary bone cancer [1]. In Canada, OS accounts for about 5% of all tumors in pediatric patients with an incidence rate of 8 cases per million each year especially in adolescents [2,3]. Despite the substantial progress in chemotherapies against OS, the mortality rate of OS patients has not been changed significantly due to chemoresistance and other factors [4].

One of those standard therapies for OS is doxorubicin (DOX), an effective anthracycline antibiotic [5]. The combination of DOX with other chemotherapeutic agents such as cisplatin, ifosfamide and methotrexate cured 60–76% of newly diagnosed non-metastatic OS [6]. Although DOX has improved survival rates in cancer patients, the effectiveness of DOX in OS is usually limited by the acquired drug resistance. This resistance is dose-dependent and may develop gradually within a month or years after the treatment initiation. Though the specific mechanism of chemoresistance associated with DOX is still unclear, several reports have demonstrated that drug inactivation, increased DNA damage repair, disturbances in intracellular drug transport and evasion of apoptosis could play a role in the chemoresistance [7].

Numerous epidemiological reports have suggested a strong association between Vitamin D (Vit-D) and cancer risk [8,9]. The deficiency of Vit-D has been reported to contribute to the development of tumors whereas, higher intake of Vit-D was accompanied by a lower incidence of cancer disease [10,11]. Experimental studies using cancer cells or tumors in mice have shown that Vit-D exhibited antitumor activities through the induction apoptosis in addition to the inhibition of cell proliferation and differentiation [12,13].

Vit-D can be classified naturally into animal-based Vit-D3 and plant-based Vitamin D2. Vit-D3, cholecalciferol, is synthesized by the mammalian skin after exposure to sunlight then metabolized into its active form, calcitriol, in the liver and kidneys [14]. Upon binding to its receptor, calcitriol activates several signaling pathways that regulate bone metabolism and calcium homeostasis. Of interest, calcitriol was shown to inhibit the growth, proliferation and differentiation of many cancer cell lines such as breast, prostate and colon cancers [15,16]. However, the anticancer activity of calcitriol was associated with significant hypercalcemia that limits its clinical utility [17]. In contrast to Vit-D3, ergocalciferol, Vit-D2, has been reported to exert a low calcemic effect and potent antitumor activity [14,18]. Ergocalciferol, Vit-D2, occurs naturally in plants and it is synthesized from proVit-D2, ergosterol, upon exposure to sunlight [14]. Of particular interest, it has been reported that Vit-D2 enhanced the cytotoxic effect of DOX on human breast and prostate cancer cell lines [19].

In light of the information described above, we hypothesized that by synthesis of DOX-Vit D, a novel DOX derivative, through conjugating Vit-D2 to DOX, the chemoresistance associated with DOX could be mitigated. For this purpose, the current study was designed to (1) investigate the antiproliferative and apoptogenic effects of DOX-Vit D in the human OS cell line, MG63 cells, and (2) explore the possible mechanism(s) involved. Our study provides substantial evidence that DOX-Vit D suppressed the growth of MG63 cells by inducing apoptosis while inhibiting cell survival and proliferative signaling pathways. Our DOX-Vit D conjugate may be of particular importance in drug delivery and may serve as a novel drug delivery approach to potentiate the delivery of DOX into the bone cancer cells.

2. Materials and Methods

2.1. Materials

Total Akt (t-Akt) rabbit polyclonal, phosphorylated-Akt (P-Akt) rabbit polyclonal, mammalian target of rapamycin C (t-mTOR) goat polyclonal and p-mTOR rabbit polyclonal were bought from Santa Cruz Biotechnology, Inc. (Santa Cruz, CA, USA). Any other materials used in the current study has been described previously [20].

2.2. Chemistry

2.2.1. Calciferol-Succinate

In a round bottom flask, 200 mg of calciferol (0.5 mmol) was dissolved in 20 mL of anhydrous dichloromethane (DCM), followed by the addition of 360 mg of succinic anhydride (3.6 mmol, 7.2 eq.) and 500 μL of triethylamine (TEA, 3.6 mmol, 7.2 eq.). The reaction mixture was stirred at ambient temperature under N_2 for 24 h in the dark. The solution was washed with water three times, and the organic layer was concentrated under reduced pressure. The compound was purified with a Combiflash RF system (hexane/ethyl acetate, 30/70) to obtain a yellowish solid with a yield of 93%. ^{1}H NMR (400 MHz, Acetone-d6) δ 6.30 (d, J = 11.2 Hz, 1H), 6.10 (dt, J = 11.3, 1.6 Hz, 1H), 5.35–5.18 (m, 2H), 5.12 (dt, J = 2.6, 1.2 Hz, 1H), 4.94 (tt, J = 7.8, 3.8 Hz, 1H), 4.84 (d, J = 2.6 Hz, 1H), 4.07 (q, J = 7.1 Hz, 1H), 2.90 (dd, J = 11.9, 4.0 Hz, 1H), 2.69–2.50 (m, 5H), 2.49–2.32 (m, 2H), 2.27–2.16 (m, 1H), 2.16–1.94 (m,

6H), 1.89 (qd, *J* = 6.9, 5.9 Hz, 1H), 1.83–1.65 (m, 4H), 1.66–1.27 (m, 7H), 1.22 (t, *J* = 7.1 Hz, 1H), 1.07 (d, *J* = 6.7 Hz, 3H), 0.96 (d, *J* = 6.9 Hz, 3H), 0.87 (t, *J* = 6.6 Hz, 6H) (Figures 1 and 2).

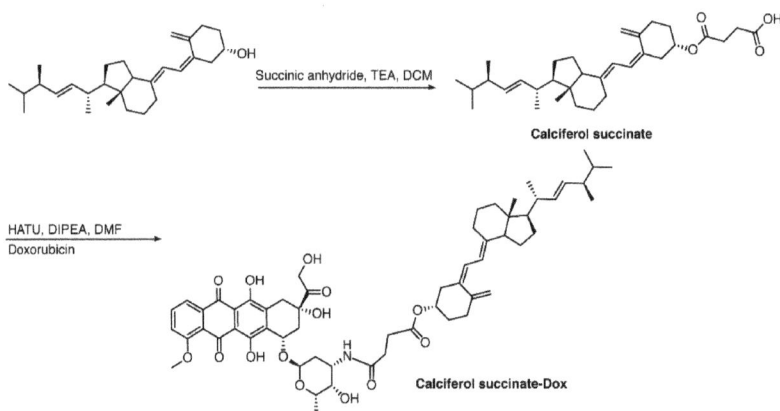

Figure 1. Chemical synthesis of DOX-Vit D.

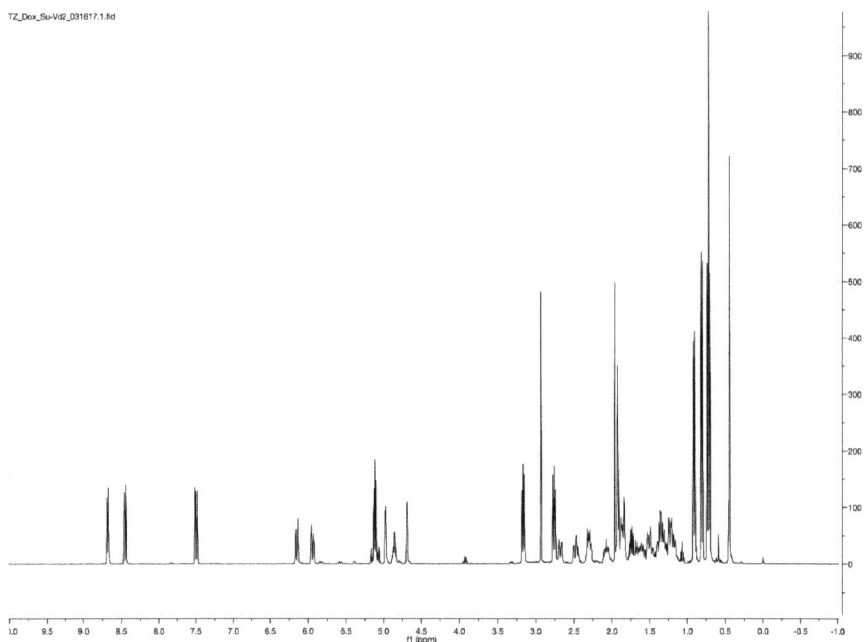

Figure 2. Schematic ¹HNMR diagram of DOX-Vit D.

2.2.2. Calciferol-Succinate-DOX

Calciferol-succinate (230 mg, 0.463 mmol), HATU, 1-[Bis(dimethylamino)methylene]-1H-1,2,3-triazolo[4,5-b]pyridinium 3-oxid hexafluorophosphate (HATU) (211 mg, 1.2 eq.), and Dox-HCl (295 mg. 1.1 eq.) were dissolved in 10 mL of anhydrous *N,N*-Dimethylformamide (DMF). To the he mixture was added 400 μL of *N,N*-Diisopropylethylamine (DIPEA). The solution was stirred at ambient temperature under N₂ for 24 h in the dark. The crude mixture was dried under reduced pressure, and purified with a Combiflash RF system (hexane/ethyl acetate, 50/50) to afford product as an orange solid with a yield of 60%. ^{1}H NMR (400 MHz, Acetone-d6) δ 8.68 (dd, J = 4.4, 1.4 Hz, 1H), 8.50–8.37 (m, 1H), 7.56–7.38 (m, 1H), 6.14 (d, J = 11.2 Hz, 1H), 5.94 (dd, J = 11.2, 1.8 Hz, 1H), 5.23–5.05 (m, 2H), 4.97 (dq, J = 3.6, 2.2, 1.8 Hz, 1H), 4.86 (tt, J = 7.7, 3.8 Hz, 1H), 4.69 (d, J = 2.7 Hz, 1H), 3.22–3.09 (m, 2H), 2.84–2.71 (m, 2H), 2.71–2.63 (m, 1H), 2.54–2.39 (m, 2H), 2.37–2.24 (m, 2H), 2.08 (dddd, J = 12.3, 5.9, 4.7, 3.5 Hz, 2H), 1.98–1.79 (m, 8H), 1.78–1.55 (m, 4H), 1.55–1.42 (m, 3H), 1.42–1.10 (m, 9H), 0.95–0.86 (m, 4H), 0.80 (d, J = 6.8 Hz, 4H), 0.72 (t, J = 6.7 Hz, 7H). ESI (m/z); calculated for $C_{59}H_{79}N_2O_{14}$ [M + NH$_4$]$^+$: 1039.5531; found: 1039.7386 (Figures 1 and 2).

2.3. Cell Culture and Treatments

The human osteosarcoma cancer cell line, MG63 cells, (ATCC, Manassas, VA, USA) was maintained according to the ATCC's instructions.

2.4. Effect of DOX and DOX-Vit D on MG63 Cell Proliferation

The effect of DOX and DOX-Vit D on MG63 cell proliferation was determined by measuring the capacity of reducing enzymes to convert 3-[4,5-dimethylthiazol-2-yl]-2,5-diphenyltetrazoliumbromide (MTT) to colored formazan crystals as described previously [21,22]. The percentage of cell proliferation was calculated relative to control wells designated as 100% viable cells using the following formula:

$$\text{cell proliferation} = (A_{\text{treated}})/(A_{\text{control}}) \times 100\% \tag{1}$$

2.5. RNA Extraction and cDNA Synthesis

Total RNA was extracted using TRIzol reagent (Invitrogen$^{\circledR}$, Carlsbad, CA, USA) as described previously [20].

2.6. Quantification of mRNA Expression by Quantitative Real-Time Polymerase Chain Reaction (Real Time-PCR)

Quantification of specific gene expression was performed by real time-PCR using ABI Prism 7500 System (Applied Biosystems, Foster City, CA, USA) as previously described [23]. Human primers sequences and probes for Caspase-3, p53, BCLxs, death receptor-4 (DR-4), heme oxygenase-1 (HO-1), NAD(P)H:quinone oxidoreductase-1 (NQO-1)and β-actin are illustrated in Table 1. These primers were purchased from Integrated DNA Technologies (IDT, Coralville, IA, USA).

Table 1. Primers sequences used for RT-PCR reactions.

Gene	Forward Primer	Reverse Primer
Caspase-3	GAGTGCTCGCAGCTCATACCT	CCTCACGGCCTGGGATTT
P53	GCCCCCAGGGAGCACTA	GGGAGAGGAGCTGGTGTTG
DR4	AGTACATCTAGGTGCGTTCCTG	GTGCTGTCCCATGGAGGTA
BCLxs	CCCAGAAAGGATACAGCTGG	GCGAT-CCGACTCACCAATAC
HO-1	ATGGCCTCCCTGTACCACATC	TGTTGCGCTCAATCTCCTCCT
NQO-1	CGCAGACCTTGTGATATTCCAG	CGTTTCTTCCATCCTTCCAGG
β-actin	CCAGATCATGTTTGAGACCTTCAA	GTGGTACGACCAGAGGCATACA

2.7. Determination of Reactive Oxygen Species (ROS) Production

ROS was measured fluorometrically using 2,7-dichlorofluorescein diacetate (DCF-DA) assay as described previously [24]. Briefly, MG63 cells were treated for 24 h with 10 µM DOX-Vit D. Thereafter, cells were washed with PBS before incubated for 30 min in fresh media containing 10 µM DCF-DA. The fluorescence was directly measured using excitation and emission wavelengths of 485 and 535 nm, respectively, the Bio-Tek Synergy H1Hybrid Multi-Mode Microplate Readers (Bio-Tek Instruments, Winooski, VT, USA).

2.8. Protein Extraction from MG63 Cells

MG63 cells were treated for 24 h with 10 µM DOX-Vit D or DOX, then the total cellular protein was extracted from the cells as described previously [20].

2.9. Immuno Blot Analysis

Cell lysates were analyzed by SDS-PAGE and immunoblotting were performed as described previously [20].

2.10. Determination of MAPKs Signaling Pathway

The protein phosphorylation of MAPKs was measured using the PhosphoTracer MAPK ELISA Kit (Abcam, Cambridge, UK) according to manufacturer's instructions and as described previously [20].

2.11. Extration of Nuclear Protein

MG63 cells were treated for 2 h with 10 µM DOX-Vit D or DOX, then the nuclear protein was extracted from the cells as described previously [20,25].

2.12. Determination of NF-κB Binding Activity

The NF-κB binding activity was determined using NF-κB Assay Chemiluminescent Kit (Millipore, Schwalbach/Ts., Germany, #70-660) according to the manufacturer's protocol as described previously [26].

2.13. Statistical Analysis

Results are shown as mean $\pm$ SEM. Statistical analysis was carried out using SigmaPlot® for Windows (Systat Software, Inc., San Jose, CA, USA). One-way analysis of variance (ANOVA) followed by Tukey-Kramer multiple comparison tests or unpaired two-sided student *t*-test was carried out. A probability value obtained less than 0.05 is considered significant.

3. Results

3.1. Physiochemical Properities of DOX-Vit D in Comaprison to DOX

Given that the main purpose of the current study is to improve the lipopilicity of DOX, we investigated the physiochemical properties of DOX-Vit D in comparison to DOX using ACD iLab and VCCLAB software (https://www.acdlabs.com/resources/ilab/) as described previously [27]. Perhaps the better predictor of lipophilicity is the distribution coefficient at pH 7.4 (LogD$_{7.4}$) since it considers the ionizable group at certain pH in addition to the estimated partition coefficient (LogP). Of interest, Table 2 shows that DOX-Vit D has clear higher predicted values of LogP and LogD$_{7.4}$ in comparison to DOX. This was consistent with a low predicted water solubility of DOX-Vit D (0.0029 µg/mL) in comparison to DOX (0.49 mg/mL) and a higher LogS value for DOX-Vit D. Together, it is reasonable to assume that our novel DOX derivative, DOX-Vit D, is more lipophilic than DOX.

Table 2. Physicochemical properties of DoxVD vs. Dox and Vitamin D2.

Compound	Doxorubicin (Free Base)	Vitamin D2	DoxVD
Structure			
Chemical Formula	$C_{27}H_{29}NO_{11}$	$C_{28}H_{44}O$	$C_{59}H_{75}NO_{14}$
Molecular Weight (g/mol)	543.53	396.65	1022.22
Log$P$ (ACD Chemsketch)	2.82 ± 1.30	9.56 ± 0.27	12.83 ± 1.32
Log$P$ (VCCLAB)	1.41	7.59	5.95
Log$P$ (experimental, Pubchem)	1.27	7.3	NA
Log $D_{7.4}$ (ACD iLab)	-0.29	7.5	8.68
Solubility H_2O (ACD iLab)	0.49 mg/mL	0.0018 mg/mL	0.0029 µg/mL
Log$S$ (VCCLAB)	-2.67	-5.96	-5.81
Solubility (experimental, drug bank)	2%	0.05 mg/mL	NA

3.2. Effect of DOX and DOX-Vit D on MG63 Cells Proliferation

To determine the cytotoxic effect of DOX and DOX-Vit D on OS, MG63 cells were exposed to 10 µM DOX and DOX-Vit D for 24 h. Thereafter, the cell proliferation was determined using MTT assay. Our results showed that a 10 µM DOX did not significantly affect cell proliferation at 24 h (Figure 3). However, 10 µM DOX-Vit D significantly decreased the cell proliferation by approximately 50% in comparison to control (Figure 3).

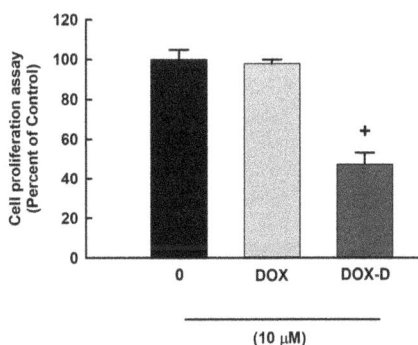

Figure 3. Effect of DOX and DOX-Vit D on MG63 cells proliferation. MG63 cells were exposed to 10 µM DOX and DOX-Vit D for 24 h. Thereafter, the cell proliferation was determined using MTT assay. The results are presented as the mean $\pm$ SEM ($n = 6$). $^+ p < 0.05$ compared to control.

3.3. Effect of DOX and DOX-Vit D on Proapoptotic Genes

To investigate whether the inhibitory effect of DOX-Vit D on MG63 cell proliferation and growth is an apoptosis-dependent mechanism, MG63 cells were treated for 24 h with 10 µM DOX and DOX-Vit

D. Thereafter, the mRNA levels of proapoptotic genes, Caspase-3, p53 and BCLxs, were determined by real time-PCR. Figure 4 shows that DOX-Vit D caused a significant induction of Caspase-3 and BCLxs genes expression by approximately 250% and 400%, respectively, in comparison to control. On the other hand, DOX significantly decreased the expression of Caspase-3, BCLxs and P53 by about 50%, 20% and 30%, respectively, in comparison to control.

In light of our findings, DOX-Vit D seems to inhibit the growth of MG63 cells through an apoptosis-dependent mechanism. Next, we questioned whether the DOX-Vit D elicited an apoptotic signal in MG63 cells is mediated extrinsically through the activation of death receptor and/or intrinsically by the induction of oxidative stress. Therefore, a series of independent experiments were conducted as follows.

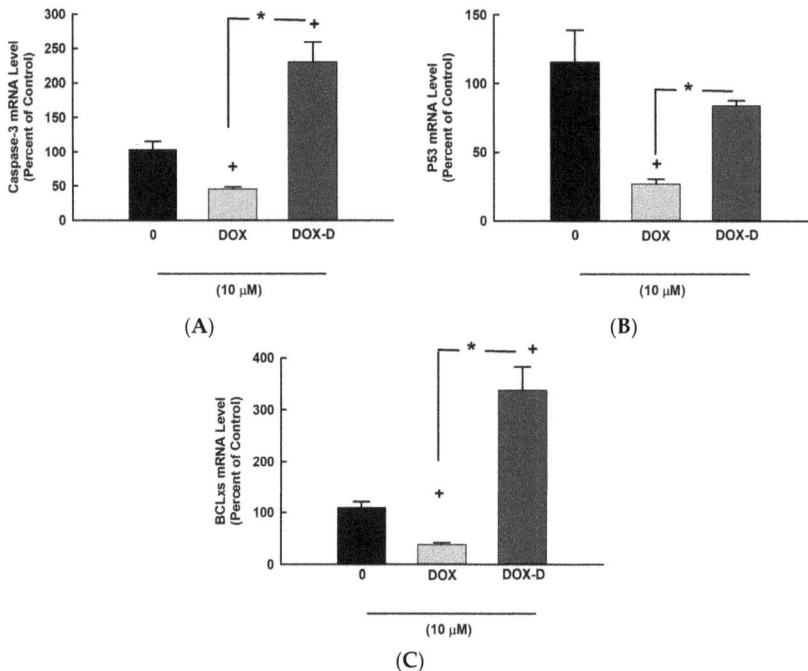

Figure 4. Effect of DOX and DOX-Vit D on proapoptotic genes. MG63 cells were treated for 24 h with 10 μM DOX and DOX-Vit D. Thereafter, total RNA was isolated using TRIzol reagent, and the mRNA levels of (**A**) Caspase-3, (**B**) p53 and (**C**) BCLxs were quantified using real time-PCR and normalized to a β-actin housekeeping gene. The results are presented as the mean ± SEM ($n = 6$). $^+ p < 0.05$ compared to control. $^* p < 0.05$ compared to DOX.

3.4. Effect of DOX and DOX-Vit D on the Expression of DR-4

In order to determine the capacity of DOX and DOX-Vit D to modulate the expression of DR-4 mRNA, MG63 cells were treated for 24 h with 10 μM DOX and DOX-Vit D. Thereafter, the mRNA levels of DR-4 was determined by real time-PCR. Figure 5 shows that treatment of MG63 cell with DOX-Vit D caused a significant induction of DR-4 by about 170% in comparison to control. On the other hand, DOX significantly inhibit the expression of DR-4 by approximately 60% in comparison to control.

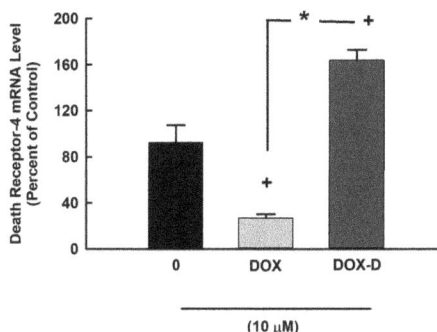

Figure 5. Effect of DOX and DOX-Vit D on the expression of DR-4. MG63 cells were treated for 24 h with 10 μM DOX and DOX-Vit D. Thereafter, total RNA was isolated using TRIzol reagent, and the mRNA level of DR-4 was quantified using real time-PCR and normalized to a β-actin housekeeping gene. The results are presented as the mean ± SEM ($n = 6$). $^+ p < 0.05$ compared to control. $^* p < 0.05$ compared to DOX.

3.5. Effect of DOX and DOX-Vit D on the Oxidative Stress

The involvement of intrinsic apoptotic pathway was addressed by two approaches. Firstly, we determined the effect of DOX-Vit D and DOX on the mRNA expression of oxidative stress markers. Figure 6 shows that treatment of cells with 10 μM DOX-Vit D caused a significant induction of NQO-1 and HO-1 by approximately 250% and 6000%, respectively, in comparison to control. In contrast, DOX significantly inhibited the expression of HO-1 by about 70%, whereas no significant changes were observed with NQO-1 (Figure 6A,B).

The second approach was to investigate the effect of DOX and DOX-Vit D on the generation of ROS using DCF assay. The incubation of MG63 cells with DOX and DOX-Vit D for 24 h caused a significant increase in the formation of ROS by about 400% and 350%, respectively, in comparison to control (Figure 6C).

Our findings suggest an involvement of both extrinsic and intrinsic pathways in the induction of proapoptotic genes by DOX-Vit D. The induction of the aforementioned pathways are known to trigger apoptosis through the activation of MAPK signaling pathway. Thus, we have investigated whether or not DOX-Vit D induces proapoptotic genes through MAPK signaling pathway.

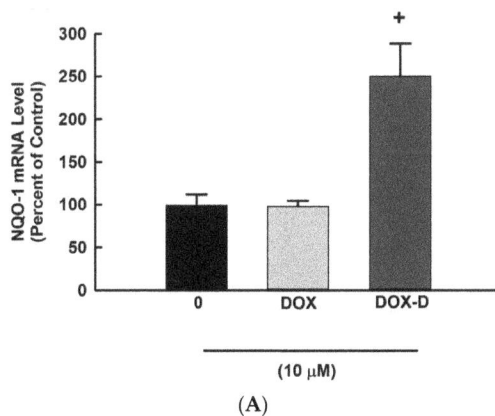

(**A**)

Figure 6. *Cont.*

(B)

(C)

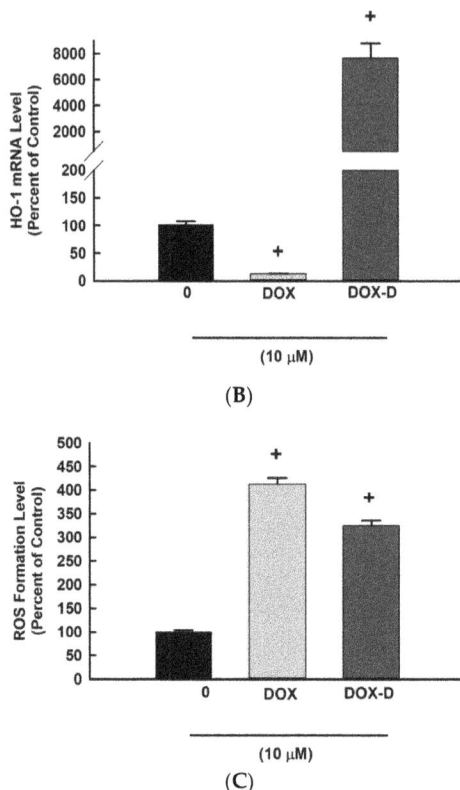

Figure 6. Effect of DOX and DOX-Vit D on the oxidative stress. MG63 cells were treated for 24 h with 10 μM DOX and DOX-Vit D. Thereafter, total RNA was isolated using TRIzol reagent, and the mRNA levels of (**A**) NQO-1 and (**B**) HO-1 were quantified using real time-PCR and normalized to a β-actin housekeeping gene. (**C**) MG63 cells were treated for 24 h with 10 μM DOX and DOX-Vit D then, cells were incubated with DCF-DA (10 μM) for 1 h. DCF formation was measured fluorometrically using excitation/emission wavelengths of 484/535 nm. The results are presented as the mean ± SEM ($n = 6$). [+] $p < 0.05$ compared to control.

3.6. Effect of DOX and DOX-Vit D on MAPK Signaling Pathway

To assess the role of MAPK signaling pathway on the DOX-Vit D mediated induction of proapoptotic genes, MG63 cells were treated with 10 μM DOX-Vit D and DOX. Thereafter, phosphorylated MAPK levels were determined using a commercially available kit. Figure 7 shows that incubation of the cells with 10 μM of DOX-Vit D but not DOX significantly induced phosphorylation of p38 and JNK by approximately 250% and 160%, respectively, in comparison to control.

To further confirm whether activation of the MAPK pathways is required for the apoptotic cell death mediated by DOX-Vit D, MG63 cells were treated with 10 μM p38 inhibitor, SB203580, and JNK inhibitor, SP600125, in the presence and absence of DOX-Vit D. Thereafter, the cells proliferation were measured using MTT assay. Figure 8 shows that DOX-Vit D alone caused a significant inhibition of MG63 cell proliferation by about 50% in comparison to control. Importantly, treatment of cells with SB203580 and SP600125 partially but significantly protects the cells against the cytotoxic effect of DOX-Vit D. Our findings suggest that the activation of MAPK is essential for the cytotoxic effect of DOX-Vit D.

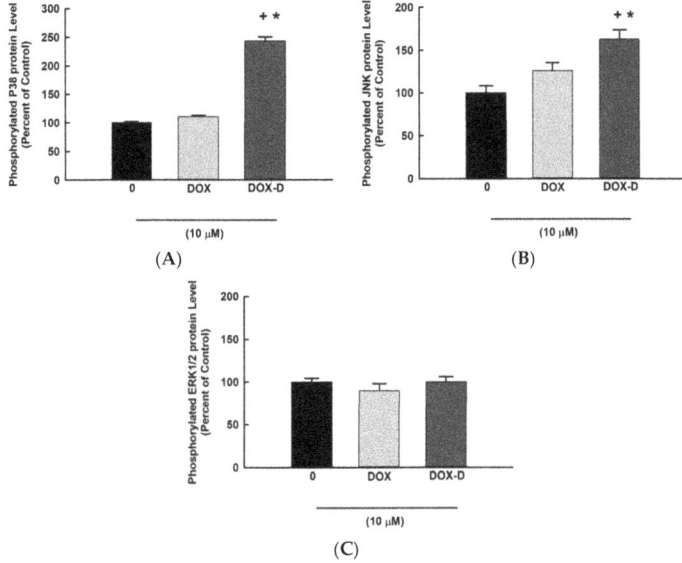

Figure 7. Effect of DOX and DOX-Vit D on MAPK signaling pathway. MG63 cells were treated for 24 h with 10 μM DOX and DOX-Vit D. Thereafter, MAPKs protein phosphorylation was determined in cytoplasmic protein extracts using the PhosphoTracer (**A**) p38 MAPK (pT180/Y182) (**B**) JNK1/2/3 (pT183/Y185) (**C**) ERK1/2 (pT202/Y204) Elisa Kit (Abcam, Cambridge, UK). The results are presented as the mean ± SEM ($n = 6$). [+] $p < 0.05$ compared to control. [*] $p < 0.05$ compared to DOX.

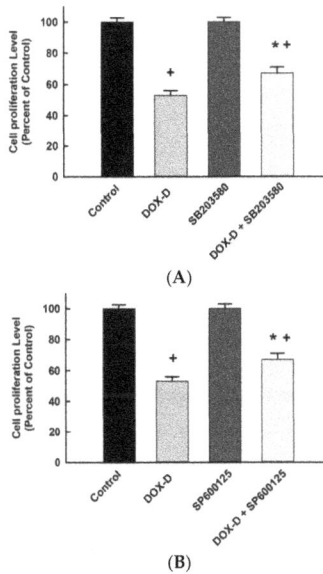

Figure 8. Effect of MAPK inhibitors on DOX-Vit D-induced cytotoxicity. MG63 cells were treated with p38 inhibitor, SB203580, and JNK inhibitor, SP600125, in the presence and absence of 10 μM DOX-Vit D. Thereafter, the cell proliferation was determined using MTT assay. The results are presented as the mean ± SEM ($n = 6$). [+] $p < 0.05$ compared to control. [*] $p < 0.05$ compared to DOX-Vit D.

In order to examine whether the inhibitory effect of DOX-Vit D on MG63 cell proliferation and growth is also attributed to the suppression of cell survival and proliferation pathways, we have determined the effect of DOX-Vit D on NF-κB, Akt and mTOR signaling pathways.

3.7. Effect of DOX and DOX-Vit D on NF-κB Signaling Pathway

The basal activity of the NF-κB transcription factor in OS seems to be crucial for their growth or resistance to chemotherapy. Therefore, we have investigated whether DOX-Vit D suppresses MG63 cell growth through the inhibition of NF-κB. For this purpose, MG63 cells were treated with 10 μM DOX and DOX-Vit D. Thereafter, NF-κB binding activity was determined using a commercially available kit. Figure 9 shows that neither DOX nor DOX-Vit D significantly affects the binding activity of NF-κB suggesting an NF-κB-independent mechanism.

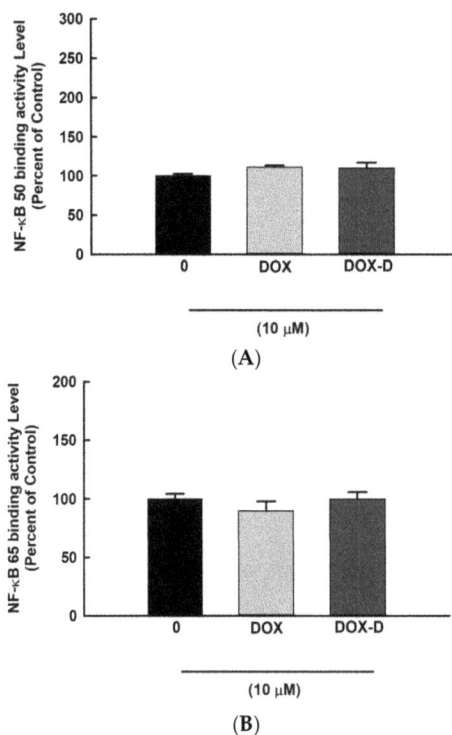

Figure 9. Effect of Effect of DOX and DOX-Vit D on NF-κB signaling pathway. MG63 cells were treated for 24 h with 10 μM DOX and DOX-Vit D. Thereafter, NF-κB binding activity was determined in nuclear extracts using a commercially available kit. The results are presented as the mean ± SEM ($n = 6$).

3.8. Effect of DOX and DOX-Vit D on Akt and mTOR Signaling Pathway

Since Akt and mTOR pathway promotes cell growth, proliferation and survival, we examined the effect of DOX-Vit D on Akt and mTOR signaling pathway. For this purpose, MG63 cells were treated with 10 μM DOX and DOX-Vit D. Thereafter, Akt and mTOR protein expression levels were determined using Western blot analysis. Figure 10 shows that DOX-Vit D caused a significant inhibition of p-Akt and p-mTOR protein expression by approximately 40% and 50%, respectively, in comparison to control suggesting an Akt/mTOR-dependent inhibition of cell growth by DOX-Vit D. In contrast, DOX did not significantly alter the expression of p-Akt and p-mTOR protein expression.

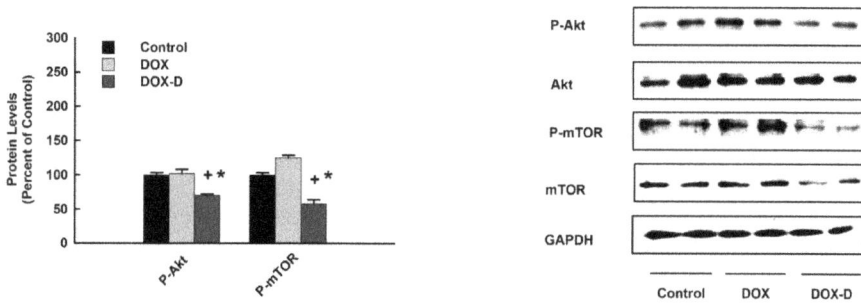

Figure 10. Effect of DOX and DOX-Vit D on Akt and mTOR signaling pathway. MG63 cells were treated for 24 h with 10 μM DOX and DOX-Vit D. Thereafter, total and phosphorylated Akt and mTOR protein expression levels were determined by Western blot analysis and detected using the enhanced chemiluminescence method. The intensity of protein bands was normalized to the signals obtained for GAPDH protein and quantified using ImageJ®. The results are presented as the mean ± SEM ($n = 6$). $^+ p < 0.05$ compared to control. $^* p < 0.05$ compared to DOX.

4. Discussion

These investigations provide strong evidence that DOX-Vit D suppresses the growth of human OS, MG63 cell line, through the induction of apoptosis and the inhibition of cell survival and proliferative signaling pathways.

One of the strategies for treating OS and minimizing the development of chemoresistance associated with chemotherapeutic agents includes the induction of apoptosis and/or the attenuation of cell survival and proliferative signaling pathways. Studies using transgenic mice provide direct evidence that overexpression of cell survival pathways and/or disruption of apoptosis promote tumorigenesis, metastasis and contribute to chemoresistance [28,29]. Therefore, the development of a new chemotherapeutic agent that is able to attenuate the proliferation of OS while inducing apoptosis is an urgently needed to overcome chemoresistance.

DOX, a broad-spectrum anthracycline antibiotic, is one of those standard therapies for the treatment of OS [5]. Unfortunately, the effectiveness of DOX in OS is usually limited by the acquired drug resistance that leads to poor prognosis and suboptimal outcomes [30]. Recently, Vit-D has been shown to suppress the growth of many human cancer cells and reverse chemotherapy drug-resistant [30,31]. Taken together, we synthesized DOX-Vit D by conjugating Vit-D to DOX in order to mitigate the chemoresistance associated with DOX. The current study was conducted to investigate the antiproliferative and apoptogenic effects of 10 μM DOX-Vit D in comparison to 10 μM DOX and the possible mechanism(s) involved using the human OS cell line, MG63 cells. The concentration of DOX used in the current study was maintained within the therapeutic range of plasma concentration reported in human. For example, human subjects given a dose of 60–75 mg/m^2 DOX for the treatment of metastatic cancer had mean plasma concentrations range from 5 and 15 μM and an average half-life of ~25 h [32,33]. In addition, several in vitro studies on human cancer cells to explore the cytotoxicity of DOX used concentrations range from 1 to 10 μM [32,33].

Initially, we have demonstrated that DOX-Vit D, but not DOX, was able to significantly suppress the MG63 cell proliferation and growth. Notably, the anticancer effect of DOX-Vit D is attributed to the induction of proapoptotic genes Caspase-3 and BCLxs. Activation of the proapoptotic genes plays a crucial role in the initiation of apoptosis through the cleavage of the key cellular proteins resulting in the irreversible commitment to cell death [29,34]. Similar to our observation, it has been reported that calciferol and its chemical derivative, MT19c, induce apoptosis and inhibit the growth and proliferation of many cancer cell lines through the activation of the Caspase-3 enzyme [35,36]. The inhibitory effect of DOX on proapoptotic genes might be attributed to the fact that our MG-63 cells are resistant to

DOX. In agreement with our results, it has been shown that 10 µM DOX was neither elicit Caspase-3 activation or apoptosis in DOX resistant MG63 cells [37]. Importantly, promoting apoptosis has been shown to overcome the chemoresistance associated with DOX during the treatment of OS [38].

Apoptosis is known to be elicited by the activation of extrinsic and/or intrinsic signals. These signals are instructing the cells to undergo programmed cell death through the activation of proapoptotic genes [29,34]. Accordingly, we investigated whether the DOX-Vit D-induced apoptosis in MG63 cells is mediated through the extrinsic and/or intrinsic apoptotic pathway. The extrinsic signals induce apoptosis through binding of cell surface death receptors such as TNF/Fas-receptor with its ligand and subsequently activates proapoptotic enzymes [39]. Hence, we have tested whether DOX-Vit D triggers extrinsic apoptotic pathway by measuring the expression of DR-4. In this current study, we found that the induction of DR-4 mRNA in response to DOX-Vit D significantly contributes to the activation of proapoptotic genes. These findings are in agreement with the observation that calciferol increases the activity of proapoptotic genes through the induction of DR [36]. On the other hand, DOX seems to decrease the expression of proapoptotic genes through the downregulation of DR-4.

The intrinsic signal is another pivotal pathway that could initiate apoptosis through the oxidative stress and ROS-dependent mechanism [40]. Oxidative stress and ROS have been considered as a potent inducer of apoptosis [41]. In this context, the involvement of the intrinsic apoptotic pathway in the cytotoxic effect of DOX-Vit D was evidenced by the induction of the oxidative stress markers, NQO-1 and HO-1, in addition to the generation of ROS. In a manner similar to our observations, it has been reported that calciferol treatment results in the accumulation of ROS and subsequently coordinates proapoptotic genes activation [36].

Our findings suggest that DOX-Vit D upregulated the expression of proapoptotic genes through the activation of both extrinsic and intrinsic apoptotic pathways. The induction of the aforementioned pathways are known to trigger apoptosis through the activation of the MAPK signaling pathway. A wealth of information suggests the involvement of MAPK cascades in cell death and survival signaling [42,43]. In particular, it has demonstrated that persistent activation of p38 and JNK promotes apoptosis and cell death [42,43]. Taken together, the possibility that DOX-Vit D would induce apoptosis through the activation of MAPKs could not be ruled out. Thus, the third objective of the current study was to explore the role of DOX-Vit D on the MAPKs signaling pathway. Our results demonstrate that DOX-Vit D, but not DOX, significantly increased the protein expression level of p-p38 and p-JNK whereas no significant changes were observed on p-ERK1/2. The direct evidence for the involvement of p-p38 and p-JNK in the DOX-Vit D-induced cytotoxicity was supported by the observation that p-p38 inhibitor, SB203580, and JNK inhibitor, SP600125, significantly protect against DOX-Vit D-induced cell death suggesting a MAPK-dependent mechanism. The premise of this observation emerges from the finding that calciferol chemical derivative, MT19c, induces apoptosis through a p38 and JNK-dependent mechanism [35].

Apoptosis-mediated cell death might induce the turn off of survival pathways, such as NF-κB and Akt/mTOR pathways, that could otherwise interfere with the apoptotic response. The aberrant activation of those proliferating proteins in OS seems to be crucial for their growth or resistance to chemotherapy [44,45]. Thus, it is imperative to investigate the effect of DOX-Vit D on the aforementioned cell survival and proliferating pathways. Although DOX-Vit D did not significantly affect the binding activity level of NF-κB in MG63 cells, DOX-Vit D significantly downregulated the protein expression level of p-Akt and p-mTOR. In addition, DOX did not significantly alter the p-Akt/p-mTOR proteins expression. These results are in agreement with previous reports on OS showing that high Akt/mTOR activity was associated with poor clinical outcome and chemoresistence associated with DOX [46,47]. On the other hand, everolimus, an mTOR inhibitor, has been shown to decrease drug-induced resistance in OS [48,49]. Our findings not only suggest an Akt/mTOR-dependent inhibition of MG63 cell growth by DOX-Vit D but also support DOX-Vit D as a promising developmental strategy for the treatment of OS resistant to chemotherapy [50].

To reiterate, our results clearly demonstrated that DOX-Vit D, a novel DOX derivative, suppresses MG63 cell growth through the induction of apoptosis and the inhibition of Akt/mTOR signaling pathways. Such observation will raise the potential of developing DOX-Vit D analogues for the treatment of OS resistant to chemotherapy. Given that DOX-VitD has a higher lipophilicity compared to DOX, it is reasonable to assume that DOX-VitD may serve as a novel drug delivery approach to minimize the first-pass effect while increasing lymphatic exposure and ultimately improving overall systemic drug exposure. Additional studies are going to test the toxicity and the kinetic of DOX-VitD in rats. Our preliminary data have shown that DOX-VitD is tolerable upon oral and intravenous administration in rats. Given that our new derivative has a stoichiometry of 1 to 1 ratio (Dox-VitD), it might not necessarily be the most active one. Therefore, we will compare our new derivative with a combination of the free drugs at different ratios in order to confirm the mechanism(s) of action and discriminate additive effects vs synergistic effects.

Author Contributions: Participated in research design: Z.H.M., N.M.D. and A.O.S.E.; Conducted experiments: Z.H.M., T.Z., M.L.F., S.A.; Performed data analysis: Z.H.M., N.M.D., M.R.D. and A.O.S.E.-K.; Wrote or contributed to the writing of the manuscript: Z.H.M., T.Z., M.L.F., N.M.D., A.O.S.E.-K.

Acknowledgments: This research was funded by a grant from the Canadian Institutes of Health Research [Grant 106665] to Ayman O.S. El-Kadi and the U.S. National Cancer Institute [Grant R01CA173292] to Marcus Laird Forrest. The authors are grateful to Erica McGinn for her helpful technical assistance.

Conflicts of Interest: The authors declare no conflict of interest.

References

1. Bacci, G.; Longhi, A.; Bertoni, F.; Bacchini, P.; Ruggeri, P.; Versari, M.; Picci, P. Primary high-grade osteosarcoma: Comparison between preadolescent and older patients. *J. Pediatr. Hematol.* **2005**, *27*, 129–134. [CrossRef]
2. Ottaviani, G.; Jaffe, N. The epidemiology of osteosarcoma. *Cancer Treat. Res.* **2009**, *152*, 3–13. [PubMed]
3. Mirabello, L.; Troisi, R.J.; Savage, S.A. International osteosarcoma incidence patterns in children and adolescents, middle ages and elderly persons. *Int. J. Cancer* **2009**, *125*, 229–234. [CrossRef] [PubMed]
4. Isakoff, M.S.; Bielack, S.S.; Meltzer, P.; Gorlick, R. Osteosarcoma: Current treatment and a collaborative pathway to success. *J. Clin. Oncol.* **2015**, *33*, 3029–3035. [CrossRef] [PubMed]
5. Waddell, A.E.; Davis, A.M.; Ahn, H.; Wunder, J.S.; Blackstein, M.E.; Bell, R.S. Doxorubicin-cisplatin chemotherapy for high-grade nonosteogenic sarcoma of bone. Comparison of treatment and control groups. *Can. J. Surg.* **1999**, *42*, 190–199. [PubMed]
6. Lewis, I.J.; Nooij, M.A.; Whelan, J.; Sydes, M.R.; Grimer, R.; Hogendoorn, P.C.; Memon, M.A.; Weeden, S.; Uscinska, B.M.; van Glabbeke, M.; et al. Improvement in histologic response but not survival in osteosarcoma patients treated with intensified chemotherapy: A randomized phase iii trial of the european osteosarcoma intergroup. *J. Natl. Cancer Inst.* **2007**, *99*, 112–128. [CrossRef] [PubMed]
7. O'Driscoll, L. Mechanisms of drug sensitivity and resistance in cancer. *Curr. Cancer Drug Targets* **2009**, *9*, 250–251. [CrossRef] [PubMed]
8. Garland, C.F.; Garland, F.C.; Gorham, E.D.; Lipkin, M.; Newmark, H.; Mohr, S.B.; Holick, M.F. The role of Vitamin D in cancer prevention. *Am. J. Public Health* **2006**, *96*, 252–261. [CrossRef] [PubMed]
9. Grant, W.B. Epidemiology of disease risks in relation to Vitamin D insufficiency. *Prog. Biophys. Mol. Biol.* **2006**, *92*, 65–79. [CrossRef] [PubMed]
10. Ma, Y.; Zhang, P.; Wang, F.; Yang, J.; Liu, Z.; Qin, H. Association between Vitamin D and risk of colorectal cancer: A systematic review of prospective studies. *J. Clin. Oncol.* **2011**, *29*, 3775–3782. [CrossRef] [PubMed]
11. Woolcott, C.G.; Wilkens, L.R.; Nomura, A.M.; Horst, R.L.; Goodman, M.T.; Murphy, S.P.; Henderson, B.E.; Kolonel, L.N.; Le Marchand, L. Plasma 25-hydroxyvitamin d levels and the risk of colorectal cancer: The multiethnic cohort study. *Cancer Epidemiol. Biomark. Prev.* **2010**, *19*, 130–134. [CrossRef] [PubMed]
12. Bouillon, R.; Eelen, G.; Verlinden, L.; Mathieu, C.; Carmeliet, G.; Verstuyf, A. Vitamin D and cancer. *J. Steroid Biochem. Mol. Biol.* **2006**, *102*, 156–162. [CrossRef] [PubMed]
13. Schwartz, G.G.; Skinner, H.G. Vitamin D status and cancer: New insights. *Curr. Opin. Clin. Nutr. Metab. Care* **2007**, *10*, 6–11. [CrossRef] [PubMed]

14. Jones, G.; Strugnell, S.A.; DeLuca, H.F. Current understanding of the molecular actions of Vitamin D. *Physiol. Rev.* **1998**, *78*, 1193–1231. [CrossRef] [PubMed]

15. Chouvet, C.; Vicard, E.; Devonec, M.; Saez, S. 1,25-dihydroxyvitamin d3 inhibitory effect on the growth of two human breast cancer cell lines (mcf-7, bt-20). *J. Steroid Biochem.* **1986**, *24*, 373–376. [CrossRef]

16. Getzenberg, R.H.; Light, B.W.; Lapco, P.E.; Konety, B.R.; Nangia, A.K.; Acierno, J.S.; Dhir, R.; Shurin, Z.; Day, R.S.; Trump, D.L.; et al. Vitamin D inhibition of prostate adenocarcinoma growth and metastasis in the dunning rat prostate model system. *Urology* **1997**, *50*, 999–1006. [CrossRef]

17. Beer, T.M.; Myrthue, A. Calcitriol in cancer treatment: From the lab to the clinic. *Mol. Cancer Ther.* **2004**, *3*, 373–381. [PubMed]

18. Knutson, J.C.; LeVan, L.W.; Valliere, C.R.; Bishop, C.W. Pharmacokinetics and systemic effect on calcium homeostasis of 1 alpha,24-dihydroxyvitamin d2 in rats. Comparison with 1 alpha,25-dihydroxyvitamin d2, calcitriol, and calcipotriol. *Biochem. Pharm.* **1997**, *53*, 829–837. [CrossRef]

19. Wigington, D.P.; Urben, C.M.; Strugnell, S.A.; Knutson, J.C. Combination study of 1,24(s)-dihydroxyvitamin d2 and chemotherapeutic agents on human breast and prostate cancer cell lines. *Anticancer Res.* **2004**, *24*, 2905–2912. [PubMed]

20. Maayah, Z.H.; Althurwi, H.N.; Abdelhamid, G.; Lesyk, G.; Jurasz, P.; El-Kadi, A.O. Cyp1b1 inhibition attenuates doxorubicin-induced cardiotoxicity through a mid-chain hetes-dependent mechanism. *Pharm. Res.* **2016**, *105*, 28–43. [CrossRef] [PubMed]

21. Maayah, Z.H.; El Gendy, M.A.; El-Kadi, A.O.; Korashy, H.M. Sunitinib, a tyrosine kinase inhibitor, induces cytochrome p450 1a1 gene in human breast cancer mcf7 cells through ligand-independent aryl hydrocarbon receptor activation. *Arch. Toxicol.* **2013**, *87*, 847–856. [CrossRef] [PubMed]

22. Liu, Y.; Peterson, D.A.; Kimura, H.; Schubert, D. Mechanism of cellular 3-(4,5-dimethylthiazol-2-yl)-2,5-diphenyltetrazolium bromide (mtt) reduction. *J. Neurochem.* **1997**, *69*, 581–593. [CrossRef] [PubMed]

23. Maayah, Z.H.; Ansari, M.A.; El Gendy, M.A.; Al-Arifi, M.N.; Korashy, H.M. Development of cardiac hypertrophy by sunitinib in vivo and in vitro rat cardiomyocytes is influenced by the aryl hydrocarbon receptor signaling pathway. *Arch. Toxicol.* **2014**, *88*, 725–738. [CrossRef] [PubMed]

24. Elbekai, R.H.; Korashy, H.M.; Wills, K.; Gharavi, N.; El-Kadi, A.O. Benzo[a]pyrene, 3-methylcholanthrene and beta-naphthoflavone induce oxidative stress in hepatoma hepa 1c1c7 cells by an ahr-dependent pathway. *Free Radic. Res.* **2004**, *38*, 1191–1200. [CrossRef] [PubMed]

25. Andrews, N.C.; Faller, D.V. A rapid micropreparation technique for extraction of DNA-binding proteins from limiting numbers of mammalian cells. *Nucleic Acids Res.* **1991**, *19*, 2499. [CrossRef] [PubMed]

26. Bhattacharya, N.; Sarno, A.; Idler, I.S.; Fuhrer, M.; Zenz, T.; Dohner, H.; Stilgenbauer, S.; Mertens, D. High-throughput detection of nuclear factor-kappab activity using a sensitive oligo-based chemiluminescent enzyme-linked immunosorbent assay. *Int. J. Cancer* **2010**, *127*, 404–411. [CrossRef] [PubMed]

27. Alrushaid, S.; Sayre, C.L.; Yanez, J.A.; Forrest, M.L.; Senadheera, S.N.; Burczynski, F.J.; Lobenberg, R.; Davies, N.M. Pharmacokinetic and toxicodynamic characterization of a novel doxorubicin derivative. *Pharmaceutics* **2017**, *9*, 35. [CrossRef] [PubMed]

28. Jiang, B.H.; Liu, L.Z. Role of mtor in anticancer drug resistance: Perspectives for improved drug treatment. *Drug Resist. Updates* **2008**, *11*, 63–76. [CrossRef] [PubMed]

29. Lowe, S.W.; Lin, A.W. Apoptosis in cancer. *Carcinogenesis* **2000**, *21*, 485–495. [CrossRef] [PubMed]

30. Yan, M.; Nuriding, H. Reversal effect of Vitamin D on different multidrug-resistant cells. *Genet. Mol. Res.* **2014**, *13*, 6239–6247. [CrossRef] [PubMed]

31. Sabzichi, M.; Mohammadian, J.; Mohammadi, M.; Jahanfar, F.; Movassagh Pour, A.A.; Hamishehkar, H.; Ostad-Rahimi, A. Vitamin D-loaded nanostructured lipid carrier (nlc): A new strategy for enhancing efficacy of doxorubicin in breast cancer treatment. *Nutr. Cancer* **2017**, *69*, 840–848. [CrossRef] [PubMed]

32. Mross, K.; Maessen, P.; van der Vijgh, W.J.; Gall, H.; Boven, E.; Pinedo, H.M. Pharmacokinetics and metabolism of epidoxorubicin and doxorubicin in humans. *J. Clin. Oncol.* **1988**, *6*, 517–526. [CrossRef] [PubMed]

33. Robert, J.; Vrignaud, P.; Nguyen-Ngoc, T.; Iliadis, A.; Mauriac, L.; Hurteloup, P. Comparative pharmacokinetics and metabolism of doxorubicin and epirubicin in patients with metastatic breast cancer. *Cancer Treat. Rep.* **1985**, *69*, 633–640. [PubMed]

34. Vecchione, A.; Croce, C.M. Apoptomirs: Small molecules have gained the license to kill. *Endoc.-Relat. Cancer* **2010**, *17*, F37–F50. [CrossRef] [PubMed]

35. Brard, L.; Lange, T.S.; Robison, K.; Kim, K.K.; Ara, T.; McCallum, M.M.; Arnold, L.A.; Moore, R.G.; Singh, R.K. Evaluation of the first ergocalciferol-derived, non hypercalcemic anti-cancer agent mt19c in ovarian cancer skov-3 cell lines. *Gynecol. Oncol.* **2011**, *123*, 370–378. [CrossRef] [PubMed]

36. Chen, W.J.; Huang, Y.T.; Wu, M.L.; Huang, T.C.; Ho, C.T.; Pan, M.H. Induction of apoptosis by Vitamin D2, ergocalciferol, via reactive oxygen species generation, glutathione depletion, and caspase activation in human leukemia cells. *J. Agric. Food Chem.* **2008**, *56*, 2996–3005. [CrossRef] [PubMed]

37. Wang, Z.; Yang, L.; Xia, Y.; Guo, C.; Kong, L. Icariin enhances cytotoxicity of doxorubicin in human multidrug-resistant osteosarcoma cells by inhibition of abcb1 and down-regulation of the pi3k/akt pathway. *Biol. Pharm. Bull.* **2015**, *38*, 277–284. [CrossRef] [PubMed]

38. Zhang, C.; Zhao, Y.; Zeng, B. Enhanced chemosensitivity by simultaneously inhibiting cell cycle progression and promoting apoptosis of drug-resistant osteosarcoma mg63/dxr cells by targeting cyclin d1 and bcl-2. *Cancer Biomark.* **2012**, *12*, 155–167. [CrossRef] [PubMed]

39. Herr, I.; Debatin, K.M. Cellular stress response and apoptosis in cancer therapy. *Blood* **2001**, *98*, 2603–2614. [CrossRef] [PubMed]

40. Tsang, W.P.; Chau, S.P.; Kong, S.K.; Fung, K.P.; Kwok, T.T. Reactive oxygen species mediate doxorubicin induced p53-independent apoptosis. *Life Sci.* **2003**, *73*, 2047–2058. [CrossRef]

41. Ravagnan, L.; Roumier, T.; Kroemer, G. Mitochondria, the killer organelles and their weapons. *J. Cell. Physiol.* **2002**, *192*, 131–137. [CrossRef] [PubMed]

42. Bian, J.; Wang, K.; Kong, X.; Liu, H.; Chen, F.; Hu, M.; Zhang, X.; Jiao, X.; Ge, B.; Wu, Y.; et al. Caspase- and p38-mapk-dependent induction of apoptosis in a549 lung cancer cells by newcastle disease virus. *Arch. Virol.* **2011**, *156*, 1335–1344. [CrossRef] [PubMed]

43. Guyton, K.Z.; Spitz, D.R.; Holbrook, N.J. Expression of stress response genes gadd153, c-jun, and heme oxygenase-1 in h2o2- and o2-resistant fibroblasts. *Free Radic. Biol. Med.* **1996**, *20*, 735–741. [CrossRef]

44. Mongre, R.K.; Sodhi, S.S.; Ghosh, M.; Kim, J.H.; Kim, N.; Sharma, N.; Jeong, D.K. A new paradigm to mitigate osteosarcoma by regulation of micrornas and suppression of the nf-kappab signaling cascade. *Dev. Reprod.* **2014**, *18*, 197–212. [CrossRef] [PubMed]

45. Hung, C.M.; Garcia-Haro, L.; Sparks, C.A.; Guertin, D.A. Mtor-dependent cell survival mechanisms. *Cold Spring Harb. Perspect. Biol.* **2012**, *4*. [CrossRef] [PubMed]

46. He, H.; Ni, J.; Huang, J. Molecular mechanisms of chemoresistance in osteosarcoma (review). *Oncol. Lett.* **2014**, *7*, 1352–1362. [CrossRef] [PubMed]

47. Bishop, M.W.; Janeway, K.A. Emerging concepts for pi3k/mtor inhibition as a potential treatment for osteosarcoma. *F1000Research* **2016**, *5*. [CrossRef] [PubMed]

48. Pignochino, Y.; Dell'Aglio, C.; Basirico, M.; Capozzi, F.; Soster, M.; Marchio, S.; Bruno, S.; Gammaitoni, L.; Sangiolo, D.; Torchiaro, E.; et al. The combination of sorafenib and everolimus abrogates mtorc1 and mtorc2 upregulation in osteosarcoma preclinical models. *Clin. Cancer Res.* **2013**, *19*, 2117–2131. [CrossRef] [PubMed]

49. O'Reilly, T.; McSheehy, P.M.; Wartmann, M.; Lassota, P.; Brandt, R.; Lane, H.A. Evaluation of the mtor inhibitor, everolimus, in combination with cytotoxic antitumor agents using human tumor models in vitro and in vivo. *Anti-Cancer Drugs* **2011**, *22*, 58–78. [CrossRef] [PubMed]

50. Ding, L.; Congwei, L.; Bei, Q.; Tao, Y.; Ruiguo, W.; Heze, Y.; Bo, D.; Zhihong, L. Mtor: An attractive therapeutic target for osteosarcoma? *Oncotarget* **2016**, *7*, 50805–50813. [CrossRef] [PubMed]

pharmaceutics

MDPI

Review

Development and Characterization of the Solvent-Assisted Active Loading Technology (SALT) for Liposomal Loading of Poorly Water-Soluble Compounds

Griffin Pauli, Wei-Lun Tang and Shyh-Dar Li *

Faculty of Pharmaceutical Sciences, University of British Columbia, Vancouver, BC V6T 1Z3, Canada
* Correspondence: shyh-dar.li@ubc.ca

Received: 22 March 2019; Accepted: 3 September 2019; Published: 9 September 2019

Abstract: A large proportion of pharmaceutical compounds exhibit poor water solubility, impacting their delivery. These compounds can be passively encapsulated in the lipid bilayer of liposomes to improve their water solubility, but the loading capacity and stability are poor, leading to burst drug leakage. The solvent-assisted active loading technology (SALT) was developed to promote active loading of poorly soluble drugs in the liposomal core to improve the encapsulation efficiency and formulation stability. By adding a small volume (~5 vol%) of a water miscible solvent to the liposomal loading mixture, we achieved complete, rapid loading of a range of poorly soluble compounds and attained a high drug-to-lipid ratio with stable drug retention. This led to improvements in the circulation half-life, tolerability, and efficacy profiles. In this mini-review, we summarize our results from three studies demonstrating that SALT is a robust and versatile platform to improve active loading of poorly water-soluble compounds. We have validated SALT as a tool for improving drug solubility, liposomal loading efficiency and retention, stability, palatability, and pharmacokinetics (PK), while retaining the ability of the compounds to exert pharmacological effects.

Keywords: liposome; water miscible solvents; remote loading; staurosporine; cancer; gambogic acid; loading gradients; mefloquine; child friendly formulation

1. Introduction

1.1. Challenges in Delivery of Poorly Water-Soluble Drugs

Clinical translation of pharmaceutical compounds is often hindered by poor solubility. Over 40% of new chemical entities are not appreciably soluble in water, often resulting in limited therapeutic use, or abandonment as drug candidates [1,2]. Methods for improving apparent solubility include salt formulations, excipients, amorphous solid dispersions, pH adjustment, co-solvents, and nanocarriers [1,2]. There has been interest in the development of systematic approaches to overcoming low soluble compounds, as outlined in the developability classification system which describes a categorization of compounds based on permeability and solubility [3]. There has also been growing interest in the use of nanocarriers as a vehicle to overcome issues of solubility, as they serve as a platform technology and provide cell targeting [4]. Lipid-based nanocarriers (LNCs) represent a class of lipid particles, including solid lipid nanoparticles, micro- or nano-emulsions, polymer-lipid hybrid nanoparticles, and liposomes.

1.1.1. Liposomes and Drug Loading

Liposomes are among the most studied class of LNCs for improving drug delivery [5]. Liposomes are nano-scale spheroid vesicles composed of one or multiple lipid bilayer(s) enclosing an aqueous core [5]. Liposomes can be manufactured using several methods, including thin-film hydration [6], reverse phase evaporation [7], and microfluidic mixing [8]. The process used to generate the liposome will impact its size and lamellarity [9–11]. Likewise, various lipids can be used to formulate liposomes with different properties, such as size, drug release kinetics, biocompatibility, surface charge, and cell targeting [12,13]. To encapsulate drugs into liposomes, two methods have been implemented; passive and active loading, both of which are briefly reviewed below.

1.1.2. Passive Loading

Passive loading describes the procedure in which liposomes are formed concurrently with drug loading (Figure 1A). In general, hydrophilic compounds are distributed homogenously in the aqueous phase (both inside and outside the liposomes), whereas hydrophobic drugs are retained inside the lipid bilayer of liposomes, respectively. Specifically, when working with poorly water-soluble drugs, the drugs are first dissolved with lipids in an organic solvent, followed by solvent evaporation to prepare a drug containing thin film, which is later hydrated with an aqueous phase to prepare liposomes. When loading water-soluble drugs, the lipid film is dispersed in a drug-containing aqueous phase.

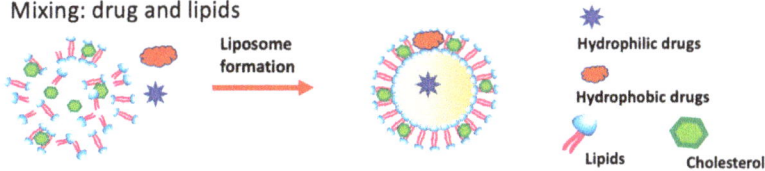
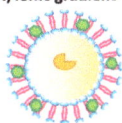
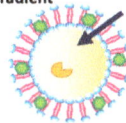

Figure 1. The two major methods for liposomal drug loading. (**A**) Passive loading involves co-current loading and liposomal formation. (**B**) In active loading, liposomes are formed containing a gradient used to load drugs.

1.1.3. Limitations of Passive Loading

The trapping efficiency of passive loading varies due to several factors, including drug solubility, vesicle size, lipid concentration, and preparation procedure. In most cases, the typical drug-to-lipid ratio (D/L) achieved by this passive loading technique is less than 0.05 (*w/w*) [14,15]. In addition, the entrapped drugs often cannot be retained stably due to weak association between the drugs and the liposomes, resulting in poor drug retention and storage stability. Passive loading often also results in a "burst release" phenomenon, whereby a large percentage of the entrapped drug is released quickly.

1.1.4. Active Loading

In active loading, liposomes are first generated containing a transmembrane gradient, i.e., the aqueous phases inside and outside the liposomes are different. Subsequently, an amphipathic drug dissolved in the exterior aqueous phase can permeate across the phospholipid bilayer(s), followed by interactions with a trapping agent in the core to lock-in the drug (Figure 1). In 1976, Deamer and Nicols [16–18] demonstrated that a pH gradient could be utilized to load catecholamine into liposomes, leading to stable retention in vitro. They formed liposomes in a low pH (pH 5) solution and then bathed the liposomes in an alkaline solution (pH 8) to create a 1000-fold difference in H_3O^+ ions across the bilayer. As the catecholamine molecules remained as the free form in the basic exterior, they freely permeated through the bilayer into the low pH core, where they were subsequently protonated. The charged catecholamine molecules were no longer membrane permeable and thus locked in. An accumulation of catecholamine molecules in the core of the liposomes was observed. Subsequent to the work of Dreamer et al. [18], Haran et al. [19] demonstrated that active loading can be achieved using an ammonium sulfate gradient. They produced liposomes with an interior containing ammonium sulfate. A concentrated solution of anthracycline was added to the liposomes and loading was then achieved after incubation at an elevated temperature. Anthracycline molecules formed aggregates in the core of the liposomes through precipitation with sulfate ions, and a high drug loading efficiency of >90% was achieved without burst drug leakage. Several other pH gradients have been established, including phosphate and calcium acetate [20]. Similar to the calcium acetate gradient, basified copper acetate gradients have also been employed for liposomal loading.

1.1.5. Limitations of Standard Active Loading

Active loading remains a powerful tool that can be used to effectively and stably retain drugs in the core of liposomes. However, traditional active loading is predicated on compounds having high solubility and membrane permeability (i.e., amphipathic), so that they can be solubilized in the exterior aqueous phase and permeable through the lipid bilayer into the liposomal core. These selection criteria exclude a large number of hydrophobic drugs currently on the market or in the development pipeline. Excipients such as ß-cyclodextrin and super saturated solutions have been also used to increase the active loading efficiency of liposomes. These techniques have helped somewhat in overcome issues of poor solubility [21]. In addition, recent approaches have focused on examining the role of nanocrystillazaion inside liposomes to improve drug retention [22]. Although discussion of these techniques is beyond the scope of this paper, these methods have significant limitations, and consequently there is a need for an improved loading technology.

2. Solvent-Assisted Active Loading Technology (SALT)

2.1. Introduction

Our laboratory has developed an innovative technology for liposomal loading of poorly water-soluble drugs. Solvent-assisted active loading technology (SALT) allows for stable and efficient loading of poorly water-soluble drugs into the aqueous core of liposomes. SALT has demonstrated efficacy across a range of poorly soluble compounds and is a versatile method for drug loading. We hypothesized that including a small volume of water-miscible solvent in the loading mixture of liposomes and a hydrophobic drug could help solubilize the drug in the aqueous phase, increase the drug penetration through the liposomal bilayer and boost active loading encapsulation efficiency. pH adjustment is among the most commonly-used methods to increase drug solubility. When compared to pH adjustment, the SALT/liposome technique enhanced the solubility of our model drugs Staurosporine, Gambogic Acid and Mefloquine from 120 µg/mL to 1 mg/mL, 5 µg/mL to 1 mg/mL, and 0.6 mg/mL to 8 mg/mL, respectively [13,23,24].

2.2. Applications and Mechanism

A drug needs to be solubilized in the free form in the exterior phase of liposomes in order to effectively penetrate the lipid bilayer. Therefore, a small amount of solvent is included in the liposomal suspension for complete solubilisation of the compound. After penetration into the liposomal core, the drug can then interact with a trapping agent inside the liposomes to form insoluble precipitates for stable a "lock-in" effect. Finally, the solvent can be removed by dialysis or gel filtration (Figure 2). To test the feasibility of this idea, we performed proof-of-principle studies [13,23,24] with a range of poorly soluble drugs. We further characterized SALT through the use of various solvents, liposomal formulations and trapping agents to demonstrate its versatility.

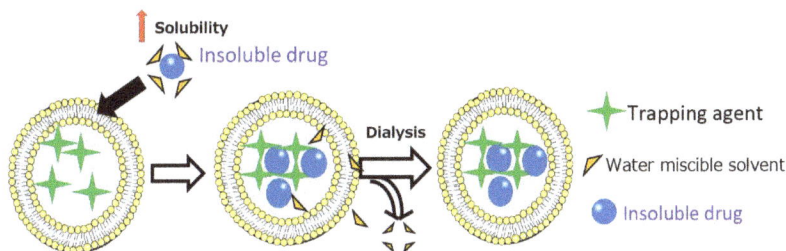

Figure 2. Solvent-assisted active loading technology (SALT) mechanism overview for liposomal loading of a poorly water-soluble drug.

2.3. Proof-of-Principle with a Weak Base Drug

Staurosporine and Liposomal Loading

Staurosporine (STS) is a broadly acting alkaloid protein kinase inhibitor with demonstrated efficacy as an antitumor agent against several cancer types [25]. STS contains a secondary amine group, but its solubility remains negligible even at a low pH [26]. Despite promising in vitro results, STS development was hindered due to poor solubility and non-specificity [26]. We hypothesized that liposomal delivery of STS would minimize its systemic toxicity through enhanced tumor targeting while also overcoming issues surrounding poor solubility. Liposomal loading of STS was previously reported by Mukthavaram et al. [27]. They utilized an active loading strategy employing a reverse pH gradient [27]. However, they were only able to achieve 70% drug encapsulation and a low drug-to-lipid ratio (D/L) of 0.09 mol/mol [27]. We hypothesized that introducing a limited amount of dimethyl sulfoxide (DMSO) to the liposomal suspension could keep STS soluble during the active loading process and would also facilitate the permeation of the drug into the inner core of liposomes. We employed an ammonium sulfate gradient to effectively trap STS via sulfate-STS nano-aggregates inside the liposomal core and achieve 100% drug encapsulation at a high D/L of 0.31/1 (mol/mol). Subsequently, gel filtration facilitated the removal of DMSO generating a final liposomal product. In the following section, we report the SALT method optimization and results of PK and efficacy studies.

Liposomal Formulation: A thin lipid film composed of 1,2-distearoyl-sn-glycero-3-phosphocholine (DSPC), cholesterol and PEGylated 1,2-distearoyl-sn-glycero-3-phosphoetholamine (DSPE-PEG2000) (55/40/5 mole ratio) (100 mg total lipids) was hydrated with 1 mL of 350 mM ammonium sulfate at 60 °C, followed by membrane extrusion to control the size ~100 nm with a polydispersity index (PDI) of <0.06. The liposomes were then dialyzed against an acetate buffer (100 mM, pH 5) to create a transmembrane gradient of ammonium sulfate.

Liposomal STS Loading Using SALT: We dissolved STS in DMSO and added into the liposomal suspension at a range of D/L with a final DMSO content of 5–60% and incubated the mixture at room temperature or 60 °C. We found that at least 5% DMSO was required to achieve complete drug loading, and that complete drug loading was maintained over a range of DMSO concentrations between 5% and 60%. Loading kinetics was impacted by the temperature. In the presence of 5% DMSO, complete loading was achieved more rapidly (5 min) with incubation at 60 °C compared to room temperature (15 min). The highest D/L achieved for complete drug loading was 0.31 w/w. The size (~100 nm) and the PDI (<0.06) of the final STS-Lipo produced with various amounts of DMSO were comparable and remained unchanged compared to the empty liposomes. We also determined that STS leakage from the liposomes was minimal (<5%) after seven days incubation in 50% fetal bovine serum (FBS) at 37 °C. Finally, the cryo-transmission electron microscopy revealed that STS molecules form spherical precipitates inside the liposomal core (Figure 3). The data support the contention that STS was actively and stably loaded into the liposomal core by the SALT.

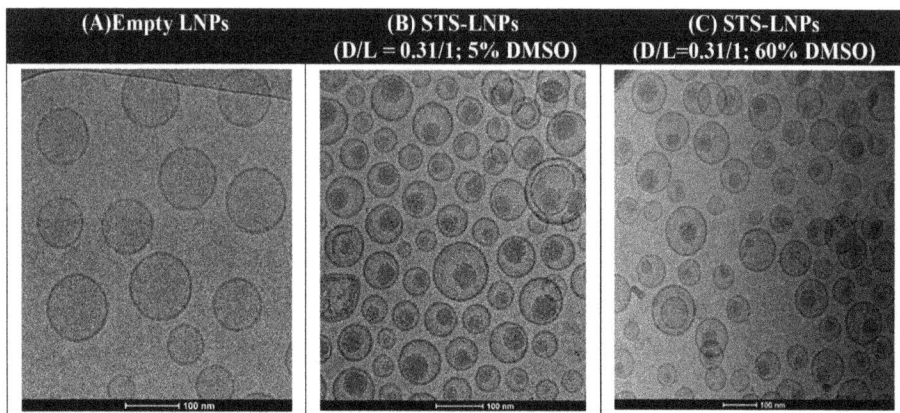

Figure 3. Cryo-TEM images of the drug free liposomes (**a**) and the staurosporine (STS)-Lipo loaded using 5% dimethyl sulfoxide (DMSO) (**b**) 60% DMSO (**c**). Scale bar represents 100 nm. Reprinted with permission from Tang et al., Pharmaceutical Research, published by Springer Nature, 2016 [24].

Safety and Efficacy Studies: It was found that the accumulated maximum tolerated dose (MTD) of STS-Lipo was 9 mg/kg relative to 3 mg/kg for free STS (dissolved in acetate buffer and 1% DMSO). STS-Lipo exhibited significantly improved efficacy against multidrug resistant EMT6-AR1 murine breast tumor in mice compared with free STS and docetaxel (DTX). The tumor growth was effectively impeded by STS-Lipo therapy and the average tumor volume was controlled by 150 mm^3 in comparison to ~800 mm^3 in the free STS group by day 18, while all tumors in the DTX and buffer treated mice all exceeded the endpoint size (1000 mm^3) before day 14. The STS-Lipo treatment did not cause significant body weight loss. On the other hand, 3 out of 3 mice in the free STS group reached humane endpoints on day 9 due to the drug toxicity, and there was ~5% body weight loss in the mice treated with DTX. The data indicate that SALT enabled liposomal delivery of STS, leading to enhanced safety and efficacy compared to free STS and the standard taxane chemotherapy.

2.4. Proof-of-Principle with a Weak Acid Drug

2.4.1. Introduction

After our initial discovery that SALT could promote loading of an insoluble weak base compound into the liposomal aqueous core to improve the drug delivery, we sought to further explore applications for SALT. We investigated whether SALT could efficiently load other drug classes. In addition, we examined whether solvents other than DMSO could be used in this technology and the role these solvents played in drug loading. To achieve this, we investigated the ability of SALT to improve loading of gambogic acid (GA). GA was selected as a model drug as it is water insoluble yet dissolvable in a range of water miscible solvents (>20 mg/mL). As such, we could investigate both the use of SALT on a different drug class (weak acid) and how SALT performs with solvents other than DMSO.

2.4.2. Gambogic Acid (GA)

GA is a naturally-derived compound found in traditional Asian medicines. GA has been reported to have anti-cancer and anti-inflammatory properties by acting through a number of cell processes [28–30]. Despite its promising results as an anti-cancer compound, its clinical development has been hindered by its poor water solubility (<5 μg/mL). Several methods have been attempted to overcome its issues of poor solubility, yet parameters relating to PK were only marginally improved [31–35]. We hypothesized that SALT could facilitate stable and efficient loading of GA into liposomes, leading to prolonged PK and tumor-targeted delivery.

SALT Promotes GA Loading into Liposomes: We began the GA study by seeking to find answers to three questions. Primarily, we were interested in determining whether SALT could be applied to load other drugs besides weak bases. Second, we sought to determine if other water miscible solvents were compatible with the SALT system. Third, we wanted to elucidate the mechanism behind the SALT system, specifically, to investigate the functions of solvents in promoting drug loading. GA is freely soluble in eight different water-miscible solvents; DMSO, DMF, EtOH, MeOH, acetonitrile, acetone, 1,4-dioxane, and NMP. We examined whether these solvents could be used to promote GA loading into liposomes. We prepared liposomes containing a gradient of basified copper acetate where $[Cu^{2+}]_{interior} > [Cu^{2+}]_{exterior}$. We rationalized that after drug solubilization and loading into the liposomal core, GA would bind cooper via coordination complex. This drug-copper conjugate forms stable complexes and effectively traps GA in the liposome. This mechanism, which parallels other trapping strategies using compounds such as calcium acetate, helps boost drug retention and minimize leakage during storage. Our results showed that all 8 solvents could facilitate GA loading and the optimal solvent content was unique to the solvent. The data suggests the solvent function is twofold. Primarily, the solvent is used to completely dissolve the drug in the liposomal exterior water phase. This is required as large drug precipitates are impermeable to the lipid bilayer. However, we discovered that just complete solubilisation did not promote complete drug loading. After the solvent has dissolved the drug, additional solvent added serves to increase the permeability of the liposomal membrane (Figure 4). However, additional solvent must not exceed the limit that induces lipid membrane instability. In these studies, we have demonstrated that SALT is capable of loading both poorly soluble weak acid and base model drugs with high efficiency and retention. SALT is a versatile platform which could be used to load an array of compounds that may have been previously non-deliverable with liposomes. In addition to STS and GA, we further demonstrated that this method could be applied for loading other drugs, including artesunate, prednisolone hemisuccinate, and quercetin [13].

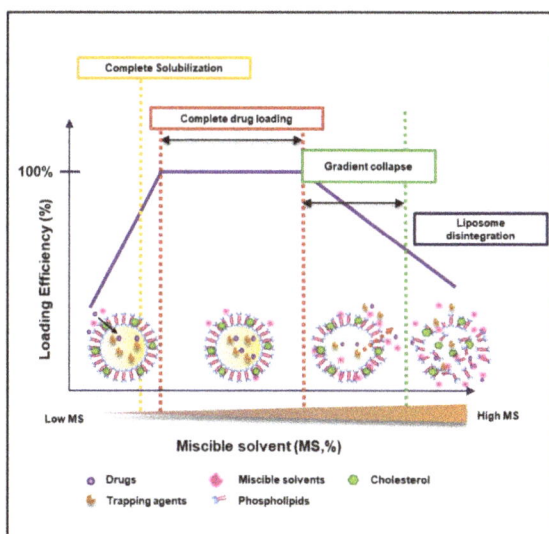

Figure 4. The solvent effect on drug loading in the SALT system. The figure is reprinted with permission from Tang et al., Biomaterials; published by Elsevier, 2018 [13].

Formulation Optimization: After confirming that SALT could facilitate active loading of GA, we next sought to optimize the formulation by modifying the loading gradient and lipid composition to improve the D/L and drug retention. We studied GA loading efficiency in 1,2-distearoyl-sn-glycero-3-phosphocholine/cholesterol/1,2-distearoyl-sn-glycero-3-phosphoethanolamine-N-[amino (polyethyleneglycol)-2000 (DSPC/Chol/DSPE-PEG2K) liposomes with a range of transmembrane gradients, including magnesium gluconate, calcium formate, and copper acetate. All gradients achieved complete drug loading in the presence of 5 vol% DMSO at a D/L of 1/5 *w/w*. However, the basified copper acetate gradient (pH 9) demonstrated the highest drug retention in 50% serum: ~45% GA retained in the liposomes after 4 h incubation. We then modified the lipid composition to further enhance GA retention. Our results show that decreasing the cholesterol content from 45% to 0% increased GA retention from 40% to 68% after 2.5 h incubation in serum. We then compared formulations prepared with an unsaturated lipid 1,2-dioleoyl-sn-glycero-3-phosphocholine (DOPC) or a saturated lipid (DSPC): the DOPC-liposomes retained >95% of the drug after 24 h incubation in serum without significant size change. This finding was unanticipated as saturated liposomes containing lipids with a high transition temperature have been shown to increase doxorubicin retention relative to unsaturated lipids [36]. However, it has been discovered that unsaturated lipids, such as DOPC, may form flexible liposomes which trap hydrophobic molecules more effectively [37].

Characterization of optimized Lipo-GA: Transmission electron cryomicroscopy (CryoTEM) imaging revealed that liposomes displayed bi-lamellar structure with an electron dense core (Figure 5). Both these features have been previously reported with liposomes containing a copper gradient for drug loading [38,39], indicative of bilayer rearrangement and copper-GA complex formation in the core. The formation of copper-GA complexes, in addition to the optimized lipid composition resulted in stable retention of GA.

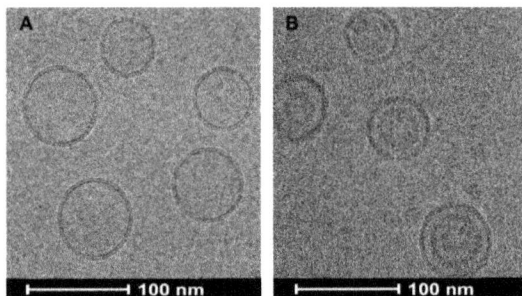

Figure 5. Cryo-TEM images of empty liposomes (DOPC/Chol/DSPE-PEG2K, 85/10/5 by mol%) (**A**) and Lipo-GA (**B**). The figure is reprinted with permission from Tang et al., Biomaterials; published by Elsevier, 2018 [13].

Safety, PK and Efficacy: GA is known to induce apoptosis of red blood cells, leading to potential toxicity in vivo [40]. While free GA was highly hemolytic, the equivalent concentration of Lipo-GA showed no activity in inducing hemolysis. Compared to free GA, the Lipo-GA formulation resulted in an 18-fold increase in plasma half-life, a 20-fold higher mean residence time, a 7.5-fold higher area under the curve ($AUC_{0-\infty}$), and 10-fold decreased clearance, confirming its prolonged circulation relative to free GA. In two murine tumor models, we observed a significant dose-dependent reduction in tumour volume after Lipo-GA therapy. In particular, in the EMT6-AR1 multidrug resistant breast tumor model, one dose of Lipo-GA completely suppressed the tumor growth, while free GA only moderately inhibited 65% tumor growth. Mice treated with Lipo-GA showed no body weight loss, suggesting good safety.

3. Pediatric Formulation

We next explored a unique application for SALT. Malaria is the world's leading parasitic disease [41], with over 200 million cases reported in 2015. Children under 5 years old are the most vulnerable population and account for over 70% of malaria associated deaths [42]. Treatment or prophylaxis of malaria often relies on the first line drug mefloquine (Mef) [43]. Despite its effectiveness as an antimalarial agent, Mef is difficult to accurately dose in children and neonates [44]. This is due to the lack of a pediatric formulation and Mef is typically administered by crushing up a portion of an adult tablet and administering with milk or food, in an effort to mask the extremely bitter taste. However, children often spit out the medicine, leading to subtherapeutic levels. Microemulsions of Mef have been developed to overcome poor solubility and increase bioavailability [45,46], yet their use in newborns is contradicted by risk of renal and liver damage caused by high concentrations of surfactants [47]. A pediatric pill of Mef is available (Artequin pediatric), yet this pill remains too large for neonatal dosing. We hypothesized that liposomal delivery of Mef could overcome many of these issues. Liposomal suspension of Mef could increase the solubility and subsequent absorption leading to higher bioavailability [48]. In addition, dosing the liquid suspension to children is easy and accurate. We hypothesized liposomes could also mask the taste of the bitter drug by shielding it from the taste buds. Liposomes can be lyophilized to prepare a powder formulation that is stable upon room temperature storage and orally dispersible for pediatric use. Lastly, liposomal formulations using only the neutral lipid DSPC and cholesterol can be considered very safe for oral administration in young children. The remaining challenge was how to prepare a stable Mef-Lipo formulation.

Liposomal Loading of Mef Using SALT: We first prepared liposomes (DSPC and cholesterol) containing an ammonium gradient, and then incubated the liposomes with Mef in the presence of 10% DMSO at a D/L concentration of 0.1 *w/w* at room temperature for 30 min to achieve complete loading. The liposomes were then purified by dialysis to remove DMSO and a formulation containing 8 mg/mL of Mef was obtained.

Lyophilizaition: The Mef-Lipo suspension was subjected to freeze drying, aiming to prepare a solid powder formulation to facilitate longer storage time and produce a rapidly dissolvable formulation. To protect the integrity of the liposomes during lyophilization we buffered the exterior with 300 mM sucrose and 20 mM phosphate. Sucrose is a known lyoprotectant and phosphate was shown to increase the stability of liposomes [49]. The lyophilization process only increased the liposome size from 110 nm to 130 nm and the PDI remained <0.1. Importantly, we observed no significant drug leakage. The lyophilized liposomes containing Mef remained stable under the storage at room temperature for >3 months. In addition, they were rapidly dissolvable in water within 10 s, indicating an orally dispersible formulation.

Drug Release: Both the liquid suspension and the lyophilized Mef-Lipo exhibited similar drug release profiles in simulated saliva, gastric, and intestinal fluids. In simulated saliva, both liquid and lyophilized Mef-Lipo demonstrated no observable drug release. Drug release was highest in the simulated stomach fluid for both formulations. There was no drug release in the simulated intestinal fluid in the absence of a bile salt for both preparations, while Mef was effectively released from the liposomes in simulated intestinal fluid supplemented with a bile salt. Our data indicate that Mef release from the liposomes was triggered by acid- or surfactant-induced destabilization of liposomes. The results also suggest that the liposomal formulation would effectively mask the drug taste in the oral cavity but rapidly release the drug in the gastrointestinal fluids.

Bitterness Masking: To determine the bitterness masking effect of the liposomal technology, we employed the Astree e-tongue technology and compared the bitterness of Mef-Lipo to 10% sucrose, Mef suspension and Infant Tylenol®. As shown in Figure 6, the bitterness of the Mef-Lipo was similar to 10% sucrose indicating a palatable formulation, while the standard Mef suspension displayed high bitterness.

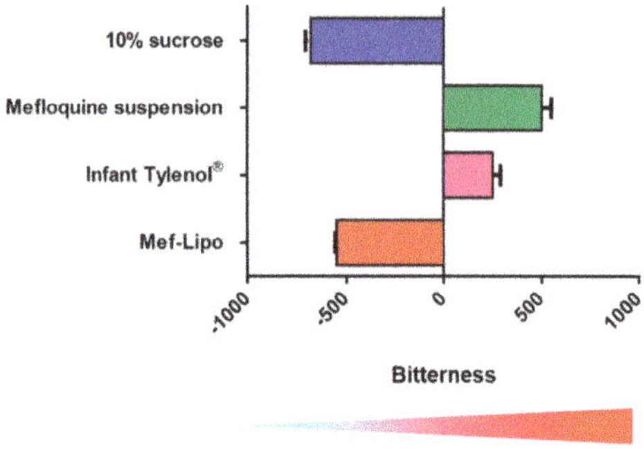

Figure 6. Quantification of bitterness of various drug formulations using the e-tongue. Data = mean ± Standard Deviation (SD) (*n* = 3). The figure is adapted with permission from Tang et al., Molecular Pharmaceutics; published by American Chemical Society, 2017 [23].

Pharmacokinetics and Bioavailability: In PK and bioavailability (BA) studies in mice, our data confirmed that the liquid and lyophilized Mef-Lipo were comparable formulations with similar C_{max}, T_{max}, area under the curve (AUC), and BA (81−86%). Mef suspension, however, displayed ~20% decreased C_{max}, AUC, and BA compared to the liposomal formulation. There was no difference in T_{max} and $t_{1/2}$ among these three formulations, suggesting that the formulations did not alter the metabolism or elimination of the drug, and the difference in BA was due to the absorption. Liposomes might improve the absorption by preventing aggregation of Mef in the gastrointestinal tract, protectiing from degradation, and promoting direct uptake by enteric cells.

4. Perspectives and Future Directions

We have demonstrated that SALT is a versatile loading technology for promoting the active loading of poorly water-soluble drugs into the liposomal core. We performed proof-of-principle studies with multiple drugs to characterize the versatility and robustness of SALT. Our studies show that SALT could improve drug solubility, liposomal loading efficiency, liposomal retention, stability, palatability, PK, and efficacy. SALT has the potential to overcome barriers currently impeding delivery of many compounds. We have also demonstrated applications of SALT in cancer therapy and child-friendly oral formulations. The application of SALT could extend beyond therapeutic drug loading and also be employed in the imaging and theranostic fields. Specifically, SALT could be useful in loading poorly soluble imaging probes into the core of liposomes. A frequent barrier to effective image-guided drug targeting is the weak association between the probe and the delivery vehicle [50]. SALT could help retain imaging probes inside the core of liposomes to improve their delivery to target tissues. Despite the high performance of SALT as a loading platform, there remain parameters that can be further optimized. As demonstrated in these specific studies, a limitation of SALT when implemented across different compounds is the requirement for optimization of the loading gradient, trapping method, and lipid composition for maximal loading efficiency and retention. However, trapping agents can be rationalized based on the drug structure and lipid composition can be empirically determined. In addition, a limitation of SALT is found in the need to find a water miscible solvent that can be used to load certain drugs, which may not always be possible. Moreover, some drugs may require large volumes of solvent that may result in degradation on the liposome itself prior to efficient drug loading although this was not found in our studies. The drugs themselves must almost possess an ionizable functional group, limiting potential candidates for this technology. Despite these limitations we anticipate that future innovations of SALT could help overcome loading of other species of drugs such as the poorly membrane-permeable yet water-soluble biomolecules such as proteins and nucleotides, both of which may benefit from liposomal delivery. SALT, however, is not limited to liposomes and will be implemented for drug loading in other bilayer-based delivery systems, further expanding the impact of this exciting technology. In addition, SALT can be used to improve current formulations that are prepared using passive loading. There exist several drugs on the market that are prepared using passive loading due to their hydrophobicity. SALT could be employed to re-formulate these existing liposomal products to improve the loading efficiency, stability and drug retention. SALT is a promising platform which may propel the field of lipid-based drug delivery to novel areas.

Funding: This research was supported by grants from Canadian Institutes of Health Research (CIHR), Natural Sciences and Engineering Research Council of Canada and Canada Foundation of Innovation. W.-L.T. was supported by the Frederick Banting and Charles Best Canada Graduate Scholarship from CIHR as well as Four Year Fellowship (4YF) Tuition Award from the UBC. S.-D.L. is a recipient of CIHR New Investigator Salary Awards and the Angiotech Professorship in Drug Delivery.

Conflicts of Interest: The authors declare no conflict of interest.

References

1. Savjani, K.T.; Gajjar, A.K.; Savjani, J.K. Drug solubility: Importance and enhancement techniques. *ISRN Pharm.* **2012**, *2012*, 195727. [CrossRef] [PubMed]
2. Kalepu, S.; Nekkanti, V. Insoluble drug delivery strategies: Review of recent advances and business prospects. *Acta Pharm. Sin. B* **2015**, *5*, 442–453. [CrossRef] [PubMed]
3. Butler, J.M.; Dressman, J.B. The developability classification system: Application of biopharmaceutics concepts to formulation development. *J. Pharm. Sci.* **2010**, *99*, 4940–4954. [CrossRef] [PubMed]
4. McNamara, K.; Tofail, S.A.M. Nanoparticles in biomedical applications. *Adv. Phys. X* **2017**, *2*, 54–88. [CrossRef]
5. Akbarzadeh, A.; Rezaei-Sadabady, R.; Davaran, S.; Joo, S.W.; Zarghami, N.; Hanifehpour, Y.; Samiei, M.; Kouhi, M.; Nejati-Koshki, K. Liposome: Classification, preparation, and applications. *Nanoscale Res. Lett.* **2013**, *8*, 102. [CrossRef] [PubMed]
6. Zhang, H. Thin-Film Hydration Followed by Extrusion Method for Liposome Preparation. *Methods Mol. Biol.* **2017**, *1522*, 17–22. [CrossRef]
7. Cortesi, R.; Esposito, E.; Gambarin, S.; Telloli, P.; Menegatti, E.; Nastruzzi, C. Preparation of liposomes by reverse-phase evaporation using alternative organic solvents. *J. Microencapsul.* **1999**, *16*, 251–256. [CrossRef] [PubMed]
8. Yu, B.; Lee, R.J.; Lee, L.J. Microfluidic methods for production of liposomes. *Methods Enzymol.* **2009**, *465*, 129–141. [CrossRef]
9. Pattni, B.S.; Chupin, V.V.; Torchilin, V.P. New Developments in Liposomal Drug Delivery. *Chem. Rev.* **2015**, *115*, 10938–10966. [CrossRef]
10. Maeki, M.; Kimura, N.; Sato, Y.; Harashima, H.; Tokeshi, M. Advances in microfluidics for lipid nanoparticles and extracellular vesicles and applications in drug delivery systems. *Adv. Drug Deliv. Rev.* **2018**, *128*, 84–100. [CrossRef]
11. Dimov, N.; Kastner, E.; Hussain, M.; Perrie, Y.; Szita, N. Formation and purification of tailored liposomes for drug delivery using a module-based micro continuous-flow system. *Sci. Rep. UK* **2017**, *7*. [CrossRef] [PubMed]
12. Anderson, M.; Omri, A. The effect of different lipid components on the in vitro stability and release kinetics of liposome formulations. *Drug Deliv.* **2004**, *11*, 33–39. [CrossRef] [PubMed]
13. Tang, W.L.; Tang, W.H.; Szeitz, A.; Kulkarni, J.; Cullis, P.; Li, S.D. Systemic study of solvent-assisted active loading of gambogic acid into liposomes and its formulation optimization for improved delivery. *Biomaterials* **2018**, *166*, 13–26. [CrossRef] [PubMed]
14. Gubernator, J. Active methods of drug loading into liposomes: Recent strategies for stable drug entrapment and increased in vivo activity. *Expert Opin. Drug Deliv.* **2011**, *8*, 565–580. [CrossRef] [PubMed]
15. Zhao, Y.C.; May, J.P.; Chen, I.W.; Undzys, E.; Li, S.D. A Study of Liposomal Formulations to Improve the Delivery of Aquated Cisplatin to a Multidrug Resistant Tumor. *Pharm. Res.* **2015**, *32*, 3261–3268. [CrossRef]
16. Bally, M.B.; Mayer, L.D.; Loughrey, H.; Redelmeier, T.; Madden, T.D.; Wong, K.; Harrigan, P.R.; Hope, M.J.; Cullis, P.R. Dopamine accumulation in large unilamellar vesicle systems induced by transmembrane ion gradients. *Chem. Phys. Lipids* **1988**, *47*, 97–107. [CrossRef]
17. Mayer, L.D.; Bally, M.B.; Cullis, P.R. Uptake of Adriamycin into Large Unilamellar Vesicles in Response to a Ph Gradient. *Biochim. Biophys. Acta* **1986**, *857*, 123–126. [CrossRef]

18. Deamer, D.W.; Prince, R.C.; Crofts, A.R. The response of fluorescent amines to pH gradients across liposome membranes. *Biochim. Biophys. Acta* **1972**, *274*, 323–335. [CrossRef]

19. Haran, G.; Cohen, R.; Bar, L.K.; Barenholz, Y. Transmembrane Ammonium-Sulfate Gradients in Liposomes Produce Efficient and Stable Entrapment of Amphipathic Weak Bases. *Biochim. Biophys. Acta* **1993**, *1151*, 201–215. [CrossRef]

20. Clerc, S.; Barenholz, Y. Loading of amphipathic weak acids into liposomes in response to transmembrane calcium acetate gradients. *Biochim. Biophys. Acta* **1995**, *1240*, 257–265. [CrossRef]

21. Bhatt, P.; Lalani, R.; Vhora, I.; Patil, S.; Amrutiya, J.; Misra, A.; Mashru, R. Liposomes encapsulating native and cyclodextrin enclosed paclitaxel: Enhanced loading efficiency and its pharmacokinetic evaluation. *Int. J. Pharm.* **2018**, *536*, 95–107. [CrossRef] [PubMed]

22. Li, T.; Cipolla, D.; Rades, T.; Boyd, B.J. Drug nanocrystallisation within liposomes. *J. Control. Release* **2018**, *288*, 96–110. [CrossRef] [PubMed]

23. Tang, W.L.; Tang, W.H.; Chen, W.C.; Diako, C.; Ross, C.F.; Li, S.D. Development of a Rapidly Dissolvable Oral Pediatric Formulation for Mefloquine Using Liposomes. *Mol. Pharm.* **2017**, *14*, 1969–1979. [CrossRef] [PubMed]

24. Tang, W.L.; Chen, W.C.; Roy, A.; Undzys, E.; Li, S.D. A Simple and Improved Active Loading Method to Efficiently Encapsulate Staurosporine into Lipid-Based Nanoparticles for Enhanced Therapy of Multidrug Resistant Cancer. *Pharm. Res.* **2016**, *33*, 1104–1114. [CrossRef] [PubMed]

25. Schwartz, G.K.; Redwood, S.M.; Ohnuma, T.; Holland, J.F.; Droller, M.J.; Liu, B.C.S. Inhibition of Invasion of Invasive Human Bladder-Carcinoma Cells by Protein-Kinase-C Inhibitor Staurosporine. *J. Natl. Cancer Inst.* **1990**, *82*, 1753–1756. [CrossRef]

26. Akinaga, S.; Gomi, K.; Morimoto, M.; Tamaoki, T.; Okabe, M. Antitumor-Activity of Ucn-01, a Selective Inhibitor of Protein-Kinase-C, in Murine and Human Tumor-Models. *Cancer Res.* **1991**, *51*, 4888–4892. [PubMed]

27. Mukthavaram, R.; Jiang, P.F.; Saklecha, R.; Simberg, D.; Bharati, I.S.; Nomura, N.; Chao, Y.; Pastorino, S.; Pingle, S.C.; Fogal, V.; et al. High-efficiency liposomal encapsulation of a tyrosine kinase inhibitor leads to improved in vivo toxicity and tumor response profile. *Int. J. Nanomed.* **2013**, *8*, 3991–4006. [CrossRef]

28. Wu, Z.Q.; Guo, Q.L.; You, Q.D.; Zhao, L.; Gu, H.Y. Gambogic acid inhibits proliferation of human lung carcinoma SPC-A1 cells in vivo and in vitro and represses telomerase activity and telomerase reverse transcriptase mRNA expression in the cells. *Biol. Pharm. Bull.* **2004**, *27*, 1769–1774. [CrossRef]

29. Li, X.F.; Liu, S.T.; Huang, H.B.; Liu, N.N.; Zhao, C.; Liao, S.Y.; Yang, C.S.; Liu, Y.R.; Zhao, C.G.; Li, S.J.; et al. Gambogic Acid Is a Tissue-Specific Proteasome Inhibitor In Vitro and In Vivo. *Cell Rep.* **2013**, *3*, 211–222. [CrossRef]

30. Ishaq, M.; Khan, M.A.; Sharma, K.; Sharma, G.; Dutta, R.K.; Majumdar, S. Gambogic acid induced oxidative stress dependent caspase activation regulates both apoptosis and autophagy by targeting various key molecules (NF-kappa B, Beclin-1, p62 and NBR1) in human bladder cancer cells. *Biochim. Biophys. Acta* **2014**, *1840*, 3374–3384. [CrossRef]

31. Cai, L.L.; Qiu, N.; Xiang, M.L.; Tong, R.S.; Yan, J.F.; He, L.; Shi, J.Y.; Chen, T.; Wen, J.L.; Wang, W.W.; et al. Improving aqueous solubility and antitumor effects by nanosized gambogic acid-mPEG(2000) micelles. *Int. J. Nanomed.* **2014**, *9*, 243–255. [CrossRef]

32. Doddapaneni, R.; Patel, K.; Owaid, I.H.; Singh, M. Tumor neovasculature-targeted cationic PEGylated liposomes of gambogic acid for the treatment of triple-negative breast cancer. *Drug Deliv.* **2016**, *23*, 1232–1241. [CrossRef] [PubMed]

33. Zhang, Z.; Qian, H.Q.; Yang, M.; Li, R.T.; Hu, J.; Li, L.; Yu, L.X.; Liu, B.R.; Qian, X.P. Gambogic acid-loaded biomimetic nanoparticles in colorectal cancer treatment. *Int. J. Nanomed.* **2017**, *12*, 1593–1605. [CrossRef] [PubMed]

34. Yin, D.K.; Yang, Y.; Cai, H.X.; Wang, F.; Peng, D.Y.; He, L.Q. Gambogic Acid-Loaded Electrosprayed Particles for Site-Specific Treatment of Hepatocellular Carcinoma. *Mol. Pharm.* **2014**, *11*, 4107–4117. [CrossRef] [PubMed]

35. Zhang, D.H.; Zou, Z.Y.; Ren, W.; Qian, H.Q.; Cheng, Q.F.; Ji, L.L.; Liu, B.R.; Liu, Q. Gambogic acid-loaded PEG-PCL nanoparticles act as an effective antitumor agent against gastric cancer. *Pharm. Dev. Technol.* **2018**, *23*, 33–40. [CrossRef] [PubMed]

36. Charrois, G.J.R.; Allen, T.M. Drug release rate influences the pharmacokinetics, biodistribution, therapeutic activity, and toxicity of pegylated liposomal doxorubicin formulations in murine breast cancer. *Biochim. Biophys. Acta* **2004**, *1663*, 167–177. [CrossRef] [PubMed]

37. Chang, H.I.; Yeh, M.K. Clinical development of liposome-based drugs: Formulation, characterization, and therapeutic efficacy. *Int. J. Nanomed.* **2012**, *7*, 49–60. [CrossRef]

38. Kheirolomoom, A.; Mahakian, L.M.; Lai, C.Y.; Lindfors, H.A.; Seo, J.W.; Paoli, E.E.; Watson, K.D.; Haynam, E.M.; Ingham, E.S.; Xing, L.; et al. Copper-Doxorubicin as a Nanoparticle Cargo Retains Efficacy with Minimal Toxicity. *Mol. Pharm.* **2010**, *7*, 1948–1958. [CrossRef]

39. Dicko, A.; Kwak, S.; Frazier, A.A.; Mayer, L.D.; Liboiron, B.D. Biophysical characterization of a liposomal formulation of cytarabine and daunorubicin. *Int. J. Pharm.* **2010**, *391*, 248–259. [CrossRef]

40. Lupescu, A.; Jilani, K.; Zelenak, C.; Zbidah, M.; Shaik, N.; Lang, F. Induction of Programmed Erythrocyte Death by Gambogic Acid. *Cell. Physiol. Biochem.* **2012**, *30*, 428–438. [CrossRef]

41. Breman, J.G.; Alilio, M.S.; Mills, A. Conquering the intolerable burden of malaria: What's new, what's needed: A summary. *Am. J. Trop. Med. Hyg.* **2004**, *71*, 1–15. [CrossRef] [PubMed]

42. Caminade, C.; Kovats, S.; Rocklov, J.; Tompkins, A.M.; Morse, A.P.; Colon-Gonzalez, F.J.; Stenlund, H.; Martens, P.; Lloyd, S.J. Impact of climate change on global malaria distribution. *Proc. Natl. Acad. Sci. USA* **2014**, *111*, 3286–3291. [CrossRef] [PubMed]

43. Schlagenhauf, P.; Adamcova, M.; Regep, L.; Schaerer, M.T.; Bansod, S.; Rhein, H.G. Use of mefloquine in children—A review of dosage, pharmacokinetics and tolerability data. *Malar. J.* **2011**, *10*. [CrossRef] [PubMed]

44. White, N.J. Antimalarial drug resistance. *J. Clin. Investig.* **2004**, *113*, 1084–1092. [CrossRef] [PubMed]

45. du Plessis, L.H.; Helena, C.; van Huysteen, E.; Wiesner, L.; Kotze, A.F. Formulation and evaluation of Pheroid vesicles containing mefloquine for the treatment of malaria. *J. Pharm. Pharmcol.* **2014**, *66*, 14–22. [CrossRef] [PubMed]

46. Mbela, T.K.M.; Deharo, E.; Haemers, A.; Ludwig, A. Submicron oil-in-water emulsion formulations for mefloquine and halofantrine: Effect of electric-charge inducers on antimalarial activity in mice. *J. Pharm. Pharmacol.* **1998**, *50*, 1221–1225. [CrossRef] [PubMed]

47. Schwartzberg, L.S.; Navari, R.M. Safety of Polysorbate 80 in the Oncology Setting. *Adv. Ther.* **2018**, *35*, 754–767. [CrossRef] [PubMed]

48. Daeihamed, M.; Dadashzadeh, S.; Haeri, A.; Akhlaghi, M.F. Potential of Liposomes for Enhancement of Oral Drug Absorption. *Curr. Drug Deliv.* **2017**, *14*, 289–303. [CrossRef] [PubMed]

49. Kannan, V.; Balabathula, P.; Thoma, L.A.; Wood, G.C. Effect of sucrose as a lyoprotectant on the integrity of paclitaxel-loaded liposomes during lyophilization. *J. Liposome Res.* **2015**, *25*, 270–278. [CrossRef] [PubMed]

50. Tang, W.L.; Tang, W.H.; Li, S.D. Cancer theranostic applications of lipid-based nanoparticles. *Drug Discov. Today* **2018**, *23*, 1159–1166. [CrossRef]

pharmaceutics

MDPI

Review

Controlled Drug Delivery Systems for Oral Cancer Treatment—Current Status and Future Perspectives

Farinaz Ketabat [1,2,3], Meenakshi Pundir [1,2,3], Fatemeh Mohabatpour [1,2,3], Liubov Lobanova [2], Sotirios Koutsopoulos [4], Lubomir Hadjiiski [5], Xiongbiao Chen [3,6], Petros Papagerakis [2,3] and Silvana Papagerakis [1,3,7,*]

1   Laboratory of Oral, Head and Neck Cancer—Personalized Diagnostics and Therapeutics, Department of Surgery—Division of Head and Neck Surgery, College of Medicine, University of Saskatchewan, Saskatoon, SK S7N 5E5, Canada
2   Laboratory of Precision Oral Health and Chronobiology, College of Dentistry, University of Saskatchewan, Saskatoon, SK S7N 5E4, Canada
3   Division of Biomedical Engineering, University of Saskatchewan, Saskatoon, SK S7K 5A9, Canada
4   Center for Biomedical Engineering, Massachusetts Institute of Technology, Cambridge, MA 02139, USA
5   Departmnet of Radiology, School of Medicine, University of Michigan, Ann Arbor, MI 48109, USA
6   Department of Mechanical Engineering, University of Saskatchewan, Saskatoon, SK S7K 5A9, Canada
7   Department of Otolaryngology-Head and Neck Surgery, School of Medicine, University of Michigan, Ann Arbor, MI 48109, USA
*   Correspondence: silvana.papagerakis@usask.ca; Tel.: +1-306-966-1960

Received: 9 June 2019; Accepted: 26 June 2019; Published: 30 June 2019

Abstract: Oral squamous cell carcinoma (OSCC), which encompasses the oral cavity-derived malignancies, is a devastating disease causing substantial morbidity and mortality in both men and women. It is the most common subtype of the head and neck squamous cell carcinoma (HNSCC), which is ranked the sixth most common malignancy worldwide. Despite promising advancements in the conventional therapeutic approaches currently available for patients with oral cancer, many drawbacks are still to be addressed; surgical resection leads to permanent disfigurement, altered sense of self and debilitating physiological consequences, while chemo- and radio-therapies result in significant toxicities, all affecting patient wellbeing and quality of life. Thus, the development of novel therapeutic approaches or modifications of current strategies is paramount to improve individual health outcomes and survival, while early tumour detection remains a priority and significant challenge. In recent years, drug delivery systems and chronotherapy have been developed as alternative methods aiming to enhance the benefits of the current anticancer therapies, while minimizing their undesirable toxic effects on the healthy non-cancerous cells. Targeted drug delivery systems have the potential to increase drug bioavailability and bio-distribution at the site of the primary tumour. This review confers current knowledge on the diverse drug delivery methods, potential carriers (e.g., polymeric, inorganic, and combinational nanoparticles; nanolipids; hydrogels; exosomes) and anticancer targeted approaches for oral squamous cell carcinoma treatment, with an emphasis on their clinical relevance in the era of precision medicine, circadian chronobiology and patient-centred health care.

Keywords: oral, head and neck squamous cell carcinoma; targeted therapies; drug delivery systems; nanoparticles; controlled drug delivery; circadian clock; chronotherapy; precision medicine

---

1. Introduction

Oral cancer refers to tumors developed in the lips, hard palate, upper and lower alveolar ridges, anterior two-thirds of the tongue, sublingual area, buccal mucosa, retromolar trigons, and floor of the mouth [1]. The majority (>90%) of oral cancer are carcinomas with squamous differentiation arising from the mucosal epithelium, thus called oral squamous cell carcinomas (OSCCs) [2,3]. In 2018, 354,864 new cases of lip and oral cavity cancer were identified, and 177,384 people died from these types of cancer worldwide [4]. According to the Canadian Cancer Society and the Canadian Dental Association, the incidence of OSCC has increased in Canada in both males and females since mid-1990s; 4700 new cases of oral cancer and 1250 oral cancer-related deaths were reported in Canada in 2017 alone [5,6]. Most often diagnosed at late stages (approximately 60% of patients present with advanced stage disease at the initial diagnosis, OSCC remains one of the most difficult challenges in head and neck oncology, and continues to be a disfiguring and deadly disease with dismal 50% to 60% five-year disease specific survival rate [7,8]. Due to its anatomic location, OSCC progression and treatment significantly impact patient quality of life, involving impairment of most vital functions (e.g., speech, swallowing, taste.), appearance and sense of self; they are associated with profound functional morbidity even when the cancer is cured [3,9].

New trends have recently emerged in the OSCC patient profile including younger patients (younger than 50 years), particularly those with human papillomavirus (HPV)-positive tumors [10,11]; a steady change in the OSCC sex ratio with a worrisome increase in OSCC incidence and mortality in females [12]; and the implications of novel, previously unrecognized factors, such as the circadian clock disruption in the initiation and progression of the OSCC [13–16].

OSCC has traditionally been associated with risk factors such as tobacco and alcohol consumption; however, HPV, a well-known cause of cervical cancer, has emerged in recent years as an etiological cause for a subset of head and neck squamous cell carcinoma (HNSCC), particularly in patients who lack the traditional risk factors [17,18]. The majority (60–80%) of HPV-driven cancers of the head and neck are oropharyngeal squamous cell carcinomas (comprising the tonsils and the base of the tongue). Recent studies have identified various types of HPV associated with both benign and malignant lesions in the oral cavity [19–22].

The HPV diagnosis is critical in planning treatment for oropharyngeal cancer (OPC) patients [23–25]. Within OPC, there is a marked difference between clinical behaviors and outcomes for patients who test positive versus negative for HPV infection. For high/late-stage patients, HPV positivity has become a significant prognostic factor that is critical for guiding the choice of treatment, with an HPV positive diagnosis resulting in lower toxicities and improved outcomes [26]. In contrast, a significant subset of early-stage OPC patients are HPV negative, their cancer rapidly progresses into advanced metastatic tumors and fails to respond to the standard of care with poor outcomes and survival. Patients with chronic exposure of the entire mucosa of the upper digestive tract (cancerization field) to carcinogenic factors (e.g., from tobacco, alcohol, and betel quid chewing) are at a higher risk for multiple primary tumors.

Oral squamous cell carcinoma is the second most common cancer in transplant patients (e.g., treated for leukemia, lymphoma, multiple myeloma, etc) [27]. The conventional approaches for oral cancer treatment involve surgery, which is the treatment of choice, ionizing radiation which is the prevalent non-surgical therapeutic approach, or a combination of radio-, chemotherapy, and surgery [28]; surgical resection leads to permanent disfigurement, altered sense of self and debilitating physiological consequences, substantial functional impairment, and morbidity, while chemo- and radio-therapies result in significant toxicities, all affecting patient wellbeing and quality of life. These treatments are efficient for the treatment of the primary tumor but are used with palliative intent in advanced cases with metastatic disease, with significant side and adverse effects [29]. Despite the advances in surgery, chemotherapy, and radiotherapy for HNSCC treatment, the prognosis for this disease has not been significantly improved over the last 50 years [8]. Thus, the development of novel therapeutic approaches or modifications of current strategies is paramount to improve individual health outcomes and survival, while early tumor detection remains a priority and significant challenge.

The oral, head, and neck cancer is an immunosuppressive disease (characterized by a lower absolute lymphocyte count and poor antigen-presenting function) that interferes with the patient's natural immune response, preventing tumor cell recognition and immune-mediated clearance [30]. Immunotherapy, a recently developed cancer treatment modality, has shown promise as an additional therapeutic option in patients having failed multiple prior therapeutic modalities, due to the success of immune-modulating agents in patients with refractory solid tumors [31,32]. The goal of immunotherapy as an anticancer approach is to either block the pathways cancer cells use to escape the immune system or to enhance the patient's immune reactions directed against tumor cells [33]. Anti-cancer immunotherapy includes: (1) systemic therapy, which is a systemic immune activation including administration of systemic cytokine, cancer vaccines, or adoptive cell transfer; (2) local-based therapy, which is based on changes in local immune status including modulation of the immunosuppressive tumor microenvironment, with immune checkpoint or small molecular inhibitors [34]. Immune-modulating approaches available for the treatment of head and neck cancer target a variety of immune processes and critical checkpoints, including cytotoxic T-lymphocyte associated antigen-4 (CTLA4), and program death (PD-1) and its ligand (PD-L1); other methods using immune modulating molecules as well as combinatorial trials evaluating these agents in the first-line setting and early-stage disease are under development [35,36].

Because HNSCC tumors have been shown to poorly present tumor antigen (TA) on the cell surface, monoclonal antibodies facilitating better TA presentation are one avenue for targeted therapeutics [37]. Nivolumab and pembrolizumab, two anti-PD-1 agents, recently approved for use as monotherapy in the second-line setting for patients with platinum-refractory recurrent/metastatic HNSCC, have shown efficacy in clinical trials [30,38]. Other targeted therapies using epidermal growth factor receptors (EGFR, highly overexpressed in 80–90% of HNSCC) inhibitors, such as cetuximab, bevacizumab, and erlotinib, have shown improvement of OSCC patient survival [39]. Despite the promise of immunotherapies, new therapeutic approaches or improvements to clinical trials design that are tailored on the tumor/patient profile are much needed in order to overcome the innate and acquired tumor resistance, as well as to address/prevent their side and adverse effects [40]. Developing novel immunotherapeutic approaches can be promising in providing long-term control of the disease in the response population, although the low efficacy and high toxicity in some patients can be a severe issue [33,34,38]. Generally, using immunotherapy can be challenging due to auto-immune side effects, variability in tumor responses rate, and financial cost [36]. A solution to enhance the efficacy of immune agents is using nano-based drug delivery systems (DDSs) through direct targeting of the cancer cells, facilitating intracellular penetration, and boosting the immunogenicity of antigens [41]. To date, there are limited studies on the utilization of DDS combined with immunotherapy for the treatment of HNSCC or OSCC. Hirabayashi et al. and Maeda et al. developed anti-EGFR antibody-conjugated microbubbles for the treatment of HNSCC and OSCC, respectively [42,43]. These studies showed promising results for future applications of combined immunotherapy with DDSs.

DDS have been developed as an alternative method aiming to enhance the benefits of the current anticancer therapies, while minimizing their undesirable toxic effects on the healthy cells. For instance, the chemotherapeutic agents have several limitations in terms of oral bioavailability, stability in natural conditions, and non-specific bio-distribution, that decrease their therapeutic efficiency [44,45]; their side effects can be severe particularly in older patients with debilitating comorbidities. For instance, the parenteral administration of chemotherapeutic drugs allows for the drug control via the bloodstream, thus affecting other non-cancerous organs/tissues in the body, besides the tumor itself; the extent and clinical consequences of these non-specific effects are hard to predict. Adverse effects such as nausea, vomiting, hair loss, infections, and diarrhea are common in patients receiving chemotherapy. Radiotherapy can be used alone or in combination with the chemotherapy to treat the primary tumor; shrink the tumor prior to surgery (neoadjuvant therapy; note: chemotherapy also can be administered in the neoadjuvant setting); as adjuvant therapy to maximize the effectiveness of the primary treatment in hopes of extending survival and reducing the risk for recurrence; or to relieve pain or control symptoms of advanced oral cancer (palliative therapy). A patient's response to neoadjuvant therapy can determine which adjuvant therapy is selected. Side effects of radiation therapy due to transient or permanent damage to healthy tissues are fatigue, sore or dry mouth and difficulty swallowing, dental problems (tooth decay), taste change, loss of appetite, nausea and vomiting, nerve damage, pain, infection, osteoradionecrosis, trismus, lymphedema, and hair loss [29]. These can affect the ability to eat and speak and can lead to other complications such as dehydration and malnutrition, social withdrawal, anxiety and depression, impacting the patient's quality of life.

Conventional therapeutic approaches need improvement in bioavailability and targeted delivery to the tumor site (for a pre-determined period) to overcome and prevent the adverse side effects of the drugs [46]. Our group has investigated the potential anticancer benefits of antacid medications, such as proton pump inhibitors and histamine 2 blockers that are commonly used in HNSCC patients to manage acid reflux, a condition that contributes to complications after surgery or during radiotherapy. Our findings in a large cohort study indicated that routine clinical usage of these two classes of antacids in HNSCC patients was correlated with enhanced survival; remarkably our analysis identified histamine 2 receptor antagonist class usage as a significant prognostic factor for recurrence-free survival in patients with oropharyngeal tumors HPV-positive [47]. Ongoing studies in our laboratory are investigating the abilities of these medications to improve the efficacy of conventional therapies, particularly in advanced HNSCC [48,49].

An innovative approach to improve the efficacy of chemotherapeutic agents is the administration of drugs in a time-specific manner (chrono-chemotherapy). It is becoming evident that administration timing is as vital as the dosing amount of chemotherapy [50]. The time of administration (morning vs. evening) influences drug toxicity and therapeutic efficacy because human body physiology is affected by the circadian clock rhythms [51]. Anticancer chemotherapeutic agents docetaxel, doxorubicin, fluorouracil, and paclitaxel have been recently recognized by the World Health Organization as drugs which target circadian clock genes (Bcl2, Top2a, Tyms, and Bcl2 respectively). Hence, they can be employed in chrono-chemotherapy for oral cancer treatment [52]. A recent study showed that chrono-chemotherapy of a combination of Docetaxel, Cisplatin, and Fluorouracil (DCF) helped to decrease the severity of the side effects of each of these drugs [53]; patients with OSCC had less vomiting, nausea, and neutropenia when treated with evening DCF dosing rather than with morning administration [53]. Thus, it seems promising that chrono-chemotherapy has the ability to reduce the severity/extent of the side effects of some chemotherapeutic drugs, which can be exploited as a novel therapeutic strategy in oral, head and neck cancer patients and beyond.

Another approach that showed promise in overcoming the complications of conventional anticancer agents while enhancing their therapeutic efficacy is the targeted drug delivery system consisting of natural and/or synthetic polymers for delivery of chemotherapeutic agents to the tumor site. Targeted drug delivery systems have the potential to increase drug bioavailability and bio-distribution at the site of the primary tumor. DDS is capable of releasing a bioactive molecule at a specific site with a specific delivery rate. Targeted DDS for oral cancer could thus improve patient compliance, enhance drug efficiency while reducing treatment duration, and consequently decrease healthcare expenses. In vivo studies have shown that targeted DDS can also improve the half-time of otherwise rapidly degradable drugs such as peptides and proteins, thus prolonging their local effects [54].

Our review of the most promising anticancer drug delivery approaches is structured in three sections as follows: first, the conventional anticancer drugs are reviewed in regard to their oral administration and potential for DDS formulation; second, a brief background of commonly used carriers in DDS for oral cancer treatment is provided; and third, the potential of different drug delivery methods for OSCC is discussed.

2. Anticancer Agents for Oral Cancer Treatment Formulated in Drug Delivery Systems

While most of the oncological treatments are traditionally administered intravenously, several anticancer drugs have recently been developed and approved by the USA Federal Drug Administration (FDA) for oral administration [55].

Administration of chemotherapeutic drugs in the form of pill or gel is an attractive approach to enhance patient compliance. This method of delivery is also desirable when the treatment requires drug exposure for prolonged periods [46]. Unfortunately, oral administration of most anticancer drugs is hindered due to the drug's physicochemical characteristics, particularly poor aqueous solubility [56,57]. However, most of the chemotherapeutic agents delivered intravenously can also be administered via other routes of delivery when incorporated in suitable carrier (bio)materials [58]. Carefully designed DDS can be used to formulate chemotherapeutic agents for local (e.g., applied to the tumor site) or intravenous delivery with higher efficacy than the standard intravenous administration. Following is an overview of the most common anticancer drugs used for the treatment of oral cavity and oropharynx cancer patients [59], which have already been investigated for their administration using controlled and/or targeted DDS with promising results.

2.1. Paclitaxel (PTX)

Paclitaxel (Taxol) is an antineoplastic agent which functions by cellular growth inhibition. Oral administration of PTX is challenging because of its low solubility and reduced permeability across the intestinal epithelium/mucosa that limit its absorption. When PTX is administered intravenously, which is the most common delivery method in the clinic, its distribution throughout the body is very extensive, causing severe side effects such as liver dysfunction [60]. To increase its absorption, Lee et al. designed a platform based on the chemical conjugation of PTX to the low molecular weight chitosan, which increased PTX's water solubility due to the presence of chitosan and its increased retention time in the gastrointestinal (GI) tract [61]. Tiwari and Amiji reported nano-emulsion formulations of PTX to improve its oral bioavailability; the nano-emulsion delivery of PTX resulted in a significant increase of the PTX concentration in systemic circulation versus control (aqueous solution of PTX), suggesting that this formulation can enhance the oral bioavailability of hydrophobic drugs such as PTX [62]. In another study, Dong and Feng added montmorillonite to poly(lactic-*co*-glycolic acid) (PLGA) in order to synthesize nanoparticles for PTX delivery; the montmorillonite-PLGA nanoparticles allowed for an enhanced cellular uptake and efficiency of PTX as compared to the PLGA nanoparticles alone, suggesting that the montmorillonite-PLGA nanoparticle formulation can extend the residence time of PTX in the GI tract [63].

2.2. Cisplatin (DDP)

Cisplatin is a chemotherapeutic agent with a recognized benefit in the treatment of various human cancers, including oral, head and neck squamous cell carcinoma, bladder, lung, ovarian, breast, and testicular cancers. Cisplatin causes apoptosis (cell death) of cancer cells due to its ability to crosslink with purine bases on DNA, interfering with DNA repair mechanism, and causing DNA damage [64,65]. Because its administration has been associated with severe side effects such as renal failure, there have been several attempts to formulate this drug in an oral sustained release system [64]. Cheng et al. exploited the ability of the low pH-responsive porous hollow nanoparticles of Fe_3O_4 to be used as a vehicle for site-specific cisplatin delivery; their system, based on the encapsulated cisplatin into porous hollow nanoparticles of Fe_3O_4, not only protected cisplatin from deactivation by plasma proteins and other biomolecules before reaching the target site, but also provided control of the release rate of cisplatin by varying the nanoparticle's pore size and pH [66]. Yan and Gemeinhart generated encapsulated cisplatin poly(acrylic acid-*co*-methyl methacrylate) micro-particles for controlled release of cisplatin, and their system enabled cisplatin to maintain its activity for prolonged periods [67]. A cisplatin analog with similar chemotherapeutic profile, Carboplatin, has also been investigated alone or as part of nanoparticle formulations in order to minimize its undesired side effects [68].

2.3. Doxorubicin

Doxorubicin (DOX) is one of the most potent anticancer agents used for the treatment of numerous cancer types, because of its ability to target rapidly dividing cells, both cancerous and non-cancerous.

Its toxicity on non-cancerous cells limits its application because it can result in cell death in major organs such as heart, brain, liver, and kidney [69–71]. Drug delivery strategies sought to minimize DOX side effects while exploiting its anticancer properties with higher therapeutic efficiency. For instance, Li et al. encapsulated DOX in dextran nanoparticles to specifically target tumor cells with the expectation that these smart nanoparticles would increase drug loading efficiency and release the drug at a particular site directly into the cancer cell's nucleus [72]. She et al. used dendronized heparin nanoparticles conjugated to DOX as a pH-responsive drug delivery vehicle for cancer treatment. These nanoparticles showed significant anti-tumor activity on a 4T1 breast tumor model without toxicity to healthy organs [73]. Collectively, this evidence showed that incorporating DOX into nanoparticles held promise for reducing toxicity on healthy cells while increasing its antitumor activity.

2.4. Docetaxel

Docetaxel (DTX), an effective anticancer drug, is most commonly administered intravenously in cancer patients because of its highly hydrophobic property, but it has low oral bioavailability due to the P-glycoprotein (P-gp)-mediated efflux and first passes effect. To address these drawbacks, Sohail et al. synthesized a chitosan scaffold in which folic acid and thiol groups were grafted to chitosan to target cancer cells and improve permeation through the gastrointestinal tract [74]. They also synthesized silver nanoclusters in situ, which allowed for the generation of core-shell nano-capsules with the hydrophobic DTX as the core and the silver nanocluster embedded chitosan as the shell; this strategy resulted in a DTX carrier system suitable for the oral delivery of DTX to cancerous tissues [30].

2.5. Methotrexate

Methotrexate (MTX), an antimetabolite agent used in anticancer chemotherapy, is a folate antagonist which inhibits the synthesis of purines and pyrimidines, thereby causing inhibition of the malignant cells' proliferation. MTX is used for the treatment of a variety of cancers, including oral, head, and neck cancer, acute lymphocytic leukemia, non-Hodgkin's lymphoma, choriocarcinoma, osteosarcoma, and breast cancer [75,76]. When administered orally, MTX systemic bioavailability is approximately 35%, which is significantly lower than when administered parenterally [77]. Oral administration of MTX is associated with significant side effects (diarrhea, ulcerative stomatitis, hemorrhagic enteritis, gastrointestinal perforation) due to inhibition of cellular proliferation. Kumar and Rao formulated MTX in proteinoid microspheres to enhance its bioavailability and targetability, with the expectation that these microspheres could deliver MTX and other pharmaceutical compounds that are prone to degradation, under gastric condition [78]. Paliwal R et al. encapsulated MTX into solid lipid nanoparticles (SLNs) consisting of stearic acid, glycerol monostearate, tristearin, and Compritol 888 ATO; the MTX loaded SLNs significantly improved the bioavailability of MTX by protecting MTX from degradation in the harsh gastric conditions [79].

2.6. Fluoropyrimidine 5-Fluorouracil

Fluoropyrimidine 5-fluorouracil (5-FU), another FDA approved anticancer drug, inhibits essential biosynthesis processes or interferes with DNA or RNA, limiting their normal function. This drug has been effective in treating various types of cancer, including oral, head and neck cancer, colorectal, and breast cancer [80]. Li et al. designed a biodegradable controlled release system composed of PLGA nanoparticles, which maintained a prolonged continuous release of 5-FU. Their results showed that these nanoparticles could enhance the oral bioavailability of 5-FU while decreasing its local gastrointestinal side effects [81]. Minhas et al. developed a pH-responsive controlled release system for 5-FU delivery, by preparing a chemically cross-linked polyvinyl alcohol-*co*-poly(methacrylic acid) hydrogel loaded with 5-FU, which enabled the release of 5-FU at pH 7.4, with the potential for being used as an oral drug delivery vehicle for 5-FU in cancer treatment, particularly colorectal cancer [38].

3. Carriers for OSCC Drug Delivery Systems

Carrier-based drug delivery systems are used for controlled release of drugs while providing improved selectivity and effectiveness, and reduced side effects compared to the chemotherapeutic agents alone. Different carrier systems based on nanoparticles, nanolipids, and hydrogels are discussed here, each with unique advantages and disadvantages (Figure 1). Additionally, exosomes have been recently introduced as potential carriers of chemotherapeutic agents for oral cancer treatment. The benefits and drawbacks of each carrier system are summarized in Table 1.

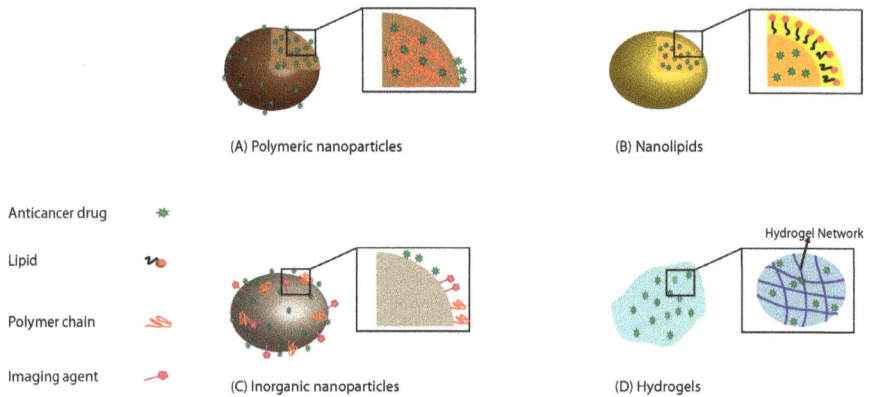

Figure 1. Different carriers used for oral cancer: (**A**) polymeric nanoparticles; (**B**) nanolipids; (**C**) inorganic nanoparticles; (**D**) hydrogels.

Table 1. Carriers for drug delivery in oral cancer treatment.

Carriers for Drug Delivery	Advantages	Disadvantages	References
Polymeric nanoparticles	<ul><li>Biodegradable and biocompatible</li><li>Suitable for controlled and sustained drugs release with increased therapeutic efficacy and reduced side effects</li></ul>	<ul><li>Difficult to handle due to particle-particle aggregation</li><li>Cytotoxic after internalization into cells</li><li>Not suitable for the release of proteins including antibodies</li><li>Associated with an immune response or local toxicity upon degradation</li></ul>	[82–86]
Inorganic nanoparticles	<ul><li>Target can be site specific by attaching the ligand to the nanoparticle (e.g., magnetic nanoparticles)</li><li>Higher photostability compared to organic dyes</li></ul>	<ul><li>Toxicity</li><li>Limited effective delivery due to limited penetration depth for photothermal therapy</li><li>Cannot deliver biomacromolecules (e.g., proteins)</li></ul>	[87–89]
Nanolipids	<ul><li>Highly stable</li><li>Provide controlled release of drugs to protect them from chemical degradation</li><li>Encapsulate and deliver drugs with low aqueous solubility</li><li>Able to penetrate deeply into tumors</li><li>Suitable for local delivery of anticancer drugs</li></ul>	<ul><li>Crystalline structure provides limited space to accommodate drugs</li><li>Solid lipid nanoparticles (SLNs) show initial burst drug release</li><li>Aggregation or gelling of nanostructured lipid carriers (NLCs) during storage</li><li>Associated with immune response</li></ul>	[83,90–93]
Hydrogels	<ul><li>Injectable to a specific site</li><li>Do not dissolve in water at physiological temperature and pH</li><li>Maintain their structural integrity and elasticity even after retaining large amounts of water</li><li>High drug loading capacity</li><li>Ability to deliver hydrophilic and hydrophobic drugs</li></ul>	<ul><li>Poor mechanical properties</li><li>Difficult to handle</li><li>Expensive</li><li>Initial burst</li></ul>	[94,95]

3.1. Nanoparticles for Drug Delivery

The use of nanotechnology in drug delivery has allowed for selective and safe methodologies for OSCC treatment [87,92]. Nanoparticles provide enhanced bioactivity due to their large surface to volume ratio [84,96]. The most common nanoparticles investigated in oral cancer treatments include gold nanoparticles, liposomes, magnetic nanoparticles, and polymeric micelles [88,97]. These nanoparticles are capable of killing cancer cells by delivering the drugs entrapped or encapsulated in them [92,97]. Utilizing nanoparticles as drug carriers have also resulted in stabilization of chemotherapeutic compounds that can be released in a controlled and sustained manner. This targeted delivery facilitates the prolonged release of a drug at a specific site, thus reducing its systemic toxicity [98].

3.1.1. Polymeric Nanoparticles for Drug Delivery System

For targeted drug delivery with improved biocompatibility and drug controlled release, nanoparticles fabricated from natural and synthetic polymer have received much attention [84]. Polymers consisting of polysaccharides, poly(lactic acid) (PLA), poly(glycolic acid) (PGA) and their copolymers, are biodegradable and thus slowly eliminated from the body after the delivery of cargo [99]. There has been substantial research into intraoral, site-specific chemoprevention using a polymeric drug delivery system. These chemopreventive agents are delivered directly to various affected sites within the oral cavity, thereby preventing the malignant conversion of oral epithelial dysplasia to frank carcinoma. Several techniques are currently employed to synthesize such nanoparticles, including nanoprecipitation, emulsifications, and self-assembly [100]. Selecting a particular method depends on the physicochemical properties of the polymer, drug solubility, and drug release behavior [100].

Endo et al. have used polymeric nanoparticles based on poly(ethylene glycol)-poly(glutamic acid) block copolymer to increase the anti-tumor effects and reduce the toxicity of cisplatin [101], the most commonly used chemotherapeutic drug in OSCC patients [102]. Cisplatin was integrated into polymeric micelles through the polymer-metal complex formation between poly(ethylene glycol)-poly(glutamic acid) block copolymers and CDDP (NC-6004). The mean particle size of polymeric micelles (NC-6004) was 30 nm. Also, static light scattering (SLS) measurement exhibited that there is no dissociation of cisplatin-loaded micelles upon dilution and the critical micelles concentration (CMC) was less than 5×10^{-7} [100,103].

The treatment of oral cancer cells with cisplatin-loaded nanoparticles (NC-6004) leads to the activation of the caspase-3 and caspase-7 pathways, which induce apoptosis [101]. In vivo results showed that the antitumor activity of NC-6004 against tumor growth in oral carcinoma-bearing mice was 4.4–6.6-fold higher compared to the control group. Additionally, the controlled release of cisplatin from these nanoparticles resulted in decreased nephrotoxicity and neurotoxicity compared with administration of cisplatin in solution [101].

Additional agents (e.g., curcumin) have been investigated for their therapeutic benefit in oral cancer based on their ability to induce apoptosis and inhibit tumor cell proliferation [91,104]. To enhance the clinical benefits of these therapeutic agents by improving their bioavailability and stability, Mazzarino et al. used a nanoprecipitation technique to generate polycaprolactone (PCL) nanoparticles coated with the polysaccharide chitosan for curcumin delivery into the oral cavity [104]. The chitosan coating on the nanoparticles was confirmed by the changes in particles size and zeta potential measurements. With the increase in concentration of chitosan, the hydrodynamic radius of nanoparticles increased for unloaded and curcumin-loaded nanoparticles (104 to 125 nm; polydispersity index (PDI) < 0.2) [101]. Additionaly, chitosan-coated nanoparticles showed increased zeta potential values (positive surface charge) compared to uncoated nanoparticles due to the presence of positively charged amino groups of chitosan molecules on the surface of the particles, thus proving that the nanoparticles were successfully coated [105]. Also, due to a strong interaction between curcumin and PCL, the core of the curcumin-loaded nanoparticles was compacted, which leads to the decrease in their size compared to the unloaded nanoparticles [101,105].

Adsorption of chitosan on PCL formed a muco-adhesive nanoemulsion, which showed an interaction between glycoprotein mucin and PCL nanoparticles. This system was evaluated by surface plasmon resonance. Better muco-adhesive properties lead to an increase in the residence time of the drug. The cytotoxic effect of these nanoparticles was evaluated in an in vitro study using an OSCC-derived cell line, SCC-9 that showed induction of apoptosis in tumoral cells. Furthermore, these polymeric nanoparticles encapsulating curcumin showed improved bioavailability [104] and improved curcumin stability by preventing its degradation in neutral solutions and upon exposure to light [106].

Interestingly, dietary substances containing bioactive compounds may also have some ability to suppress cancer. Studies indicated that ellagic acid (a polyphenolic chemopreventive agent) has anti-cancerous, antioxidant, and antiviral properties. However, its usage is limited due to low oral bioavailability and water solubility [107]. Bio-polymeric nanoparticles may overcome these drawbacks, increasing the drug efficiency by preventing the degradation of unstable chemotherapeutic biomolecules. Arulmozhi et al. developed chitosan nanoparticles encapsulating ellagic acid using the ionotropic gelation technique, which enhanced the anticancer properties of ellagic acid, thus, making this formulation a promising platform for oral cancer treatment [100,108].

3.1.2. Inorganic Nanoparticles for Drug Delivery System

Inorganic nanoparticles have been extensively used in treatments due to their lower toxicity, higher tolerance towards organic solvents, and better bioavailability compared with the free drug [88]. Inorganic nanoparticles based on noble metals (e.g., gold) have been used in diagnostic and imaging processes and received much attention due to their highly controlled optical properties [109,110]. Such nanoparticles are potential photo-thermal agents with high efficacy in therapeutic applications. Sayed et al. prepared anti-epithelial growth factor receptor (EGFR) antibody-conjugated gold (Au) nanoparticles (with an average particle size of 40 nm characterized by transmission electron microscopy (TEM) and incubated them with OSCC cell lines and a control benign epithelial cell line [110]. Continuous wave (CW) argon ion laser was used to produce photothermal destruction. These in vitro results showed that the malignant cells with anti-EGFR/Au conjugates required less energy to produce photothermal destruction due to the targeting of the Au nanoparticles on the surface of EGFR-overexpressing malignant cells but not on benign cells. In clinical applications, near-infrared (NIR) laser light with deep penetration allowed for effective delivery of anti-EGFR/Au conjugates to the cells. Furthermore, the surface plasmon absorption of Au nanoparticles can be finely tuned by modifying the nanoparticles' size to allow for better absorption of this NIR laser light, thus maximizing their therapeutic benefit [110].

Recently, other therapeutic techniques, including photodynamic therapy (PDT), have been employed to increase the penetration of drugs deeper into tissues, required for the treatment of advanced and recurrent oral cancer [111]. Lucky et al. developed up-conversion nanoparticles (UCN) loaded with PEGylated titanium dioxide (TiO_2) to increase tissue penetration using NIR; these nanoparticles were used for targeting EGFRs on the surface of cancer cells using anti-EGFR-antibody conjugated with PEGylated TiO_2-UCNs to inhibit tumor proliferation, invasion, angiogenesis, and metastasis. Anti-EGFR-PEG-TiO_2-UCNs nanoparticles were characterized by TEM and a well-defined core-shell structure was observed with approximately 50 nm in diameter. Further, the composition of nanoparticles was confirmed by Energy-dispersive X-ray (EDX) spectroscopy showing formation of Na (Sodium), Y (Yttrium), F (Fluorine), Yb (Ytterbium), and Tm (Thulium) from the core nanocrystals and Ti (Titanium), Si (Silicon) and O (Oxygen) from the shell [107]. In vivo studies investigating anti-EGFR-PEG-TiO_2-UCNs showed no toxic side effects, whereas in vitro studies showed enhanced apoptosis and tumor growth inhibition [111,112].

Drug delivery using nanoparticles allowed for increased concentration of therapeutic agents at the tumor site, which resulted in cancer cell inhibition with reduced toxicity on the surrounding non-cancerous healthy cells. Nevertheless, there are still challenges linked to carriers stability and fate in the human body, and their limited effective delivery remains problematic. To overcome some of these drawbacks, Eguchi et al. prepared innovative magnetic nanoparticles consisting of μ-oxo *N,N'*-bis (salicylidene) ethylenediamine iron (Fe(Salen)) for targeted delivery of anticancer agents. Since these particles were difficult to solubilize, they were suspended in water or saline after sonication. Iron–salen particles were characterized using DLS and TEM, showing size ranged 1.2–3 μm for unsonicated particles and 60–800 nm for sonicated particles. The sonication for approximately 6 hours reduced particle size (confirmed by TEM) with smooth edges of the particles as compared to the unsonicated particles. The sonicated Fe(Salen) particles showed zeta potential value of −24.1 mV, thus confirming the stability of the colloidal dispersion [113].

Sato et al. used Fe(Salen) nanoparticles with average size of 200 nm for targeted delivery of anticancer agents. These nanoparticles were sonicated for 30 min and were suspended in normal saline. Alternating magnetic field (AMF) combining chemotherapy and hyperthermia was used to heat Fe(Salen) nanoparticles and resulted in increased induction of cancer cell apoptosis and better carrier stability, as compared to individual chemotherapy or magnetic guided delivery. Fe(Salen) nanoparticles were useful for controlled drug delivery and hyperthermia therapy, with an increase in anti-cancer therapeutic efficacy and reduced toxicity [89].

Other inorganic nanoparticles systems, such as mesoporous silica nanoparticles (MSNP), showed promise for cancer therapy. These nanoparticles' advantages include high porosity, biocompatibility, and amenability for surface functionalization [114]. The porous nature of MSNPs provides much free space for antitumor drugs to be incorporated. These nanoparticles, combined with polymers, can carry drugs with high efficiency in targeting OSCC cells [114,115], but additional investigations are required for the routine implementation of these systems into clinical practice.

3.1.3. Combinational (Polymeric-Inorganic) Nanoparticles

Combinational drug treatment is recognized for its increased therapeutic benefits. Targeted drug delivery offers improved therapeutic efficacy with reduced toxicity. Quinacrine (QC) is an anticancer agent that is also used as an antimalarial drug; it has shown therapeutic benefits in breast, lung, colon, and renal cell carcinoma. Despite these positive outcomes, QC clinical applications are limited due to its poor bioavailability and various side effects, including skin rash and pigmentation, and immunological complications [116]. Inorganic silver-based nanoparticles (AgNPs) also have potential as anticancer agents due to their ability to induce tumor cell apoptosis. Combinational approaches have been employed to address AgNP's limitation of toxicity to healthy cells at higher doses, which resulted in the enhanced anticancer activity of AgNPs [100,116]. Satapathy et al. prepared highly stable PLGA based quinacrine (QC)–silver hybrid nanoparticles (QAgNP) using an oil-in-water emulsion solvent evaporation technique. The TEM analysis determined the size and morphology of QAgNP with size ranging 50–100 nm. Average particle size of 382.4 ± 0.11 nm was obtained by DLS with a positive zeta potential of 0.523 ± 0.09 mV [111]. These nanoparticles were allowed to interact with various oral cancer cell lines and OSCC-derived stem cells and evaluated for their antitumor activity. PLGA/quinacrine/silver nanoparticles showed high cytotoxicity against cancer cells with improved ability to destroy specifically the OSCC-derived stem cells. The study also confirmed that PLGA/quinacrine/silver nanoparticles not only inhibited proliferation of OSCC but also reduced neo-angiogenesis, suggesting that this hybrid nanoparticle drug delivery system can be a promising platform for the treatment of OSCC [100,116].

3.2. Nanolipids

Polymeric nanoparticles' cytotoxicity, due to low internalization into the tumor cells, restricts their therapeutic efficiency [85,86]. Solid lipid-based nanoparticles (SLNs) have overcome this problem because they can penetrate cancer cells. Furthermore, their high stability provides controlled drug release, drug protection from chemical degradation, and they can serve as carriers for drugs with low aqueous solubility [92,117]. Therefore, these nanoparticles seem suitable for local delivery of drugs and chemopreventive agents [118,119].

One limitation of nanoparticles prepared from solid lipids is their crystalline structure, which allows for only limited space to accommodate drugs. Nanostructured lipid carriers (NLCs) have been designed and tested in cancer therapy to overcome this limitation. These NLCs consist of both solid and liquid lipids in a core matrix, thereby distorting the crystal structure and providing space for drugs to be encapsulated in amorphous clusters [120,121]. Thus, NLCs addressed the issues of poor solubility, low bioavailability, and instability of anticancer drugs and therapeutic agents [93,121]. A recent study by Fang et al. reported the enhanced bioavailability of curcumin loaded into nanostructured lipid particles, an emerging method for treating OSCC [122]. Other studies reported the fabrication of nanostructured lipids with other therapeutic agents, such as docetaxel and etoposide, which have shown promise in treating oral cancer [123–125].

3.3. Hydrogel-Based Drug Delivery Systems

Hydrogels are three-dimensional (3D) mesh structures of hydrophilic fibers that contain a large amount of water or biological fluids. Hydrogels resemble the soft body tissues and are capable of encapsulating drugs and biomolecules such as proteins and genetic materials [126]. Depending on the mechanism used for their gelation, there are two types of hydrogels, physical and chemical. Physical gelation is not inherently permanent, but reversible whereas chemical gelation is reversible because it involves chemical bonds, and thus results in permanent or very stable hydrogels [127–129].

Hydrogels act as localized, targeted drug delivery systems and offer some advantages when juxtaposed with active and passive targeting by using nanocarriers [130]. For instance, a limitation of nanoparticle-based systems is the swift elimination from blood circulation due to their small size and renal clearance. Also, the tumor microvascular morphology, characterized by increased interstitial fluid pressure, results in low intra-tumoral penetration of the drug-loaded nanocarriers, which in turn results in decreased therapeutic efficiency [130–133]. In contrast, hydrogels can provide sustained administration of both hydrophilic and hydrophobic drugs, proteins and other biomolecules independently of the microvascular system of the tumor, allowing for high drug loading capacity, as high as the drug's solubility in water [134,135]. Hydrogels can also control the release of the drug for short or long periods (up to several months) by altering the density of the nanofibers in the hydrogel [136].

Moreover, hydrogels allow for co-administration of multiple drugs with synergistic anti-cancer effects and decreased drug resistance [46,130]. In one study, a thermosensitive physical hydrogel composed of poly(ethylene glycol)-poly(ε-caprolactone)-poly(ethylene glycol) (PEG-PCL-PEG, PECE) showed great potential as an in situ controlled delivery system for suberoylanilide hydroxamic acid (SAHA), a histone deacetylase (HDAC) inhibitor in combination with cisplatin (DDP). When injected intratumorally in a OSCC mouse model, the PECE hydrogel provided sustained release of the loaded SAHA and DDP for more than 14 days, enhanced therapeutic effects, and reduced side effects [137].

3.4. Exosomes

Exosomes are membranous vesicles with sizes between 40–120 nm that are secreted by different cells, such as dendritic cells, macrophages, mesenchymal stem cells, endothelial, and epithelial cells, into the extracellular space [138–141]. Due to their nanosized dimensions and natural formation, exosomes have received much attention and are involved in many biological and pathological processes. Exosomes are secreted when the multivesicular body (MVB) fuses with the plasma membrane. Exosomes can contain many types of biomolecules and play an essential role in inter-cellular communication [142]. Their ability to bind to the cell membrane through adhesion proteins and ligands has made them a sound carrier system for targeted drug delivery applications [139,141]. They have been used as a vehicle for chemotherapeutic agents such as curcumin, DOX, and PTX, helping to reduce their side effects while increasing their therapeutic efficiency [139,143,144]. Tian et al. used targeted exosomes as a targeted delivery system for DOX to treat breast cancer cells; when injected intravenously in mice, these exosomes delivered DOX targeted to tumor tissues, which resulted in inhibition of tumor growth without overt toxicity [144]. Despite their promising preclinical evidence for cancer therapy, several limitations prevent exosomes utilization as an efficient drug delivery system in the clinical practice, mainly due to their limited capacity to deliver high doses of therapeutic agents. Also, the separation of exosomes with high purity is a long and demanding process that usually generate low amounts. Finally, studies showed that exosome administration in patients might lead to adverse immune reactions [140]. Conclusively, exosomes can be a useful tool for the treatment of oral cancer, but their purification, analysis, and administration are still challenging [145].

4. Controlled Drug Delivery Approaches for Oral Cancer

The treatment options for advanced OSCC are limited and suboptimal. Conventional therapeutic approaches (i.e., surgery, chemotherapy, and radiotherapy) significantly impact patient' wellbeing and quality of life. Thus, there is an imperative requirement for new therapeutic methods with reduced side effects and systemic toxicity. Several controlled drug delivery and release strategies have been developed to overcome the current challenges associated with the parenteral (intravenous, IV) administration of chemotherapeutic agents. These strategies include: the administration of chemotherapeutics via intra-tumoral injection; local delivery; photo-thermal administration using drug-loaded nanoparticles; and ultra-sonoporation using microbubbles (Figure 2). These approaches are reviewed and discussed herein regarding oral cancer.

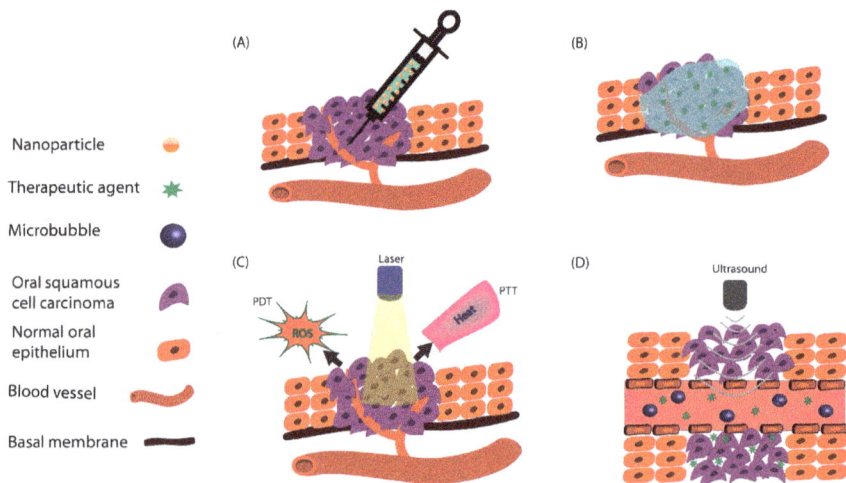

Figure 2. Different controlled drug delivery approaches: (**A**) Intra-tumoral drug delivery; (**B**) local drug delivery; (**C**) photo-thermal therapies combined to drug delivery systems; (**D**) ultrasound-mediated microbubble.

4.1. Intra-Tumoral Drug Delivery in Oral Cancer

One approach is local intra-tumoral administration [146,147]. Li et al. developed a controlled release system that optimized the combined therapeutic benefits of two anticancer drugs while minimizing their side effects, by using suberoylanilide hydroxamic acid (SAHA) and cisplatin (DDP) loaded into PECE hydrogel for the OSCC treatment. Six mice groups were comparatively analyzed (1st group was injected with normal saline (NS); the 2nd was injected with blank hydrogel; the 3rd with SAHA; the 4th with DDP; the 5th with SAHA-DDP; and the 6th with SAHA-DDP/PECE; the mice in the sixth group had the smallest tumor volume with no noticeable systemic cytotoxicity compared to other groups at the end of the study [137]. Intra-tumoral delivery of chemotherapeutic drugs incorporated in a hydrogel is considered as a promising approach for further exploration of OSCC treatment [137].

4.2. Local Drug Delivery in Oral Cancer

Local drug delivery is a tumor-targeted approach that delivers the drug to the proximity of the tumor. With this approach, the drugs enter the systemic circulation to a lesser extent compared with other administration routes, thus limiting the adverse side effects of the drugs on healthy cells [148]. For example, locally delivered drugs formulated inside nanoparticles can reach cancer cells passively or through active targeting. In the case of passive targeting, the nanoparticles reach cancer cells by diffusion and enter the cytoplasm by endocytosis, while in the case of active targeting the nanoparticles are functionalized to identify specific receptors on the cancer cell surface resulting in increased drug delivery inside the cancer cell, leaving the majority of the healthy cells unaffected (Figure 3) [149].

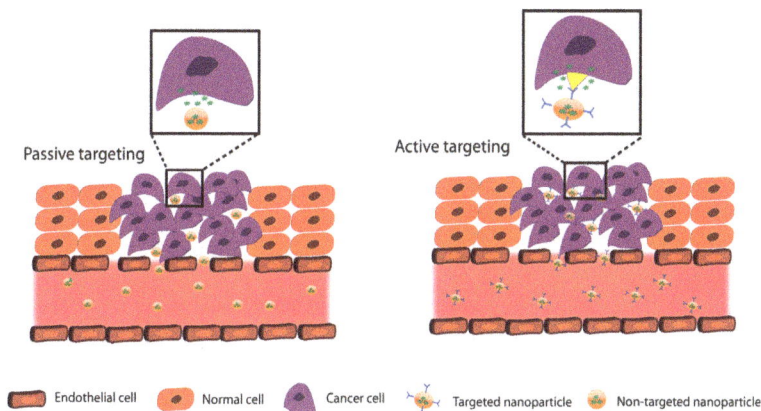

Figure 3. Tumor targeting approaches in oral drug delivery.

Local delivery of anticancer drugs to the oral cavity provides a convenient and safe local administration, with benefit of rapid turnover of the oral mucosa; this allows for a rapid self-repair after given damage and is a significant advantage that helps alleviate the adverse effects caused by long-term local drug delivery [150]. The majority of studies that employed local drug delivery for OSCC treatment used chitosan as a mucoadhesive polymer. To highlight the promise of local delivery in oral cancer treatment, a remarkable study authored by Arulmozhi et al. reported the encapsulation of ellagic acid (EA, an anticancer drug with poor water solubility and oral bioavailability) inside chitosan nanoparticles, which were then evaluated for their therapeutic efficacy in a human oral cancer-derived cell line (i.e., KB cells). The significant cytotoxicity exhibited by the EA nanoparticles suggested that this system has the potential to overcome the limitations of any drug with poor oral bioavailability via targeted local delivery to cancer cells by enhancing its local therapeutic benefits while reducing its systemic side effects [108].

4.3. Phototherapy Approaches in Drug Delivery

Phototherapy is a minimally invasive method that is commonly used in the treatment of neoplastic disease. The first phototherapeutic technique is photodynamic therapy (PDT), consisting of administration of a photosensitizing agent followed by irradiation, which is absorbed by the agent at a specific wavelength. The photosensitizer generates reactive oxygen species (ROS) following the utilization of near-infrared (NIR) light, which results in the apoptosis of cancer cells. This process has proven to be efficient in killing the cancer cells, with the limitation that accumulation of the photosensitizer in the tumor is relatively low [151,152]. Photo-thermal therapy (PTT) is another method of phototherapy, which employs light absorbing agents to generate heat, that damages cancer cells and consequently eliminates the tumor [153]. However, PTT is not considered for clinical applications because the laser power density is high and can also damage the surrounding normal tissue [152].

Drug delivery systems (DDS) can improve the phototherapy techniques and address their limitations. Recent studies have focused on incorporating chemotherapeutic agents and photosensitizers or light absorbing agents into nanocarriers. After delivery of these agents at the tumor site, local irradiation has resulted in the killing of the cancerous cells and tumor shrinkage. Current research studies are focused on the use of magnetic nanoparticles for targeting or tracking cancer cells by magnetic resonance imaging (MRI) [154–157].

He et al. combined photodynamic therapy (PDT) with chemotherapy to simultaneously release anticancer and photosensitizer drugs at the tumor site for the treatment of resistant head and neck cancer. Coordination polymer (NCP)-based core-shell nanoparticles were prepared and loaded with cisplatin and the photosensitizer pyrolipid. They performed in vivo studies, where mice were treated with a combination of nanoparticles loaded with cisplatin and pyrolipid; a remarkable tumor reduction (83%) occurred in cisplatin-resistant SQ20B subcutaneous xenograft murine HNSCC model after the combined treatment of loaded nanoparticles and irradiation. This system of delivery allowed for high loadings of cisplatin and pyrolipid to be locally released after irradiation at the tumor site, with increased anticancer effects as compared to monotherapy [158].

4.4. Microbubbles Mediated Ultrasound in Drug Delivery

Microbubbles are micrometer-sized (1–2 μm) gas bubbles that are used as ultrasound contrast agents. The injection of microbubbles into blood circulation improves the contrast of ultrasound images. In addition to their diagnostic usage, the combination of microbubbles and ultrasound can be used in local drug delivery for the treatment of cancer. Microbubbles can be targeted to specific tumor sites by incorporation of ligands or monoclonal antibodies binding to receptors expressed on cancer cell membranes. The combination of chemotherapeutic agents with microbubble-mediated ultrasound therapy increases drug uptake in targeted tissues through so-called 'sonoporation', improves the drugs' biodistribution and decreases their systemic toxicity [159–161]. Sonoporation is defined as a drug delivery system that uses ultrasound for intracellular delivery of agents that cannot move into cancerous cells under normal conditions [42].

One crucial strategy for the treatment of HNSCC is the inhibition of EGFR signaling, but current methods cannot suppress this signaling completely. EGFR inhibition can occur through RNA interference by using microbubbles as nucleic acid delivery vectors. Microbubbles delivered to the site get ruptured by ultrasound-targeted microbubble destruction (UTMD) resulting in drug release from the microbubbles' shell to the insonified area [162].

Recently, Hirabayashi et al. developed anti-EGFR antibody-conjugated microbubbles for colon squamous cell carcinoma treatment. In in vivo studies, anti-EGFR-microbubbles were injected directly into the tumor, while the anticancer drug bleomycin (BLM) was injected via the tail vein. The findings of this study showed that anti-EGFR-microbubbles bound to EGFR on Ca9-22 cells, and the BLM uptake was increased following anti-EGFR-microbubbles binding to cancerous cells. This system is promising to enable effective targeted delivery of anticancer drugs into oral cancer cells [42].

Carson et al. highlighted the potential use of microbubbles as carriers of anti-EGFR siRNA along with ultrasound-targeted microbubble destruction (UTMD) in SCC-VII-induced murine squamous cell carcinoma model. Delivery of microbubbles to the tumor site, where they were ruptured by UTMD and resulted in drug release from the microbubbles to the insonified area, led to tumor growth suppression in mice with OSCC [162]. Recent studies on drug delivery for oral cancer are summarized in Table 2.

A novel immunotherapy strategy involves using small molecules as monotherapy or combined with other anticancer therapies [163]. The main advantages of these small molecules are good oral bioavailability, ability to penetrate the physiological barriers, precise formulations and dosing options, and lower cost to produce and administer [163,164]. A summary of the small molecules designed for HNSCC and/or OSCC treatment are provided in Table 3.

Table 2. Drug delivery studies for the treatment of oral cancer. OSCC: oral squamous cell carcinoma; PLA: poly(lactic acid); SAHA: suberoylanilide hydroxamic acid; DDP: cisplatin; EGFR: epithelial growth factor receptor.

Study	Outcomes	Material	Anticancer Drug/Small Molecules	Target Cells/ Target Tumor	Delivery Approach	Type of Study	Sex/Species	Reference
Microbranchytherapy for intratumoral injection of holmium-166 microspheres into 13 cats with inoperable OSCC	• Local response rate: 55% • Mean survival time: 113 days overall and 296 days for the cases with local response	PLA microspheres loaded with holmium acetylacetonate and then suspended in Pluronic F-68 solution	Holmium-166 microspheres	Tumors located in the: tongue/sublingual ($n = 10$); gingiva of the mandible ($n = 1$); gingiva or the maxilla ($n = 2$)	Intratumoral injection of radioactive agents	In vivo	Eight male and five female cats	[165]
Injection of drug loaded gels into tumors (up to 6 weeks treatments), at dosage: 0.25 mL of active or placebo gel per cm^3 of the tumor up to 10 mL total	• The tumor response noted in 29% of patients, including 19% cases with complete responses in the drug-loaded gel group versus 2% for placebo ($P < 0.001$).	Purified bovine collagen/gel	Cisplatin/ Epinephrine	Head and neck tumors	Intratumoral	Clinical study (178 patients pretreated with recurrent or refractory HNSCC); prospective, double-blind placebo-controlled phase III trials	Male and female humans	[147]
SAHA and DDP were loaded into a biodegradable and thermosensitive hydrogel (PECE)	• Mice treated with SAHA-DDP/PECE had the smallest tumor volume (62.43 mm^3) compared to other groups tumor volume.	PECE	Cisplatin (DDP)/SAHA	In vitro: HSC-3 and HOK16-F6E7 cells. In vivo: 2×10^6 HSC-3 cells were injected subcutaneously into the right flank regions	Intratumoral	In vitro and in vivo	Female mice	[137]
Synthesizing DTX encapsulated PLGA nanoparticles for in situ delivery to the tumor site	• The slow release profile of the drug (60% of DTX released in 9 days) • Higher cytotoxic effect against SCC-9 cells compared to free drug	PLGA	Docetaxel (DTX)	Human tongue squamous carcinoma derived cell line SCC-9	Intratumoral	In vitro	N/A	[166]
Irradiation following intra-tumoral injection of gold nanorods (GNRs) conjugated with rose bengal (RB)	• The tumor inhibition rate was significant (95.5%) on the 10th day after treatment for (f).	Gold nanorods (GNRs)/ Rose Bengal	-	Tumors induced in hamster cheek pouches	Intratumoral combined with photo-dynamic (PDT) and photothermal (PTT) therapy	In vitro and in vivo	Male hamsters	[167]

Table 2. *Cont.*

Study	Outcomes	Material	Anticancer Drug/Small Molecules	Target Cells/Target Tumor	Delivery Approach	Type of Study	Sex/Species	Reference
Synthesizing and drug encapsulation of EA loaded chitosan nanoparticles	• Sustain drug release by 48 h • Decreased proliferation of human oral cancer KB cell lines (in vitro)	Chitosan	Ellagic acid (EA)	Human oral cancer KB cell line	local	In vitro	N/A	[108]
Curcumin-loaded in PCL nanoparticles and coated with chitosan as a mucoadhesive polymer	• Reduced viability of SCC-9 human oral cancer cell line • Decreased toxicity of curcumin incorporated in nanoparticles compared to its free state	Chitosan	Curcumin	SCC-9 human oral squamous carcinoma cell; for permeation studies: esophageal mucosa of at least two different animals	local	In vitro	N/A	[104]
Nano-emulsions loaded with Gen and coated with chitosan in the form of tablets	• Controlled release profile • Anticancer activity against two oropharyngeal carcinoma-derived cell lines • Both formulations showed equivalent cell kill ratio within 48 h	Nanoemulsion, chitosan, cellulose microcrystalline, dextrose	Genistein (Gen)	SCC-4 cells, FaDu cells, and murine connective tissue fibroblasts (L929) (in vitro)/ porcine buccal Mucosa (ex vivo)	local	In vitro and ex vivo	N/A	[168]
Using MTX loaded liposomes to prepare the mucoadhesive film	• Increased apoptosis rate in HSC-3 cells by three fold in M-LP-F7 • The pro-oxidant effect in HSC-3 cells by M-LP-F7	Liposomes, chitosan (CH), poly(vinyl alcohol) (PVA), hydroxypropyl methylcellulose (HPMC)	Methotrexate (MTX)	HSC-3 cells	local	In vitro	N/A	[169]

Table 2. *Cont.*

Study	Outcomes	Material	Anticancer Drug/Small Molecules	Target Cells/ Target Tumor	Delivery Approach	Type of Study	Sex/Species	Reference
Preparation of a targeted nanoparticle platform combing Pc 4 with IO and a cancer targeting ligand, then intravenous injection of non-formulated Pc4 and two nanoparticle formulations: targeted (Fmp-IO-Pc4) and non-targeted (IO-Pc4) were administered to mice	• Significant tumor inhibition in both Fmp-IO-Pc4 and IO-Pc4 compared to free Pc4 • Significant reduction in tumor volume in targeted nanoparticles (Fmp-IO-Pc 4) compared to IO-Pc4	Iron oxide (IO) nanoparticles	PDT drug (Pc 4)	In vitro: M4E, M4E-15, 686LN, and TU212 cell lines	PDT	In vitro and in vivo	Female mice	[170]
Preparation of gold nanoparticles conjugated with anti-EGFR antibody, then evaluation of the effect of PDT combined with administration of anti-EGFR antibody conjugated Au nanoparticles on two OSCC lines and one epithelial cell line	• No photothermal destruction was seen in any of the cell lines in the absence of Au nanoparticles, but one-quarter of this energy was enough to kill the tumor cells in the presence of anti-EGFR/Au nanoparticles	Anti-EGFR antibody conjugated gold nanoparticles	-	Two OSCC cell lines (HSC 313 and HOC 3 Clone 8); one benign epithelial cell line (HaCaT)	PDT	In vitro	N/A	[110]
Preparation of self-assembled core-shell nanoparticles loaded with cisplatin and pyrolipid for treatment of resistant head and neck cancers.	• Reduced the tumor volume only in NCP@pyrolipid plus irradiation group in cisplatin-resistant SQ20B tumors by 83% • No tumor growth inhibition was observed in NCP@pyrolipid without irradiation	1,2-dioleoyl-sn-glycero-3-phosphate sodium salt (DOPA) coated nanoscale coordination polymer (NCP)-based core-shell Nanoparticles with PEG	Cisplatin and pyrolipid (as photosensitizer)	In vitro: cisplatin-sensitive HNSCC135 and SCC61 as well as cisplatin-resistant JSQ3 and SQ20B In vivo: SQ20B subcutaneous xenograft murine models	PDT	In vitro and in vivo	Female Mice	[158]

Table 2. *Cont.*

Study	Outcomes	Material	Anticancer Drug/Small Molecules	Target Cells/ Target Tumor	Delivery Approach	Type of Study	Sex/Species	Reference
Injection of anti-EGFR-microbubbles into the tumor site, with intravenous injection of BLM 5 min after microbubble injection	• Increased BLM uptake after sonoporation with anti-EGFR-microbubbles • The greater anti-tumor effect in anti-EGFR-microbubbles compared to microbubbles alone • Improved BLM cytotoxicity in Ca9-22 cells in vitro and in vivo	Liposomes with PEG chains	Bleomycin (BLM)	In vitro: Ca9-22 cells In vivo: Ca9-22 cells injected into the back of mice	Local using microbubbles and ultrasound	In vitro and in vivo	Male Mice	[42]
Sonoporation using microbubbles with anti-EGFR antibody and administration of BLM to assess its effect on Ca9-22 growth	• Remarkable inhibition of Ca9-22 cells growth • Surface deformation of Ca9-22 after sonoporation in the presence of antibody • Increased number of apoptotic cells with using a low dosage of BLM and the Fab fragment of an anti-EGFR antibody	SonoVue as microbubble agent	BLM	Ca9-22 cell line	Local using microbubbles and ultrasound	In vitro	N/A	[43]

Table 3. Monoclonal antibodies-based therapies for the treatment of head and neck cancer.

Drugs	Mechanism of Action	Reference
Cetuximab, panitumumab, zalutumumab and nimotuzumab	EGFR inhibitors	[171]
Gefitinib, erlotinib, lapatinib, afatinib and dacomitinib	EGFR tyrosine kinase inhibitors	[171]
Bevacizumab	VEGF inhibitors	[171]
Sorafenib, sunitinib and vandetanib	VEGFR inhibitors	[171]
Rapamycin, temsirolimus, everolimus, torin1, PP242 and PP30, BYL719	PI3K/AKT/mTOR pathway inhibitors	[171,172]
Pembrolizumab and nivolumab	Anti-PD-1 antibodies	[171]
Motolimond (VTX-2337)	TLR8 agonist	[173]
AZD1775 (Adavosertib)	Elective small molecule inhibitor of WEE1 G2 checkpoint serin/threoin/protein kinase	[174]
Abemaciclib (LY2835219)	Cyclin-dependent kinase inhibitor	[175]
TPST-1120	Selective antagonist of PPARα	[176]
Sitravatinib (MGCD516)	RTK inhibitor	[177]
Nintedanib (BIBF1120)	Triple receptor tyrosine kinase inhibitor (PDGFR/FGFR and VEGFR)	[178]
Durvalumab (Imfinzi, MEDI4736)	(IgG1κ) monoclonal antibody	[179,180]
Tremelimumab	Anti-CTLA4 antibody	[170,181]

Abbreviations: EGFR, epidermal growth factor receptor; VEGF, vascular endothelial growth factor; VEGFR, VEGF receptor; PI3K, phosphatidylinositol 3-kinase; AKT, serine/threonine-specific protein kinase; mTOR, mammalian target of rapamycin; PD-1, program death receptor 1; TLR8, a selective toll-like receptor 8; PPARα, peroxisome proliferator-activated receptor alpha; RTK, receptor tyrosine kinase; PDGF-R, Platelet-derived growth factor receptor; CTLA4, cytotoxic T-lymphocyte associated antigen-4; IgG1κ, human immunoglobulin G1 kappa; WEE1, Wee1-like protein kinase.

There is also promising burgeoning research on immunotherapy and gene therapy for oral cancer treatment, and these therapies can also benefit from DDS [162,182,183]. A significant advantage of DDS is their clinical potential for oral cancer diagnostics and treatment simultaneously. Therefore, designing theranostic systems containing both imaging and anticancer agents will significantly improve the diagnosis and treatment of OSCC at early stages.

5. Conclusions and Future Perspective

The major challenge in the management of HNSCC patients today is the development of the evasive cancer cell resistance to conventional therapies. Drug delivery systems employed for the administration of chemotherapeutic agents have shown promise in the abilities to overcome the limitations of the conventional anticancer therapeutic approaches. Drug delivery systems for oral cancer consist of three major components: the anticancer agents (single or multiple); carriers to encapsulate the agents; and the methods of delivering the agents to the tumor site. The carriers can be chosen among natural, synthetic, or a combination of materials. They can be prepared in the form of hydrogels or nanocarriers, including nanoparticles and nanolipids. New drug delivery approaches in oral cancer focused on intratumoral or local drug delivery, photothermal therapies combined with DDS, and delivery using ultrasound-mediated microbubbles. Even though controlled drug delivery systems have been around for more than 30 years, improving clinical efficiency and release profiles of anti-cancer drugs as well as lowering their side effects remains a challenge. One of the main hindrances for the commercialization of these systems is the low production reproducibility. Currently, most research investigations are still focused on in vitro or in vivo studies, whereas only a few systems have been implemented into the clinic (Table 2). A nano-formulation of DOX (liposomal-encapsulated formulation of DOX, DOXIL®) was approved by the USA Food and Drug Administration (FDA) in 1995 [184] and is used for breast and ovarian cancer treatment [185]. Similar or novel formulations and delivery methods are required to address unmet needs for the treatment of oral cancer. A personalized, reliable drug delivery system explicitly tailored on the unique genetic, molecular, histological, and circadian profile of a given tumor and a given patient seems the ideal approach in treating patients with oral cancer and beyond.

Author Contributions: S.P. and F.K. conceived the outline of the manuscript. F.K. wrote the manuscript under the guidance of Drs. S.P. (supervisor), P.P. (co-supervisor) and X.C. (co-supervisor) with input from all authors. F.M. and M.P. designed the figures. L.L. and Drs. S.K. and L.H. commented on the manuscript. Drs. S.P., P.P., and X.C. edited the complete manuscript. All authors contributed to the final manuscript.

Funding: University of Saskatchewan Dean's scholarships (F.K. and F.M.); University of Saskatchewan College of Graduate and Postgraduate Studies Scholarship (M.P.); University of Saskatchewan, College of Medicine and College of Dentistry start-up funds; Saskatchewan Centre for Patient Oriented Research SCPOR/CIHR–Saskatchewan Health Research Foundation (SHRF) Patient-Oriented Research Leadership Grant; Saskatoon Royal University Hospital Foundation Grant; Alpha Omega Foundation of Canada Awards; American Cancer Society Research Scholar Grant RSG-13-103-01-CCE. NIH R21 DE027169-01 to Silvana Papagerakis, Petros Papagerakis and Nikos Chronis (Multiple-PI grant).

Acknowledgments: The authors would like to thank Janice Michael, MBA, CPA, CGA, Research Facilitator of the Colleges of Dentistry and Public Health at the University of Saskatchewan, for editing the manuscript.

Conflicts of Interest: The authors declare no conflict of interest.

References

1. Vogel, D.W.T.; Zbaeren, P.; Thoeny, H.C. Cancer of the oral cavity and oropharynx. *Cancer Imaging* **2010**, *10*, 62.
2. Manikandan, M.; Rao, A.K.D.M.; Arunkumar, G.; Manickavasagam, M.; Rajkumar, K.S.; Rajaraman, R.; Munirajan, A.K. Oral squamous cell carcinoma: microRNA expression profiling and integrative analyses for elucidation of tumourigenesis mechanism. *Mol. Cancer* **2016**, *15*, 28. [CrossRef] [PubMed]
3. Rivera, C. Essentials of oral cancer. *Int. J. Clin. Exp. Pathol.* **2015**, *8*, 11884. [PubMed]
4. Bray, F.; Ferlay, J.; Soerjomataram, I.; Siegel, R.L.; Torre, L.A.; Jemal, A. Global cancer statistics 2018: GLOBOCAN estimates of incidence and mortality worldwide for 36 cancers in 185 countries. *CA. Cancer J. Clin.* **2018**, *68*, 394–424. [CrossRef] [PubMed]
5. Oral Cavity Cancer Statistics-Canadian Cancer Society. Available online: https://www.cancer.ca/en/cancer-information/cancer-type/oral/statistics/?region=on (accessed on 28 June 2019).
6. Denise, M.; Laronde, T.G.; Hislop, J.M.; Elwood, M.R. Oral Cancer: Just the Facts-Canadian Dental Association. Available online: https://cda-adc.ca/jcda/vol-74/issue-3/269.pdf (accessed on 28 June 2019).
7. Marur, S.; Forastiere, A.A. Head and neck cancer: Changing epidemiology, diagnosis, and treatment. *Mayo Clin. Proc.* **2008**, *83*, 489–501. [CrossRef]
8. Nör, J.E.; Gutkind, J.S. Head and neck cancer in the new era of precision medicine. *J. Dent. Res.* **2018**, *97*, 601–602. [CrossRef] [PubMed]
9. Prince, V.M.; Papagerakis, S.; Prince, M.E. Oral Cancer and Cancer Stem Cells: Relevance to Oral Cancer Risk Factors, Premalignant Lesions, and Treatment. *Curr. Oral Heal. Rep.* **2016**, *3*, 65–73. [CrossRef]
10. Heck, J.E.; Berthiller, J.; Vaccarella, S.; Winn, D.M.; Smith, E.M.; Shan'gina, O.; Schwartz, S.M.; Purdue, M.P.; Pilarska, A.; Eluf-Neto, J. Sexual behaviours and the risk of head and neck cancers: a pooled analysis in the International Head and Neck Cancer Epidemiology (INHANCE) consortium. *Int. J. Epidemiol.* **2009**, *39*, 166–181. [CrossRef]
11. Majchrzak, E.; Szybiak, B.; Wegner, A.; Pienkowski, P.; Pazdrowski, J.; Luczewski, L.; Sowka, M.; Golusinski, P.; Malicki, J.; Golusinski, W. Oral cavity and oropharyngeal squamous cell carcinoma in young adults: a review of the literature. *Radiol. Oncol.* **2014**, *48*, 1–10. [CrossRef]
12. Pickard, R.K.L.; Xiao, W.; Broutian, T.R.; He, X.; Gillison, M.L. The prevalence and incidence of oral human papillomavirus infection among young men and women, aged 18–30 years. *Sex. Transm. Dis.* **2012**, *39*, 559–566. [CrossRef]
13. Nirvani, M.; Khuu, C.; Utheim, T.P.; Sand, L.P.; Sehic, A. Circadian clock and oral cancer. *Mol. Clin. Oncol.* **2018**, *8*, 219–226. [PubMed]
14. Hsu, C.; Lin, S.; Lu, C.; Lin, P.; Yang, M. Altered expression of circadian clock genes in head and neck squamous cell carcinoma. *Tumor Biol.* **2012**, *33*, 149–155. [CrossRef] [PubMed]
15. Adeola, H.A.; Papagerakis, P.; Papagerakis, S. System Biology approaches and Precision Oral Health: A Circadian Clock Perspective. *Front. Physiol.* **2019**, *10*, 399. [CrossRef] [PubMed]
16. Cancer Tomorrow. Available online: https://gco.iarc.fr/tomorrow/home (accessed on 28 June 2019).

17. Adams, A.K.; Hallenbeck, G.E.; Casper, K.A.; Patil, Y.J.; Wilson, K.M.; Kimple, R.J.; Lambert, P.F.; Witte, D.P.; Xiao, W.; Gillison, M.L. DEK promotes HPV-positive and-negative head and neck cancer cell proliferation. *Oncogene* **2015**, *34*, 868. [CrossRef] [PubMed]

18. Hübbers, C.U.; Akgül, B. HPV and cancer of the oral cavity. *Virulence* **2015**, *6*, 244–248. [CrossRef] [PubMed]

19. Bouda, M.; Gorgoulis, V.G.; Kastrinakis, N.G.; Giannoudis, A.; Tsoli, E.; Danassi-Afentaki, D.; Foukas, P.; Kyroudi, A.; Laskaris, G.; Herrington, C.S. "High risk" HPV types are frequently detected in potentially malignant and malignant oral lesions, but not in normal oral mucosa. *Mod. Pathol.* **2000**, *13*, 644. [CrossRef] [PubMed]

20. Kojima, A.; Maeda, H.; Sugita, Y.; Tanaka, S.; Kameyama, Y. Human papillomavirus type 38 infection in oral squamous cell carcinomas. *Oral Oncol.* **2002**, *38*, 591–596. [CrossRef]

21. Feller, L.; Wood, N.H.; Khammissa, R.A.G.; Lemmer, J. Human papillomavirus-mediated carcinogenesis and HPV-associated oral and oropharyngeal squamous cell carcinoma. Part 2: Human papillomavirus associated oral and oropharyngeal squamous cell carcinoma. *Head Face Med.* **2010**, *6*, 15. [CrossRef]

22. Feller, L.; Wood, N.H.; Khammissa, R.A.G.; Lemmer, J. Human papillomavirus-mediated carcinogenesis and HPV-associated oral and oropharyngeal squamous cell carcinoma. Part 1: Human papillomavirus-mediated carcinogenesis. *Head Face Med.* **2010**, *6*, 14. [CrossRef]

23. Pinatti, L.M.; Walline, H.M.; Carey, T.E. Human papillomavirus genome integration and head and neck cancer. *J. Dent. Res.* **2018**, *97*, 691–700. [CrossRef]

24. Gillison, M.L.; Broutian, T.; Pickard, R.K.L.; Tong, Z.; Xiao, W.; Kahle, L.; Graubard, B.I.; Chaturvedi, A.K. Prevalence of oral HPV infection in the United States, 2009–2010. *Jama* **2012**, *307*, 693–703. [CrossRef] [PubMed]

25. Kreimer, A.R.; Villa, A.; Nyitray, A.G.; Abrahamsen, M.; Papenfuss, M.; Smith, D.; Hildesheim, A.; Villa, L.L.; Lazcano-Ponce, E.; Giuliano, A.R. The epidemiology of oral HPV infection among a multinational sample of healthy men. *Cancer Epidemiol. Prev. Biomark.* **2011**, *20*, 172–182. [CrossRef] [PubMed]

26. Ang, K.K.; Harris, J.; Wheeler, R.; Weber, R.; Rosenthal, D.I.; Nguyen-Tân, P.F.; Westra, W.H.; Chung, C.H.; Jordan, R.C.; Lu, C. Human papillomavirus and survival of patients with oropharyngeal cancer. *N. Engl. J. Med.* **2010**, *363*, 24–35. [CrossRef] [PubMed]

27. Rabinovics, N.; Mizrachi, A.; Hadar, T.; Ad-El, D.; Feinmesser, R.; Guttman, D.; Shpitzer, T.; Bachar, G. Cancer of the head and neck region in solid organ transplant recipients. *Head Neck* **2014**, *36*, 181–186. [CrossRef] [PubMed]

28. Neville, B.W.; Day, T.A. Oral cancer and precancerous lesions. *CA. Cancer J. Clin.* **2002**, *52*, 195–215. [CrossRef] [PubMed]

29. Furness, S.; Glenny, A.-M.; Worthington, H.V.; Pavitt, S.; Oliver, R.; Clarkson, J.E.; Macluskey, M.; Chan, K.K.; Conway, D.I. The CSROC Expert Panel Interventions for the treatment of oral cavity and oropharyngeal cancer: chemotherapy. In *The Cochrane Database of Systematic Reviews*; Furness, S., Ed.; John Wiley & Sons, Ltd: Chichester, UK, 2010; p. CD006386.

30. Moskovitz, J.; Moy, J.; Ferris, R.L. Immunotherapy for head and neck squamous cell carcinoma. *Curr. Oncol. Rep.* **2018**, *20*, 22. [CrossRef] [PubMed]

31. Rapidis, A.D.; Wolf, G.T. Immunotherapy of head and neck cancer: Current and future considerations. *J. Oncol.* **2009**, *2009*, 346345. [CrossRef] [PubMed]

32. Yamaguchi, Y. Overview of Current Cancer Immunotherapy. In *Immunotherapy of Cancer: An Innovative Treatment Comes of Age*; Yamaguchi, Y., Ed.; Springer Japan: Tokyo, Japan, 2016; pp. 3–17. ISBN 978-4-431-55031-0.

33. Cheng, C.-T.; Castro, G.; Liu, C.-H.; Lau, P. Advanced nanotechnology: An arsenal to enhance immunotherapy in fighting cancer. *Clin. Chim. Acta* **2019**, *492*, 12–19. [CrossRef] [PubMed]

34. Song, W.; Musetti, S.N.; Huang, L. Nanomaterials for cancer immunotherapy. *Biomaterials* **2017**, *148*, 16–30. [CrossRef]

35. Khalil, D.N.; Budhu, S.; Gasmi, B.; Zappasodi, R.; Hirschhorn-Cymerman, D.; Plitt, T.; De Henau, O.; Zamarin, D.; Holmgaard, R.B.; Murphy, J.T. The new era of cancer immunotherapy: manipulating T-cell activity to overcome malignancy. In *Advances in Cancer Research*; Elsevier: Amsterdam, The Netherlands, 2015; Volume 128, pp. 1–68. ISBN 0065-230X.

36. Lubek, J.E. Head and Neck Cancer Research and Support Foundations. *Oral Maxillofac. Surg. Clin.* **2018**, *30*, 459–469. [CrossRef]

37. Colevas, A.D.; Yom, S.S.; Pfister, D.G.; Spencer, S.; Adelstein, D.; Adkins, D.; Brizel, D.M.; Burtness, B.; Busse, P.M.; Caudell, J.J. NCCN guidelines insights: Head and neck cancers, version 1.2018. *J. Natl. Compr. Cancer Netw.* **2018**, *16*, 479–490. [CrossRef] [PubMed]

38. Sim, F.; Leidner, R.; Bell, R.B. Immunotherapy for Head and Neck Cancer. *Oral Maxillofac. Surg. Clin.* **2019**, *31*, 85–100. [CrossRef] [PubMed]

39. Ling, D.C.; Bakkenist, C.J.; Ferris, R.L.; Clump, D.A. Role of immunotherapy in head and neck cancer. *Semin. Radiat. Oncol.* **2018**, *28*, 12–16. [CrossRef] [PubMed]

40. Moskovitz, J.M.; Ferris, R.L. Tumor Immunology, Immunotherapy and Its Application to Head and Neck Squamous Cell Carcinoma (HNSCC). In *Critical Issues in Head and Neck Oncology*; Springer International Publishing: Cham, Switzerland, 2018; pp. 341–355.

41. Chowdhury, M.M.H.; Kubra, K.; Kanwar, R.K.; Kanwar, J.R. Nanoparticles Advancing Cancer Immunotherapy. In *Biomedical Applications of Graphene and 2D Nanomaterials*; Elsevier: Amsterdam, The Netherlands, 2019; pp. 283–304.

42. Hirabayashi, F.; Iwanaga, K.; Okinaga, T.; Takahashi, O.; Ariyoshi, W.; Suzuki, R.; Sugii, M.; Maruyama, K.; Tominaga, K.; Nishihara, T. Epidermal growth factor receptor-targeted sonoporation with microbubbles enhances therapeutic efficacy in a squamous cell carcinoma model. *PLoS ONE* **2017**, *12*, e0185293. [CrossRef] [PubMed]

43. Maeda, H.; Tominaga, K.; Iwanaga, K.; Nagao, F.; Habu, M.; Tsujisawa, T.; Seta, Y.; Toyoshima, K.; Fukuda, J.; Nishihara, T. Targeted drug delivery system for oral cancer therapy using sonoporation. *J. Oral Pathol. Med.* **2009**, *38*, 572–579. [CrossRef] [PubMed]

44. Masood, F. Polymeric nanoparticles for targeted drug delivery system for cancer therapy. *Mater. Sci. Eng. C* **2016**, *60*, 569–578. [CrossRef] [PubMed]

45. Pérez-Herrero, E.; Fernández-Medarde, A. Advanced targeted therapies in cancer: drug nanocarriers, the future of chemotherapy. *Eur. J. Pharm. Biopharm.* **2015**, *93*, 52–79. [CrossRef] [PubMed]

46. Karavasili, C.; Andreadis, D.A.; Katsamenis, O.L.; Panteris, E.; Anastasiadou, P.; Kakazanis, Z.; Zoumpourlis, V.; Markopoulou, C.K.; Koutsopoulos, S.; Vizirianakis, I.S.; et al. Synergistic Antitumor Potency of a Self-Assembling Peptide Hydrogel for the Local Co-delivery of Doxorubicin and Curcumin in the Treatment of Head and Neck Cancer. *Mol. Pharm.* **2019**, *16*, 2326–2341. [CrossRef] [PubMed]

47. Papagerakis, S.; Bellile, E.; Peterson, L.A.; Pliakas, M.; Balaskas, K.; Selman, S.; Hanauer, D.; Taylor, J.M.G.; Duffy, S.; Wolf, G. Proton pump inhibitors and histamine 2 blockers are associated with improved overall survival in patients with head and neck squamous carcinoma. *Cancer Prev. Res.* **2014**, *7*, 1258–1269. [CrossRef]

48. Desiderio, V.; Papagerakis, P.; Tirino, V.; Zheng, L.; Matossian, M.; Prince, M.E.; Paino, F.; Mele, L.; Papaccio, F.; Montella, R. Increased fucosylation has a pivotal role in invasive and metastatic properties of head and neck cancer stem cells. *Oncotarget* **2015**, *6*, 71. [CrossRef]

49. Matossian, M.; Vangelderen, C.; Papagerakis, P.; Zheng, L.; Wolf, G.T.; Papagerakis, S. In silico modeling of the molecular interactions of antacid medication with the endothelium: novel therapeutic implications in head and neck carcinomas. *Int. J. Immunopathol. Pharmacol.* **2014**, *27*, 573–583. [CrossRef] [PubMed]

50. Lévi, F.; Okyar, A. Circadian clocks and drug delivery systems: Impact and opportunities in chronotherapeutics. *Expert Opin. Drug Deliv.* **2011**, *8*, 1535–1541. [CrossRef] [PubMed]

51. Lévi, F.; Schibler, U. Circadian rhythms: Mechanisms and therapeutic implications. *Annu. Rev. Pharmacol. Toxicol.* **2007**, *47*, 593–628. [CrossRef] [PubMed]

52. Zhang, R.; Lahens, N.F.; Ballance, H.I.; Hughes, M.E.; Hogenesch, J.B. A circadian gene expression atlas in mammals: Implications for biology and medicine. *Proc. Natl. Acad. Sci.* **2014**, *111*, 16219–16224. [CrossRef] [PubMed]

53. Tsuchiya, Y.; Ushijima, K.; Noguchi, T.; Okada, N.; Hayasaka, J.; Jinbu, Y.; Ando, H.; Mori, Y.; Kusama, M.; Fujimura, A. Influence of a dosing-time on toxicities induced by docetaxel, cisplatin and 5-fluorouracil in patients with oral squamous cell carcinoma; a cross-over pilot study. *Chronobiol. Int.* **2018**, *35*, 289–294. [CrossRef] [PubMed]

54. Parveen, S.; Misra, R.; Sahoo, S.K.; Misra, R.; Sahoo, S.K. Nanoparticles: A Boon to Drug Delivery, Therapeutics, Diagnostics and Imaging. In *Nanomedicine in Cancer*; Pan Stanford: Singapore, 2017; pp. 47–98.

55. FDA Approved Drugs in Oncology|CenterWatch. Available online: https://www.centerwatch.com/drug-information/fda-approved-drugs/therapeutic-area/12/oncology (accessed on 28 June 2019).

56. O'neill, V.J.; Twelves, C.J. Oral cancer treatment: developments in chemotherapy and beyond. *Br. J. Cancer* **2002**, *87*, 933. [CrossRef]

57. Chidambaram, M.; Manavalan, R.; Kathiresan, K. Nanotherapeutics to overcome conventional cancer chemotherapy limitations. *J. Pharm. Pharm. Sci.* **2011**, *14*, 67–77. [CrossRef]

58. Pridgen, E.M.; Alexis, F.; Farokhzad, O.C. Polymeric nanoparticle drug delivery technologies for oral delivery applications. *Expert Opin. Drug Deliv.* **2015**, *12*, 1459–1473. [CrossRef]

59. Chemotherapy for Oral Cavity and Oropharyngeal Cancer. Available online: https://www.cancer.org/cancer/oral-cavity-and-oropharyngeal-cancer/treating/chemotherapy.html (accessed on 28 June 2019).

60. Choi, J.-S. Pharmacokinetics of paclitaxel in rabbits with carbon tetrachloride-Induced hepatic failure. *Arch. Pharm. Res.* **2002**, *25*, 973–977. [CrossRef]

61. Lee, E.; Lee, J.; Lee, I.-H.; Yu, M.; Kim, H.; Chae, S.Y.; Jon, S. Conjugated chitosan as a novel platform for oral delivery of paclitaxel. *J. Med. Chem.* **2008**, *51*, 6442–6449. [CrossRef]

62. Tiwari, S.B.; Amiji, M.M. Improved oral delivery of paclitaxel following administration in nanoemulsion formulations. *J. Nanosci. Nanotechnol.* **2006**, *6*, 3215–3221. [CrossRef] [PubMed]

63. Dong, Y.; Feng, S.-S. Poly(D,L-lactide-*co*-glycolide)/montmorillonite nanoparticles for oral delivery of anticancer drugs. *Biomaterials* **2005**, *26*, 6068–6076. [CrossRef] [PubMed]

64. Nakano, K.; Ike, O.; Wada, H.; Hitomi, S.; Amano, Y.; Ogita, I.; Nakai, N.; Takada, K. Oral sustained-release cisplatin preparation for rats and mice. *J. Pharm. Pharmacol.* **1997**, *49*, 485–490. [CrossRef] [PubMed]

65. Dasari, S.; Tchounwou, P.B. Cisplatin in cancer therapy: Molecular mechanisms of action. *Eur. J. Pharmacol.* **2014**, *740*, 364–378. [CrossRef] [PubMed]

66. Cheng, K.; Peng, S.; Xu, C.; Sun, S. Porous hollow Fe_3O_4 nanoparticles for targeted delivery and controlled release of cisplatin. *J. Am. Chem. Soc.* **2009**, *131*, 10637–10644. [CrossRef] [PubMed]

67. Yan, X.; Gemeinhart, R.A. Cisplatin delivery from poly(acrylic acid-*co*-methyl methacrylate) microparticles. *J. Control. Release* **2005**, *106*, 198–208. [CrossRef] [PubMed]

68. Nanjwade, B.K.; Singh, J.; Parikh, K.A.; Manvi, F.V. Preparation and evaluation of carboplatin biodegradable polymeric nanoparticles. *Int. J. Pharm.* **2010**, *385*, 176–180. [CrossRef] [PubMed]

69. Tacar, O.; Sriamornsak, P.; Dass, C.R. Doxorubicin: An update on anticancer molecular action, toxicity and novel drug delivery systems. *J. Pharm. Pharmacol.* **2013**, *65*, 157–170. [CrossRef]

70. Astra, L.I.; Hammond, R.; Tarakji, K.; Stephenson, L.W. Doxorubicin-Induced Canine CHF: Advantages and Disadvantages 1. *J. Card. Surg.* **2003**, *18*, 301–306. [CrossRef]

71. Christiansen, S.; Autschbach, R. Doxorubicin in experimental and clinical heart failure. *Eur. J. Cardio-Thoracic Surg.* **2006**, *30*, 611–616. [CrossRef]

72. Li, Y.; Zhu, L.; Liu, Z.; Cheng, R.; Meng, F.; Cui, J.; Ji, S.; Zhong, Z. Reversibly stabilized multifunctional dextran nanoparticles efficiently deliver doxorubicin into the nuclei of cancer cells. *Angew. Chem.* **2009**, *121*, 10098–10102. [CrossRef]

73. She, W.; Li, N.; Luo, K.; Guo, C.; Wang, G.; Geng, Y.; Gu, Z. Dendronized heparin– doxorubicin conjugate based nanoparticle as pH-responsive drug delivery system for cancer therapy. *Biomaterials* **2013**, *34*, 2252–2264. [CrossRef] [PubMed]

74. Sohail, M.F.; Hussain, S.Z.; Saeed, H.; Javed, I.; Sarwar, H.S.; Nadhman, A.; Rehman, M.; Jahan, S.; Hussain, I.; Shahnaz, G. Polymeric nanocapsules embedded with ultra-small silver nanoclusters for synergistic pharmacology and improved oral delivery of Docetaxel. *Sci. Rep.* **2018**, *8*, 13304. [CrossRef] [PubMed]

75. Jolivet, J.; Cowan, K.H.; Curt, G.A.; Clendeninn, N.J.; Chabner, B.A. The pharmacology and clinical use of methotrexate. *New Engl. J. Med.* **1983**, *309*, 1094–1104. [CrossRef] [PubMed]

76. McLean-Tooke, A.; Aldridge, C.; Waugh, S.; Spickett, G.P.; Kay, L. Methotrexate, rheumatoid arthritis and infection risk—What is the evidence? *Rheumatology* **2009**, *48*, 867–871. [CrossRef] [PubMed]

77. Campbell, M.A.; Perrier, D.G.; Dorr, R.T.; Alberts, D.S.; Finley, P.R. Methotrexate: bioavailability and pharmacokinetics. *Cancer Treat. Rep.* **1985**, *69*, 833–838. [PubMed]

78. Kumar, A.B.M.; Rao, K.P. Preparation and characterization of pH-sensitive proteinoid microspheres for the oral delivery of methotrexate. *Biomaterials* **1998**, *19*, 725–732. [CrossRef]

79. Paliwal, R.; Rai, S.; Vaidya, B.; Khatri, K.; Goyal, A.K.; Mishra, N.; Mehta, A.; Vyas, S.P. Effect of lipid core material on characteristics of solid lipid nanoparticles designed for oral lymphatic delivery. *Nanomed. Nanotechnol. Biol. Med.* **2009**, *5*, 184–191. [CrossRef]

80. Longley, D.B.; Harkin, D.P.; Johnston, P.G. 5-Fluorouracil: mechanisms of action and clinical strategies. *Nat. Rev. Cancer* **2003**, *3*, 330. [CrossRef]
81. Li, X.; Xu, Y.; Chen, G.; Wei, P.; Ping, Q. PLGA nanoparticles for the oral delivery of 5-Fluorouracil using high pressure homogenization-emulsification as the preparation method and in vitro/in vivo studies. *Drug Dev. Ind. Pharm.* **2008**, *34*, 107–115. [CrossRef]
82. Jawahar, N.; Meyyanathan, S. Polymeric nanoparticles for drug delivery and targeting: A comprehensive review. *Int. J. Heal. Allied Sci.* **2012**, *1*, 217. [CrossRef]
83. Sim, R.B.; Wallis, R. Surface properties: Immune attack on nanoparticles. *Nat. Nanotechnol.* **2011**, *6*, 80. [CrossRef] [PubMed]
84. Rizvi, S.A.A.; Saleh, A.M. Applications of nanoparticle systems in drug delivery technology. *Saudi Pharm. J.* **2018**, *26*, 64–70. [CrossRef] [PubMed]
85. Kulkarni, S.A.; Feng, S.S. Effects of particle size and surface modification on cellular uptake and biodistribution of polymeric nanoparticles for drug delivery. *Pharm. Res.* **2013**, *30*, 2512–2522. [CrossRef] [PubMed]
86. Hoshyar, N.; Gray, S.; Han, H.; Bao, G. The effect of nanoparticle size on in vivo pharmacokinetics and cellular interaction. *Nanomedicine* **2016**, *11*, 673–692. [CrossRef]
87. Huang, H.; Barua, S.; Sharma, G.; Dey, S.K.; Rege, K. Inorganic nanoparticles for cancer imaging and therapy. *J. Control. Release* **2011**, *155*, 344–357. [CrossRef]
88. Subramani, K.; Ahmed, W. *Nanoparticulate Drug Delivery Systems for Oral Cancer Treatment*, 1st ed.; Elsevier Inc.: Amsterdam, The Netherlands, 2012; ISBN 9781455778621.
89. Sato, I.; Umemura, M.; Mitsudo, K.; Fukumura, H.; Kim, J.H.; Hoshino, Y.; Nakashima, H.; Kioi, M.; Nakakaji, R.; Sato, M.; et al. Simultaneous hyperthermia-chemotherapy with controlled drug delivery using single-drug nanoparticles. *Sci. Rep.* **2016**, *6*, 1–12. [CrossRef]
90. Ghasemiyeh, P.; Mohammadi-samani, S. Solid lipid nanoparticles and nanostructured lipid carriers as novel drug delivery systems: Applications, advantages and disadvantages. *Res. Pharm. Sci.* **2018**, *13*, 288–303.
91. Khosa, A.; Reddi, S.; Saha, R.N. Biomedicine & Pharmacotherapy Nanostructured lipid carriers for site-specific drug delivery. *Biomed. Pharmacother.* **2018**, *103*, 598–613.
92. Calixto, G.; Bernegossi, J.; Fonseca-Santos, B.; Chorilli, M. Nanotechnology-based drug delivery systems for treatment of oral cancer: A review. *Int. J. Nanomed.* **2014**, *9*, 3719–3735. [CrossRef]
93. Sun, M.; Su, X.; Ding, B.; He, X.; Liu, X.; Yu, A.; Lou, H.; Zhai, G. Advances in nanotechnology-based delivery systems for curcumin. *Nanomedicine* **2012**, *7*, 1085–1100. [CrossRef] [PubMed]
94. Coelho, J.F.; Ferreira, P.C.; Alves, P.; Cordeiro, R.; Fonseca, A.C.; Góis, J.R.; Gil, M.H. Drug delivery systems: Advanced technologies potentially applicable in personalized treatments. *EPMA J.* **2010**, *1*, 164–209. [CrossRef] [PubMed]
95. Singh, G.; Lohani, A.; Bhattacharya, S.S. Hydrogel as a novel drug delivery system: A review. *J. Fundam. Pharm. Res.* **2014**, *2*, 35–48.
96. Buzea, C.; Pacheco, I.I.; Robbie, K. Nanomaterials and nanoparticles: Sources and toxicity. *Biointerphases* **2007**, *2*, MR17–MR71. [CrossRef] [PubMed]
97. Poonia, M.; Ramalingam, K.; Goyal, S.; Sidhu, K.S. Nanotechnology in oral cancer: A comprehensive review. *J. Oral Maxillofac. Pathol.* **2017**, *3*, 407–414.
98. Brannon-Peppas, L.; Blanchette, J.O. Nanoparticle and targeted systems for cancer therapy. *Adv. Drug Deliv. Rev.* **2012**, *64*, 206–212. [CrossRef]
99. Brewer, E.; Coleman, J.; Lowman, A. Emerging technologies of polymeric nanoparticles in cancer drug delivery. *J. Nanomater.* **2011**, *2011*, 10. [CrossRef]
100. Desai, K.G.H. Polymeric drug delivery systems for intraoral site-specific chemoprevention of oral cancer. *J. Biomed. Mater. Res. Part B Appl. Biomater.* **2018**, *106*, 1383–1413. [CrossRef]
101. Endo, K.; Ueno, T.; Kondo, S.; Wakisaka, N.; Murono, S.; Ito, M.; Kataoka, K.; Kato, Y.; Yoshizaki, T. Tumor-targeted chemotherapy with the nanopolymer-based drug NC-6004 for oral squamous cell carcinoma. *Cancer Sci.* **2013**, *104*, 369–374. [CrossRef]
102. Madhulaxmi, M.; Iyer, K.; Periasamy, R.; Gajendran, P.; Lakshmi, T. Role of cisplatin in oral squamous cell carcinoma—A review. *J. Adv. Pharm. Educ. Res.* **2017**, *7*, 39–42.
103. Uchino, H.; Matsumura, Y.; Negishi, T.; Koizumi, F.; Hayashi, T.; Honda, T.; Nishiyama, N.; Kataoka, K.; Naito, S.; Kakizoe, T. Cisplatin-incorporating polymeric micelles (NC-6004) can reduce nephrotoxicity and neurotoxicity of cisplatin in rats. *Br. J. Cancer* **2005**, *93*, 678. [CrossRef] [PubMed]

104. Mazzarino, L.; Loch-neckel, G.; Bubniak, S.; Mazzucco, S.; Santos-silva, M.C.; Borsali, R.; Lemos-senna, E. Curcumin-loaded chitosan-coated nanoparticles as a new approach for the local treatment of oral cavity cancer. *J. Nanosci. Nanotechnol.* **2015**, *15*, 781–791. [CrossRef] [PubMed]

105. Mazzarino, L.; Travelet, C.; Ortega-Murillo, S.; Otsuka, I.; Pignot-Paintrand, I.; Lemos-Senna, E.; Borsali, R. Elaboration of chitosan-coated nanoparticles loaded with curcumin for mucoadhesive applications. *J. Colloid Interface Sci.* **2012**, *370*, 58–66. [CrossRef] [PubMed]

106. Cardoso, S.G.; Mazzarino, L.; Dora, C.L.; Bellettini, I.C.; Minatti, E.; Lemos-Senna, E. del documento: Curcumin-loaded polymeric and lipid nanocapsules: Preparation, characterization and chemical stability evaluation. *Indizada en Chem. Abstr. Serv. Int. Pharm. Abstr. Serv. Biosci. Inf. Serv. (Biol. Abstr. Period. Int. Pharm. Technol. Prod. Manuf. Abstr. Ref. Zhurnal EMBAS)* **2002**, *29*, 933–940.

107. Weisburg, J.H.; Schuck, A.G.; Reiss, S.E.; Wolf, B.J.; Fertel, S.R.; Zuckerbraun, H.L.; Babich, H. Ellagic acid, a dietary polyphenol, selectively cytotoxic to HSC-2 oral carcinoma cells. *Anticancer Res.* **2013**, *33*, 1829–1836. [PubMed]

108. Arulmozhi, V.; Pandian, K.; Mirunalini, S. Ellagic acid encapsulated chitosan nanoparticles for drug delivery system in human oral cancer cell line (KB). *Colloids Surf. B Biointerfaces* **2013**, *110*, 313–320. [CrossRef] [PubMed]

109. Senapati, S.; Mahanta, A.K.; Kumar, S.; Maiti, P. Controlled drug delivery vehicles for cancer treatment and their performance. *Signal Transduct. Target. Ther.* **2018**, *3*, 7. [CrossRef]

110. El-Sayed, I.H.; Huang, X.; El-Sayed, M.A. Selective laser photo-thermal therapy of epithelial carcinoma using anti-EGFR antibody conjugated gold nanoparticles. *Cancer Lett.* **2006**, *239*, 129–135. [CrossRef]

111. Lucky, S.S.; Idris, N.M.; Huang, K.; Kim, J.; Li, Z.; Thong, P.S.P.; Xu, R.; Soo, K.C.; Zhang, Y. In vivo biocompatibility, biodistribution and therapeutic efficiency of titania coated upconversion nanoparticles for photodynamic therapy of solid oral cancers. *Theranostics* **2016**, *6*, 1844–1865. [CrossRef]

112. Marcazzan, S.; Varoni, E.M.; Blanco, E.; Lodi, G.; Ferrari, M. Nanomedicine, an emerging therapeutic strategy for oral cancer therapy. *Oral Oncol.* **2018**, *76*, 1–7. [CrossRef]

113. Eguchi, H.; Umemura, M.; Kurotani, R.; Fukumura, H.; Sato, I.; Kim, J.-H.; Hoshino, Y.; Lee, J.; Amemiya, N.; Sato, M. A magnetic anti-cancer compound for magnet-guided delivery and magnetic resonance imaging. *Sci. Rep.* **2015**, *5*, 9194. [CrossRef] [PubMed]

114. Wang, D.; Xu, X.; Zhang, K.; Sun, B.; Wang, L.; Meng, L.; Liu, Q.; Zheng, C.; Yang, B.; Sun, H. Codelivery of doxorubicin and MDR1-siRNA by mesoporous silica nanoparticles-polymerpolyethylenimine to improve oral squamous carcinoma treatment. *Int. J. Nanomed.* **2018**, *13*, 187–198. [CrossRef] [PubMed]

115. Shi, X.L.; Li, Y.; Zhao, L.M.; Su, L.W.; Ding, G. Delivery of MTH1 inhibitor (TH287) and MDR1 siRNA via hyaluronic acid-based mesoporous silica nanoparticles for oral cancers treatment. *Colloids Surf. B Biointerfaces* **2019**, *173*, 599–606. [CrossRef] [PubMed]

116. Satapathy, S.R.; Siddharth, S.; Das, D.; Nayak, A.; Kundu, C.N. Enhancement of Cytotoxicity and Inhibition of Angiogenesis in Oral Cancer Stem Cells by a Hybrid Nanoparticle of Bioactive Quinacrine and Silver: Implication of Base Excision Repair Cascade. *Mol. Pharm.* **2015**, *12*, 4011–4025. [CrossRef] [PubMed]

117. Rana, V. Therapeutic Delivery. *Ther. Deliv* **2016**, *7*, 117–138.

118. Sah, A.K.; Vyas, A.; Suresh, P.K.; Gidwani, B. Application of nanocarrier-based drug delivery system in treatment of oral cancer. *Artif. Cells Nanomed. Biotechnol.* **2018**, *46*, 650–657. [CrossRef] [PubMed]

119. Reddy, R.S.; Dathar, S. Nano drug delivery in oral cancer therapy: An emerging avenue to unveil. *J. Med. Radiol. Pathol. Surg.* **2015**, *1*, 17–22. [CrossRef]

120. Beloqui, A.; Solinís, M.Á.; Rodríguez-Gascón, A.; Almeida, A.J.; Préat, V. Nanostructured lipid carriers: Promising drug delivery systems for future clinics. *Nanomed. Nanotechnol. Biol. Med.* **2016**, *12*, 143–161. [CrossRef] [PubMed]

121. Fang, C.-L.; Al-Suwayeh, S.A.; Fang, J.-Y. Nanostructured Lipid Carriers (NLCs) for Drug Delivery and Targeting. *Recent Pat. Nanotechnol.* **2012**, *7*, 41–55. [CrossRef]

122. Zlotogorski, A.; Dayan, A.; Dayan, D.; Chaushu, G.; Salo, T.; Vered, M. Nutraceuticals as new treatment approaches for oral cancer-I: Curcumin. *Oral Oncol.* **2013**, *49*, 187–191. [CrossRef]

123. Liu, D.; Liu, Z.; Wang, L.; Zhang, C.; Zhang, N. Nanostructured lipid carriers as novel carrier for parenteral delivery of docetaxel. *Colloids Surf. B Biointerfaces* **2011**, *85*, 262–269. [CrossRef] [PubMed]

124. Iida, S.; Shimada, J.; Sakagami, H. Cytotoxicity induced by docetaxel in human oral squamous cell carcinoma cell lines. *In Vivo (Brooklyn)* **2013**, *27*, 321–332.

125. Zhang, T.; Chen, J.; Zhang, Y.; Shen, Q.; Pan, W. Characterization and evaluation of nanostructured lipid carrier as a vehicle for oral delivery of etoposide. *Eur. J. Pharm. Sci.* **2011**, *43*, 174–179. [CrossRef] [PubMed]
126. Li, J.; Mooney, D.J. Designing hydrogels for controlled drug delivery. *Nat. Rev. Mater.* **2016**. [CrossRef] [PubMed]
127. Maitra, J.; Kumar Shukla, V. Cross-linking in Hydrogels—A Review. *Am. J. Polym. Sci.* **2014**, *2014*, 25–31.
128. Ketabat, F.; Khorshidi, S.; Karkhaneh, A. Application of minimally invasive injectable conductive hydrogels as stimulating scaffolds for myocardial tissue engineering. *Polym. Int.* **2018**, *67*, 975–982. [CrossRef]
129. Ketabat, F.; Karkhaneh, A.; Mehdinavaz Aghdam, R.; Hossein Ahmadi Tafti, S. Injectable conductive collagen/alginate/polypyrrole hydrogels as a biocompatible system for biomedical applications. *J. Biomater. Sci. Polym. Ed.* **2017**, *28*, 794–805. [CrossRef]
130. Sepantafar, M.; Maheronnaghsh, R.; Mohammadi, H.; Radmanesh, F.; Hasani-sadrabadi, M.M.; Ebrahimi, M.; Baharvand, H. Engineered Hydrogels in Cancer Therapy and Diagnosis. *Trends Biotechnol.* **2017**, *35*, 1074–1087. [CrossRef]
131. *Multifunctional Nanoparticles for Drug Delivery Applications*; Svenson, S.; Prud'homme, R.K. (Eds.) Nanostructure Science and Technology; Springer US: Boston, MA, USA, 2012; ISBN 978-1-4614-2304-1.
132. Svenson, S.; Prud'homme, R.K. *Multifunctional Nanoparticles for Drug Delivery Applications: Imaging, Targeting, and Delivery*; Springer Science & Business Media: Berlin, Germany, 2012; ISBN 146142304X.
133. Wilhelm, S.; Tavares, A.J.; Dai, Q.; Ohta, S.; Audet, J.; Dvorak, H.F.; Chan, W.C.W. Analysis of nanoparticle delivery to tumours. *Nat. Rev. Mater.* **2016**, *1*, 16014. [CrossRef]
134. Karavasili, C.; Panteris, E.; Vizirianakis, I.S.; Koutsopoulos, S.; Fatouros, D.G. Chemotherapeutic Delivery from a Self-Assembling Peptide Nanofiber Hydrogel for the Management of Glioblastoma. *Pharm. Res.* **2018**, *35*, 166. [CrossRef]
135. Koutsopoulos, S.; Unsworth, L.D.; Nagai, Y.; Zhang, S. Controlled release of functional proteins through designer self-assembling peptide nanofiber hydrogel scaffold. *Proc. Natl. Acad. Sci.* **2009**, *106*, 4623–4628. [CrossRef] [PubMed]
136. Koutsopoulos, S.; Zhang, S. Two-layered injectable self-assembling peptide scaffold hydrogels for long-term sustained release of human antibodies. *J. Control. Release* **2012**, *160*, 451–458. [CrossRef] [PubMed]
137. Li, J.; Gong, C.; Feng, X.; Zhou, X.; Xu, X.; Xie, L.; Wang, R.; Zhang, D.; Wang, H.; Deng, P. Biodegradable thermosensitive hydrogel for SAHA and DDP delivery: therapeutic effects on oral squamous cell carcinoma xenografts. *PLoS ONE* **2012**, *7*, e33860. [CrossRef] [PubMed]
138. Mathivanan, S.; Ji, H.; Simpson, R.J. Exosomes: Extracellular organelles important in intercellular communication. *J. Proteomics* **2010**, *73*, 1907–1920. [CrossRef] [PubMed]
139. Batrakova, E.V.; Kim, M.S. Using exosomes, naturally-equipped nanocarriers, for drug delivery. *J. Control. Release* **2015**, *219*, 396–405. [CrossRef] [PubMed]
140. Ha, D.; Yang, N.; Nadithe, V. Exosomes as therapeutic drug carriers and delivery vehicles across biological membranes: current perspectives and future challenges. *Acta Pharm. Sin. B* **2016**, *6*, 287–296. [CrossRef] [PubMed]
141. Jiang, X.-C.; Gao, J.-Q. Exosomes as novel bio-carriers for gene and drug delivery. *Int. J. Pharm.* **2017**, *521*, 167–175. [CrossRef] [PubMed]
142. Wang, W.; Wang, G.; Bunggulawa, E.J.; Wang, N.; Yin, T.; Wang, Y.; Durkan, C. Recent advancements in the use of exosomes as drug delivery systems. *J. Nanobiotechnol.* **2018**, *16*, 1–13.
143. Sun, D.; Zhuang, X.; Xiang, X.; Liu, Y.; Zhang, S.; Liu, C.; Barnes, S.; Grizzle, W.; Miller, D.; Zhang, H.G. A novel nanoparticle drug delivery system: The anti-inflammatory activity of curcumin is enhanced when encapsulated in exosomes. *Mol. Ther.* **2010**, *18*, 1606–1614. [CrossRef]
144. Tian, Y.; Li, S.; Song, J.; Ji, T.; Zhu, M.; Anderson, G.J.; Wei, J.; Nie, G. A doxorubicin delivery platform using engineered natural membrane vesicle exosomes for targeted tumor therapy. *Biomaterials* **2014**, *35*, 2383–2390. [CrossRef]
145. Luan, X.; Sansanaphongpricha, K.; Myers, I.; Chen, H.; Yuan, H.; Sun, D. Engineering exosomes as refined biological nanoplatforms for drug delivery. *Acta Pharmacol. Sin.* **2017**, *38*, 754–763. [CrossRef] [PubMed]
146. Dehari, H.; Ito, Y.; Nakamura, T.; Kobune, M.; Sasaki, K.; Yonekura, N.; Kohama, G.; Hamada, H. Enhanced antitumor effect of RGD fiber-modified adenovirus for gene therapy of oral cancer. *Cancer Gene Ther.* **2003**, *10*, 75. [CrossRef] [PubMed]

147. Wenig, B.L.; Werner, J.A.; Castro, D.J.; Sridhar, K.S.; Garewal, H.S.; Kehrl, W.; Pluzanska, A.; Arndt, O.; Costantino, P.D.; Mills, G.M. The role of intratumoral therapy with cisplatin/epinephrine injectable gel in the management of advanced squamous cell carcinoma of the head and neck. *Arch. Otolaryngol. Neck Surg.* **2002**, *128*, 880–885. [CrossRef]

148. Minko, T.; Dharap, S.S.; Pakunlu, R.I.; Wang, Y. Molecular targeting of drug delivery systems to cancer. *Curr. Drug Targets* **2004**, *5*, 389–406. [CrossRef] [PubMed]

149. Wu, T.-T.; Zhou, S.-H. Nanoparticle-based targeted therapeutics in head-and-neck cancer. *Int. J. Med. Sci.* **2015**, *12*, 187. [CrossRef]

150. Sankar, V.; Hearnden, V.; Hull, K.; Juras, D.V.; Greenberg, M.S.; Kerr, A.R.; Lockhart, P.B.; Patton, L.L.; Porter, S.; Thornhill, M. Local drug delivery for oral mucosal diseases: Challenges and opportunities. *Oral Dis.* **2011**, *17*, 73–84. [CrossRef]

151. Agostinis, P.; Berg, K.; Cengel, K.A.; Foster, T.H.; Girotti, A.W.; Gollnick, S.O.; Hahn, S.M.; Hamblin, M.R.; Juzeniene, A.; Kessel, D. Photodynamic therapy of cancer: An update. *CA. Cancer J. Clin.* **2011**, *61*, 250–281. [CrossRef]

152. Guo, R.; Peng, H.; Tian, Y.; Shen, S.; Yang, W. Mitochondria-targeting magnetic composite nanoparticles for enhanced phototherapy of cancer. *Small* **2016**, *12*, 4541–4552. [CrossRef]

153. Huang, X.; Jain, P.K.; El-Sayed, I.H.; El-Sayed, M.A. Plasmonic photothermal therapy (PPTT) using gold nanoparticles. *Lasers Med. Sci.* **2008**, *23*, 217. [CrossRef]

154. Feng, Q.; Zhang, Y.; Zhang, W.; Hao, Y.; Wang, Y.; Zhang, H.; Hou, L.; Zhang, Z. Programmed near-infrared light-responsive drug delivery system for combined magnetic tumor-targeting magnetic resonance imaging and chemo-phototherapy. *Acta Biomater.* **2017**, *49*, 402–413. [CrossRef]

155. Feng, Q.; Zhang, Y.; Zhang, W.; Shan, X.; Yuan, Y.; Zhang, H.; Hou, L.; Zhang, Z. Tumor-targeted and multi-stimuli responsive drug delivery system for near-infrared light induced chemo-phototherapy and photoacoustic tomography. *Acta Biomater.* **2016**, *38*, 129–142. [CrossRef] [PubMed]

156. Liu, J.; Detrembleur, C.; De Pauw-Gillet, M.; Mornet, S.; Jérôme, C.; Duguet, E. Gold Nanorods Coated with Mesoporous Silica Shell as Drug Delivery System for Remote Near Infrared Light-Activated Release and Potential Phototherapy. *Small* **2015**, *11*, 2323–2332. [CrossRef] [PubMed]

157. Einafshar, E.; Asl, A.H.; Nia, A.H.; Mohammadi, M.; Malekzadeh, A.; Ramezani, M. New cyclodextrin-based nanocarriers for drug delivery and phototherapy using an irinotecan metabolite. *Carbohydr. Polym.* **2018**, *194*, 103–110. [CrossRef] [PubMed]

158. He, C.; Liu, D.; Lin, W. Self-assembled core–shell nanoparticles for combined chemotherapy and photodynamic therapy of resistant head and neck cancers. *ACS Nano* **2015**, *9*, 991–1003. [CrossRef] [PubMed]

159. Sennoga, C.A.; Kanbar, E.; Auboire, L.; Dujardin, P.-A.; Fouan, D.; Escoffre, J.-M.; Bouakaz, A. Microbubble-mediated ultrasound drug-delivery and therapeutic monitoring. *Expert Opin. Drug Deliv.* **2017**, *14*, 1031–1043. [CrossRef]

160. Ibsen, S.; Schutt, C.E.; Esener, S. Microbubble-mediated ultrasound therapy: a review of its potential in cancer treatment. *Drug Des. Devel. Ther.* **2013**, *7*, 375. [CrossRef] [PubMed]

161. Sorace, A.G.; Warram, J.M.; Umphrey, H.; Hoyt, K. Microbubble-mediated ultrasonic techniques for improved chemotherapeutic delivery in cancer. *J. Drug Target.* **2012**, *20*, 43–54. [CrossRef]

162. Carson, A.R.; McTiernan, C.F.; Lavery, L.; Grata, M.; Leng, X.; Wang, J.; Chen, X.; Villanueva, F.S. Ultrasound-targeted microbubble destruction to deliver siRNA cancer therapy. *Cancer Res.* **2012**, *72*, 6191–6199. [CrossRef]

163. Kerr, W.G.; Chisholm, J.D. The Next Generation of Immunotherapy for Cancer: Small Molecules Could Make Big Waves. *J. Immunol.* **2019**, *202*, 11–19. [CrossRef]

164. Zhu, H.-F.; Li, Y. Small-molecule targets in tumor immunotherapy. *Nat. Prod. Bioprospect.* **2018**, *8*, 297–301. [CrossRef]

165. Van Nimwegen, S.A.; Bakker, R.C.; Kirpensteijn, J.; van Es, R.J.J.; Koole, R.; Lam, M.; Hesselink, J.W.; Nijsen, J.F.W. Intratumoral injection of radioactive holmium (^{166}Ho) microspheres for treatment of oral squamous cell carcinoma in cats. *Vet. Comp. Oncol.* **2018**, *16*, 114–124. [CrossRef] [PubMed]

166. Gupta, P.; Singh, M.; Kumar, R.; Belz, J.; Shanker, R.; Dwivedi, P.D.; Sridhar, S.; Singh, S.P. Synthesis and in vitro studies of PLGA-DTX nanoconjugate as potential drug delivery vehicle for oral cancer. *Int. J. Nanomed.* **2018**, *13*, 67. [CrossRef] [PubMed]

167. Wang, B.; Wang, J.-H.; Liu, Q.; Huang, H.; Chen, M.; Li, K.; Li, C.; Yu, X.-F.; Chu, P.K. Rose-bengal-conjugated gold nanorods for in vivo photodynamic and photothermal oral cancer therapies. *Biomaterials* **2014**, *35*, 1954–1966. [CrossRef] [PubMed]
168. Gavin, A.; Pham, J.T.H.; Wang, D.; Brownlow, B.; Elbayoumi, T.A. Layered nanoemulsions as mucoadhesive buccal systems for controlled delivery of oral cancer therapeutics. *Int. J. Nanomedicine* **2015**, *10*, 1569. [PubMed]
169. Jin, B.; Dong, X.; Xu, X.; Zhang, F. Development and in vitro evaluation of mucoadhesive patches of methotrexate for targeted delivery in oral cancer. *Oncol. Lett.* **2018**, *15*, 2541–2549. [CrossRef]
170. Wang, D.; Fei, B.; Halig, L.V.; Qin, X.; Hu, Z.; Xu, H.; Wang, Y.A.; Chen, Z.; Kim, S.; Shin, D.M. Targeted iron-oxide nanoparticle for photodynamic therapy and imaging of head and neck cancer. *ACS Nano* **2014**, *8*, 6620–6632. [CrossRef]
171. Kozakiewicz, P.; Grzybowska-Szatkowska, L. Application of molecular targeted therapies in the treatment of head and neck squamous cell carcinoma. *Oncol. Lett.* **2018**, *15*, 7497–7505. [CrossRef]
172. Razak, A.R.A.; Ahn, M.-J.; Yen, C.-J.; Solomon, B.J.; Lee, S.-H.; Wang, H.-M.; Munster, P.N.; Van Herpen, C.M.L.; Gilbert, J.; Pal, R.R.; et al. Phase Ib/II study of the PI3Kα inhibitor BYL719 in combination with cetuximab in recurrent/metastatic squamous cell cancer of the head and neck (SCCHN). *J. Clin. Oncol.* **2014**, *32*, 6044. [CrossRef]
173. Dietsch, G.N.; Lu, H.; Yang, Y.; Morishima, C.; Chow, L.Q.; Disis, M.L.; Hershberg, R.M. Coordinated Activation of Toll-Like Receptor8 (TLR8) and NLRP3 by the TLR8 Agonist, VTX-2337, Ignites Tumoricidal Natural Killer Cell Activity. *PLoS ONE* **2016**, *11*, e0148764. [CrossRef]
174. WEE1 Inhibitor With Cisplatin and Radiotherapy: A Trial in Head and Neck Cancer. Available online: https://clinicaltrials.gov (accessed on 23 May 2019).
175. Clinical Trial of Abemaciclib in Combination with Pembrolizumab in Patients with Metastatic or Recurrent Head and Neck Cancer. Available online: https://clinicaltrials.gov (accessed on 23 May 2019).
176. TPST-1120 as Monotherapy and in Combination with (Nivolumab, Docetaxel or Cetuximab) in Subjects with Advanced Cancers. Available online: https://clinicaltrials.gov (accessed on 23 May 2019).
177. Sitravatinib (MGCD516) and Nivolumab in Oral Cavity Cancer Window Opportunity Study. Available online: https://clinicaltrials.gov (accessed on 23 May 2019).
178. Trial of BIBF1120 (Nintedanib) in Patients with Recurrent or Metastatic Salivary Gland Cancer of the Head and Neck. Available online: https://clinicaltrials.gov (accessed on 23 May 2019).
179. Azacitidine, Durvalumab, and Tremelimumab in Recurrent and/or Metastatic Head and Neck Cancer Patients. Available online: https://clinicaltrials.gov (accessed on 23 May 2019).
180. Safety and Efficacy of MEDI0457 and Durvalumab in Patients with HPV Associated Recurrent/Metastatic Head and Neck Cancer. Available online: https://clinicaltrials.gov (accessed on 23 May 2019).
181. Phase III Open Label Study of MEDI 4736 With/Without Tremelimumab Versus Standard of Care (SOC) in Recurrent/Metastatic Head and Neck Cancer. Available online: https://clinicaltrials.gov (accessed on 23 May 2019).
182. Shah, J.P.; Gil, Z. Current concepts in management of oral cancer–surgery. *Oral Oncol.* **2009**, *45*, 394–401. [CrossRef]
183. Okunaga, S.; Takasu, A.; Meshii, N.; Imai, T.; Hamada, M.; Iwai, S.; Yura, Y. Ultrasound as a method to enhance antitumor ability of oncolytic herpes simplex virus for head and neck cancer. *Cancer Gene Ther.* **2015**, *22*, 163. [CrossRef] [PubMed]
184. Drugs@FDA: FDA Approved Drug Products. Available online: https://www.accessdata.fda.gov/scripts/cder/daf/ (accessed on 23 May 2019).
185. Muggia, F.M. Randomized phase III trial of pegylated liposomal doxorubicin versus vinorelbine or mitomycin C plus vinblastine in women with taxane-refractory advanced breast cancer. *Breast Dis. a YB Q.* **2005**, *16*, 186–187. [CrossRef]

pharmaceutics

MDPI

Review

A Snapshot of Transdermal and Topical Drug Delivery Research in Canada

Mahdi Roohnikan, Elise Laszlo, Samuel Babity and Davide Brambilla *

Faculty of Pharmacy, University of Montreal, Montreal, QC H3T 1J4, Canada;
mahdi.roohnikan@umontreal.ca (M.R.); elise.laszlo@umontreal.ca (E.L.); samuel.babity@umontreal.ca (S.B.)
* Correspondence: davide.brambilla@umontreal.ca; Tel.: +1-514-343-6111

Received: 1 May 2019; Accepted: 30 May 2019; Published: 1 June 2019

Abstract: The minimally- or non-invasive delivery of therapeutic agents through the skin has several advantages compared to other delivery routes and plays an important role in medical care routines. The development and refinement of new technologies is leading to a drastic expansion of the arsenal of drugs that can benefit from this delivery strategy and is further intensifying its impact in medicine. Within Canada, as well, a few research groups have worked on the development of state-of-the-art transdermal delivery technologies. Within this short review, we aim to provide a critical overview of the development of these technologies in the Canadian environment.

Keywords: transdermal drug delivery; Canada; skin; permeation enhancers

1. Introduction

The skin is the most accessible organ of the body, stretching over a surface area of 1.7 m^2 and making up roughly 10–16% of the total mass of the body [1,2]. Its primary function is to act as a protective layer against environmental hazards such as chemicals, heat, and toxins, as well as to defend the body against invading microorganisms and allergens. Moreover, the skin plays a key role in homeostasis and body temperature regulation, and acts as a sensory organ exposed to the environment, detecting external stimuli such as temperature, pressure, and pain [2].

While the skin might appear to be an ideal route of administration for local and systemic therapeutics, it actually represents a challenging barrier against the penetration of most compounds [3]. It is composed of three main layers—the epidermis, the dermis, and the hypodermis—with a total thickness ranging from 0.5 to 4 mm, dictated by various factors such as body region, age, and sex [1,3]. The epidermis is a non-vascularized, multilayered stratum whose cells receive nutrients via diffusion from the lower layers, with the outermost portion, the stratum corneum, acting as the main barrier against the passage of drugs [4]. The stratum corneum has a wall-like structure with corneocytes—non-nucleated keratinocytes composed of 70–80% keratin and 20% lipids—acting as "bricks" in a network of intercellular lipids, while desmosomes act as structural links between the "bricks" [5]. Beneath the epidermis is the dermis, which contains a network of blood vessels, which provide nutrients, regulate body temperature, and remove waste products; as well as a network of lymphatic vessels, which are important in regulating interstitial pressure and clearing large molecules [5]. These networks are critical for the distribution of molecules crossing the epidermis into the systemic circulation. When a skin-permeable chemical is applied to the surface of the skin, this process creates a concentration gradient between the surface and the dermis which helps to drive the drug into the skin over time. Thus, the capillary network embedded in the dermis is the main target for transdermal delivery strategies. Finally, the innermost layer, the hypodermis, is mostly composed of adipose tissue and its primary roles are to provide thermal insulation, protect against physical shock, and serve as an energy reserve.

Though transdermal and topical drug delivery has important advantages, including bypassing the harsh conditions of the gastrointestinal tract, as well as the pain and requirement for trained personnel associated with parenteral administration, the skin represents a significant physical barrier and only a very limited number of drugs are compatible with this route of administration without the use of permeation promoters. An important effort has thus been focused on the design of more efficient strategies to facilitate the permeation of drugs across the skin, while avoiding tissue damage. These strategies have drastically improved small molecule delivery for cosmetic, dermatological, and other localized applications, and have allowed the delivery of macromolecules and vaccines in clinical trials, along with a few other systemic applications [6–8].

The industrial sector for topical and transdermal drug delivery has grown as the global market continues to prosper. In addition to established transdermal formulations developed by larger manufacturing companies, there are a number of initiatives focusing on new systems, and numerous start-ups have been successfully seeded in this expanding market [9]. The general acceptance of transdermal products by patients is very high, which is also evident in their increasing market share, worth $20.5 billion in 2010 and currently estimated at over $32 billion [9,10]. Canada is the world's 10th largest pharmaceutical market with a 1.9% share of the global market, and annual domestic pharmaceutical manufacturing was valued at $9.6 billion in 2017 [11]. Hence, research on topical and transdermal drug delivery systems is of great importance in Canada. In this brief review, we present and discuss the works of Canadian research groups in the field, critically comparing them to similar research performed outside the country, as well as highlighting instances where research has translated to the private sector. We categorized the proposed systems based on their nominal nature and/or performance mechanisms, namely: chemically enhanced, physically enhanced, and nanoparticle-based delivery systems (Table 1). We believe that this short review suitably fits the overall topic of this Special Issue, entitled "Drug Delivery Technology Development in Canada", and will be useful for the community in positioning Canada within this important research field.

2. Chemical Permeation Enhancer-Based Systems

Chemical permeation enhancers (CPEs)—generally defined as molecular compounds able to destabilize the stratum corneum and facilitate the passage of drugs while, ideally, limiting or avoiding deeper tissue damage—have long been used in transdermal and topical formulations [12]. In the past, several compounds, such as alcohols, surfactants, terpenes, and fatty acids, have been shown to enhance the permeation of therapeutics, however, only a few have been adopted in commercial transdermal and topical products [13,14].

A remarkable example of the use of CPEs to help deliver larger molecules across the stratum corneum has been proposed by a group at the University of Waterloo. The system, initially described in a patent in 1998 [15], consists of multilamellar biphasic lipid vesicles (BPVs) ranging in size from 0.1–10 µm. These vesicles were proposed to contain a lipophilic dispersion within an aqueous core, surrounded by over 15 concentric phospholipid bilayers separated by additional aqueous phases. This allows molecules of interest (otherwise too large to cross the stratum corneum) to be loaded throughout the vesicle, enabling local or systemic delivery, with permeation enhanced by the incorporation of solvents and surfactants within the vesicle structure. Importantly, the works demonstrated that the relative proportions of lipids and permeation enhancers used in these formulations can vary significantly, possibly as a function of the molecule being delivered. Two of the most abundant enhancers, namely propylene glycol and oleic acid, were included in a study of the mechanisms of permeation enhancement by CPEs [16]. Though both were defined as solvents, oleic acid was demonstrated to increase fluidity and disorder in the structural lipids of the stratum corneum, whereas propylene glycol was speculated to primarily enhance permeation through solubilization of the molecule being delivered. Early uses of this technology were described in the context of the transdermal delivery of insulin and vaccines [17,18], highlighting the initial promise of the delivery method, though neither application has thus far led to clinical or commercial translation. Nonetheless, research into this topic appears to be ongoing, with more recent articles describing mechanistic studies of vesicle delivery and the use of BPVs for the topical (rather than transdermal) delivery of interferon (IFN)-α in the context of human papilloma virus (HPV) treatment. It was demonstrated in a guinea pig model that, when applied topically, the treatment effectively delivered IFN-α to the dermis, reaching a maximum local concentration of ~100,000 IU/100 cm^2 within 6 h, and remaining at a therapeutically effective level for up to 72 h [19]. However, more comprehensive studies of the half-life of IFN-α in the skin were not conducted. Additionally, when compared to an intradermal injection (the standard delivery method for IFN-α), the vesicle-based topical application never generated elevated systemic concentrations (<100 IU/mL), while the intradermal administration resulted in rapid systemic absorption of the drug and consequently an increased likelihood of adverse side-effects. Overall, dermal application of BPV-IFN-α led to a sustained local release of IFN-α with minimal systemic exposure [19]. The authors reported that the topical IFN-α was generally well tolerated by animals, as no significant difference in body weight, apparent pain, or visual appearance of the skin was observed between animals treated with the vesicle-based IFN-α, those treated with a placebo formulation, and animals that did not receive treatment. In cases where irritation or redness did occur (3.5%), it was minor and resolved itself within 2 h of application. The transdermal delivery mechanism of IFN-α was investigated using confocal microscopy, differential scanning calorimetry (DSC), and small- and wide-angle X-ray scattering (SAXS/WAXS) [20]. Confocal microscopy analysis indicated that encapsulated IFN-α was delivered across the stratum corneum to the viable epidermis and dermis, at a depth of roughly 70 μm. Data extracted from DSC and SAXS/WAXS analyses suggest that a three-dimensional cubic Pn3m polymeric phase rearrangement of intercellular lipids is induced by the interaction between stratum corneum lipids and the biphasic vesicles, possibly explaining the improved delivery of IFN-α. More importantly, in another work, the authors investigated the delivery of BPV-IFN-α in humans [21]. The particles were demonstrated to be between 1000 and 1100 nm in size and their zeta potentials were measured between 70 and 78 mV. Following the application of 5, 15, and 40 MIU/g formulations of encapsulated IFN-α to the upper, inner arm of healthy volunteers as a topical patch, dermal levels of IFN-α were measured to be 120 ± 30, 380 ± 60, and 400 ± 80 IU/mg respectively, suggesting a limit to the local concentration deliverable through this system. These local concentrations indicated a delivery of between 3% and 5% of the dose contained in the patch to the skin, comparable to many topical formulations of small molecules [22]. In a pilot study of 12 patients with external Condylomata acuminata warts (a topical manifestation of HPV infection in the genitals) the application of a significantly lower dose of encapsulated IFN-α (1 MIU/g) twice daily for two weeks resulted in a decrease in lesion size and 2′,5′-oligoadenylate synthetase activity (a marker for viral infection), as well as a significant decrease in systemic HPV viral load [21]. In light of this demonstrated potential for IFN-α delivery, multilamellar BPVs could be envisioned as suitable candidates for the non-invasive topical delivery of other therapeutic macromolecules to the skin. Despite these promising initial results, the most recent publication concerning these biphasic vesicles was released in 2013, likely owing to the acquisition of the technology by the Vancouver-based company Altum Pharmaceuticals Inc. Branded as BiPhasix™, Altum appears to have further developed the technology in the context of IFN-α delivery, and has conducted clinical trials of a vesicle-containing cream for treatment of HPV in females, with Phase III trials slated to begin in Q2 2019, according to their website [23,24]. While the specific use of multilamellar BPVs for topical or transdermal delivery was quite unique (likely owing to the proprietary nature of the technology), no other examples of work on similar lipid-based transdermal delivery systems were found in Canada. For the simplest forms of lipid-based carriers (namely single-layer liposomes), this can be attributed to their limited success at breaching the stratum corneum without additional permeation enhancers, and ambiguity regarding their mechanism of transdermal delivery [25]. Despite this, over the past few years there has been a growing interest in the concept of nanostructured lipid carriers (NLCs) as a means of overcoming some of the difficulties associated with traditional liposomes [26]. Similar to some of the principles used in the design of multilamellar biphasic vesicles, NLCs consist of a solid lipid nanoparticle with a variable percentage of liquid lipids and surfactants included within, resulting in a disordered internal structure, and allowing increased drug loading and improved skin permeation. While many groups have been studying this delivery route [27–29], none are based in Canada, highlighting a potential topic of interest for groups studying topical or transdermal delivery. In particular, since the systemic delivery of insulin, vaccines, and other larger therapeutic compounds through BPVs never reached clinical stages, NLCs could present another avenue for investigation, especially as they have been primarily studied in the context of small molecule delivery.

Another important class of CPEs is ionic liquids (ILs); low melting point salts that have drawn interest for their uses in green chemistry, but are also being investigated based on their potential for transdermal drug delivery [30–32]. Indeed, ILs have shown the ability to facilitate transdermal transport, bypassing the physical barrier of the stratum corneum through disruption of cellular structure, lipid bilayer fluidization, and generation of permeation routes, all of which facilitate the diffusion of drugs to the dermis [33–36]. For instance, Zakrewsky et al. have demonstrated that choline-based ILs can enhance the transdermal delivery of mannitol, a small hydrophilic molecule with low skin permeability [35]. Importantly, this interest in ILs has also extended to the design of ILs based on active pharmaceutical ingredients (API-ILs); the combination of an API with an optimal counterion results in a new chemical entity with improved pharmaceutical properties, including better solubility and ADME characteristics [37]. Although ionized forms of a drug typically cross biological membranes to a lesser extent (owing to their charged nature), limiting their transport through lipid membranes, recent reports have demonstrated that additional interactions (e.g., hydrogen bonding) with a counterion can promote their co-transport to a greater extent than the free ionic drug [38–42]. However, despite the advantages they offer, research into the transdermal applications of API-ILs remains in early stages. Recently, a group at McGill University described the development of ILs to improve the delivery of poorly soluble drugs through the skin. Zavgorodnya et al. studied the effects of various counterions on the membrane permeability of salicylate-based APIs-ILs [43]. Specifically, they paired three counterions (choline, tributylammonium, and triethylene glycol monomethyl ether tributylammonium) with a salicylate anion to generate three different API-ILs and assessed their impact on transdermal diffusion using a silicone membrane as a skin mimic. Remarkably, each of the ILs showed increased transmembrane diffusion relative to sodium salicylate dissolved in triethylene glycol monomethyl ether, suggesting that the counterions play an important role in the permeation of salicylate, beyond simple solubilization. In particular, the polyethylene glycol (PEG)-functionalized counterion (triethylene glycol monomethyl ether tributylammonium) enhanced transdermal transport up to 2.5-fold relative to the non-PEGylated tributylammonium. The improved permeation observed with the PEGylated counterion was potentially due to capacity of triethylene glycol to act as a CPE while also improving the dissolution of the whole IL complex [44,45]. Although further evaluations, both ex vivo and in vivo, are needed to confirm the actual potentials of these salicylate formulations, the same group proceeded to evaluate the in vivo transdermal delivery of lidocaine—a common local anesthetic often selected as a model compound due to its limited transdermal permeability—using a similar strategy [46,47]. For this study, the authors prepared two API-IL pairs: lidocainium docusate ([Lid][Doc]) and Lidocaine·Ibuprofen (Lid·Ibu)—which have previously been reported to generate strong hydrogen bonds, promoting transport across a synthetic membrane [39]—and compared them to lidocainium chloride ([Lid]Cl). To perform in the in vivo tests, each form was dissolved in a commercially-available moisturizing vehicle cream (LUBRIDERM®), topically applied to shaved rats, and the plasma profile of concentration vs. time was assessed by ELISA assay. Among the different forms, Lid·Ibu demonstrated the greatest and most rapid systemic exposure of lidocaine (4 h AUC of 12, 602, and 1763 µM·h for [Lid][Doc], [Lid]Cl, and Lid·Ibu). Interestingly, the [Lid][Doc] API-IL displayed a drastically lower plasma concentration compared to the salt API ([Lid]Cl), which could be due to the strong hydrophobicity of the ionic salt and the high molecular weight of the counterion. Importantly, after application of Lid·Ibu, the authors also measured the plasma profile of ibuprofen and observed a higher plasma concentration relative to lidocaine, in contrast with previous observations in synthetic membranes, where the two drugs showed the same kinetics [47].

This phenomenon requires more investigation; a possible explanation could be that the two compounds permeate the stratum corneum in paired form, and become dissociated within the complex skin matrix, leading to different plasma absorption kinetics. Despite the relative novelty of the API-IL strategy for transdermal and topical delivery, early works appear to indicate that it is a promising method for the enhancement of transdermal and topical delivery, which could be applied to a vast range of low molecular-weight drugs [47,48]. Nonetheless, more systematic studies are needed, to investigate different counterions and their effects on drug absorption, cytotoxicity, and skin irritation in order to better understand their mechanism of action, and to compare them with other transdermal and topical delivery systems.

3. Physical Permeation Enhancer-Based Systems

Physical permeation enhancers use a physical process to promote the passage of drugs through the superficial layers of the skin, avoiding damage to the deeper layers. While some of these methods permit the delivery of drugs across the stratum corneum without damage—for instance iontophoresis, which uses an electrical field to promote the electrophoretic mobility of a drug—others cause only superficial physical disruption to the skin. Indeed, one conceptually simple and effective way to bypass the stratum corneum without the use of chemical compounds is to physically pierce the superficial layers of the skin and inject the active compound. However, classical hypodermic needles are usually too large to do this without damaging the deeper layers, potentially causing pain and tissue damage. Thus, microneedles (MNs), pointed microstructures with a submillimeter length, have been developed as an alternative technique to deliver vaccines and drugs. Their potential clinical use presents the substantial advantage of being painless (as the MNs are not long enough to reach skin nociceptors) and could potentially be self-administered, similar to other topical formulations [49]. Designs for transdermal drug delivery include solid, dissolving, and hollow MNs, which can be arranged as in-plane or out-of-plane arrays. Among these, hollow MNs have the primary advantage of being able to deliver large doses (comparable to hypodermic needles) directly to the dermis and can be used with any drug without the need for optimization of the formulation, or post-manufacturing sterilization. Although different methods have been proposed for the manufacture of hollow MNs, including femtosecond laser two-photon polymerization [50] and microinjection molding [51], their commercial use has been curbed by their high costs of fabrication [52]. One important player in this field is the Stoeber group, located at the University of British Columbia in Vancouver. They introduced a new, allegedly more cost-effective process based on solvent casting, to manufacture hollow out-of-plane clay-reinforced polyimide MNs with lengths up to 250 µm [53]. To do so, they used photolithography to manufacture re-usable micromolds containing pillar-shaped MNs using an epoxy-based photoresist. The mold was then spin-coated with polydimethylsiloxane and treated with O_2 plasma to improve its surface wetting. Using these molds, they optimized the manufacture of hollow MNs by casting a montmorillonite nanoclay powder mixed with *N*-methyl-2-pyrrolidone (NMP) dispersed in a solution of polyimide PI-2611 (85–95% *N*-methyl-2-pyrrolidone, 10–20% S-biphenyldianhydride/p-phenylendiamine) onto the mold structures. Following a 2 h evaporation of NMP at 65 °C, 250 µm MNs were formed and removed from the molds. The tips of the MNs were then opened at an aperture of 50 µm using 3 µm aluminium oxide polishing film. Mechanical tests indicated that the MNs were robust enough to penetrate rabbit skin (an in vivo model generally recognized as a suitable mimic of the thicknesses and elasticity of human skin) and efficiently deliver a 0.0025 wt % suspension of 0.21 µm fluorescent polystyrene beads when attached to a standard syringe. The same group adjusted the manufacturing process to allow the preparation of metallic MNs with the same hollow, out-of-plane geometry [54]. To do so, they used a MN mold coated with a layer of poly(methyl methacrylate) seeded with carbon black (a conductive polymer) as a cathode, and a pure nickel anode, both immersed in an electroplating solution consisting of nickel sulfate, nickel chloride, and boric acid. After the application of a 2 mA current for 150 min, a 70 μm thick nickel backing layer was obtained.

Following this electroplating process, the tips of the MNs were opened using O_2/CF_4 plasma etching. Subsequently, the outer surface of the MNs was covered with a 20 nm thick gold layer to avoid any dermal allergic reaction that could be caused by the nickel. Mechanical compression tests indicated that the MNs were strong enough to pierce human skin without breaking, with a measured fracture force of 4.2 ± 0.61 N. Moreover, the delivery of 2.28 μm fluorescent beads into pig skin was demonstrated, using 500 μm hollow metallic MNs with a tip lumen diameter of 40 μm. The significant advantage of this new hollow out-of-plane MN preparation process is that MNs with a wide range of heights, spacing, and lumen sizes could be prepared. Importantly, this group has funded the start-up company Microdermics®, and has begun clinical testing, with the primary goal of evaluating the safety of these MNs in humans [55]. Their aim was to evaluate the biocompatibility and inertness of gold- and silver-coated MNs, relative to uncoated MNs when applied to the skin, as nickel is known to cause skin irritation. Though this clinical trial was completed in 2015, no results have yet been disclosed.

4. Nanoparticle-Based Systems

The last decade has witnessed a remarkable rise of nanomaterials in drug delivery research, which has also translated into promising results in the field of transdermal delivery [56]. Indeed, micro- and nano-carriers are among the most sought-after methods that have been extensively studied as potential delivery systems for the transport of non-permeable molecules across the stratum corneum [57]. Specifically, microemulsions are thermodynamically stable colloidal systems containing oil and water, stabilized by a combination of surfactants and co-surfactants, that have attracted significant attention for topical and transdermal delivery purposes [58]. These systems have been studied and developed for the delivery of a vast range of compounds to and across the skin for dermatological, cosmetic, and systemic applications. In comparison with conventional emulsions, microemulsions have been claimed to enhance skin delivery primarily by virtue of their reduced droplet size and the disruption of the stratum corneum by their constituents [58]. For instance, microemulsion-based formulations for lidocaine delivery generally have a longer-lasting effect than emulsion-based ones and result in 1.5–2 times greater permeation of lidocaine than the emulsion-based EMLA® cream [59,60].

At the University of Toronto, the Acosta group has investigated the design and optimization of microemulsion-based systems for transdermal drug delivery applications [61]. To do this, they used a donor-skin-receiver mass balance model to study the effects of the concentration of surfactant used to generate the microemulsions on the transdermal delivery of lidocaine. Among different classes of components, lecithin-based microemulsions have attracted attention thanks to the generally recognized as safe status of their main constituent [62]. However, lecithin (a mixture of amphiphilic substances) tends to form lamellar and other liquid-crystal phases, and the addition of co-surfactants—generally medium-chain alcohols (e.g., butanol and pentanol)—is thus necessary for the formation of stable microemulsions [63].

While presence of these co-surfactants results in the low interfacial tension and small particle size observed in the emulsions [63], they are known to have skin-irritation properties [64]. To solve this problem, the group investigated other classes of additives as co-surfactants, which led to the selection of sorbitan monooleate (Span 80) as a lipophilic linker, and a mixture of caprylic acid (CA) and sodium caprylate (SC) as a hydrophilic linker for the fabrication of stable oil-in-water (type I), water-in-oil (type II), and bicontinuous (type III or IV) microemulsions (classified by studying their phase behavior), based on an isopropyl myristate oil phase [64,65]. The ratio of Span 80 to lecithin was kept constant at 3:1, while the ratio of CA to lecithin was maintained at 0.75:1. Using these emulsions, transdermal delivery performance was assessed as a function of lecithin concentration. The droplet size (radius) was measured by dynamic light scattering and was found to be constant at 6 nm regardless of surfactant concentration. It was shown that in lecithin-linker microemulsions, an increase in surfactant concentration was associated with an increased quantity of lidocaine delivered across porcine ear skin using a MatTek permeation device [61,66]. In addition, the authors demonstrated the superior lidocaine delivery of their lecithin-linker-based formulations relative to a pentanol-based formulation, with the type II microemulsions being the most effective, with a flux of up to 0.4 mg/(cm^2·h) compared to the maximum of 0.12 mg/(cm^2·h) for type I microemulsions.

In vitro cytotoxicity studies using a (3-(4,5-dimethylthiazol-2-yl)-2,5-diphenyltetrazolium bromide) tetrazolium cell viability assay on human reconstructed skin showed that these lecithin-linker microemulsions had a reduced toxicity profile compared to medium-chain alcohol-based microemulsions [64]. Despite having been extensively explored, the actual mechanism by which nano- and micro-formulations can promote the delivery of compounds through the epidermis remains controversial [57]. To explain the observed permeation results with their optimized formulations, the group proposed a dominant transport mechanism: due to their small size, the microemulsion droplets migrate to the lower epidermis and upper dermis, creating a depot-like effect and release the drug into the deeper layers of the skin. However, the observed increase in permeation as a function of surfactant concentration might suggest a combined mechanism in which the surfactants destabilize the lipid structure within the stratum corneum (acting in a CPE-like manner), leading to the diffusion of the nano-droplets into the deeper layers. Given this possible mechanism of action, this technology could also be considered as a CPE-based system.

It should be noted that the proposed system is likely to require further investigation. Indeed, it has been shown that when a lecithin-related, naturally-derived monoacyl phosphatidycholine (MAPL) surfactant was used, crystal-like structures formed at the surface of the skin, acting as an additional barrier and further limiting drug diffusion. This observation, combined with the limited permeation enhancement of lecithin surfactants, might raise concerns regarding the overall efficacy of this strategy for the topical or transdermal delivery of drugs [67].

Regardless of the mechanism, after establishing the transdermal delivery potential of their lecithin-based microemulsion system, the same group went on to tackle a classical problem associated with topical formulations, namely that the low viscosity of microemulsions makes them challenging to apply and localize on a designated area of skin [65–68]. As a result, a longer-releasing formula using gelatin was developed to enhance the viscosity of the system. The selected formulation containing 20% gelatin had a zero-shear viscosity close to 3 Pa·s, an order of magnitude higher than the original microemulsions and within the range of commercial topical creams (1–10 Pa·s) [69]. The authors reported that their microemulsion-based gels (MBGs) performed similarly to their lecithin microemulsions, though with a slightly lower loading and release of lidocaine.

To make these lecithin-linker microemulsions, the authors modified their previous formulation, replacing the hydrophilic linkers sodium octanoate and octanoic acid with a milder combination of PEG-6-caprylic/capric glycerides and decaglycerol monocaprylate/caprate, less irritating to human skin [70]. Permeability experiments studying passage through a synthetic membrane made of silicone, as well as transdermal delivery to and through pig ear skin, showed a comparable efficiency for these newly-formulated MBGs and the parent microemulsions (permeability coefficients = $6 \pm 1 \times 10^{-3}$ and $6.3 \pm 0.4 \times 10^{-3}$ cm/h, respectively). Though the addition of a gelling agent improves the rheological behavior of the formulation for clinical use, previous studies by other groups have suggested that the addition of a gelling agent reduces transdermal delivery (by roughly 1.5-fold) for hydrophobic nonsteroidal anti-inflammatory drugs, highlighting a potential limitation of the formulation which would need more evaluation [71]. Compared to commercial microemulsion-based formulations such as Topicaine® (ESBA Laboratories Inc., Jupiter, FL, USA, 30–60 mg/(cm^2·h)), these lecithin-based formulations, with the same loading of 4% *w/w*, presented a much slower release profile (3 mg/cm^2 in 18 h) for the local delivery of lidocaine, potentially beneficial for a longer-release formulation [72,73]. In addition to the reported works, studies of transdermal delivery in vivo and biocompatibility will be necessary to determine the clinical potential of this system.

Recently, the use of archaeosomes, liposome-like structures composed of archaeal lipids, have generated interest in drug delivery applications, with the Krishnan group at the National Research

Council of Canada being a very active player in the field. Although most of their research is focused on the design of archaeosomes as immune adjuvants and delivery systems for parenteral administration, in 2017, Jia et al. investigated the transdermal permeability potentials of these structures. The authors screened the capacity of a pool of archaeosomes composed of archaeal total polar lipids, as well as semi-synthetic glycosylarchaeol, to diffuse through the skin and deliver ovalbumin (a common reference protein for vaccination experiments) and compared it to a standard DPPC/DPPG liposome formulation with similar size (100–300 nm) and comparable ovalbumin loading capacities. Using pig ear skin, the authors showed that all the tested particles generated from the total archaeal lipids had remarkably improved (up to 5 times) their capacity to cross the SC and deliver ovalbumin to the dermal layer compared to the liposomes composed of standard lipids or semi-synthetic glycosylarchaeol. While the authors observed that this improved permeability at least partially correlated with the fluid character of these archaeal vesicles, as well as with their negative surface charge, no other insights on the actual mechanism behind these activities has yet been provided [74]. While the small particle size is self-explanatory when it comes to permeation-enhancing properties of this system, a few previous works investigated similar systems for transdermal drug delivery applications and showed that their physical deformability is an important factor behind their significant SC permeability [75]. Nevertheless, further systematic structure-activity investigations will be needed to fully understand by which mechanism these compounds are able to cross the SC and promote the transdermal delivery of large hydrophilic molecules, which might lead to the design of ideal synthetic delivery methods. Aside from the mechanistic investigation, in vivo studies will be necessary to confirm the observed results and to assess the safety of such archaea-derived compounds.

During the last decade, carbon nanotubes (CNTs) have gained popularity as potential drug and gene delivery vehicles for two main reasons: (1) their large inner volume, which allows the loading of either large pharmaceutical molecules, or larger quantities of smaller drugs, and (2) their observed capacity to operate as "nano-needles" which are able to effectively cross biological membranes, via a diffusion-like mechanism [76,77]. CNTs have thus been described for the delivery of several chemotherapeutic and antifungal agents such as cisplatin, doxorubicin, methotrexate, taxol, and amphotericin B, by parenteral administration [78]. However, their intrinsic hydrophobicity strongly limits their medical applications and consequently, different strategies have been described to overcome this limitation. Among them, surface decoration with polar or charged groups (such as the cationic polymer polyethylenimine) has been widely used to functionalize CNTs, leading to increased solubility and permitting the effective delivery of therapeutically active compounds [79]. Moreover, this functionalization with cationic residues has been shown to drastically improve the loading of nucleic acid-based therapeutics such as siRNA [80]. By virtue of their ability to cross biological membranes, a work from Western University in London, Ontario in 2014, investigated for the first time the use of single-walled carbon nanotubes (swCNTs) non-covalently functionalized with succinated polyethylenimine to topically deliver pharmaceutically active siRNA for the management of melanoma [81]. Functionalized swCNTs were loaded with an siRNA targeting Braf (a kinase involved in tumor growth via the MAPK pathway) at a remarkable *w/w* ratio of 2:1. These swCNTs demonstrated the selective downregulation of the targeted gene in melanoma cells (B16-F10), although no comparison with standard transfecting agents (i.e., cationic liposomes) was performed by the authors. Following these promising in vitro results, and based on a previous work demonstrating the ability of swCNTs to deliver low molecular-weight drugs transdermally, in association with an iontophoresis system [82]; the authors investigated the capacity of the particles to deliver siRNA across the epidermal layer, and to transfect melanoma cells in vivo after topical administration. Fluorescent microscopy of frozen sections of skin following the administration of swCNTs loaded with Cy3-labelled siRNA demonstrated the capacity of the nucleic acid sequence to cross the epidermis and reach the dermal layer (Figure 1a), while the same formulation without swCNTs did not cross the stratum corneum. Though these experiments highlighted the key role played by swCNTs in helping the siRNA to cross the epidermis, is should be noted that 10% DMSO, a well-known permeation enhancer, was added to the formulation. The authors then investigated the actual pharmaceutical potential of the swCNTs loaded with an anti-Braf siRNA in a mouse model of melanoma (intradermal inoculation of B16-F10 cells). Down-regulation experiments in the tumor cells indicated a remarkable 70% knockdown of Braf when delivering siRNA loaded in the functionalized swCNTs, 24 h after a single topical administration. Moreover, multiple administrations every 2 days for 25 days resulted in drastic inhibition of tumor growth (Figure 1b). Overall, these experiments by Siu et al. highlighted for the first time the potential of swCNTs to deliver pharmaceutics, even relatively large and hydrophilic molecules, across the epidermis, though the actual role of DMSO should be clarified. Nevertheless, prior to any actual clinical translation, the reported experiments will need to be validated in a more quantitative manner and in a more relevant animal model for transdermal studies, such as newborn pigs. Above all, while the described protocol used a functionalization step to increase the solubility of the CNTs, which has been reported to reduce their toxic accumulation and retention within the body, the actual fate of the CNTs as well as the non-degradable functionalization polymer will have to be carefully investigated [83,84].

Table 1. Transdermal drug delivery systems covered in the review.

Category	Technology	In Vivo Evaluation	Clinical Trial	Pros	Cons	Ref.
Chemical permeation enhancer (CPE)-based systems	Biphasic vesicles	Yes (guinea pig)	Phase II	Sustained release Versatility (small and large molecules)	Control of the delivered dose	[15–24]
	Ionic liquids (ILs) and active pharmaceutical ingredient-ionic liquids (API-ILs)	Yes (rats)	N/A	APIs with enhanced skin permeation properties of ionic liquids Properties can be fine-tuned	Requires specific choice of counter-ions Limited to small molecules	[42,43,47]
Physical enhancer-based systems	Hollow microneedles	Ex vivo: rabbit ear skin	Completed	Large doses Versatility	Manufacturing cost Potential clogging Skilled personnel	[53,54]
Nanoparticle-based systems	Lecithin-based microemulsions	N/A	N/A	Low skin irritation Sustained release and higher permeation compared to standard emulsions	Lecithin could lead to skin permeation complications	[61,64,65,69]
	Archaeosomes	N/A	N/A	Versatility Sustained release	Biocompatibility unclear Permeation mechanism unclear	[74]
	Carbon nanotubes (CNTs)	Yes (mice)	N/A	Effective skin permeation without CPEs High drug loading	Complexity Biocompatibility unclear	[81]

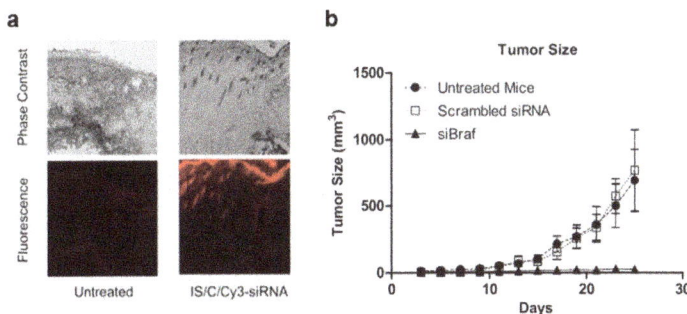

Figure 1. Carbon nanotube (CNT)-mediated delivery of siRNA in mice. (**a**) Representative images of tumor-bearing mouse skin treated with Cy3-labeled siRNA loaded CNTs; (**b**) tumor size evolution after topical administration of siRNA-loaded CNTs. Three days after the mice were injected with tumor cells, siRNA-loaded CNT solution was applied every 2 days for 25 days. Adapted with permission from [81]; published by Elsevier, 2014.

5. Conclusions

In this brief review, we summarized and discussed the main topical and transdermal drug delivery technologies that have been developed in recent years in Canada. Overall, although the number of works describing new technologies appears limited at first glance, it is interesting to note that they cover a wide range of strategies, from nanotechnology to CPEs, to hollow MNs. Furthermore, the research spans from the development of innovative technologies based on new materials that have shown a remarkable capacity to enhance the transdermal delivery of high molecular weight drugs (e.g., functionalized CNTs), but require intensive and more quantitative studies to better identify their clinical potentials and biocompatibility profiles; to more fundamental studies with the broader goal of identifying superior enhancement compounds based on well-known materials. Finally, advanced strategies have been developed over the past decade, and are currently being further studied by private companies in clinical settings (e.g., the BiPhasix™ technology) and could ideally soon reach the market, highlighting the favorable environment for the development of new medical and pharmaceutical technologies in the country.

Nonetheless, it should be noted that some transdermal drug delivery technologies undergoing rapid development are not currently represented in Canada. A notable example of this can be seen in polymeric MNs, two varieties of which have seen increasing use for drug delivery. The first of these are dissolving MNs made of soluble polymers, where a drug is loaded within the soluble polymeric matrix of the MNs, allowing release across the skin upon dissolution of the tips following application [85]. The Prausnitz group at Georgia Tech helped pioneer this technology in the context of non-invasive vaccination and drug delivery [86–88] and continue to work on this topic [89]. While research is also active globally, particularly in the context of delivering peptide-based drugs and macromolecules [90–92], no Canadian groups are currently investigating this class of MNs.

The other class of polymeric MNs being studied for transdermal drug delivery is swellable hydrogel MNs, typically made of crosslinked hydrophilic polymers able to swell by absorbing fluid from the skin. While these have also been used for sampling biological fluids from the skin [93], the Donnelly group at Queen's University Belfast have primarily studied them for drug delivery purposes, as their swelling properties also allow drug molecules contained within the MNs to flow into the skin after application [94,95]. This research appears to have progressed significantly, with recent studies focusing on optimizing the system for clinical applications [96–98]. Iontophoresis is another transdermal drug delivery strategy currently experiencing worldwide growth [99]. By passing an electrical current through the skin, this technique serves to enhance skin permeability, as well as allow positively charged compounds to be transported into the skin by the resulting electric field. Though various groups are developing iontophoresis-based transdermal delivery methods for both small molecule- and peptide-based drugs [100,101], the topic is seemingly not undergoing active research within Canada.

These methods, alongside the ones discussed previously, serve to reveal the current state of transdermal drug delivery technology worldwide. Though Canada has generated meaningful contributions within the past decade, it remains clear that many opportunities for further work exist, if groups within Canada wish to further the progression of this field.

Author Contributions: M.R., E.L., and S.B. performed the primary literature search and wrote the manuscript. D.B. corrected and implemented the manuscript.

Funding: The authors acknowledge financial support from the Fonds de Recherche du Québec, the Bourses d'excellence TransMedTech (M.R.), and the Canadian Generic Pharmaceutical Association and Biosimilars Canada (D.B.).

Conflicts of Interest: The authors declare no conflict of interest.

References

1. Babity, S.; Roohnikan, M.; Brambilla, D. Advances in the design of transdermal microneedles for diagnostic and monitoring applications. *Small* **2018**, *14*, 1803186. [CrossRef] [PubMed]

2. Benson, H.A.E. *Transdermal and Topical Drug Delivery: Principles and Practice*; Benson, H.A.E., Watkinson, A.C., Eds.; Wiley: Hoboken, NJ, USA, 2012; Chapter 1, pp. 1–22.

3. Menon, G.K.; Cleary, G.W.; Lane, M.E. The structure and function of the stratum corneum. *Int. J. Pharm.* **2012**, *435*, 3–9. [CrossRef] [PubMed]

4. Eckhart, L.; Lippens, S.; Tschachler, E.; Declercq, W. Cell death by cornification. *Biochim. Biophys. Acta Mol. Cell Res.* **2013**, *1833*, 3471–3480. [CrossRef]

5. Michaels, A.S.; Chandrasekaran, S.K.; Shaw, J.E. Drug permeation through human skin: Theory and in vitro experimental measurement. *AIChE J.* **1975**, *21*, 985–996. [CrossRef]

6. Prausnitz, M.R.; Mitragotri, S.; Langer, R. Current status and future potential of transdermal drug delivery. *Nat. Rev. Drug Discov.* **2004**, *3*, 115–124. [CrossRef]

7. Arora, A.; Prausnitz, M.R.; Mitragotri, S. Micro-scale devices for transdermal drug delivery. *Int. J. Pharm.* **2008**, *364*, 227–236. [CrossRef]

8. Prausnitz, M.R.; Langer, R. Transdermal drug delivery. *Nat. Biotechnol.* **2008**, *26*, 1261–1268. [CrossRef] [PubMed]

9. Watkinson, A.C.; Kearney, M.C.; Quinn, H.L.; Courtenay, A.J.; Donnelly, R.F. Future of the transdermal drug delivery market—Have we barely touched the surface? *Expert Opin. Drug Deliv.* **2016**, *13*, 523–532. [CrossRef]

10. Paudel, K.S.; Milewski, M.; Swadley, C.L.; Brogden, N.K.; Ghosh, P.; Stinchcomb, A.L. Challenges and opportunities in dermal/transdermal delivery. *Ther. Deliv.* **2010**, *1*, 109–131. [CrossRef]

11. Pharmaceutical Industry Profile. Available online: http://www.webcitation.org/query?url=https%3A%2F%2Fwww.ic.gc.ca%2Feic%2Fsite%2Flsg-pdsv.nsf%2Feng%2Fh_hn01703.html&date=2019-02-02 (accessed on 2 February 2019).

12. Chen, Y.; Quan, P.; Liu, X.; Wang, M.; Fang, L. Novel chemical permeation enhancers for transdermal drug delivery. *Asian J. Pharm.* **2014**, *9*, 51–64. [CrossRef]

13. Karande, P.; Jain, A.; Ergun, K.; Kipersky, V.; Mitragotri, S. Design principles of chemical penetration enhancers for transdermal drug delivery. *Proc. Natl. Acad. Sci. USA* **2005**, *102*, 4688–4693. [CrossRef]

14. Williams, A.C.; Barry, B.W. Penetration enhancers. *Adv. Drug Deliv. Rev.* **2004**, *56*, 603–618. [CrossRef] [PubMed]

15. Foldvari, M. Biphasic Multilamellar Lipid Vesicles. U.S. Patent 5,853,755, 29 December 1998.

16. Moghadam, S.H.; Saliaj, E.; Wettig, S.D.; Dong, C.; Ivanova, M.V.; Huzil, J.T.; Foldvari, M. Effect of chemical permeation enhancers on stratum corneum barrier lipid organizational structure and interferon alpha permeability. *Mol. Pharm.* **2013**, *10*, 2248–2260. [CrossRef] [PubMed]

17. King, M.J.; Badea, I.; Solomon, J.; Kumar, P.; Gaspar, K.J.; Foldvari, M. Transdermal delivery of insulin from a novel biphasic lipid system in diabetic rats. *Diabetes Technol. Ther.* **2002**, *4*, 479–488. [CrossRef]

18. Baca-Estrada, M.E.; Foldvari, M.; Ewen, C.; Badea, I.; Babiuk, L.A. Effects of IL-12 on immune responses induced by transcutaneous immunization with antigens formulated in a novel lipid-based biphasic delivery system. *Vaccine* **2000**, *18*, 1847–1854. [CrossRef]

19. King, M.; Kumar, P.; Michel, D.; Batta, R.; Foldvari, M. In vivo sustained dermal delivery and pharmacokinetics of interferon alpha in biphasic vesicles after topical application. *Eur. J. Pharm. Biopharm.* **2014**, *84*, 532–539. [CrossRef]

20. Foldvari, M.; Badea, I.; Wettig, S.; Baboolal, D.; Kumar, P.; Creagh, A.L.; Haynes, C.A. Topical delivery of interferon alpha by biphasic vesicles: Evidence for a novel nanopathway across the stratum corneum. *Mol. Pharm.* **2010**, *7*, 751–762. [CrossRef]

21. Foldvari, M.; Badea, I.; Kumar, P.; Wettig, S.; Batta, R.; King, M.; Shear, N. Biphasic vesicles for topical delivery of interferon alpha in human volunteers and treatment of patients with human papillomavirus infections. *Curr. Drug Deliv.* **2011**, *8*, 307–319. [CrossRef]

22. Rougier, A.; Lotte, C. Predictive approaches I: The stripping technique. In *Topical Drug Bioavailability, Bioequivalence, and Penetration*; Shah, V.P., Maibach, H.I., Eds.; Springer: Boston, MA, USA, 1993; pp. 163–181.

23. Kurzeja, R.; Böhmer, G.; Schneider, A. Clinical outcome of topical interferon Alpha-2b cream in phase II trial for LSIL/CIN 1 patients. *J. Cancer Ther.* **2011**, *2*, 203–208. [CrossRef]

24. Altum Pharmaceuticals Pipeline. Available online: http://www.webcitation.org/query?url=https%3A%2F%2Fwww.altumpharma.com%2Fpipeline%2F&date=2019-02-02 (accessed on 2 February 2019).

25. Ashtikar, A.; Nagarsekar, K.; Fahr, A. Transdermal delivery from liposomal formulations—Evolution of the technology over the last three decades. *J. Control. Release* **2016**, *242*, 126–140. [CrossRef] [PubMed]

26. Beloqui, A.; Solinís, M.Á.; Rodríguez-Gascón, A.; Almeida, A.J.; Préat, V. Nanostructured lipid carriers: Promising drug delivery systems for future clinics. *Nanomedicine* **2016**, *12*, 143–161. [CrossRef]

27. Mennini, N.; Cirri, M.; Maestrelli, F.; Mura, P. Comparison of liposomal and NLC (nanostructured lipid carrier) formulations for improving the transdermal delivery of oxaprozin: Effect of cyclodextrin complexation. *Int. J. Pharm.* **2016**, *515*, 684–691. [CrossRef] [PubMed]

28. Pinto, M.F.; Moura, C.C.; Nunes, C.; Segundo, M.A.; Costa Lima, S.A.; Reis, S. A new topical formulation for psoriasis: Development of methotrexate-loaded nanostructured lipid carriers. *Int. J. Pharm.* **2014**, *477*, 519–526. [CrossRef]

29. Puglia, C.; Sarpietro, M.G.; Bonina, F.; Castelli, F.; Zammataro, M.; Chiechio, S. Development, characterization, and in vitro and in vivo evaluation of benzocaine- and lidocaine-loaded nanostructured lipid carriers. *J. Pharm. Sci.* **2011**, *100*, 1892–1899. [CrossRef]

30. Das, R.N.; Roy, K. Advances in QSPR/QSTR models of ionic liquids for the design of greener solvents of the future. *Mol. Divers.* **2013**, *17*, 151–196. [CrossRef]

31. Gadilohar, B.L.; Shankarling, G.S. Choline based ionic liquids and their applications in organic transformation. *J. Mol. Liquids* **2017**, *227*, 234–261. [CrossRef]

32. Egorova, K.S.; Gordeev, E.G.; Ananikov, V.P. Biological activity of ionic liquids and their application in pharmaceutics and medicine. *Chem. Rev.* **2017**, *117*, 7132–7189. [CrossRef] [PubMed]

33. Kundu, N.; Roy, S.; Mukherjee, D.; Maiti, T.K.; Sarkar, N. Unveiling the interaction between fatty-acid-modified membrane and hydrophilic imidazolium-based ionic liquid: Understanding the mechanism of ionic liquid cytotoxicity. *J. Phys. Chem. B* **2017**, *121*, 8162–8170. [CrossRef]

34. Lim, G.S.; Jaenicke, S.; Klähn, M. How the spontaneous insertion of amphiphilic imidazolium-based cations changes biological membranes: A molecular simulation study. *Phys. Chem. Chem. Phys.* **2015**, *17*, 29171–29183. [CrossRef] [PubMed]

35. Zakrewsky, M.; Lovejoy, K.S.; Kern, T.L.; Miller, T.E.; Le, V.; Nagy, A.; Goumas, A.M.; Iyer, R.S.; Del Sesto, R.E.; Koppisch, A.T.; et al. Ionic liquids as a class of materials for transdermal delivery and pathogen neutralization. *Proc. Natl. Acad. Sci. USA* **2014**, *111*, 13313–13318. [CrossRef]

36. Lane, M.E. Skin penetration enhancers. *Int. J. Pharm.* **2013**, *447*, 12–21. [CrossRef]

37. Sidat, Z.; Marimuthu, T.; Kumar, P.; du Toit, L.C.; Kondiah, P.P.D.; Choonara, Y.E.; Pillay, V. Ionic liquids as potential and synergistic permeation enhancers for transdermal drug delivery. *Pharmaceutics* **2019**, *11*, 96. [CrossRef]

38. Kelley, S.P.; Narita, A.; Holbrey, J.D.; Green, K.D.; Reichert, W.M.; Rogers, R.D. Understanding the effects of ionicity in salts, solvates, co-crystals, ionic co-crystals, and ionic liquids, rather than nomenclature, is critical to understanding their behavior. *Cryst. Growth Des.* **2013**, *13*, 965–975. [CrossRef]

39. Wang, H.; Gurau, G.; Shamshina, J.L.; Cojocaru, O.A.; Janikowski, J.; MacFarlane, D.R.; Davis, J.H.J.; Rogers, R.D. Simultaneous membrane transport of two active pharmaceutical ingredients by charge assisted hydrogen bond complex formation. *Chem. Sci.* **2014**, *5*, 3449–3456. [CrossRef]

40. Raza, K.; Kumar, P.; Ratan, S.; Malik, R.; Arora, S. Polymorphism: The phenomenon affecting the performance of drugs. *SOJ Pharm. Pharm. Sci.* **2014**, *1*, 10. [CrossRef]

41. Balk, A.; Holzgrabe, U.; Meinel, L. 'Pro et contra' ionic liquid drugs—Challenges and opportunities for pharmaceutical translation. *Eur. J. Pharm. Biopharm.* **2015**, *94*, 291–304. [CrossRef]

42. Hough, W.L.; Smiglak, M.; Rodríguez, H.; Swatloski, R.P.; Spear, S.K.; Daly, D.T.; Pernak, J.; Grisel, J.E.; Carliss, R.D.; Soutullo, M.D.; et al. The third evolution of ionic liquids: Active pharmaceutical ingredients. *New J. Chem.* **2007**, *31*, 1429–1436. [CrossRef]

43. Zavgorodnya, O.; Shamshina, J.L.; Mittenthal, M.; McCrary, P.D.; Rachiero, G.P.; Titi, H.M.; Rogers, R.D. Polyethylene glycol derivatization of the non-active ion in active pharmaceutical ingredient ionic liquids enhances transdermal delivery. *New J. Chem.* **2017**, *41*, 1499–1508. [CrossRef]

44. Stoimenovski, J.; MacFarlane, D.R. Enhanced membrane transport of pharmaceutically active protic ionic liquids. *Chem. Commun.* **2011**, *47*, 11429–11431. [CrossRef]
45. Thong, H.Y.; Zhai, H.; Maibach, H.I. Percutaneous penetration enhancers: An overview. *Skin Pharmacol. Physiol.* **2007**, *20*, 272–282. [CrossRef]
46. Lloyd, R.S.; Blythe, J.O.J. Clinical experiences with lidocaine as a local anesthetic. *J. Am. Dent. Assoc.* **1949**, *39*, 296–298. [CrossRef]
47. Berton, P.; Di Bona, K.R.; Yancey, D.; Rizvi, S.A.A.; Gray, M.; Gurau, G.; Shamshina, J.L.; Rasco, J.F.; Rogers, R.D. Transdermal bioavailability in rats of lidocaine in the forms of ionic liquids, salts, and deep eutectic. *ACS Med. Chem. Lett.* **2017**, *8*, 498–503. [CrossRef]
48. Park, H.J.; Prausnitz, M.R. Lidocaine-ibuprofen ionic liquid for dermal anesthesia. *AIChE J.* **2015**, *61*, 2732–2738. [CrossRef]
49. Van der Maaden, K.; Jiskoot, W.; Bouwstra, J. Microneedle technologies for (trans)dermal drug and vaccine delivery. *J. Control. Release* **2012**, *161*, 645–655. [CrossRef]
50. Ovsianikov, A.; Chichkov, B.; Mente, P.; Monteiro-Riviere, N.A.; Doraiswamy, A.; Narayan, R.J. Two photon polymerization of polymer-ceramic hybrid materials for transdermal drug delivery. *Int. J. Appl. Ceram. Technol.* **2007**, *4*, 22–29. [CrossRef]
51. Yung, K.L.; Xu, Y.; Kang, C.; Liu, H.; Tam, K.F.; Ko, S.M.; Kwan, F.Y.; Lee, T.M.H. Sharp tipped plastic hollow microneedle array by microinjection moulding. *J. Micromech. Microeng.* **2012**, *22*, 015016. [CrossRef]
52. Lhernould, M.S.; Deleers, M.; Delchambre, A. Hollow polymer microneedles array resistance and insertion tests. *Int. J. Pharm.* **2015**, *480*, 152–157. [CrossRef]
53. Mansoor, I.; Häfeli, U.O.; Stoeber, B. Hollow out-of-plane polymer microneedles made by solvent casting for transdermal drug delivery. *J. Microelectromech. Syst.* **2012**, *21*, 44–52. [CrossRef]
54. Mansoor, I.; Liu, Y.; Häfeli, U.O.; Stoeber, B. Arrays of hollow out-of-plane microneedles made by metal electrodeposition onto solvent cast conductive polymer structures. *J. Micromech. Microeng.* **2013**, *23*, 085011. [CrossRef]
55. University of British Columbia. Safety Demonstration of Microneedle Insertion, NCT02995057. Available online: https://clinicaltrials.gov/ct2/show/NCT02995057 (accessed on 31 May 2019).
56. Takeuchi, I.; Shimamura, Y.; Kakami, Y.; Kameda, T.; Hattori, K.; Miura, S.; Okumura, M.; Inagi, T.; Terada, H.; Makino, K. Transdermal delivery of 40-nm silk fibroin nanoparticles. *Colloids Surf. B Biointerfaces* **2018**, *175*, 564–568. [CrossRef] [PubMed]
57. Nastiti, C.; Ponto, T.; Abd, E.; Grice, J.E.; Benson, H.A.E.; Roberts, M.S. Topical nano and microemulsions for skin delivery. *Pharmaceutics* **2017**, *9*, 37. [CrossRef] [PubMed]
58. Kogan, A.; Garti, N. Microemulsions as transdermal drug delivery vehicles. *Adv. Colloid Interface Sci.* **2006**, *123–126*, 369–385. [CrossRef]
59. Kreilgaard, M.; Pedersen, E.J.; Jaroszewski, J.W. NMR characterisation and transdermal drug delivery potential of microemulsion systems. *J. Control. Release* **2000**, *69*, 421–433. [CrossRef]
60. Lee, M.-W.C. Topical triple-anesthetic gel compared with 3 topical anesthetics. *Cosmet. Dermatol.* **2003**, *16*, 35–37.
61. Yuan, J.S.; Yip, A.; Nguyen, N.; Chu, J.; Wen, X.Y.; Acosta, E.J. Effect of surfactant concentration on transdermal lidocaine delivery with linker microemulsions. *Int. J. Pharm.* **2010**, *392*, 274–284. [CrossRef] [PubMed]
62. Van Nieuwenhuyzen, W.; Tomás, M.C. Update on vegetable lecithin and phospholipid technologies. *Eur. J. Lipid Sci. Technol.* **2008**, *110*, 472–486. [CrossRef]
63. Heuschkel, S.; Goebel, A.; Neubert, R.H. Microemulsions—Modern colloidal carrier for dermal and transdermal drug delivery. *J. Pharm. Sci.* **2008**, *97*, 603–631. [CrossRef]
64. Yuan, J.S.; Ansari, M.; Samaan, M.; Acosta, E.J. Linker-based lecithin microemulsions for transdermal delivery of lidocaine. *Int. J. Pharm.* **2008**, *349*, 130–143. [CrossRef]
65. Xuan, X.Y.; Cheng, Y.L.; Acosta, E.J. Lecithin-linker microemulsion gelatin gels for extended drug delivery. *Pharmaceutics* **2012**, *4*, 104–129. [CrossRef] [PubMed]
66. Jacobi, U.; Kaiser, M.; Toll, R.; Mangelsdorf, S.; Audring, H.; Otberg, N.; Sterry, W.; Lademann, J. Porcine ear skin: An in vitro model for human skin. *Skin Res. Technol.* **2007**, *13*, 19–24. [CrossRef]
67. Hoppel, M.; Juric, S.; Ettl, H.; Valenta, C. Effect of monoacyl phosphatidylcholine content on the formation of microemulsions and the dermal delivery of flufenamic acid. *Int. J. Pharm.* **2015**, *479*, 70–76. [CrossRef] [PubMed]

68. Hajjar, B.; Zier, K.I.; Khalid, N.; Azarmi, S.; Löbenberg, R. Evaluation of a microemulsion-based gel formulation for topical drug delivery of diclofenac sodium. *J. Pharm. Investig.* **2018**, *48*, 351–362. [CrossRef]
69. Adeyeye, M.C.; Jain, A.C.; Ghorab, M.K.; Reilly, W.J.J. Viscoelastic evaluation of topical creams containing microcrystalline cellulose/sodium carboxymethyl cellulose as stabilizer. *AAPS Pharm. Sci. Technol.* **2002**, *3*, 16–25. [CrossRef]
70. Acosta, E.; Chung, O.; Xuan, X.Y. Lecithin-linker microemulsions in transdermal delivery. *J. Drug Deliv. Sci. Technol.* **2011**, *21*, 77–87. [CrossRef]
71. Kriwet, K.; Müller-Goymann, C.C. Diclofenac release from phospholipid drug systems and permeation through excised human stratum corneum. *Int. J. Pharm.* **1995**, *135*, 231–242. [CrossRef]
72. Young, K.D. What's new in topical anesthesia. *Clin. Pediatr. Emerg. Med.* **2007**, *8*, 232–239. [CrossRef]
73. Friedman, P.M.; Mafong, E.A.; Friedman, E.S.; Geronemus, R.G. Topical anesthetics update: EMLA and beyond. *Dermatol. Surg.* **2001**, *27*, 1019–1026. [CrossRef] [PubMed]
74. Jia, Y.; McCluskie, M.J.; Zhang, D.; Monette, R.; Iqbal, U.; Moreno, M.; Sauvageau, J.; Williams, D.; Deschatelets, L.; Jakubek, Z.J.; et al. In vitro evaluation of archaeosome vehicles for transdermal vaccine delivery. *J. Liposome Res.* **2018**, *28*, 305–314. [CrossRef] [PubMed]
75. Higa, L.H.; Arnal, L.; Vermeulen, M.; Perez, A.P.; Schilrreff, P.; Mundiña-Weilenmann, C.; Yantorno, O.; Vela, M.E.; Morilla, M.J.; Romero, E.L. Ultradeformable archaeosomes for needle free nanovaccination with Leishmania braziliensis antigens. *PLoS ONE* **2016**, *11*, e0150185. [CrossRef] [PubMed]
76. Jain, K.K. Advances in use of functionalized carbon nanotubes for drug design and discovery. *Expert Opin. Drug Discov.* **2012**, *7*, 1029–1037. [CrossRef]
77. Mehra, N.K.; Jain, K.; Jain, N.K. Pharmaceutical and biomedical applications of surface engineered carbon nanotubes. *Drug Discov. Today* **2015**, *20*, 750–759. [CrossRef]
78. Karimi, M.; Solati, N.; Ghasemi, A.; Estiar, M.A.; Hashemkhani, M.; Kiani, P.; Mohamed, E.; Saeidi, A.; Taheri, M.; Avci, P.; et al. Carbon nanotubes part II: A remarkable carrier for drug and gene delivery. *Expert Opin. Drug Deliv.* **2015**, *12*, 1089–1105. [CrossRef]
79. Zintchenko, A.; Philipp, A.; Dehshahri, A.; Wagner, E. Simple modifications of branched PEI lead to highly efficient siRNA carriers with low toxicity. *Bioconj. Chem.* **2008**, *19*, 1448–1455. [CrossRef] [PubMed]
80. Wang, L.; Shi, J.; Zhang, H.; Li, H.; Gao, Y.; Wang, Z.; Wang, H.; Li, L.; Zhang, C.; Chen, C.; et al. Synergistic anticancer effect of RNAi and photothermal therapy mediated by functionalized single-walled carbon nanotubes. *Biomaterials* **2013**, *34*, 262–274. [CrossRef]
81. Siu, K.S.; Chen, D.; Zheng, X.; Zhang, X.; Johnston, N.; Liu, Y.; Yuan, K.; Koropatnick, J.; Gillies, E.R.; Min, W.P. Non-covalently functionalized single-walled carbon nanotube for topical siRNA delivery into melanoma. *Biomaterials* **2014**, *35*, 3435–3442. [CrossRef]
82. Degim, I.T.; Burgess, D.J.; Papadimitrakopoulos, F. Carbon nanotubes for transdermal drug delivery. *J. Microencapsul.* **2010**, *27*, 669–681. [CrossRef] [PubMed]
83. Kesharwani, P.; Gajbhiye, V.; Jain, N.K. A review of nanocarriers for the delivery of small interfering RNA. *Biomaterials* **2012**, *33*, 7138–7150. [CrossRef]
84. Principi, E.; Girardello, R.; Bruno, A.; Manni, I.; Gini, E.; Pagani, A.; Grimaldi, A.; Ivaldi, F.; Congiu, T.; De Stefano, D.; et al. Systemic distribution of single-walled carbon nanotubes in a novel model: Alteration of biochemical parameters, metabolic functions, liver accumulation, and inflammation in vivo. *Int. J. Nanomed.* **2016**, *11*, 4299–4316. [CrossRef]
85. Lee, J.W.; Park, J.-H.; Prausnitz, M.R. Dissolving microneedles for transdermal drug delivery. *Biomaterials* **2008**, *29*, 2113–2124. [CrossRef] [PubMed]
86. Henry, S.; McAllister, D.V.; Allen, M.G.; Prausnitz, M.R. Microfabricated microneedles: A novel approach to transdermal drug delivery. *J. Pharm. Sci.* **1998**, *87*, 922–925. [CrossRef]
87. Prausnitz, M.R.; Mikszta, J.A.; Cormier, M.; Andrianov, A.K. Microneedle-based vaccines. *Curr. Top. Microbiol. Immunol.* **2009**, *333*, 369–393. [CrossRef]
88. Sullivan, S.P.; Koutsonanos, D.G.; del Pilar Martin, M.; Lee, J.-W.; Zarnitsyn, V.; Murthy, N.; Compans, R.W.; Skountzou, I.; Prausnitz, M.R. Dissolving polymer microneedle patches for influenza vaccination. *Nat. Med.* **2010**, *16*, 915–920. [CrossRef] [PubMed]

89. Joyce, J.C.; Carroll, T.D.; Collins, M.L.; Chen, M.H.; Fritts, L.; Dutra, J.C.; Rourke, T.L.; Goodson, J.L.; McChesney, M.B.; Prausnitz, M.R.; et al. A microneedle patch for measles and rubella vaccination is immunogenic and protective in infant rhesus monkeys. *J. Infect. Dis.* **2018**, *218*, 124–132. [CrossRef] [PubMed]

90. Dillon, C.; Hughes, H.; O'Reilly, N.J.; Allender, C.J.; Barrow, D.A.; McLoughlin, P. Dissolving microneedle based transdermal delivery of therapeutic peptide analogues. *Int. J. Pharm.* **2019**, *565*, 9–19. [CrossRef]

91. Tian, Z.; Cheng, J.; Liu, J.; Zhu, Y. Dissolving graphene/poly(acrylic acid) microneedles for potential transdermal drug delivery and photothermal therapy. *J. Nanosci. Nanotechnol.* **2019**, *19*, 2453–2459. [CrossRef] [PubMed]

92. Chen, X.; Wang, L.; Yu, H.; Li, C.; Feng, J.; Haq, F.; Khan, A.; Khan, R.U. Preparation, properties and challenges of the microneedles-based insulin delivery system. *J. Control. Release* **2018**, *288*, 173–188. [CrossRef]

93. Chang, H.; Zheng, M.; Yu, X.; Than, A.; Seeni, R.Z.; Kang, R.; Tian, J.; Khanh, D.P.; Liu, L.; Chen, P.; et al. A swellable microneedle patch to rapidly extract skin interstitial fluid for timely metabolic analysis. *Adv. Mater.* **2017**, *29*. [CrossRef]

94. Donnelly, R.F.; Singh, T.R.R.; Garland, M.J.; Migalska, K.; Majithiya, R.; McCrudden, C.M.; Kole, P.L.; Mahmood, T.M.T.; McCarthy, H.O.; Woolfson, A.D. Hydrogel-forming microneedle arrays for enhanced transdermal drug delivery. *Adv. Funct. Mater.* **2012**, *22*, 4879–4890. [CrossRef]

95. Migdadi, E.M.; Courtenay, A.J.; Tekko, I.A.; McCrudden, M.T.C.; Kearney, M.C.; McAlister, E.; McCarthy, H.O.; Donnelly, R.F. Hydrogel-forming microneedles enhance transdermal delivery of metformin hydrichloride. *J. Control. Release* **2018**, *285*, 142–151. [CrossRef]

96. Donnelly, R.F.; Moffatt, K.; Alkilani, A.Z.; Vicente-Pérez, E.M.; Barry, J.; McCrudden, M.T.; Woolfson, A.D. Hydrogel-forming microneedle arrays can be effectively inserted in skin by self-application: A pilot study centred on pharmacist intervention and a patient information leaflet. *Pharm. Res.* **2014**, *31*, 1989–1999. [CrossRef]

97. McCrudden, M.T.; Alkilani, A.Z.; Courtenay, A.J.; McCrudden, C.M.; McCloskey, B.; Walker, C.; Alshraiedeh, N.; Lutton, R.E.; Gilmore, B.F.; Woolfson, A.D.; et al. Considerations in the sterile manufacture of polymeric microneedle arrays. *Drug Deliv. Transl. Res.* **2015**, *5*, 3–14. [CrossRef] [PubMed]

98. Vicente-Pérez, E.M.; Quinn, H.L.; McAlister, E.; O'Neill, S.; Hanna, L.A.; Barry, J.G.; Donnelly, R.F. The use of a pressure-indicating sensor film to provide feedback upon hydrogel-forming microneedle array self-application in vivo. *Pharm. Res.* **2016**, *33*, 3072–3080. [CrossRef] [PubMed]

99. Ita, K. Transdermal iontophoretic drug delivery: Advances and challenges. *J. Drug Target.* **2016**, *24*, 386–391. [CrossRef]

100. Cordery, S.F.; Husbands, S.M.; Bailey, C.P.; Guy, R.H.; Delgado-Charro, M.B. Simultaneous transdermal delivery of buprenorphine hydrochloride and naltrexone hydrochloride by iontophoresis. *Mol. Pharm.* **2019**. [CrossRef]

101. Pawar, K.; Kolli, C.S.; Rangari, V.K.; Babu, R.J. Transdermal iontophoretic delivery of lysine-proline-valine (KPV) peptide across microporated human skin. *J. Pharm. Sci.* **2017**, *106*, 1814–1820. [CrossRef]

pharmaceutics

MDPI

Review

Plant/Bacterial Virus-Based Drug Discovery, Drug Delivery, and Therapeutics

Esen Sokullu, Hoda Soleymani Abyaneh and Marc A. Gauthier *

Institut National de la Recherche Scientifique (INRS), EMT Research Center, Varennes, QC J3X 1S2, Canada;
sokullu@emt.inrs.ca (E.S.); hoda.soleymani@emt.inrs.ca (H.S.A.)
* Correspondence: gauthier@emt.inrs.ca; Tel.: +1-514-228-6932

Received: 22 March 2019; Accepted: 25 April 2019; Published: 3 May 2019

Abstract: Viruses have recently emerged as promising nanomaterials for biotechnological applications. One of the most important applications of viruses is phage display, which has already been employed to identify a broad range of potential therapeutic peptides and antibodies, as well as other biotechnologically relevant polypeptides (including protease inhibitors, minimizing proteins, and cell/organ targeting peptides). Additionally, their high stability, easily modifiable surface, and enormous diversity in shape and size, distinguish viruses from synthetic nanocarriers used for drug delivery. Indeed, several plant and bacterial viruses (e.g., phages) have been investigated and applied as drug carriers. The ability to remove the genetic material within the capsids of some plant viruses and phages produces empty viral-like particles that are replication-deficient and can be loaded with therapeutic agents. This review summarizes the current applications of plant viruses and phages in drug discovery and as drug delivery systems and includes a discussion of the present status of virus-based materials in clinical research, alongside the observed challenges and opportunities.

Keywords: virus; plant; bacteriophage; phage display; drug discovery; encapsulation; drug delivery

1. Introduction

In the 18th century, the term "virus" was defined as a morbid principle or poisonous substance produced in the body as the result of some disease [1]. This was due to the initial identification of viruses as infectious agents that could be transferred to other humans or animals, similarly to bacteria but of different size. Since the discovery of the first virus by Ivanovsky in 1892, our understanding of the properties of viruses has changed significantly [2]. Owing to advances in virology, their definition was changed to small, non-cellular obligate parasites carrying non-host genetic information [3]. In fact, these advances not only improved our knowledge about the nature of viruses, but also contributed to change their negative connotation. After the discovery of bacterial viruses (bacteriophages; phages), the French Canadian Felix d'Herelle recognized their ability to replicate exponentially and kill bacteria [4]. These observations suggested potential clinical applications for bacteriophages such as their use as antibacterial agents for the treatment of infectious disease. Although many large pharmaceutical companies marketed phage products in the 1920s and 1930s, clinical failures and theoretical concerns led to their abandonment [5]. In parallel, the discovery of antibiotics equally contributed to the loss of interest in phage therapy around this time [6]. Nonetheless, in the 1940s bacteriophages regained attention in the field of molecular biology. They were notably used as model organisms to understand the genetic basis of virus–host interactions in addition to enabling the discovery of several genetic processes such as transcription, translation, recombination, and regulation of gene expression [7]. While these discoveries reshaped the paradigm of virology, in 1985 the invention of the so-called "phage display" technique greatly broadened the field to new areas of application including drug discovery, vaccine development, antibody engineering, enzyme evolution, and gene therapy [8,9].

Thirty-three years after its development, the contribution of phage display technology to the selection of peptides and antibodies was recognized by a Nobel Prize in Chemistry in 2018 [10]. Another milestone was reached around the year 2000, when viruses were recognized as nanomaterials with the ability to encapsulate molecules within their capsids, to template biomineralization of inorganic materials, and to form self-assembled 2D/3D nanostructures. In this regard, viruses have become promising materials for several applications ranging from biosensing, imaging, and targeted drug/gene delivery to energy/electronic applications (memory devices, batteries, and light-harvesting systems). This review summarizes the current applications of plant and bacterial viruses in medicine, with a focus on virus-based drug discovery approaches and drug delivery systems. Human/animal viruses used e.g., for gene delivery, have been reviewed elsewhere and will not be discussed herein [11,12]. The review begins with an overview of the structure and chemistry of the most common plant and bacterial viruses used in nanomedicine and is concluded by a discussion of the present status of virus-based materials in clinical research, alongside current existing challenges and opportunities.

2. Viruses as Nanomedicine

From a material point of view, viruses can be considered as protein-based supramolecular assemblies composed of multiple copies of coat proteins assembled into shell structures of different shapes/sizes ranging from tens to hundreds of nanometers. The protein outer shell (i.e., the so-called capsid) encapsulates the genomic material that contains all essential genes to replicate within a host [13]. The primary function of the capsid is to protect the genomic material and this feature makes viruses stable under conditions such as extreme temperature and pH [14]. Although animal viruses are widely recognized as a delivery vehicle, or 'vector', for gene therapy, their use as a nanocarrier has remained relatively limited due to safety concerns. Indeed, while it has been demonstrated that the administration of 10^{11} plant viruses or phage to mice showed no sign of toxicity, the same dose of animal viruses caused severe hepatotoxicity [15–18]. Therefore, plant and bacteria viruses have received considerably more attention than animal viruses in nanomedicine for applications other than gene delivery. The most studied bacterial virus is M13 phage, whereas the tobacco mosaic virus (TMV), cowpea chlorotic mottle virus (CCMV), and cowpea mosaic virus (CMV) are the most extensively studied plant viruses [19]. Viruses possess precise, nanoscale structures and dimensions that are difficult replicate using chemical synthesis or top-down fabrication methods [20]. The diversity in shape and size of viruses provides a wealth of possibilities to researchers, who can choose the most appropriate system for a given application. For instance, while viruses with higher aspect ratios are more suitable to target diseased vessel walls, viruses with flexible rod shapes have been shown to penetrate better into tumor tissue [21,22]. In addition, the surface properties of viruses can be controlled using chemical and genetic approaches without destroying their structural integrity. This feature enables spatial control on the position of functional moieties, such as targeting ligands, drug molecules, and contrast agents on the virus surface and allows for the design of multifunctional systems bearing combinations of the above [23]. One of the most interesting properties of viruses, in terms of material synthesis, is that they can be produced in large quantities by infecting host cells, and can be purified inexpensively on a large scale. Therefore, virus-based materials are a niche nanomaterial with several unique features compared to synthetic nanomaterials.

2.1. Bacterial Viruses (Bacteriophages)

2.1.1. Filamentous Bacteriophages (M13 and fd)

M13 and fd phages belong to a group of filamentous phages (Ff) that specifically infect *Escherichia coli* bacteria. As their genomes are more than 98% identical and their gene products are interchangeable, they are usually collectively referred to as Ff phage [24]. Thus, only the properties of M13 phage are discussed herein as a representative example of filamentous phages. The relatively simple structure of the M13 virion has been extensively studied and is very well known. M13 is 65 Å in diameter and its length depends on the size of enclosed genome (9300 Å in the case of the wild-type M13) (Figure 1A). The flexible filamentous structure contains a circular, 6407 base-pair single-stranded DNA genome coated with 2700 copies of the major coat protein p8 (Figure 2A). The major coat proteins form a tube around the DNA, in an overlapping helical array. The N-terminus of the p8 protein extends towards the exterior of the capsid while the C-terminus interacts with the DNA inside. The hydrophobic domain located in the central part of p8 protein stabilizes the viral particle by interlocking the coat proteins with their neighbors. Additionally, four other minor coat proteins are present, at five copies per particle. p7 and p9 are located at one end of the capsid, while p3 and p6 are located at the other end. p3 is the largest and most complex coat protein and is responsible for the host cell recognition and infection [25–27].

M13 phage engages in a chronic infection life cycle where the propagated phage particles are slowly released from the host cell by secretion through the outer membrane, a process that does not lead to bacteria lysis. Phage infection starts with the attachment of p3 protein to the F pilus of bacteria. The phage genome enters the cell and is converted into double-stranded DNA. Afterwards, the synthesis of all M13 phage proteins starts, and the double-stranded DNA is amplified in a process involving p2 and p10 proteins to produce plus-strand copies of the phage DNA. Protein p5 is employed in coating the amplified DNA molecules while the coat proteins p8, p7, p9, p6, and p3 are inserted into the inner bacterial membrane. A small uncovered hairpin of single-stranded DNA is captured by a complex of integral membrane proteins p1, p4, and p9. This complex is described as a membrane pore where the phage is assembled and extruded from the bacterium. As the release of mature M13 virions occurs right after phage assembly, they do not accumulate inside the bacteria and the infected cell continue to grow, albeit at a reduced rate [26,28–31].

2.1.2. T4 Bacteriophage

The T4 phage is a double-stranded DNA virus that is known as one of the largest viruses to infect bacteria. It belongs to the Myoviridae family and infects *Escherichia coli* and the closely related *Shigella*. Like other members of Myoviridae family, T4 has a prolate icosahedral head, a collar with whiskers, and a contractile tail terminating in a baseplate that is attached to six long tail fibers (Figure 1B). While the fibers recognize the host cell surface and attach to the bacterium during infection, the baseplate binds to specific surface receptors and degrades the bacterial wall with its enzymes. This process enables the introduction of DNA into the cell. The virion consists of several components including DNA, proteins, and a few non-protein constituents such as polyamines associated with DNA (putrescine, spermidine, cadaverine), ATP and Ca^{2+} associated with the tail sheath, and dihydropteroylhexaglutamate associated with the baseplate [32].

The DNA of T4 phage is tightly packed inside the protein capsid, which has a length of 120 nm and a width of 86 nm. The capsid is built from three essential proteins: the major capsid protein gp23* (49 kDa, *: final form within capsid, following enzymatic processing) present at 930 copies, the vertex protein gp24* (47 kDa) present at 55 copies, and the portal protein gp20 (61 kDa) present at 12 copies (Figure 2C). Additionally, there are two outer capsid proteins that are nonessential and bind to the capsid after assembly. The highly antigenic outer capsid protein (Hoc, 39.1 kDa) occupies the center of the gp23 capsomers and is present in up to 155 copies per capsid particle. The rod-like small outer capsid protein (Soc, 9.7 kDa) binds to the capsid surface between the gp23* capsomers (up to 810 copies per capsid) and form a nearly continuous mesh on the surface encircling the gp23* hexamers. Interaction of a Soc protein with two gp23* proteins of adjacent capsomers, as well as trimerization of Soc proteins through C-terminal interactions, stabilize the gp23* hexameric capsomers. Although Soc protein is not essential, the assembly of Soc proteins on the surface of T4 improves stability towards pH (up to pH 11), temperature (<60 °C), osmotic shock, and denaturants. Nevertheless, deletion of either one or both Hoc and Soc genes does not affect phage viability or infectivity under standard laboratory conditions [33–35].

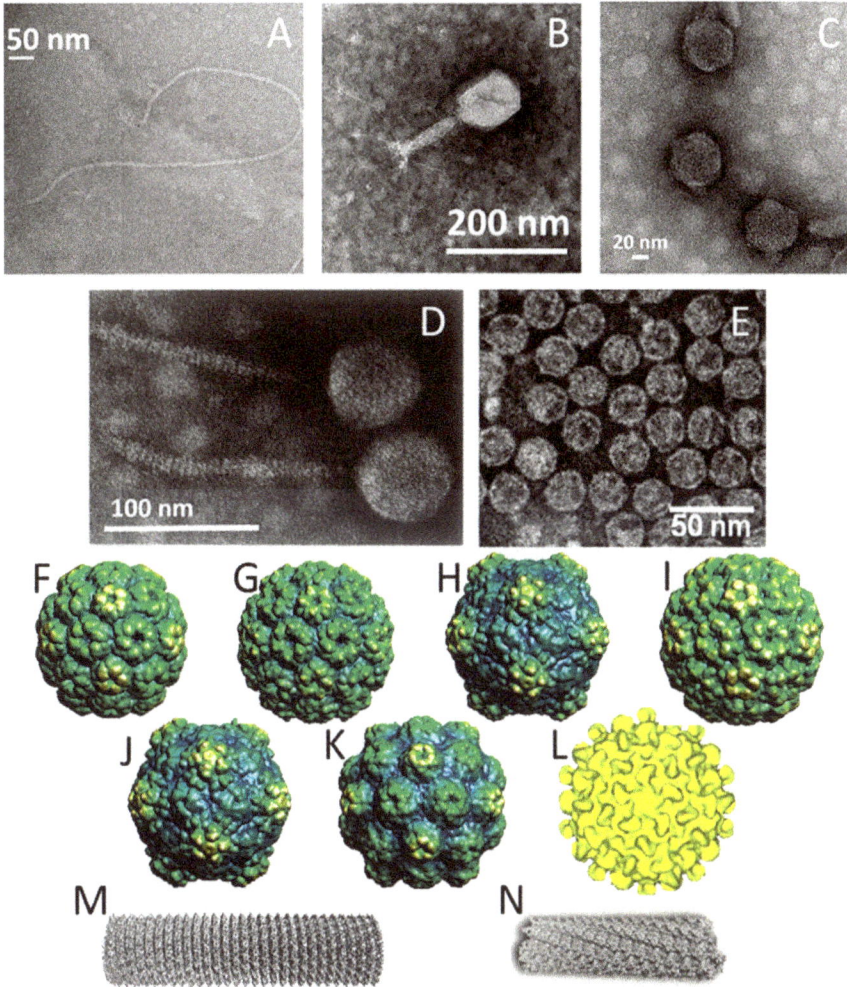

Figure 1. Structures of the viruses discussed in this review. Transmission electron microscopy (TEM) images of (**A**) M13 phage, (**B**) T4 phage, (**C**) T7 phage, (**D**) λ (lambda) phage, and (**E**) MS2 phage. (TEM Images were acquired by the authors, except for λ phage (reprinted with permission from [36], Copyright Elsevier, 1968) and TEM image of MS2 phage (reprinted with permission from [37], Copyright The Royal Society of Chemistry, conveyed through Copyright Clearance Center, Inc., 2011). Structures of plant viruses (**F**) brome mosaic virus (BMV), (**G**) cowpea chlorotic mottle virus (CCMV), (**H**) cowpea mosaic virus (CPMV), (**I**) cucumber mosaic virus (CMV), (**J**) red clover necrotic mosaic virus (RCNMV), (**K**) turnip yellow mosaic virus (TYMV), (**L**) hibiscus chlorotic ringspot virus (HCRSV), (**M**) tobacco mosaic virus (TMV), and (**N**) PVX. (Images of the following viruses were obtained from the VIPERdb (http://viperdb.scripps.edu/) [38]: BMV, CCMV, CPMV, CMV, RCNMV, TYMV. The image of HCRSV was reprinted with permission from [39], Copyright Elsevier, 2003. The image of TMV was reprinted with permission from [40], Copyright Elsevier, 2007. The image of PVX was reprinted with permission from [41], Copyright Elsevier, 2017).

2.1.3. T7 Bacteriophage

T7 bacteriophage belongs to the genus of T7-like bacteriophages, which are characterized by their isometric capsid and non-contractile tail (Figure 1C). T7 phage contains a short tail and a 60 nm symmetrical polyhedral capsid with a conspicuous core (composed of proteins gp14, 15, and 16) containing a 40,000 base-pair double-stranded DNA. Phage assembly begins with the formation of a prohead composed of the major head protein gp10A (36.4 kDa), the minor head protein gp10B (41.7 kDa, derived from a read-through of gp10A), and the scaffolding gp9 protein (Figure 2B). During the process of DNA packaging, the prohead interacts with DNA through the terminase proteins (gp18 and 19) and loses the scaffolding protein once the encapsidation of DNA is complete. Afterwards, the connector protein (gp8) that attaches the core to the tail is incorporated into the capsid structure. The function of core structure is believed to be essential for infectivity, but not for the stability of the prohead structure [42,43].

2.1.4. λ (lambda) Phage

λ phage is a temperate *Escherichia coli* virus composed of a flexible helical tail and a 62 nm diameter icosahedral capsid containing a 48,500 base-pair double-stranded DNA genome (Figure 1D) [44]. As a part of its temperate life cycle, λ phage initially integrates its DNA into the bacterial genome where it is replicated with bacterial chromosomes and transmitted to new cells. Thereafter, the lytic cycle begins and phage proteins are synthesized to form the phage particles. The lytic cycle takes ~40 min and produces ~100 phage, and the ends with cell lysis, which releases the phages. The generation of empty procapsids is the first step of phage assembly. Four hundred and five copies of protein E (gpE), one of the major capsid proteins, are organized into hexameric and pentameric capsomers (Figure 2D). The phage DNA is then packed into the procapsid, which provokes a reconfiguration of gpE as well as the expansion of the procapsid. The expansion is followed by the attachment of protein gpD, the head decoration protein, to form mature capsid. Four hundred and twenty copies of protein gpD are present within the capsid and stabilize the expanded capsid structure to prevent DNA release [44,45].

2.1.5. MS2

MS2 is an RNA-containing *Escherichia coli* bacteriophage with a 27 nm diameter icosahedral capsid (Figure 1E). The phage capsid is composed of 180 copies of coat protein and a single copy of maturation protein (A protein) that is responsible for attachment to the host bacterial cell during infection [46]. During the assembly of phage particles, coat proteins initially form dimers and attachment of the dimer to an RNA hairpin produces a complex initiating the growth of the capsid [47]. As the complete RNA sequence is not necessary for initiation of the capsid formation, the self-assembly of purified coat proteins can be achieved with only the RNA hairpin loop to form empty virus-like particles [46,48]. Empty capsids can be also produced by removing the RNA genome in alkaline conditions (pH ~11.8), conditions that are suspected to degrade the RNA molecule through phosphate hydrolysis and reduce its affinity for the capsid proteins. In addition to the ability to produce phage capsids without genetic material, the presence of 32 pores, each with a 1.8 nm diameter, on the capsid surface enables the use this virus as a nanocarrier [49].

2.2. Plant Viruses

Like bacteriophages, plant viruses also possess many shapes, sizes, and surface properties, which offer a great diversity of possibilities for medical applications. Plant viruses can be conveniently produced from infected leaves, where infection is achieved by exposure to purified virus particles, infected leaf samples, or simple genomic products of the virus such as cDNA and RNA transcripts [50]. Moreover, plant viruses demonstrate remarkable stability over a wide pH range (3.5–9), temperatures up to 90 °C, and towards a variety of organic solvents (e.g., ethanol, dimethyl sulfoxide) [51–54]. As the shape of the viruses can significantly affect their performance in a given biomedical application, in the following sections they are categorized based on their shape. There are two main groups of plant viruses: icosahedral viruses with spherical shapes and rod-shaped viruses with high aspect ratios (Table 1).

Table 1. Plant viruses used in drug delivery systems.

Virus	Size	Symmetry	Family	Genome	Locations on Coat Proteins for Genetic Modification	Ref
BMV	30 nm	Icosahedral	*BromoViridae*	ssRNA	Valine 168	[55,56]
CCMV	28 nm	Icosahedral	*BromoViridae*	ssRNA	Lysine 42 Serine 102/130	[57,58]
CPMV	30 nm	Icosahedral	*Comoviridae*	ssRNA	βB-βC loop of the small subunit/ βE-βF loop of the large subunit	[53,59]
CMV	29 nm	Icosahedral	*Bromoviridae*	ssRNA		[60]
HCRSV	30 nm	Icosahedral	*Tombusviridae*	ssRNA		[61]
RCNMV	36 nm	Icosahedral	*Tombusviridae*	ssRNA		[62]
TYMV	28 nm	Icosahedral	*Tymoviridae*	ssRNA	Threonine 44 Lysine 45 Lysine 53/68	[63]
TMV	300 × 18 nm	Rod-like	*Tobamoviridae*	ssRNA	Threonine 104/158 Serine 123 N/C-terminal of coat protein	[64–66]
PVX	515 × 13 nm	Rod-like	*Potexviridae*	ssRNA	N-terminal of coat protein	[67]

2.2.1. Icosahedral Plant Viruses

The icosahedral structure is the most common shape for plant viruses. The members of this group most commonly used in medicine are cowpea mosaic virus (CPMV, Figure 1H), cowpea chlorotic mottle virus (CCMV, Figure 1G), brome mosaic virus (BMV, Figure 1F), cucumber mosaic virus (CMV, Figure 1I), hibiscus chlorotic ringspot virus (HCRSV, Figure 1L), red clover necrotic mosaic virus (RCNMV, Figure 1J), and turnip yellow mosaic virus (TYMV, Figure 1K). They are composed of 180 copies of coat protein that form capsid structures with sizes ranging from 28–36 nm (Table 1), each containing a single-stranded RNA genome [51,55,60–62,68]. In addition, RNA-free capsids can be generated artificially using pressure, basic/acidic environments, denaturing agents (ca. urea), ribonucleases, or repeated freeze–thaw cycles [51,60,61,68]. In the case of CPMV particles, empty capsids are efficiently produced by co-expressing the fused small/large subunits of coat protein (VP60) along with the 24K proteinase in insect cells and plants [69].

2.2.2. Rod-Shaped Plant Viruses

Tobacco mosaic virus (TMV, Figure 1M) and potato virus x (PVX, Figure 1N) are the rod shaped plant viruses that have been most widely used for drug delivery systems. While both viruses contain single stranded RNA genetic material, they are formed of different coat proteins with different copy numbers, resulting in different lengths. TMV consists of 2130 copies of coat proteins, helically arranged around RNA and forming a hollow nanorod with a 4 nm wide interior channel that is 300 nm in length. It is also possible to produce them as disks in the absence of RNA, which is another nanostructure consisting of 34 coat proteins. The most interesting feature of TMV particles is their ability to form RNA-free spherical nanoparticles from the rod-shaped virus due to thermal processing. The size of the spherical particles can be tuned between 100–800 nm [70–72]. On their side, PVX particles have a rod-like shape with sizes similar to that of TMV, and are 515 × 13 nm in size [67,73].

3. Virus-Based Drug Discovery

3.1. Phage Display Platforms

Various types of phage have been employed to create phage display platforms and have been used for the purpose of drug discovery. This section will present the properties of the most popular of these, with emphasis on their relative advantages and shortcomings.

3.1.1. M13 Phage Display Platform

M13 has been the most widely used phage for phage display since the invention of this technology in 1985 by George P. Smith [74,75]. Combinatorial libraries of polypeptides fused to coat proteins have been used to screen interaction partners towards several targets as well as to study structure–function relations in proteins [25]. In phage display, the genome of M13 phage is manipulated by inserting a DNA sequence into the gene encoding a coat protein. Generally, diverse combinatorial libraries of short peptide sequences (8–12 amino acids) are displayed as fusions to these. Any modification of the phage genome is reflected in a corresponding modification in the coat proteins of the phage, which provides a link between the phenotype and genotype. Selection of the best peptide binding sequence for a given target material is performed through an enrichment process called "biopanning". Initially, the phage are allowed to bind to the target then, after washing away the non-bound phage, the bound phage are eluted and amplified through host bacterial infection. This artificial evolutionary process to select the best binding peptide sequence is repeated several times for enrichment of the best binding partners. Finally, the selected binding peptides are identified by DNA sequencing of the phage genome [76–78]. Phage display technology has been extensively used to identify specific binding peptides for many biological molecules including toxins, bacteria, organs, and tumor-associated antigens [79–82]. Although all five coat proteins have been used to display foreign proteins, the most common approach is to fuse foreign sequences to the N-terminus of p3 and p8 coat proteins [83–85].

There are three different strategies to display proteins as fusions to p3 and p8 coat proteins, which are categorized as phage, phagemid, and hybrid systems. In the phage system, the gene encoding the foreign protein is directly inserted into the phage genome and results in fusion proteins displayed on every copy of chosen coat protein. As a general rule, larger proteins are more efficiently displayed on p3, and the p8 protein is limited to displaying short peptides (~6–10 amino acids). Nonetheless, p3 remains limited in what it can display, and proteins larger than 50 amino acids cannot be displayed on all five copies. Thus, it can be necessary to decrease the copy number of fusion proteins to efficiently display them on the desired coat protein. The phagemid system is used to overcome this limitation. A phagemid is a plasmid carrying the viral gene encoding the fusion coat protein, phage origin of replication, and a phage-packaging signal. The genes required for phage assembly, including the wild type coat protein, are provided by packaging-defective 'helper' phage. Upon coinfection of bacteria by phagemid and helper phage, wild type proteins and fusion coat proteins are synthesized and preferentially-assembled around the phagemid DNA. This results in hybrid phages displaying only a few copies of the fusion coat protein. The hybrid system was also invented with a similar motivation to the phagemid system: to enable the display of large protein sequences on the phage's surface. However, unlike the phagemid system, it only employs the phage genome, which carries both the gene encoding wild type coat protein and the gene encoding fusion protein. Smith defined these three systems using the terms"3", "3+3", and "33" respectively (Figure 2F). Number "3" indicates p3 coat protein whereas formats "8", "8+8", and "88" are used for phage display on p8 coat protein [26,86–88].

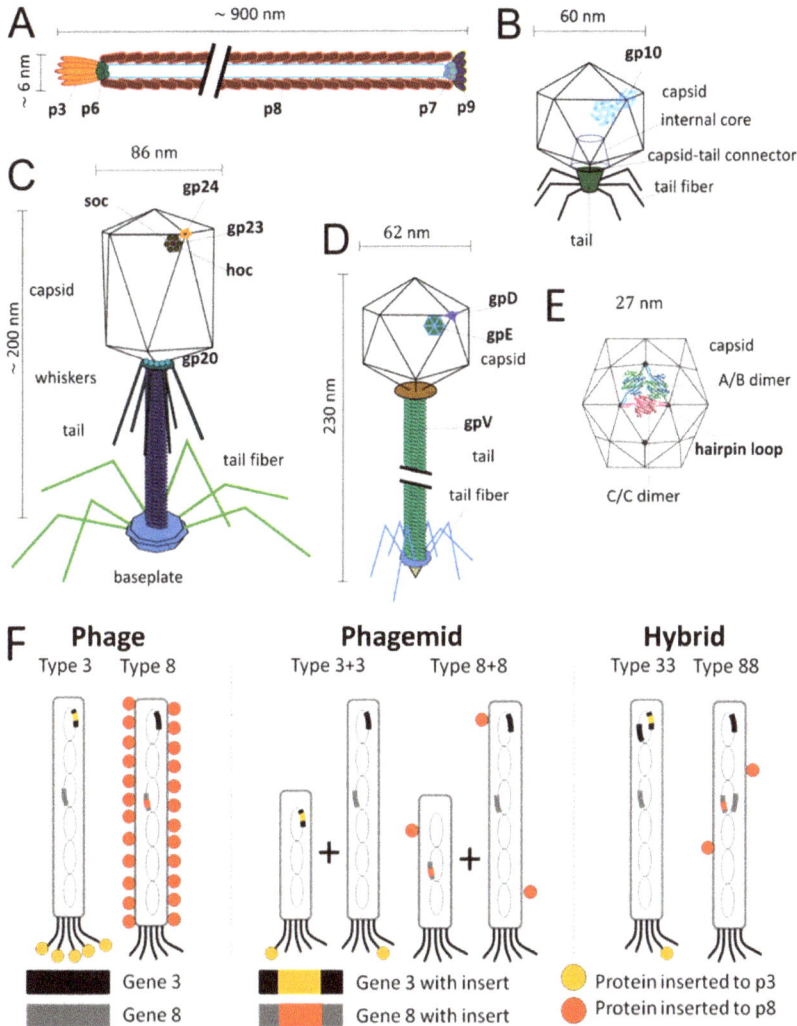

Figure 2. Assembly of coat proteins on bacteriophage (**A**) M13, (**B**) T7, (**C**) T4, (**D**) λ (lambda), and (**E**) MS2 (Images of M13, T7, T4, and λ (lambda) phages were adapted with permission from [89], Copyright American Chemical Society, 2015. The image of MS2 phage was adapted with permission from [90], Copyright the PCCP Owner Societies, 2010). (**F**) Schematic of M13 phage display systems; phage system (type 3/8), phagemid system (type 3+3/8+8), and hybrid system (type 33/88) (The image was adapted with permission from [88], Copyright Elsevier, 1993).

3.1.2. T4 Phage Display Platform

The display of fusion peptides/proteins on T4 phage has emerged as a promising tool to overcome the limitations of phage display platforms employing filamentous phages. For instance, one of the drawbacks of filamentous phage display is the small size of the peptides displayed on the major coat protein (6–10 amino acid residues). Larger polypeptides can only be displayed on minor coat proteins but at very low copy numbers. Moreover, during phage amplification, synthesized coat proteins are inserted into the inner cell membrane where virion assembly and export occur. Due to the membrane-mediated nature of this process, fusion proteins that cannot cross the cell membrane will not be displayed on the phage surface. The secretion system of *E. coli* may also prevent the display of some peptides that are toxic to bacteria and may also create the problem of achieving correct folding of the displayed protein [91–94]. However, T4 phage uses a lytic life cycle for reproduction in which phage assembly takes place inside the infected cell and, afterwards, progeny phages are released by cell lysis. This feature of T4 enables the display of a broader range of proteins with different size, structure, and biological functions that may not be possible with filamentous phage display [32,95].

In T4 phage display, Soc and Hoc are used for the fusion of foreign proteins (Figure 2C). Because both Soc and Hoc sites can be used simultaneously if desired, it has been shown that higher copy numbers of fusions on the phage surface can be achieved by display on both sites (Soc and Hoc) [96,97]. Foreign proteins can be displayed on T4 phage by in vivo and in vitro approaches. The in vivo approach can be performed in different ways, one of which is based on the integration of a modified *soc* gene into a *soc*-deleted T4 genome through a modified positive selection plasmid. In this type of plasmid, the *soc* gene is flanked on its 5′ side by a 3′ portion of the T4 lysozyme gene (e′), and on its 3′ side by a 5′ part of another T4 gene (denV′), which allows homologous recombination between the phage and the plasmid. Integration of the *soc* fusion gene into the T4 genome allows the expression and in vivo binding of fusion proteins to the phage capsid [93]. Alternatively, Soc and Hoc fusion proteins can be incorporated to the phage capsid through a natural assembly process in host bacteria expressing the fusion proteins from a designed expression vector. Upon infection of bacteria with T4 phage strains having defective *soc* or *hoc* genes, fusion proteins are expressed and assembled onto the phage capsid [98,99].

Although in vivo approaches have been widely used to display different proteins on the surface of T4 phage, they are limited to the display of single components, such as a peptide, a domain, or a protein. Limitations in displaying multiple components and large domains arise from the fact that the expression and assembly of the foreign proteins occur during phage infection. This lytic phage cycle leads to problems such as the loss of critical epitopes due to nonspecific proteolysis, low and variable copy number of displayed proteins due to the variations in intracellular expression, structural heterogeneity due to aggregation of the expressed proteins, insolubility, and improper folding [93,100]. Thus an in vitro approach has been developed to overcome these drawbacks for efficient and controlled display of large proteins on the phage surface. In this approach, foreign proteins fused to Soc and Hoc proteins are overexpressed in bacteria and purified. The high affinity interactions between Hoc/Soc proteins and the phage capsid enable in vitro assembly of purified proteins on Hoc- and Soc-defective phage, which is performed by simply mixing the components. Therefore, the in vitro approach results in a phenotype no longer connected to the genotype of the engineered phage, which contrasts to the in vivo approach. An attractive feature of the in vitro approach is that the expression of Hoc/Soc fusion proteins is not restricted to *E.coli* or another specific host, thus any expression system can be used for production of fusions. Consequently, functionally well-characterized and conformationally homogenous fusion proteins are produced and displayed on phage capsids. Additionally, the copy number of displayed proteins can be controlled by changing the ratio of protein to capsid binding sites. It is worthy to note that the in vitro approach also allows customized engineering of T4 phage to display multiple proteins on the same capsid [101–103].

3.1.3. T7 Phage Display Platform

The T7 phage capsid is composed of gp10A and gp10B, which have been employed to display peptide moieties (Figure 2B). While high display numbers can be achieved for peptide sequences shorter than 50 amino acids, only a few copies of larger proteins (<1,200 amino acids) are displayed per capsid [95]. Therefore, the T7 phage display platform becomes favorable for the display of large proteins. As for T4, T7 phage display vectors also overcome certain limitations of filamentous phage display platforms. The lytic nature of T7 phage eliminates the need for protein export and enables the display of a broad diversity of proteins on the phage's surface [104,105]. Moreover, the lytic life cycle shortens the time required for biopanning steps and thus accelerates the selection from the phage library [106]. Unlike many other phage display systems, the coat proteins of T7 phage are anchored to the phage through their N-termini, which makes their C-termini available to display the peptide moieties. This feature makes T7 phage attractive to develop recognition moieties for targeting protein domains that preferentially interact at their N-termini [107].

3.1.4. λ (lambda) Phage Display Platform

Lambda phage libraries are used as another lytic phage display platform to overcome the limitations of filamentous phage display systems, and employ either the gpV tail protein or the gpD decorative capsid protein (Figure 2D). The tail tube consists of 32 disks, each containing six subunits of gpV protein. The small C-terminal domain of gpV protein is exposed and allows the expression of protein moieties as fusions, however, the fusion proteins are only displayed at low levels (ca. one molecule per phage particle) [108]. On the other hand, the gpD protein enables the display of fusions on both its N- and C-termini [109]. While the level of display depends on the size of the fusion, large tetrameric proteins can be displayed at lower levels compared to the small protein domains [110]. In order to overcome low display levels of lambda phage, two different approaches have been investigated. The first approach uses a two-gene system, where both wild type gpD protein and gpD fusion coat proteins are co-packaged into the lambda's head and generate mosaic phage particles expressing both proteins [111,112]. In the second approach, nonsense suppression is used to control the level of the fusion protein. A stop codon is introduced between the gene of gpD protein and its fusion partner. As the gene of gpD protein cannot be fully suppressed, some wild type protein will be also expressed and displayed on phage surface [110,113].

3.1.5. MS2 Phage Display Platform

The major coat protein of MS2 can be used to display foreign proteins/peptides on its surface with high copy number. While the insertion of peptide sequences can be accomplished at different regions of the coat protein, the short hairpin loop between two β-strands (βA and βB) of the protein subunits has been the most commonly used part, as it allows the display of the peptide sequences on the outer surface of the phage (Figure 2E) [114–119]. Small peptide sequences can be displayed as N-terminal extensions of the coat protein subunits, which protrude from the phage surface as well. However, some deletions and base-substitutions can be observed due to the lack of genetic stability. The genetic stability of the insert highly depends on the structure of the RNA hairpin loop encoding the insert and it is determined by the choice of the nucleic acid sequence [120].

3.2. Phage Display-Derived Therapeutics

Phage display is one of today's most important drug discovery technologies. It allows the identification of a broad range of potential therapeutic peptides and antibodies, as well as polypeptides with a variety of functions (protease inhibitors [121,122], minimizing proteins [123], novel scaffolds [124], and DNA binding proteins [125]). Amongst all of these, monoclonal antibodies (mAb) have received considerable attention. The generation of mAbs started with the discovery of the hybridoma technology by Köhler and Milstein in 1975, in which hybrid cells were developed by the fusion of B-cells from immunized animals with myeloma cells to produce antibodies [126]. 10 years later, the first approved mAb, muromonab- CD (Orthoclone OKT3®) [127], was produced using this technology. However, due to the non-human origin of this mAb, a significant percentage of patients developed immune responses, which called into question the safety and efficacy of the non-human mAb therapy [128]. In the late 1980s, recombinant DNA technologies allowed the humanization of non-human mAbs to make them more similar to antibodies within the human body [128]. Starting in the 1990s, human antibodies were produced by in vivo immunization and hybridoma technology using transgenic mice or rats containing the human antibody gene repertoire, or parts of it [129–131]. However, the immunization of transgenic animals could not be used for the production of in vivo antibodies for all types of antigens (e.g., unstable, conserved, and toxic antigens). These limitations impose the use of other alternatives such as in vitro selection technologies, which can be used to discover antibodies towards almost every type of antigen, as they do not depend on immunization.

As such, in vitro display technologies such as phage display, yeast display, ribosome display, bacterial display, mammalian cell-surface display, mRNA display, and DNA display have been used for antibody discovery [132]. Among these, phage display is the most widely used for antibody selection [133]. Its use has resulted in the discovery of over 80 mAbs that have entered clinical trials [134]. In 2002, adalimumab became the first phage display-derived mAb to have been granted market approval, and was also the first approved human mAb [131]. Small recombinant antibody fragments (e.g., scFv) [135,136] or fragment antigen-binding (Fab) [137,138] are also commonly selected by phage display, in addition to full antibodies [139]. A selection of FDA-approved phage display-derived antibody therapeutics are summarized in Table 2. Four exceptions in this table are the non-antibody peptides ecallantide (Kalbitor®), Romiplostim (Nplate®), albigutide (Tanzeum®), and Xyntha purification peptide. Antibody libraries are huge collections ($>10^{10}$) of antibody genes encoding antibodies with unknown properties. These are an essential resource for antibody discovery by phage display and other in vitro selection technologies [133]. Depending on their source of origin, antibody libraries are classified as immune libraries and universal libraries. Immune libraries containing affinity-matured antibodies [133] are constructed using donors (humans or animals) who have received immunization, have been infected or a chronically-diseased, or those suffering from cancer [140]. Affinity maturation is achieved by mutation and clonal selection in which mutated antibodies with higher antigen-binding affinity are enriched [141]. However, extensive or hypermutation may increase the risk of immunogenicity [133]. Of course, it is not possible to construct immune libraries for each disease due to ethical issues, high cost, and laborious procedures. One solution to this problem is to use universal libraries that are generated by a source for which the immune system had not been activated to recognize a specific antigen (naïve) [132]. Moreover, due to the lack of affinity maturation, these universal libraries have low risk of immunogenicity [133]. In principle, a universal library can be applicable to mAb selection of any type of antigen. This is because the library comprises a high variability of antibody genes and comprises antibody genes from many donors [132]. Universal libraries are further sub-classified as naïve, semi-synthetic, or fully-synthetic. Naïve antibody libraries are constructed from the natural human IgM repertoire (i.e., from not intentionally immunized donors) [132,133]. Examples of naïve universal libraries are the human Fab library constructed by de Haard and colleagues at Dyax (now Shire) [137], the scFv libraries from Cambridge Antibody Technology (CAT) [142,143], scFv and Fab libraries from XOMA59, and the HAL scFv libraries [135,144]. Fully synthetic libraries are constructed to include synthetic genes derived from known (human) antibody frameworks with the capacity to generate a large diversity in appropriate regions [145]. Semi-synthetic libraries are a combination of natural (i.e., donor-derived antibody) and synthetic antibody sequences [138]. A combination of naïve and synthetic repertoires was used for the Dyax FAB310 library. Fully-synthetic libraries were developed by MorphoSys [133]. In addition, a particular type of antibody library is generated during guided phage display selection of human antibody using a non-human original antibody sequence. This strategy has been used for humanization and the discovery of fully human antibodies with similar properties to the murine antibody template, such as adalimumab [146].

Overall, the identification of mAbs and mAb derivatives by phage display technology was a breakthrough that has enabled the isolation of human antibodies towards many types of antigen without immunization. Since then, it is one of the main platforms for generation of human therapeutic antibodies together with transgenic immunized mice, antibody humanization techniques, and single B-cell expression cloning [169].

Table 2. Selected phage display-derived antibodies/peptide therapeutics approved by FDA, as of Dec 06 2018.

Product, Trade Name®, Manufacturer, FDA Approval	Type	Target	Phage Display Type	Phage Display Technology	Indication	Ref
Antibodies						
Adalimumab (D2E7), Humira®, Abbott Laboratories, 2002	IgGκ	TNFα	Humanization by guided selection, scFv	CAT	Rheumatoid arthritis, juvenile idiopathic arthritis, psoriatic arthritis, ankylosing spondylitis, Crohn's disease, ulcerative colitis, plaque psoriasis	[131,133,147]
Ranibizumab, Lucentis®, Genentech, 2006	Fab fragment	VEGF-A	Affinity maturation of bevacizumab by phage display	Genentech	Age-related macular degeneration, macular edema after retinal vein occlusion, diabetic macular edema, diabetic retinopathy	[133,148]
Belimumab, Benlysta® GSK, 2011	IgG1λ	BLyS	Naïve, scFv	CAT	Autoantibody-positive, systemic lupus	[133,149,150]
Raxibacumab, Abthrax®, GSK, 2012	IgG1λ	Protective antigen	Naïve, scFv	CAT	Anthrax	[133,151]
Ramucirumab (IMC-1121B), Cyramza® ImClone/ Lilly, 2014	IgG1	VEGF-R2	Naïve, Fab	Dyax	Gastric cancer, colorectal cancer, non-small cell lung cancer	[133,152]
Necitumumab (IMC-11F8), Portrazza® ImClone/ Lilly, 2015	IgG1κ	EGFR	Naïve, Fab	Dyax	Squamous non-small cell lung cancer	[133,153]
Avelumab, Bavencio® EMD Serono/ Pfizer, 2017	IgG1λ	PD-L1	Naïve, Fab	Dyax	Metastatic Merkel cell carcinoma	[133,154]
Guselkumab, Tremfya®, Janssen, 2017	IgG1λ	P19 subunit of Interleukin 23	Synthetic, Fab	MorphoSys	Psoriasis	[133,155]
Lanadelumab (DX-2930), Takhzyro®, Shire, 2018	IgG1	Kalikrein	Naïve, Fab	Dyax	Types I and II hereditary angioedema	[133,156,157]
Moxetumomab Pasudotox (HA22 or CAT-8015) Lumoxiti®, AstraZeneca, 2018	Antibody-fusion protein	CD22	Affinity maturation of BL22 (CAT-3888) by phage display	CAT	Relapsed or refractory hairy cell leukemia	[158,159]
Emapalumab-lzsg (NI-0501), Gamifant®, Novimmune/ Serono, 2018	IgG1	Interferon-gamma	Naïve, scFv	CAT	Primary hemophagocytic lymphohistiocytosis	[133,134]

Table 2. *Cont.*

Product, Trade Name®, Manufacturer, FDA Approval	Type	Target	Phage Display Type	Phage Display Technology	Indication	Ref
		Peptides				
Ecallantide, Kalbitor®, Shire, 2009	60 AA polypeptide	Plasma kallikrein	-	Dyax	Hereditary angioedema	[160,161]
Romiplostim, Nplate®, Amgen, 2008	Peptide Fc fusion	Thrombopoietin receptor (TPOR)	-	Affymax	Immune thrombocytopenic purpura	[162,163]
Affinity ligand for Xyntha, Wyeth Pharmaceuticals	Polypeptide	Factor VIII	-	Dyax	Hemophilia A	[162,164,165]
Albiglutide, Tanzeum®, GlaxoSmithKline, 2014	Peptide albumin fusion	GLP-1	-	AlbudAb	Type 2 diabetes mellitus, glycemia	[166–168]

AA, amino acid; BLyS, B-lymphocyte stimulator; CAT, Cambridge Antibody Technology; EGFR, epidermal growth factor receptor; FDA, Food and Drug Administration; Fab, fragment antigen-binding; GSK, GlaxoSmithKline; IgG, immunoglobulin G; k, kappa light chain; λ, lambda light chain; PD-L1, programmed death–ligand 1; scFv, single chain fragment variable; VEGFA, vascular endothelial growth factor A; VEGF-R2, vascular endothelial growth factor receptor-2; GLP-1, glucagon-likepeptide-1 receptor; AlbudAb, domain antibodies to serum albumin.

3.3. Phage Display Selection of Peptide Binders for Biomineralization/Self-Assembly of Inorganic Materials

Reports showing that peptides sequences displayed on the outer surface of *E.coli* could recognize and specifically-bind to metal/metal oxide surfaces (e.g., gold, iron oxide, and chromium) inspired research to extend this concept to phage display [170,171]. The first application of phage display libraries to evolve peptide sequences binding to inorganic substrates was performed for a range of semiconductor surfaces with the motivation of directing nanoparticles to specific locations on semiconductor structures for the fabrication of complex, sophisticated electronic materials [172]. This achievement led to research in phage display selection of material-binding peptides. Several peptide sequences with affinity to different materials (e.g., platinum, palladium, titanium, silicon, silver, gold, zinc sulfide, cadmium sulfide, graphite, calcite, indium phosphide, chlorine-doped polypyrrole (PPyCl), and carbon nanotubes) have been identified [78,173–187].

Peptides selected by phage display have not only been used to achieve binding, but have also been employed to direct the mineralization of nanomaterials. This approach is inspired by the process of biomineralization of materials in Nature by living organisms. Several biominerals are formed in a biologically-controlled manner under mild conditions and include calcium phosphate minerals in teeth and bone, silica in sponges, and magnetic particles of magnetite (Fe_3O_4) or greigite (Fe_3S_4). Recent interest in biomineralization has grown as it offers a greener and cheaper alternative to inorganic synthesis of materials, which usually requires high temperatures and harsh chemical reagents [188]. Peptide sequences that can facilitate biomineralization have been identified by phage display selection against target materials, considering that some selected peptides with binding affinity to the target material can nucleate or promote the formation of these materials. M13 phage was used to select several peptide sequences capable of recognition and nucleation of different materials like zinc sulfide (ZnS), cadmium sulfide (CdS) nanocrystals, iridium oxide (IrO_2), cobalt platinum (CoPt), and iron platinum (FePt) materials [189–193]. Although the specific interactions between the peptide sequences and the ions are crucial for the nucleation and growth of these materials, the uniform conformation of the displayed peptides on phage surface is also mentioned to be important for controlled crystallization of the materials with single-crystal nature. While not directly used as therapeutics per se, the ability of phage display technology to identify peptides that bind with affinity to inorganic substrates has led to the use of viruses as building blocks of functional structures for drug delivery applications. In the following section, the techniques used for fabrication of virus-based drug carriers and the approaches to design virus-based drug delivery systems are presented.

4. Virus-Based Drug Delivery Systems

4.1. Encapsulation/Decoration with External Cargos

One of the most attractive features of plant viruses is that the coat proteins can self-assemble around synthetic materials, offering a stable and biodegradable delivery platform for various compounds. The self-assembly of virus coat proteins around cargo molecules can be achieved in different manners, depending on the virus. The interaction of coat proteins with a specific sequence of viral RNA (RCNMV) or a negatively charged material (BMV, CCMV, HCRSV) to replace the negatively charged RNA is crucial for some viral platforms to encapsulate foreign materials [61,194–196]. However, some virus particles can form empty capsids in the absence of any genetic/external material (CPMV) by simple co-expression of essential viral proteins in plant cells [197]. Moreover, the pores present on the viral capsids can be employed to encapsulate small molecules [60]. However, the size of capsid pores may not be sufficiently large for diffusion of small molecules in all virus platforms. In this case, pore formation can be induced by depletion of capsid-integrated divalent ions for some viral particles (e.g., RCNMV) [198]. Like many plant viruses, the relatively large interior volume of MS2 phage capsid provides an interesting platform to load molecules or materials. Spontaneous assembly of MS2 coat proteins in the presence of RNA sequences enables the loading of RNA-conjugated functional materials inside the phage capsid. Encapsulation of phage RNA is mediated by a 19-nucleotide RNA stem-loop, the so-called pac site, which specifically interacts with coat proteins. Inspired by this mechanism, molecules of interest, such as antisense oligodeoxynucleotides, antisense RNAs, quantum dots, drugs, and toxins, have been conjugated with a pac site to initiate the assembly of coat proteins and to achieve packaging within the capsid [199–201]. The reassembly of coat protein dimers around functional moieties is another strategy employed to encapsulate external cargos within the MS2 capsid. In this approach, phage capsids are first disassembled into dimers with acetic acid. Then, reassembly is initiated with negatively charged DNA/polymer/amino acid tags conjugated to a cargo molecule, due to their electrostatic interactions with the interior positively-charged capsid dimers. While the presence of highly-negatively charged cargo (ca. DNA-coated gold nanoparticles) is sufficient to initiate the reassembly, it might be necessary to add protein stabilizing osmolytes to increase the yield of reassembly around protein cargos, such as enzymes [202,203].

The ability to empty the T7 phage capsid of its genetic material opens up the opportunity to use these hollow phage heads as cages for encapsulation of foreign materials. It has been demonstrated that the phage DNA can be released from the capsid by applying osmotic pressure, and the empty capsids then filled with precursor ions for mineralization [204]. The self-assembly of nanoparticles on the capsid of T7 has also been reported in the literature. This was accomplished by introducing functional moieties (biotinylation peptide, gold binding peptide) onto the coat protein through genetic engineering, and were used to assemble quantum dots and gold nanoparticles [205,206]. M13 and MS2 phage capsids have also been used as templates to assemble a variety of materials (e.g., ^{64}Cu, Gd^{3+}, drugs, single wall carbon nanotubes, iron oxide nanoparticles and fluorophores) with high copy number by means of genetic/chemical modification of the coat proteins for site-specific material conjugation [84,207–211].

The capsid of T4 phage has also used as a bio-scaffold to fabricate functional materials. The fabrication of such nanostructures typically relies on the reduction of metal nanoparticles on the capsid surface, and is generally performed in two steps. T4 phage are initially incubated in a solution of metal salts and then the metal ions that interact with phage coat proteins are reduced by dimethylaminoborane [212]. Although this approach has been applied to synthesize different phage-templated metal nanostructures (e.g., gold, platinum, rhodium, cobalt, iron, palladium, and nickel), the mechanism of metal nanoparticle formation on phage coat protein remains unknown [213]. It has been suggested that the side-chains of the surface-exposed amino acids on coat proteins interact with the metal ions and mediate the nucleation and organization of the nanoparticles on T4 capsid [214].

4.2. Drug Delivery Systems

The extensive use of phage in drug discovery and for the identification of binding partners to various targets have also led researchers to consider using the virus itself as a targeting probe/nanocarrier in medicine. Genetically/chemically-modified viruses that display targeting peptides or synthetic functional molecules have been used as building blocks to design self-assembled nanostructures for drug delivery and the treatment of diseases. In particular, the reported ability of filamentous phages to penetrate to the central nervous system, which is difficult for most of the drug molecules and drug delivery systems due to the relative impermeability of blood-brain barrier, may contribute to making viruses promising drug delivery platforms [215].

4.2.1. Anti-Cancer Drugs

The most common virus-based drug delivery platforms are based on chemotherapeutics, and the viral particles are employed as carriers for small drug molecules. In this manner, doxorubicin, which is a clinically approved anticancer agent, has been extensively studied. Viral particles have been employed to improve the drug efficacy and reduce systemic toxicity. In order to create a linker with high serum stability and sensitivity to enzymatic hydrolysis by cysteine protease (cathepsin-B; present within the lysosomes of target cells), cathepsin-B cleavable DFK peptides were displayed on the p8 coat proteins of filamentous phage and employed to attach doxorubicin with a high copy number (~3500) via carbodiimide coupling chemistry. Although, direct conjugation of the drug to the coat proteins yielded higher numbers of drug molecules per phage (~10,000), the engineered drug-release mechanism significantly improved the potency of the carrier as a result of release of the drug at the targeted cells [216]. Considering that all five coat proteins of M13 can be used for phage display, it is possible to introduce more than one functionality onto the surface of the phage to achieve new properties. By using three different coat proteins of M13, a phage-based therapeutic platform was designed for simultaneous prostate cancer imaging and targeted drug delivery [84]. In this system, doxorubicin was conjugated to the major coat protein while other coat proteins (p3 and p9) were used to display cancer-targeting peptides and fluorophores (imaging agent), respectively (Figure 3). The phage-based therapeutic enabled cancer cell targeting, imaging, and drug delivery.

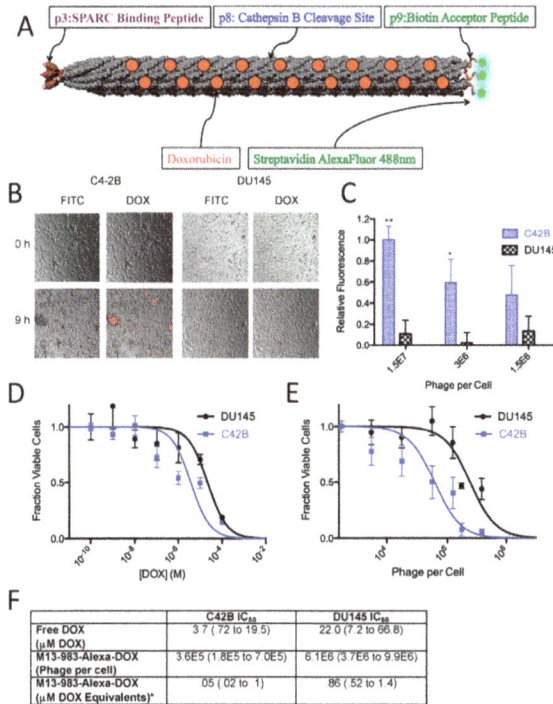

Figure 3. (**A**) Schematic of M13-983 Phage. The red dots along the phage coat represent doxorubicin (DOX) attached to p8. p3 displays a peptide with affinity for SPARC (Secreted Protein, Acidic and Rich in Cysteine), and p9 can be enzymatically biotinylated and loaded with streptavidin-functionalized fluorophores AlexaFluor 488 nm. (**B**) Overlay of brightfield and fluorescent images of SPARC positive C42B cells (first and second column) and less expressing SPARC DU145 cells (third and fourth column) incubated with M13-983-Alexa-DOX at 0 h (top row) and 9 h post-treatment (bottom row). FITC channel represents fluorescence from Alexa Fluor 488, and DOX is designated by red fluorescence from DOX uptake. C42B samples showed increased fluorescence of phage uptake, indicated by green fluorescence (bottom row, first column) and DOX uptake (bottom row, second column) as compared to DU145 cells after 9 h. (**C**) Targeted uptake measured by quantifying fluorescence intensity (** $p < 0.001$; * $p < 0.01$). C42B consistently shows higher fluorescence intensity than DU145, confirming the observations in panel B. Higher phage concentrations report larger differences between C42B and DU145 fluorescence. (**D**) Cell viability of C42B and DU145 as a function of free DOX. (**E**) Cell viability of C42B and DU145 cell lines as a function of increasing M13-983-Alexa-DOX. All samples were run in triplicate and error bars represent standard deviations. (**F**) IC50 values for C42B and DU145 are given with the 95% confidence interval given in parenthesis. * Based on 257 DOX particles per phage (Adapted with permission from [84], Copyright American Chemical Society, 2012).

As an alternative approach, rather than loading drug molecules onto viral particles, drug-loaded materials can be modified with viruses for targeting purposes. For instance, M13 phage modified with folic acid to target cancer cells, was attached onto a biodegradable polymer (poly(caprolactone-b-2-vinylpyridine, PCL–P2VP) particles that were loaded with doxorubicin [217]. By providing a large surface area with control over the spacing and orientation, phage particles enabled multivalent target-receptor interaction and improved targeting. Indeed, in vitro studies with human nasopharyngeal cells showed that doxorubicin-loaded phage coated polymer particles had significantly higher cellular uptake and selectivity in comparison to free drug.

Encapsulation of doxorubicin within the cavity at the center of MS2 has been an alternative strategy to improve the delivery of the drugs to target cells. It has been demonstrated that reduced intracellular accumulation of doxorubicin in human hepatocellular carcinoma cells (Hep3B) due to the moderate P-glycoprotein levels can be overcome by loading empty MS2 virus capsids with drug molecules [199]. As targeted drug-loaded viral particles are internalized via receptor-mediated endocytosis, they can circumvent efflux mechanisms of P-glycoproteins and can kill cancer cells at lower drug concentrations (20-fold improvement) compared to free drugs. Moreover, encapsulation of doxorubicin inside the MS2 viral capsid demonstrated significantly different time-dependent cytotoxicity. Indeed, while the free drug was highly toxic to all studied cell types exposed to the drug for 24 h, encapsulation of the drugs within the targeted viral particles showed a high degree of specificity towards Hep3B cells, with an >80% reduction in cell viability. In contrast, the viability of non-targeted cells after a 7 days exposure to the encapsulated drug remained relatively unaffected. Similar results were observed for doxorubicin-loaded cancer targeted RCNMV, CMV, and HCRSV particles, which reduced the cytotoxicity of the drug in non-targeted cells, due to the specific cell uptake in target cells [60,62,218]. CMV drug carriers exhibited a sustained in vitro drug release profile over 5 days. The efficacy of doxorubicin-loaded viral particles in tumor-bearing animal models have been studied, as well. Regarding doxorubicin-loaded TMV particles, treatment showed significant delay in tumor growth and increased survival due to the efficient tumor accumulation of the drug carrier platform while free doxorubicin had no effect on tumor burden or survival [219]. PEGylated doxorubicin-loaded PVX particles also showed similar efficacy in animal models. Tumor growth rates were significantly lower compared to free doxorubicin [220]. Cardiotoxicity of doxorubicin was also studied in viral drug delivery systems in order to demonstrate the efficiency of targeted drug delivery along with the prevented drug release within the heart. Johnson grass chlorotic stripe mosaic virus (JgCSMV, a member of the family *Tombusviridae*), another plant virus recently gained attention as an alternative viral platform, was loaded with doxorubicin and drug delivery efficiency of doxorubicin loaded JgCSMV has been investigated in MCF-7 tumor-bearing athymic mouse models [221,222]. The study showed that tumor volume of the mice treated with doxorubicin loaded JgCSMV was 2.22 times smaller than the control group which was not treated with any drug. More interestingly, hearts of the mice treated with doxorubicin loaded JgCSMV and untreated negative control mice showed no significant pathological changes while thrombi were observed in hearts of the mice treated with free doxorubicin. In addition to rod-like filamentous structures, TMV coat proteins can be self-assembled into stable disk-shaped particles, which expands the shape library of the protein-based nanomaterials for drug delivery. Delivery of doxorubicin molecules to glioblastoma cells in vitro demonstrated that TMV disks could provide a promising nanocarrier platform resulting in significant cell death after 72 h of incubation while the cells incubated with TMV disks alone showed ~100% viability [64].

The formation of hollow mesoporous silica nanocapsules around CPMV viral templates has been another approach to control drug delivery from a viral scaffold. APTES ((3-aminopropyl) triethoxysilane) and TEOS (tetraethyl-orthosilicate) were used to form a silicate network around viral capsids due to electrostatic interactions between the carboxyl/carbonyl groups of CPMV and amine groups of silicates (Figure 4A) [223]. In order to create a hollow cavity, the capsid proteins of viruses were denatured by increasing the temperature to 40 °C for a period of 24 h, which enabled their diffusion out from the core. The pores within the silica nanocapsules enabled the loading of doxorubicin into its cavity and their release by diffusion (Figure 4B). The fabrication of surfactant free, hollow mesoporous silica nanocapsules provided high drug stability as a result of slow decomposition of the drug molecules, which were protected inside the capsules and their regulated sustained release from the pores. The in vitro efficacy of the virus-templated mesoporous silica nanocapsules were investigated by utilizing Hek293 and HepG2 cell lines, where the raise in drug dosage resulted in an increase in the cell survival (Figure 4C–D) and showed the efficiency of the platform.

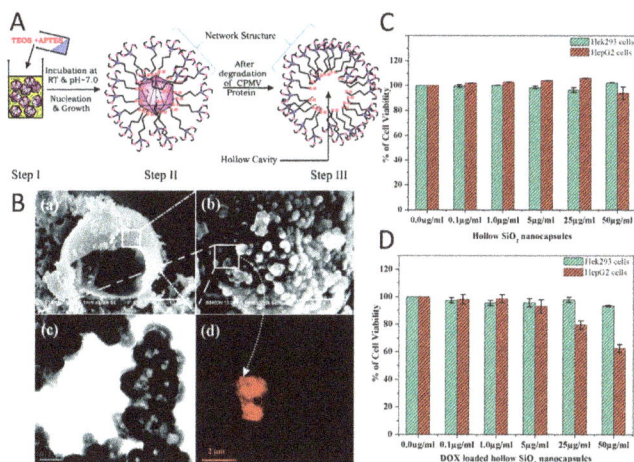

Figure 4. (**A**) Schematic representing the proposed synthesis mechanism of the CPMV-templated mesoporous silica nanocapsules in three steps (Step I, Step II and Step III). (**B**) (a) SEM image of single hollow capsules formed through the self-assembly of hollow SiO_2 nanocapsules synthesized in the presence of CPMV, (b) surface textures of the same formed by the self-assembly of nanoparticles. (Scale bar 10.0 μm and 1.0 μm, respectively), (c) TEM of hollow SiO_2 nanocapsules (shown in (b), scale bar 0.2 μm), and (d) confocal microscopy image of hollow SiO_2 nanocapsules loaded with Rh6G (a small fluorescent molecule) (scale bar 2 μm). Cytotoxicity assay of Hek293 (Human embryonic kidney cell line) and HepG2 cells (Human carcinoma cell line) (**C**) with mesoporous SiO_2 nanocapsules free from drugs and (**D**) nanoformulated hollow SiO_2 nanocapsules (doxorubicin (DOX)-loaded hollow SiO_2 nanocapsules) of different doses (Adapted with permission from [223], Copyright The Royal Society of Chemistry, 2015).

In addition to doxorubicin, other anticancer pro-drug/drug molecules have also been successfully loaded inside the viral nanoparticles and their drug delivery efficiencies in cancer cells have been evaluated. Proflavine, mostly known as a bacteriostatic, is one such compound that has shown antiproliferative activity in cancer cells and tumors due to its intercalation into DNA [224]. Loading of proflavin within the CPMV particles has been achieved through the diffusion of drug molecules into the viral capsids. The loading mechanism of the drug within the capsid was explained by the genetic material inside the capsid acting as a sponge that absorbs the drug molecules. Drug delivery studies have shown that the interaction of proflavine with viral RNA was reversible and enabled the release of the drug molecules in several cancer cell lines (HeLa (cervical cancer cells), HT-29 (colon cancer cells), and PC-3 (prostate cancer cells)) while no cargo release was observed in cell-free medium.

Cisplatin (cis-$[Pt(NH_3)_2Cl_2]$) is another anticancer drug that has been efficiently delivered to cancer cells via viral particles. TMV particles have been decorated with mannose and lactose moieties to specifically target the galectin-rich human breast cancer cell line MCF-7 and asialoglycoprotein receptor (ASGPR) over-expressing hepatocellular carcinoma cell line HepG2, respectively. For this purpose, alkyne modified TMV particles were modified with azido sugar derivatives via the copper(I)-catalyzed azide–alkyne cycloaddition. The interior capsid surface of the TMV particles was covalently modified with cisplatin molecules through a stable chelate structure with the carboxyl groups of glutamate residues, and the drug later slowly released and resulted in enhanced apoptosis efficiency in specific targeted cell lines [225]. A similar drug delivery approach has also been employed for the treatment of ovarian cancer cells with platinum resistance, which may appear at the onset of disease or develop in response to platinum-based chemotherapy. As cisplatin offers greater efficacy than its analogue carboplatin, it is crucial to develop alternative drug delivery platforms employing cisplatin with reduced toxicity. It has been reported that the delivery of drug molecules conjugated to the interior surface of TMV provided a platform capable of circumventing the resistance mechanisms in platinum resistant ovarian cancer cells and restoring efficacy of cisplatin treatment at low concentrations. Loading of TMV particles with cisplatin was achieved through electrostatic interactions between the deprotonated interior glutamic acid residues of the capsid proteins with positively-charged cisplatin molecules, which were produced via a reaction with silver nitrate. The enhanced efficacy of cisplatin-loaded TMV particles in ovarian cancer cells suggests that encapsulation of cisplatin in viral particles increased the rate of drug uptake/retention as well as DNA damage inside the cells [226].

4.2.2. Protein Therapeutics

The plant hormone indole-3-acetic acid (IAA) is a prodrug used in a virus-based drug carrier to treat human prostate cancer cells. IAA generates a radical upon reaction with horseradish peroxidase and produces radical-dependent cytotoxicity as well as cell death. The viral drug carrier was designed by engineering M13 phage to display a short peptide to enhance prostate cancer cell recognition/penetration. Moreover, a NeutrAvidin–horseradish peroxidase conjugate was attached to the p9 phage coat protein of M13 fused with a biotinylated peptide [227]. The treatment of cancer cells with this system led to a significant reduction in cell viability due to intracellular delivery. This virus-based prodrug activation approach has also been investigated for tamoxifen, which is one of the most widely used prodrug in the treatment of hormone-dependent breast cancer [228,229]. Tamoxifen is mainly metabolised in the liver by cytochrome P450 (CYPs) enzymes, resulting in the active drug. Encapsulation of CYPs inside cancer-targeting CCMV particles was suggested as a pro-drug activation strategy to increase the drug efficiency as well as to reduce the severe side-effects of the drug in normal cells. In order to encapsulate the CYPs, viral particles were disassembled and then reassembled in the presence of the enzyme molecules. The electrostatic interactions between the negatively charged CYPs and the positively charged interior of the viral capsid was used as a driving force to internalize the enzyme molecules within the viruses. Preliminary studies have shown that CYPs encapsulated within the viral particles maintained their activity, though the catalytic activity was one order of magnitude lower compared to the activity of the free CYPs. The decrease in enzyme activity was attributed to deleterious effects of crowding inside the capsid cavity, diffusion of the substrate into the virus capsid, and improper orientation of the active site of the enzyme.

4.2.3. Antibiotics

Antibiotics are another group of drugs that have been loaded in viral-based drug carriers. It has been reported that the conjugation of many copies of the drug molecule onto the phage's major coat proteins increased potency by creating a microenvironment around bacterial cell with a locally high drug concentration. Treatment of gram positive pathogenic bacteria *Staphylococcus aureus* with the low potency antibiotic chloramphenicol, well-known for its toxicity to blood cells, conjugated to the fd phage retarded the growth of bacterial cells ~20-fold more efficiently than free chloramphenicol [230].

4.2.4. Photodynamic and Photothermal Therapy

Efficient delivery of drug molecules to targeted cells is also important for photodynamic therapy (PDT) and photothermal therapy (PTT). PDT relies on the activation of a photosensitizer by light in the presence of oxygen, which produces reactive oxygen species (ROS). As cell damage occurs due to the reaction of ROS with cellular components, it is important to deliver the photosensitizers to the targeted cells and avoid their nonspecific delivery to the healthy cells. The nonspecific dispersal of the photosensitizer throughout the body creates sensitivity in patients. For instance, sunlight must be avoided for several weeks following treatment. Moreover, the insolubility of many photosensitizers in physiological solutions is another problem encountered in PDT-related drug delivery systems and it is necessary to develop new platforms to address this challenge. For this purpose, icosahedral CCMV particles have been explored for PDT by dual labelling with both cell-targeting moieties and photosensitizers. The surface coat protein of the virus was genetically-modified to display a cysteine residue, used to attach the photosensitizer Ru(bpy)$_2$-5-iodoacetoamino-1,10-phenanthroline(phen-IA). An antibody specific for *Staphylococcus aureus*, was chemically conjugated to the lysine residues of the coat proteins [57]. PDT studies using this system showed that the photosensitizer-labelled CCMV particles were more efficient in killing *S. aureus* cells compared to free photosensitizer due to the enhanced cell targeting ability. On the other hand, the cell killing efficiency of the photosensitizer-labelled CCMV particles was approximately the same as that achieved using and anti-*S.aureus* antibody-photosensitizer conjugate. While the number of photosensitizers per binding event for the CCMV platform (~70) was significantly higher than the antibody-photosensitizer conjugate (~2), it was suggested that the large size of the CCMV particles significantly reduced the proximity of photosensitizer to the cell surface, which is an important factor in efficacy due to the very short diffusion length of singlet oxygen. The encapsulation of photosensitizers within the capsid cavity of CCMV particles has been another approach for PDT. Water-soluble zinc phthalocyanine (ZnPc) was encapsulated inside CCMV by two different routes: i) self-assembly of coat proteins dimers around aggregated ZnPc molecules at neutral pH; and ii) diffusion of ZnPc molecules into empty CCMV particles through capsid pores at acidic pH [231]. Although the potential use of ZnPc-loaded CCMV particles as PDT delivery system was tested on macrophages and resulted in cell death, further studies are required to evaluate the efficiency of this platform as targeted PDT delivery system for cancer therapy. Zn-EpPor (5-(4-ethynylphenyl)-10,15,20-tris(4-methylpyridin-4-ium-1-yl)porphyrin-zinc(II) triiodide) is another photosensitizer used in viral PDT systems. The interior channel of TMV particles were loaded with the photosensitizer by exploiting electrostatic interactions between the negatively-charged amino acid residues of virus and the positive charge of Zn-EpPor [232]. Zn-EpPor-loaded TMV particles were stable and possessed a good shelf-life (drug release was not observed during one-month storage at 4°C in 0.01 M potassium phosphate buffer, pH 7.0). Cellular uptake and drug efficacy studies were performed with melanoma cancer cells. The release of the photosensitizer was suggested to occur inside the acidic endolysosomal compartment, which caused protonation of TMV's interior carboxylic acid groups resulting in drug release. Zn-EpPor-loaded TMV particles showed enhanced cell killing efficacy compared to free Zn-EpPor molecules, which was attributed to the increased cellular uptake of photosensitizer as a result of their delivery within the TMV particles. Recently, the effect of photosensitizer's charge on drug loading efficiency of TMV particles was investigated in a study where different zinc porphyrin (Zn-Por) formulations (monocationic (Zn-Por^{1+}), dicationic (Zn-Por^{2+}), tricationic (Zn-Por^{3+}), and tetracationic (Zn-Por^{4+})) were employed [233]. While the tricationic formulation demonstrated the highest loading efficiency (~600 molecules/TMV), the results were attributed to the combined effect of electrostatic and hydrophobic/hydrophilic interactions; higher positive charge was suggested to result in better stabilization of the photosensitizers inside TMV particles due to the electrostatic interactions with the deprotonated carboxylate residue of glutamic acid at pH 7.8. Moreover, the increased hydrophobic nature of the monocationic and dicationic Zn-Por formulations with their electrostatic properties resulted in aggregation and reduced the loading efficiency. In order to develop a targeted drug delivery system, F3 peptide was conjugated to the surface proteins of TMV particles to target a shuttle protein, nucleolin, overexpressed on HeLa cells. Zn-Por^{+3} loaded TMV-F3 particles accumulated on the cell membrane and showed a fivefold increase in cell killing efficacy compared to the free drug. The results are explained with some possible mechanisms which are the cell toxicity through cell membrane disruption by light activation, the release of the photosensitizers at the cell surface and favored cell uptake of the Zn-Por molecules due to their positive charge. The self-assembly of drug-loaded liposomes on M13 phage has been another PDT delivery approach using phage as a nanocarrier [234]. Cationic liposomes were loaded with zinc phthalocyanine (ZnPc) and were assembled on M13 phage displaying eight glutamic acid residues on the p8 major coat protein, via electrostatic interactions. Phage-templated drug-loaded liposomes had enhanced excited singlet oxygen generation efficiency and were able to internalize in breast cancer cells. These two properties make phage-liposome complexes a promising tool for targeted drug delivery given that this property can be introduced by displaying targeting peptides on the minor coat proteins of phage. Moreover, the phage template can stabilize the liposomes in biological media against flocculation and can help the delivery of the content loaded inside the liposome to specific targets.

Photothermal therapy (PTT), so-called hyperthermal therapy, employs gold nanoparticles as a heat source for inducing cell damage as a result of a light-to-heat conversion process. As gold nanoparticles generate local heating under light illumination, efficient delivery of gold nanoparticles to targeted cells is desirable for selective cell killing. Clusters of gold nanoparticles on T7 phage particles have been fabricated as a PTT delivery platform to treat prostate cancer cells in vitro [206]. The assembly of gold nanoparticles on the capsid surface was achieved via the display of a gold-binding peptide and a prostate cancer cell-binding peptide, in tandem (Figure 5A). Phage-templated gold clusters maintained their cell-targeting functionality and promoted delivery of the system to the cancer cells. Clusters were localized within vesicular organelles (e.g., endosomes), generating even larger clusters with a diameter of up to few hundred nanometers, which suggested receptor-mediated endocytosis as a possible internalization mechanism (Figure 5C,D). Irradiation of the prostate cancer cells resulted in cell death in a very selective manner, whereas no remarkable cell death was observed in both healthy cells and non-targeted cancer cells (Figure 5B).

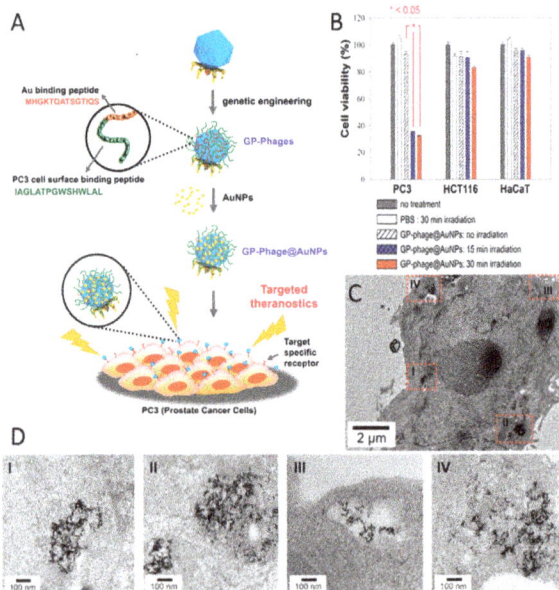

Figure 5. (**A**) Schematic illustration of cancer-selective photothermal therapy via prostate cancer-targeted intracellular delivery of T7-templated AuNP nanoclusters, where T7 phages are genetically modified to display gold-binding and prostate cancer cell-targeting peptides on the viral surface. (**B**) The viability of each cell line (prostate cancer cell (PC3), human colorectal carcinoma cells (HCT116), and normal cells (HaCat)) by photothermal effects of T7-templated AuNP nanoclusters. (**C,D**) TEM images of T7-templated AuNP nanoclusters internalized within PC3 cells in ultrathin section specimens. The cells were treated with T7-templated AuNP nanoclusters for 5 h, followed by medium replacement and additional incubation for 20 h (Adapted with permission from [206], Copyright American Chemical Society, 2015).

4.2.5. Incorporation into Polymer Matrices

Polymeric materials are widely used as drug delivery systems because of their ability to be use as a matrix that protects drug molecules and controls drug release via e.g., its rate of degradation or swelling profile. Moreover, in some cases, the high water content and soft structures of polymer matrices make them similar to natural extracellular matrices, which contributes to minimizing tissue irritation and cell adhesion and makes them promising drug delivery systems. However, an initial burst, or very fast release of drug molecules remains a challenge that can limit their use to certain types of drugs. The incorporation of viruses into polymeric matrices has been proposed as a solution to overcome this limitation and to better control the release of the drugs molecules. The affinity-based polymeric drug release system is one such platforms in which M13 phage are embedded inside a polymer matrix to suppress the release of drug proteins due to their specific interactions with phage particles. For this purpose, the p3 coat protein of M13 phage was genetically modified to display peptides that bind to antibodies and mixed with a gelatin solution containing antibodies to form hybrid hydrogels [235]. While the antibodies were gradually released from M13-free gelatin hydrogels within 48 h, their release from the M13-gelatin hybrid hydrogels was ~1% after 144 h, indicating suppression of the release due to phage-antibody interactions.

Stimuli-responsive polymers are another group of material where M13 phage has been introduced to create polymer–protein bioconjugates combining both bioactivities of the protein molecules and the responsive properties of the polymers. A polymer–M13 bioconjugate was produced by grafting a boronic acid containing polymer (poly(NIPAM*co*- phenylboronic acid (PBA)) onto M13 surface through amino groups of the coat proteins [236]. By doing so, around 400 poly(NIPAM*co*-PBA) chains were conjugated per phage particle. Due to the temperature-responsive gelation behavior of the polymer, the polymer–M13 phage bioconjugate was mixed with insulin at 4 °C and could then be converted into hydrogels by injecting into PBS buffer at 37 °C. The insulin-encapsulated hydrogels demonstrated glucose-responsive release behavior, which can be an interesting strategy in the regulation of diabetes. During hydrogel formation, the virus particles interact with poly(NIPAM*co*-PBA) polymer, which is in a collapsed hydrophobic state. However, in the presence of glucose, the boronic acid moieties of poly(NIPAM*co*-PBA) couple to glucose to yield hydrophilic boronates that are substantially more hydrophilic. This changes the structure of hydrogel matrix and enables faster release of encapsulated insulin molecules. While 70% of insulin was released over 24 h, the release rate of insulin increased in the presence of glucose with a peak value of 90% at 10 h. Similar carbohydrate-responsive polymers have also been prepared by cross-linking phenylboronic acid (PBA)–M13 bioconjugates with poly(vinyl alcohol), resulting in hydrogels with excellent injectability and self-healing behavior (Figure 6A) [237]. The self-healing nature of the hydrogels was investigated by placing two pieces of hydrogels adjacent to each other (Figure 6B). It was observed that the interface became smeared after 15 min and then completely disappeared after 1 h. Additionally, insulin molecules could be introduced into both the PBA–M13 bioconjugate suspension and the PVA solution, resulting in insulin-loaded hydrogels. Glucose-responsive insulin release studies showed that an initial burst release, which was observed in less than 15 h, was similar in the presence and absence of glucose (Figure 6C). At a later stage of release, the hydrogel in the presence of glucose released most of the insulin with a speed faster than that without glucose as a result of swelling of the gel matrix caused by the diffusion of glucose into the hydrogel, which partially disrupts the physical crosslinking between the diols of PVA and the boronic acids.

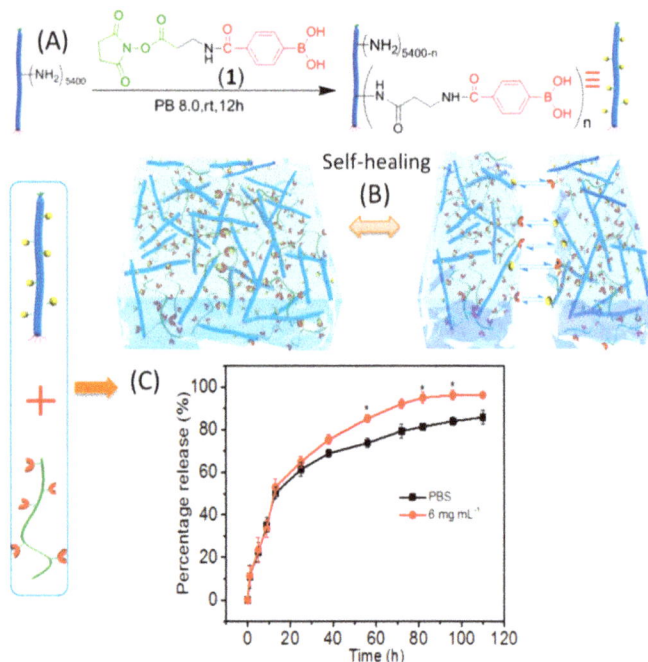

Figure 6. Schematic representing (**A**) the preparation of viral bioconjugates of M13 viruses with low-pKa phenylboronic acid derivative (PBA-M13) and (**B**) fabrication of dynamic hydrogels via binding of PBA-M13 with multiple diol-containing polymers and demonstration of self-healing behavior. (**C**) Glucose responsiveness-regulated insulin release behaviors. FITC-insulin was loaded into the hydrogel and then placed into PBS buffer with or without glucose. The released FITC-insulin was monitored by fluorescent measurements. The asterisks represent the statistical significance, which is calculated by multiple t tests-one per row: * $p < 0.05$, ** $p < 0.01$, *** $p < 0.001$. The hydrogels consist of 1 wt % PBA-M13 and 0.15 wt % PVA (Adapted with permission from [237], Copyright American Chemical Society, 2018).

RCNMV particles has been another virus incorporated into polymeric drug delivery systems. The ability to trigger the opening and closing of RCNMV capsid pores by changing divalent ion (Ca^{2+} and Mg^{2+}) concentrations and pH enables their use as an additional release mechanism in drug delivery systems. Doxorubicin-loaded RCNMV particles could be incorporated into two different polymeric matrices (poly(lactic acid)(PLA) and poly(lactic acid):polyethylene oxide (PLA:PEO)) by electrospinning into fibers [238]. Two approaches were employed to incorporate viral particles in these matrices; i) the direct mixing of viral particles and polymer solution before fiber formation; and ii) physisorption of the viral particles onto the pre-formed fibers by immersion (Figure 7A). It has been demonstrated that electrospinning of drug-loaded viral particles with polymer solutions resulted in nanofibers with significantly lower release profiles (Figure 7B). A slightly lower release profile was also observed for PLA polymer compared to PLA–PEO polymer, which was attributed to the more hydrophilic nature of PEO. Upon appropriate stimuli (reducing divalent ion concentrations or making pH more basic), the release of the drug molecules from virus-polymer matrix, produced via direct processing followed a two-phase kinetic profile. While the first phase included the diffusion of the mobile phase through the polymer to trigger the opening of the virus capsid pores, which enables the release of the drug, in the second phase the drug molecules diffuse through the polymer matrix.

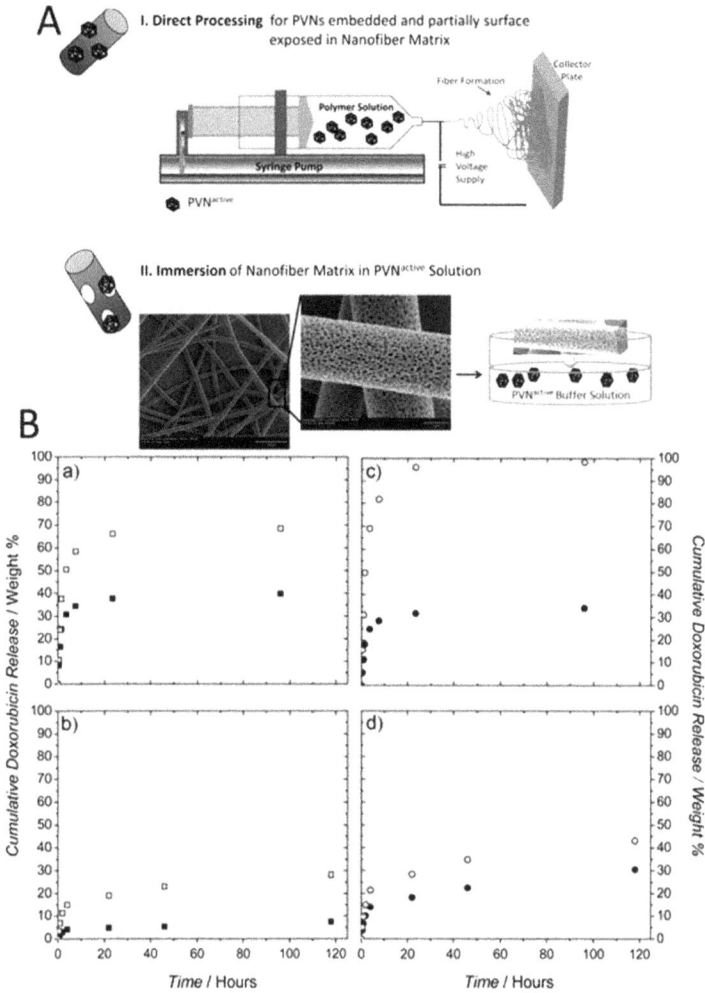

Figure 7. (**A**) Schematic of processes for incorporating plant viral nanoparticles (PVN) in nanofiber matrices. In the direct processing method (I), the PVN active are electrospun in situ with the polymer solution. In the immersion process (II), the nanofiber matrices are dipped into a specific volume and concentration of PVN active particles. (Scale bars of the images from left to right 50.0 μm and 2.0 μm, respectively) (**B**) Cumulative doxorubicin release over time of PLA nanofiber matrices combined with PVN Dox (■) or free doxorubicin (□) where PLA was combined with the active either a) post mat fabrication (dipping method) or b) prior to electrospinning (co-spinning method) and of 70:30 PLA:PEO nanofiber matrices with PVN dox (o) or free doxorubicin (●) where PLA:PEO was combined with the active either c) post mat fabrication (dipping method) or d) prior to electrospinning (co-spinning method) (Adapted with permission from [238], Copyright WILEY-VCH Verlag GmbH & Co., 2013).

5. Challenges to Clinical Applications

Pharmacokinetics and toxicity are critical factors that determine the viability of materials intended to be used as drug delivery system for medical applications. Therefore, several studies investigating the circulation, clearance, blood half-life, stability, immunogenicity, and organ biodistribution of viral platforms have been performed and have provide useful information. While not numerous, some fundamental studies on the interaction of plant viruses and phage with mammalian cells and pathways that they used to enter into cells have been performed. For instance, while M13 phage tends to only bind to the cell membrane of epithelial cells, it shows cell-type dependent interactions and internalization mechanism: clathrin-mediated endocytosis and macropinocytosis for HeLa cells; vesicular transport; clathrin-mediated endocytosis, and macropinocytosis for a human breast cancer cell line (MCF-7); and caveolae-mediated endocytosis for human dermal microvascular endothelial cell (HDMEC) [239]. On the other hand, CPMV particles naturally interact with mammalian cells, including endothelial cells and particularly tumor neovascular endothelium *in vivo*, via a surface exposed cell protein, vimentin [240].

After their administration into the body, viral particles, like many other nanoparticles, are recognized as foreign agents by the cells of the host immune system and are eliminated from the blood. B-cells are one type of these immune cells and B-cell dependent immunoglubins have been shown to play an important role in rapid neutralization of T7 phage in murine blood, resulting in a short half-life (<5 min) [241]. The reticuloendothelial system (RES) is another component of the immune system that is highly involved in the clearance of viral particles. Indeed, clearance by macrophages and accumulation in organs such as the liver and spleen have been reported for many viruses (ca. M13, MS2, CPMV, PVX, and TMV) [22,242–245]. Nevertheless, there are significant differences between the efficiency of the viral particles in avoiding clearance by phagocytosis, which results in different blood circulation times. For instance, filamentous M13 phage can effectively avoid rapid clearance from the RES with a plasma half-time of 3.6 h [246]. In contrast, another rod-like virus TMV has much shorter blood clearance time (~3 min) [243]. Different plasma clearance half-lives have also been observed for spherical viruses. For instance, while CPMV and CCMV particles are rapidly cleared from the blood with short circulation times (CPMV, 4–7 min), MS2 viral particles show longer plasma half-life [16,208]. It has been demonstrated that certain rod-like particles can effectively evade phagocytosis since the larger contact angles with macrophages do not favor engulfment and internalization. The surface charge of the viral capsid is another parameter that influences their fate in tissues. Due to the negative charge of the surface of mammalian cells resulting from abundant proteoglycans, positively-charged viral particles are expected to interact electrostatically with cells, which has been reported to increase their intracellular delivery and tumor penetration. The short blood circulation of CPMV particles is attributed to their negative surface charge, hence their cellular uptake in HeLa cells could be improved by conjugation of cationic peptide sequences [247].

The studies above suggest that the fast blood clearance of viral particles are at least partially due to the host immune response and highlight a need for a coating to tailor the properties of the viral particles for enhanced blood circulation times. For this, the modification of the viral capsid with passivating agents, such as polymers or proteins, has been investigated. For instance, the modification of the CPMV capsid with poly(ethylene glycol) (PEG) effectively blocked CPMV–cell interactions and shielded the viral particles from inducing a primary antibody response [248,249]. The effect of PEGylation has been also observed on plasma circulation times. The half-lives of PEGylated CPMV and TMV particles were 20.8 min and 6.6 min, respectively, which were longer than the half-lives mentioned above [22,250]. In addition to PEG, poly(2-oxazoline) (POx) and serum albumin have also been employed as alternative coatings agents [251]. In comparison to PEGylated TMV, POx-coated viruses showed lower cellular uptake rates indicating less favored TMV–cell interactions due to the higher polymer grafting density of Pox relative to PEG for the tested samples. High polymer density was associated with a reduction in cellular recognition as well as protein adhesion to the polymer-coated particles. On the other hand, pharmacokinetic studies showed a two-phase decay for the polymer-coated TMV particles. During the initial clearance period, POxylated TMV particles had longer plasma half-life (11.3 min) than PEGylated TMV particles (0.01 min) indicating better screening capability of POx polymer. Serum albumin (SA) has been another coating material investigated to shield TMV particles from the immune system. Similar to POx, in cellular uptake and pharmacokinetic studies of the SA coating also outperformed the PEG [252]. While the rates of macrophage uptake of PEG- and SA-coated TMV particles were comparable, showing a 4-fold reduction in macrophage-particle interactions relative to uncoated TMV particles (Figure 8A), their circulation half-lives in Balb/C mice were 10 min and 100 min, respectively (Figure 8B). According to the authors, the enhanced pharmacokinetics of SA coating was associated with its relatively high molecular weight. It was suggested that compared to flexible PEG chains with low molecular weight, the globular and rigid structure of SA provided enhanced steric hindrance, resulting in better stealth properties.

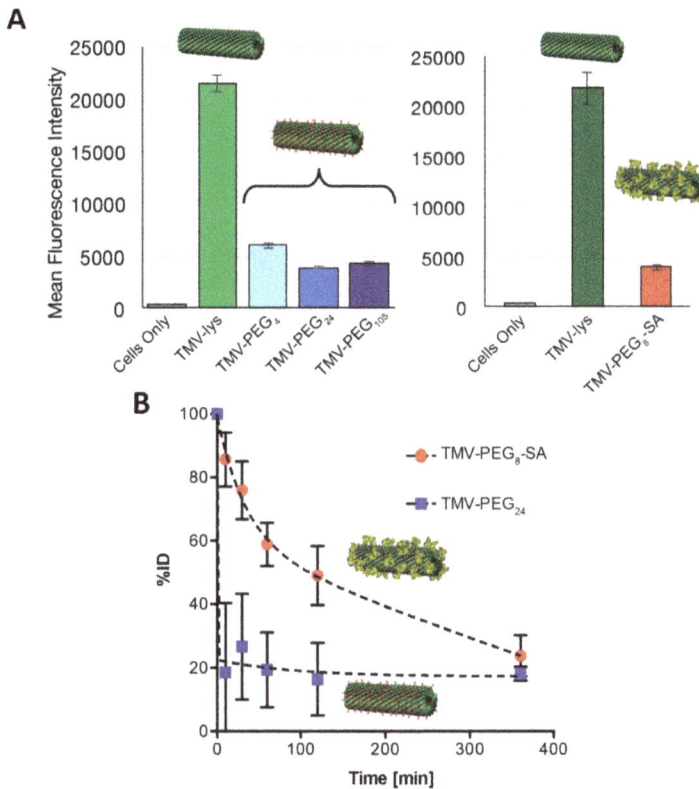

Figure 8. (**A**) In vitro recognition of SA- and PEG-coated TMV vs. 'naked' TMV by RAW264.7 macrophages. Quantitative flow cytometry analysis of the interactions between 'naked' and 'stealth' TMV formulations and RAW264.7 cells. (**B**) Pharmacokinetics of the TMV-PEG24 and TMV-PEG8-SA particles in Balb/C mouse model. The particles were administered intravenously at the amount of 400 mg/mouse. Blood was collected before injection at $t = 0$ and after injection at $t = 10$ min, $t = 30$ min, $t = 60$ min, $t = 120$ min, and $t = 360$ min (the experiments were completed at an $n = 3$ per group) (Adapted with permission from [252], Copyright Elsevier, 2016).

6. Conclusions

Overall, a unique feature of viruses highlighted in this review is that they are perfectly-defined nanomaterials that can be conveniently modified, either at the genetic level or using bioconjugate chemistry, to yield interesting tools for drug discovery or drug delivery. Indeed, the structural diversity of viruses as well as the connection between genotype and phenotype has provided several complementary platforms for the discovery of different classes of therapeutic polypeptides by phage display, some of which have made it to the clinic. While in vitro selection procedures by phage display are the most commonplace, a better understanding of the health risks posed by plant viruses and bacteriophages to Humans may unbridle more complex in vivo selection procedures towards targets that cannot be emulated in vitro. Indeed, the available yet limited number of studies examining the application of in vivo phage screening in human patients have shown promising results [253,254]. In contrast to the use of viruses for drug discovery, most studies employing viruses or virus-like particles as drug delivery systems are limited to in vitro proof-of-concepts. Little information is available regarding in vivo toxicity or biodistribution. While some immunohistochemistry studies suggest that virus-like particles do not induce adverse effects in animal tissues, in terms of no overt signs toxicity (tissue degeneration, cell apoptosis, and necrosis), it is necessary to support these results with organ–function studies in the future. This information will be important for guiding future developments of plant and bacterial viruses as drug delivery systems or components thereof. On the other hand, some studies on bacterial viruses has shown that food-ingested phage DNA, like any foreign DNA, can get inside the cells of mouse and, even on rare occasions, can covalently link to mouse DNA [255]. While this information speculates some medically-relevant implications in terms of mutagenesis and carcinogenesis, it has been one of the major concerns preventing the implementation of viral particles in medicine. In this manner, viruses, which enable the removal of genetic material and form empty capsids, could reduce DNA-related concerns and become more interesting as drug carrier platform.

Funding: This research was funded by the Natural Sciences and Engineering Council of Canada (NSERC), grant number RGPIN-2015-04254.

Acknowledgments: H.S.A acknowledges a postdoctoral scholarship from NSERC.

Conflicts of Interest: The authors declare no conflict of interest.

References

1. Van der Want, J.P.H.; Dijkstra, J. A history of plant virology. *Arch. Virol.* **2006**, *151*, 1467–1498. [CrossRef] [PubMed]
2. Brown, J.C. Virology. In *eLS*; John Wiley & Sons, Ltd: Chichester, UK, 2001. [CrossRef]
3. Bos, L. 100 years of virology: From vitalism via molecular biology to genetic engineering. *Trends Microbiol.* **2000**, *8*, 82–87. [CrossRef]
4. Kropinski, A.M. Phage Therapy—Everything Old is New Again. *Can. J. Infect. Dis. Med. Microbiol.* **2006**, *17*, 297–306. [CrossRef]
5. Merril, C.R.; Scholl, D.; Adhya, S.L. The prospect for bacteriophage therapy in Western medicine. *Nat. Rev. Drug Discov.* **2003**, *2*, 489. [CrossRef]
6. Sunderland, K.S.; Yang, M.; Mao, C. Phage-Enabled Nanomedicine: From Probes to Therapeutics in Precision Medicine. *Angew. Chem.* **2017**, *56*, 1964–1992. [CrossRef]
7. Henry, M.; Debarbieux, L. Tools from viruses: Bacteriophage successes and beyond. *Virology* **2012**, *434*, 151–161. [CrossRef]
8. Domingo-Calap, P.; Georgel, P.; Bahram, S. Back to the future: Bacteriophages as promising therapeutic tools. *HLA* **2016**, *87*, 133–140. [CrossRef]
9. Méthot, P.-O. Writing the history of virology in the twentieth century: Discovery, disciplines, and conceptual change. *Stud. Hist. Philos. Sci. Part C* **2016**, *59*, 145–153. [CrossRef] [PubMed]
10. The Royal Swedish Academy of Sciences. Press Release: The Nobel Prize in Chemistry 2018. Available online: https://www.nobelprize.org/prizes/chemistry/2018/press-release/ (accessed on 29 April 2019).
11. Nayerossadat, N.; Maedeh, T.; Ali, P.A. Viral and nonviral delivery systems for gene delivery. *Adv. Biomed. Res.* **2012**, *1*, 27. [CrossRef] [PubMed]
12. Lundstrom, K. Viral Vectors in Gene Therapy. *Diseases* **2018**, *6*, 42. [CrossRef] [PubMed]
13. Liu, Z.; Qiao, J.; Niu, Z.; Wang, Q. Natural supramolecular building blocks: From virus coat proteins to viral nanoparticles. *Chem. Soc. Rev.* **2012**, *41*, 6178–6194. [CrossRef]
14. Pokorski, J.K.; Steinmetz, N.F. The Art of Engineering Viral Nanoparticles. *Mol. Pharm.* **2011**, *8*, 29–43. [CrossRef]

15. Kaiser, C.R.; Flenniken, M.L.; Gillitzer, E.; Harmsen, A.L.; Harmsen, A.G.; Jutila, M.A.; Douglas, T.; Young, M.J. Biodistribution studies of protein cage nanoparticles demonstrate broad tissue distribution and rapid clearance in vivo. *Int. J. Nanomed.* **2007**, *2*, 715–733.
16. Singh, P.; Prasuhn, D.; Yeh, R.M.; Destito, G.; Rae, C.S.; Osborn, K.; Finn, M.G.; Manchester, M. Bio-distribution, toxicity and pathology of cowpea mosaic virus nanoparticles in vivo. *J. Control. Release* **2007**, *120*, 41–50. [CrossRef]
17. Green, N.K.; Herbert, C.W.; Hale, S.J.; Hale, A.B.; Mautner, V.; Harkins, R.; Hermiston, T.; Ulbrich, K.; Fisher, K.D.; Seymour, L.W. Extended plasma circulation time and decreased toxicity of polymer-coated adenovirus. *Gene Ther.* **2004**, *11*, 1256. [CrossRef]
18. Vaks, L.; Benhar, I. In vivo characteristics of targeted drug-carrying filamentous bacteriophage nanomedicines. *J. Nanobiotechnol.* **2011**, *9*, 58. [CrossRef]
19. Flynn, C.E.; Lee, S.-W.; Peelle, B.R.; Belcher, A.M. Viruses as vehicles for growth, organization and assembly of materials11The Golden Jubilee Issue—Selected topics in Materials Science and Engineering: Past, Present and Future, edited by *S. Suresh*. *Acta Mater.* **2003**, *51*, 5867–5880. [CrossRef]
20. Lee, S.-Y.; Lim, J.-S.; Harris, M.T. Synthesis and application of virus-based hybrid nanomaterials. *Biotechnol. Bioeng.* **2012**, *109*, 16–30. [CrossRef] [PubMed]
21. Wen, A.M.; Wang, Y.; Jiang, K.; Hsu, G.C.; Gao, H.; Lee, K.L.; Yang, A.C.; Yu, X.; Simon, D.I.; Steinmetz, N.F. Shaping bio-inspired nanotechnologies to target thrombosis for dual optical-magnetic resonance imaging. *J. Mater. Chem. B* **2015**, *3*, 6037–6045. [CrossRef]
22. Shukla, S.; Ablack, A.L.; Wen, A.M.; Lee, K.L.; Lewis, J.D.; Steinmetz, N.F. Increased Tumor Homing and Tissue Penetration of the Filamentous Plant Viral Nanoparticle Potato virus X. *Mol. Pharm.* **2013**, *10*, 33–42. [CrossRef] [PubMed]
23. Rong, J.; Niu, Z.; Lee, L.A.; Wang, Q. Self-assembly of viral particles. *Curr. Opin. Colloid Interface Sci.* **2011**, *16*, 441–450. [CrossRef]
24. Russel, M.; Lowman, H.B.; Clackson, T. Introduction to Phage Biology and Phage Display. In *Phage Display: A Practical Approach*; Clackson, T., Lowman, H.B., Eds.; Oxford University Press: Oxford, UK, 2004; pp. 1–26.
25. Hemminga, M.A.; Vos, W.L.; Nazarov, P.V.; Koehorst, R.B.M.; Wolfs, C.J.A.M.; Spruijt, R.B.; Stopar, D. Viruses: Incredible nanomachines. New advances with filamentous phages. *Eur. Biophys. J.* **2010**, *39*, 541–550. [CrossRef] [PubMed]
26. Kehoe, J.W.; Kay, B.K. Filamentous Phage Display in the New Millennium. *Chem. Rev.* **2005**, *105*, 4056–4072. [CrossRef]
27. Sidhu, S.S. Engineering M13 for phage display. *Biomol. Eng.* **2001**, *18*, 57–63. [CrossRef]
28. Clokie, M.R.J.; Millard, A.D.; Letarov, A.V.; Heaphy, S. Phages in nature. *Bacteriophage* **2011**, *1*, 31–45. [CrossRef]
29. Salmond, G.P.C.; Fineran, P.C. A century of the phage: Past, present and future. *Nat. Rev. Microbiol.* **2015**, *13*, 777–786. [CrossRef] [PubMed]
30. Onodera, K. Molecular Biology and Biotechnology of Bacteriophage. In *Nano/Micro Biotechnology*; Endo, I., Nagamune, T., Eds.; Springer: Berlin/Heidelberg, Germany, 2010; pp. 17–43. [CrossRef]
31. Tzagoloff, H.; Pratt, D. The initial steps in infection with coliphage M13. *Virology* **1964**, *24*, 372–380. [CrossRef]
32. Kurzępa, A.; Dąbrowska, K.; Świtała-Jeleń, K.; Górski, A. Molecular modification of T4 bacteriophage proteins and its potential application—Review. *Folia Microbiol.* **2009**, *54*, 5–15. [CrossRef]
33. Sathaliyawala, T.; Islam, M.Z.; Li, Q.; Fokine, A.; Rossmann, M.G.; Rao, V.B. Functional Analysis of the Highly Antigenic Outer Capsid Protein, Hoc, a Virus Decoration Protein from T4-like Bacteriophages. *Mol. Microbiol.* **2010**, *77*, 444–455. [CrossRef]
34. Sathaliyawala, T.; Rao, M.; Maclean, D.M.; Birx, D.L.; Alving, C.R.; Rao, V.B. Assembly of Human Immunodeficiency Virus (HIV) Antigens on Bacteriophage T4: A Novel In Vitro Approach To Construct Multicomponent HIV Vaccines. *J. Virol.* **2006**, *80*, 7688–7698. [CrossRef] [PubMed]
35. Fokine, A.; Bowman, V.D.; Battisti, A.J.; Li, Q.; Chipman, P.R.; Rao, V.B.; Rossmann, M.G. Cryo-electron microscopy study of bacteriophage T4 displaying anthrax toxin proteins. *Virology* **2007**, *367*, 422–427. [CrossRef]
36. Kemp, C.L.; Howatson, A.F.; Siminovitch, L. Electron microscope studies of mutants of lambda bacteriophage: I. General description and quantitation of viral products. *Virology* **1968**, *36*, 490–502. [CrossRef]

37. Nguyen, T.H.; Easter, N.; Gutierrez, L.; Huyett, L.; Defnet, E.; Mylon, S.E.; Ferri, J.K.; Viet, N.A. The RNA core weakly influences the interactions of the bacteriophage MS2 at key environmental interfaces. *Soft Matter* **2011**, *7*, 10449–10456. [CrossRef]
38. Carrillo-Tripp, M.; Shepherd, C.M.; Borelli, I.A.; Venkataraman, S.; Lander, G.; Natarajan, P.; Johnson, J.E.; Brooks, C.L., 3rd; Reddy, V.S. VIPERdb2: An enhanced and web API enabled relational database for structural virology. *Nucleic Acids Res.* **2009**, *37*, D436–D442. [CrossRef] [PubMed]
39. Doan, D.N.P.; Lee, K.C.; Laurinmäki, P.; Butcher, S.; Wong, S.M.; Dokland, T. Three-dimensional reconstruction of hibiscus chlorotic ringspot virus. *J. Struct. Biol.* **2003**, *144*, 253–261. [CrossRef]
40. Sachse, C.; Chen, J.Z.; Coureux, P.D.; Stroupe, M.E.; Fändrich, M.; Grigorieff, N. High-resolution Electron Microscopy of Helical Specimens: A Fresh Look at Tobacco Mosaic Virus. *J. Mol. Biol.* **2007**, *371*, 812–835. [CrossRef] [PubMed]
41. Le, D.H.T.; Hu, H.; Commandeur, U.; Steinmetz, N.F. Chemical addressability of potato virus X for its applications in bio/nanotechnology. *J. Struct. Biol.* **2017**, *200*, 360–368. [CrossRef] [PubMed]
42. Agirrezabala, X.; Martín-Benito, J.; Castón, J.R.; Miranda, R.; Valpuesta, J.M.; Carrascosa, J.L. Maturation of phage T7 involves structural modification of both shell and inner core components. *EMBO J.* **2005**, *24*, 3820. [CrossRef]
43. Cerritelli, M.E.; Studier, W.F. Assembly of T7 Capsids from Independently Expressed and Purified Head Protein and Scaffolding Protein. *J. Mol. Biol.* **1996**, *258*, 286–298. [CrossRef] [PubMed]
44. Fuller, D.N.; Raymer, D.M.; Rickgauer, J.P.; Robertson, R.M.; Catalano, C.E.; Anderson, D.L.; Grimes, S.; Smith, D.E. Measurements of single DNA molecule packaging dynamics in bacteriophage lambda reveal high forces, high motor processivity, and capsid transformations. *J. Mol. Biol.* **2007**, *373*, 1113–1122. [CrossRef] [PubMed]
45. Lander, G.C.; Evilevitch, A.; Jeembaeva, M.; Potter, C.S.; Carragher, B.; Johnson, J.E. Bacteriophage lambda stabilization by auxiliary protein gpD: Timing, location, and mechanism of attachment determined by cryo-EM. *Structure* **2008**, *16*, 1399–1406. [CrossRef]
46. Valegård, K.; Liljas, L.; Fridborg, K.; Unge, T. The three-dimensional structure of the bacterial virus MS2. *Nature* **1990**, *345*, 36. [CrossRef]
47. Fu, Y.; Li, J. A novel delivery platform based on Bacteriophage MS2 virus-like particles. *Virus Res.* **2016**, *211*, 9–16. [CrossRef]
48. Ma, Y.; Nolte, R.J.M.; Cornelissen, J.J.L.M. Virus-based nanocarriers for drug delivery. *Adv. Drug Deliv. Rev.* **2012**, *64*, 811–825. [CrossRef] [PubMed]
49. Hooker, J.M.; Kovacs, E.W.; Francis, M.B. Interior Surface Modification of Bacteriophage MS2. *J. Am. Chem. Soc.* **2004**, *126*, 3718–3719. [CrossRef]
50. Wen, A.M.; Steinmetz, N.F. Design of virus-based nanomaterials for medicine, biotechnology, and energy. *Chem. Soc. Rev.* **2016**, *45*, 4074–4126. [CrossRef] [PubMed]
51. Barnhill, H.N.; Reuther, R.; Ferguson, P.L.; Dreher, T.; Wang, Q. Turnip Yellow Mosaic Virus as a Chemoaddressable Bionanoparticle. *Bioconj. Chem.* **2007**, *18*, 852–859. [CrossRef] [PubMed]
52. Wang, Q.; Lin, T.; Tang, L.; Johnson, J.E.; Finn, M.G. Icosahedral Virus Particles as Addressable Nanoscale Building Blocks. *Angew. Chem. Int. Ed.* **2002**, *41*, 459–462. [CrossRef]
53. Wang, Q.; Kaltgrad, E.; Lin, T.; Johnson, J.E.; Finn, M.G. Natural Supramolecular Building Blocks: Wild-Type Cowpea Mosaic Virus. *Chem. Biol.* **2002**, *9*, 805–811. [CrossRef]
54. Narayanan, K.B.; Han, S.S. Icosahedral plant viral nanoparticles—Bioinspired synthesis of nanomaterials/nanostructures. *Adv. Colloid Interface Sci.* **2017**, *248*, 1–19. [CrossRef] [PubMed]
55. Yildiz, I.; Tsvetkova, I.; Wen, A.M.; Shukla, S.; Masarapu, M.H.; Dragnea, B.; Steinmetz, N.F. Engineering of Brome mosaic virus for biomedical applications. *RSC Adv.* **2012**, *2*, 3670–3677. [CrossRef] [PubMed]
56. Dixit, S.K.; Goicochea, N.L.; Daniel, M.-C.; Murali, A.; Bronstein, L.; De, M.; Stein, B.; Rotello, V.M.; Kao, C.C.; Dragnea, B. Quantum Dot Encapsulation in Viral Capsids. *Nano Lett.* **2006**, *6*, 1993–1999. [CrossRef] [PubMed]
57. Suci, P.A.; Varpness, Z.; Gillitzer, E.; Douglas, T.; Young, M. Targeting and Photodynamic Killing of a Microbial Pathogen Using Protein Cage Architectures Functionalized with a Photosensitizer. *Langmuir* **2007**, *23*, 12280–12286. [CrossRef]
58. Zlotnick, A.; Aldrich, R.; Johnson, J.M.; Ceres, P.; Young, M.J. Mechanism of Capsid Assembly for an Icosahedral Plant Virus. *Virology* **2000**, *277*, 450–456. [CrossRef]

59. Huynh, N.T.; Hesketh, E.L.; Saxena, P.; Meshcheriakova, Y.; Ku, Y.-C.; Hoang, L.T.; Johnson, J.E.; Ranson, N.A.; Lomonossoff, G.P.; Reddy, V.S. Crystal Structure and Proteomics Analysis of Empty Virus-like Particles of Cowpea Mosaic Virus. *Structure* **2016**, *24*, 567–575. [CrossRef] [PubMed]

60. Zeng, Q.; Wen, H.; Wen, Q.; Chen, X.; Wang, Y.; Xuan, W.; Liang, J.; Wan, S. Cucumber mosaic virus as drug delivery vehicle for doxorubicin. *Biomaterials* **2013**, *34*, 4632–4642. [CrossRef]

61. Ren, Y.; Wong, S.-M.; Lim, L.-Y. In vitro-reassembled plant virus-like particles for loading of polyacids. *J. Gen. Virol.* **2006**, *87*, 2749–2754. [CrossRef]

62. Lockney, D.M.; Guenther, R.N.; Loo, L.; Overton, W.; Antonelli, R.; Clark, J.; Hu, M.; Luft, C.; Lommel, S.A.; Franzen, S. The Red clover necrotic mosaic virus Capsid as a Multifunctional Cell Targeting Plant Viral Nanoparticle. *Bioconj. Chem.* **2011**, *22*, 67–73. [CrossRef] [PubMed]

63. Dreher, T.W. Turnip yellow mosaic virus: Transfer RNA mimicry, chloroplasts and a C-rich genome. *Mol. Plant Pathol.* **2004**, *5*, 367–375. [CrossRef]

64. Finbloom, J.A.; Han, K.; Aanei, I.L.; Hartman, E.C.; Finley, D.T.; Dedeo, M.T.; Fishman, M.; Downing, K.H.; Francis, M.B. Stable Disk Assemblies of a Tobacco Mosaic Virus Mutant as Nanoscale Scaffolds for Applications in Drug Delivery. *Bioconj. Chem.* **2016**, *27*, 2480–2485. [CrossRef]

65. Shukla, S.; Eber, F.J.; Nagarajan, A.S.; DiFranco, N.A.; Schmidt, N.; Wen, A.M.; Eiben, S.; Twyman, R.M.; Wege, C.; Steinmetz, N.F. The Impact of Aspect Ratio on the Biodistribution and Tumor Homing of Rigid Soft-Matter Nanorods. *Adv. Healthc. Mater.* **2015**, *4*, 874–882. [CrossRef]

66. Bazzini, A.A.; Hopp, H.E.; Beachy, R.N.; Asurmendi, S. Infection and coaccumulation of tobacco mosaic virus proteins alter microRNA levels, correlating with symptom and plant development. *Proc. Natl. Acad. Sci. USA* **2007**, *104*, 12157. [CrossRef]

67. Shukla, S.; Dickmeis, C.; Nagarajan, A.S.; Fischer, R.; Commandeur, U.; Steinmetz, N.F. Molecular farming of fluorescent virus-based nanoparticles for optical imaging in plants, human cells and mouse models. *Biomater. Sci.* **2014**, *2*, 784–797. [CrossRef]

68. Kwak, M.; Minten, I.J.; Anaya, D.-M.; Musser, A.J.; Brasch, M.; Nolte, R.J.M.; Müllen, K.; Cornelissen, J.J.L.M.; Herrmann, A. Virus-like Particles Templated by DNA Micelles: A General Method for Loading Virus Nanocarriers. *J. Am. Chem. Soc.* **2010**, *132*, 7834–7835. [CrossRef] [PubMed]

69. Saunders, K.; Sainsbury, F.; Lomonossoff, G.P. Efficient generation of cowpea mosaicvirus empty virus-like particles by the proteolytic processing of precursors in insect cells and plants. *Virology* **2009**, *393*, 329–337. [CrossRef]

70. Marín-Caba, L.; Chariou, P.L.; Pesquera, C.; Correa-Duarte, M.A.; Steinmetz, N.F. Tobacco Mosaic Virus-Functionalized Mesoporous Silica Nanoparticles, a Wool-Ball-like Nanostructure for Drug Delivery. *Langmuir* **2019**, *35*, 203–211. [CrossRef] [PubMed]

71. Bruckman, M.A.; Hern, S.; Jiang, K.; Flask, C.A.; Yu, X.; Steinmetz, N.F. Tobacco mosaic virus rods and spheres as supramolecular high-relaxivity MRI contrast agents. *J. Mater. Chem. B* **2013**, *1*, 1482–1490. [CrossRef]

72. Niehl, A.; Appaix, F.; Boscá, S.; van der Sanden, B.; Nicoud, J.-F.; Bolze, F.; Heinlein, M. Fluorescent Tobacco mosaic virus-Derived Bio-Nanoparticles for Intravital Two-Photon Imaging. *Front. Plant Sci.* **2016**, *6*, 1244. [CrossRef]

73. Sánchez, F.; Sáez, M.; Lunello, P.; Ponz, F. Plant viral elongated nanoparticles modified for log-increases of foreign peptide immunogenicity and specific antibody detection. *J. Biotechnol.* **2013**, *168*, 409–415. [CrossRef]

74. Smith, G.P. Filamentous fusion phage: Novel expression vectors that display cloned antigens on the virion surface. *Science* **1985**, *228*, 1315–1317. [CrossRef]

75. Huang, J.X.; Bishop-Hurley, S.L.; Cooper, M.A. Development of Anti-Infectives Using Phage Display: Biological Agents against Bacteria, Viruses, and Parasites. *Antimicrob. Agents Chemother.* **2012**, *56*, 4569–4582. [CrossRef]

76. Seker, U.O.S.; Demir, H.V. Material Binding Peptides for Nanotechnology. *Molecules* **2011**, *16*, 1426–1451. [CrossRef] [PubMed]

77. Merzlyak, A.; Lee, S.-W. Phage as templates for hybrid materials and mediators for nanomaterial synthesis. *Curr. Opin. Chem. Biol.* **2006**, *10*, 246–252. [CrossRef] [PubMed]

78. Kriplani, U.; Kay, B.K. Selecting peptides for use in nanoscale materials using phage-displayed combinatorial peptide libraries. *Curr. Opin. Biotechnol.* **2005**, *16*, 470–475. [CrossRef]

79. Garet, E.; Cabado, A.G.; Vieites, J.M.; González-Fernández, Á. Rapid isolation of single-chain antibodies by phage display technology directed against one of the most potent marine toxins: Palytoxin. *Toxicon* **2010**, *55*, 1519–1526. [CrossRef] [PubMed]

80. Nanduri, V.; Bhunia, A.K.; Tu, S.-I.; Paoli, G.C.; Brewster, J.D. SPR biosensor for the detection of L. monocytogenes using phage-displayed antibody. *Biosens. Bioelectron.* **2007**, *23*, 248–252. [CrossRef] [PubMed]

81. Pasqualini, R.; Ruoslahti, E. Organ targeting In vivo using phage display peptide libraries. *Nature* **1996**, *380*, 364–366. [CrossRef]

82. Deutscher, S.L. Phage Display in Molecular Imaging and Diagnosis of Cancer. *Chem. Rev.* **2010**, *110*, 3196–3211. [CrossRef]

83. Govarts, C.; Somers, K.; Stinissen, P.; Somers, V. Frameshifting in the P6 cDNA Phage Display System. *Molecules* **2010**, *15*, 9380–9390. [CrossRef]

84. Ghosh, D.; Kohli, A.G.; Moser, F.; Endy, D.; Belcher, A.M. Refactored M13 Bacteriophage as a Platform for Tumor Cell Imaging and Drug Delivery. *ACS Synth. Biol.* **2012**, *1*, 576–582. [CrossRef]

85. Gao, C.; Mao, S.; Lo, C.-H.L.; Wirsching, P.; Lerner, R.A.; Janda, K.D. Making artificial antibodies: A format for phage display of combinatorial heterodimeric arrays. *Proc. Natl. Acad. Sci. USA* **1999**, *96*, 6025–6030. [CrossRef] [PubMed]

86. Pande, J.; Szewczyk, M.M.; Grover, A.K. Phage display: Concept, innovations, applications and future. *Biotechnol. Adv.* **2010**, *28*, 849–858. [CrossRef]

87. Makowski, L. Phage display: Structure, assembly and engineering of filamentous bacteriophage M13. *Curr. Opin. Struct. Biol.* **1994**, *4*, 225–230. [CrossRef]

88. Smith, G.P. Preface. *Gene* **1993**, *128*, 1–2. [CrossRef]

89. Molek, P.; Bratkovič, T. Bacteriophages as Scaffolds for Bipartite Display: Designing Swiss Army Knives on a Nanoscale. *Bioconjugate Chemistry* **2015**, *26*, 367–378. [CrossRef] [PubMed]

90. Morton, V.L.; Burkitt, W.; O'Connor, G.; Stonehouse, N.J.; Stockley, P.G.; Ashcroft, A.E. RNA-induced conformational changes in a viral coat protein studied by hydrogen/deuterium exchange mass spectrometry. *Phys. Chem. Chem. Phys.* **2010**, *12*, 13468–13475. [CrossRef]

91. Wilson, D.R.; Finlay, B.B. Phage display: Applications, innovations, and issues in phage and host biology. *Can. J. Microbiol.* **1998**, *44*, 313–329. [CrossRef]

92. Smith, G.P.; Petrenko, V.A. Phage Display. *Chem. Rev.* **1997**, *97*, 391–410. [CrossRef]

93. Ren, Z.J.; Black, L.W.; Lewis, G.K.; Wingfield, P.T.; Locke, E.G.; Steven, A.C. Phage display of intact domains at high copy number: A system based on SOC, the small outer capsid protein of bacteriophage T4. *Protein Sci.* **1996**, *5*, 1833–1843. [CrossRef]

94. Skerra, A.; Plückthun, A. Secretion and in vivo folding of the Fab fragment of the antibody McPC603 in Escherichia coli: Influence of disulphides and cis-prolines. *Protein Eng.* **1991**, *4*, 971–979. [CrossRef]

95. Bratkovič, T. Progress in phage display: Evolution of the technique and its applications. *Cell. Mol. Life Sci.* **2010**, *67*, 749–767. [CrossRef]

96. Gao, J.; Wang, Y.; Liu, Z.; Wang, Z. Phage display and its application in vaccine design. *Ann. Microbiol.* **2010**, *60*, 13–19. [CrossRef]

97. Wu, J.; Tu, C.; Yu, X.; Zhang, M.; Zhang, N.; Zhao, M.; Nie, W.; Ren, Z. Bacteriophage T4 nanoparticle capsid surface SOC and HOC bipartite display with enhanced classical swine fever virus immunogenicity: A powerful immunological approach. *J. Virol. Methods* **2007**, *139*, 50–60. [CrossRef]

98. Ren, Z.-j.; Black, L.W. Phage T4 SOC and HOC display of biologically active, full-length proteins on the viral capsid. *Gene* **1998**, *215*, 439–444. [CrossRef]

99. Oślizło, A.; Miernikiewicz, P.; Piotrowicz, A.; Owczarek, B.; Kopciuch, A.; Figura, G.; Dąbrowska, K. Purification of phage display-modified bacteriophage T4 by affinity chromatography. *BMC Biotechnol.* **2011**, *11*, 59. [CrossRef]

100. Jiang, J.; Abu-Shilbayeh, L.; Rao, V.B. Display of a PorA peptide from Neisseria meningitidis on the bacteriophage T4 capsid surface. *Infect. Immun.* **1997**, *65*, 4770–4777.

101. Li, Q.; Shivachandra, S.B.; Zhang, Z.; Rao, V.B. Assembly of the Small Outer Capsid Protein, Soc, on Bacteriophage T4: A novel system for high density display of multiple large anthrax toxins and foreign proteins on phage capsid. *J. Mol. Biol.* **2007**, *370*, 1006–1019. [CrossRef] [PubMed]

102. Rao, V.B.; Black, L.W. Structure and assembly of bacteriophage T4 head. *Virol. J.* **2010**, *7*, 356. [CrossRef]

103. Gamkrelidze, M.; Dąbrowska, K. T4 bacteriophage as a phage display platform. *Arch. Microbiol.* **2014**, *196*, 473–479. [CrossRef] [PubMed]

104. Danner, S.; Belasco, J.G. T7 phage display: A novel genetic selection system for cloning RNA-binding proteins from cDNA libraries. *Proc. Natl. Acad. Sci. USA* **2001**, *98*, 12954–12959. [CrossRef] [PubMed]

105. Krumpe, L.R.H.; Atkinson, A.J.; Smythers, G.W.; Kandel, A.; Schumacher, K.M.; McMahon, J.B.; Makowski, L.; Mori, T. T7 lytic phage-displayed peptide libraries exhibit less sequence bias than M13 filamentous phage-displayed peptide libraries. *Proteomics* **2006**, *6*, 4210–4222. [CrossRef] [PubMed]

106. Caberoy, N.B.; Zhou, Y.; Jiang, X.; Alvarado, G.; Li, W. Efficient identification of tubby-binding proteins by an improved system of T7 phage display. *J. Mol. Recognit.* **2010**, *23*, 74–83. [CrossRef]

107. Sharma, S.C.; Memic, A.; Rupasinghe, C.N.; Duc, A.-C.E.; Spaller, M.R. T7 phage display as a method of peptide ligand discovery for PDZ domain proteins. *Pept. Sci.* **2009**, *92*, 183–193. [CrossRef] [PubMed]

108. Maruyama, I.N.; Maruyama, H.I.; Brenner, S. Lambda foo: A lambda phage vector for the expression of foreign proteins. *Proc. Natl. Acad. Sci. USA* **1994**, *91*, 8273–8277. [CrossRef]

109. Gi Mikawa, Y.; Maruyama, I.N.; Brenner, S. Surface Display of Proteins on Bacteriophage λ Heads. *J. Mol. Biol.* **1996**, *262*, 21–30. [CrossRef]

110. Hoess, R.H. Bacteriophage Lambda as a Vehicle for Peptide and Protein Display. *Curr. Pharm. Biotechnol.* **2002**, *3*, 23–28. [CrossRef]

111. Zanghi, C.N.; Lankes, H.A.; Bradel-Tretheway, B.; Wegman, J.; Dewhurst, S. A simple method for displaying recalcitrant proteins on the surface of bacteriophage lambda. *Nucleic Acids Res.* **2005**, *33*, e160. [CrossRef]

112. Beghetto, E.; Gargano, N. Lambda-Display: A Powerful Tool for Antigen Discovery. *Molecules* **2011**, *16*, 3089–3105. [CrossRef] [PubMed]

113. Nicastro, J.; Sheldon, K.; Slavcev, R.A. Bacteriophage lambda display systems: Developments and applications. *Appl. Microbiol. Biotechnol.* **2014**, *98*, 2853–2866. [CrossRef]

114. Heal, K.G.; Hill, H.R.; Stockley, P.G.; Hollingdale, M.R.; Taylor-Robinson, A.W. Expression and immunogenicity of a liver stage malaria epitope presented as a foreign peptide on the surface of RNA-free MS2 bacteriophage capsids. *Vaccine* **1999**, *18*, 251–258. [CrossRef]

115. Peabody, D.S.; Manifold-Wheeler, B.; Medford, A.; Jordan, S.K.; do Carmo Caldeira, J.; Chackerian, B. Immunogenic Display of Diverse Peptides on Virus-like Particles of RNA Phage MS2. *J. Mol. Biol.* **2008**, *380*, 252–263. [CrossRef] [PubMed]

116. Mastico, R.A.; Talbot, S.J.; Stockley, P.G. Multiple presentation of foreign peptides on the surface of an RNA-free spherical bacteriophage capsid. *J. Gen. Virol.* **1993**, *74*, 541–548. [CrossRef] [PubMed]

117. Chackerian, B.; Caldeira, J.d.C.; Peabody, J.; Peabody, D.S. Peptide Epitope Identification by Affinity Selection on Bacteriophage MS2 Virus-Like Particles. *J. Mol. Biol.* **2011**, *409*, 225–237. [CrossRef]

118. Brown, W.L.; Mastico, R.A.; Wu, M.; Heal, K.G.; Adams, C.J.; Murray, J.B.; Simpson, J.C.; Lord, J.M.; Taylor-Robinson, A.W.; Stockley, P.G. RNA Bacteriophage Capsid-Mediated Drug Delivery and Epitope Presentation. *Intervirology* **2002**, *45*, 371–380. [CrossRef]

119. Peabody, D.S. Subunit Fusion Confers Tolerance to Peptide Insertions in a Virus Coat Protein. *Arch. Biochem. Biophys.* **1997**, *347*, 85–92. [CrossRef]

120. Van Meerten, D.; Olsthoorn, R.C.L.; van Duin, J.; Verhaert, R.M.D. Peptide display on live MS2 phage: Restrictions at the RNA genome level. *J. Gen. Virol.* **2001**, *82*, 1797–1805. [CrossRef]

121. Nixon, A.E. Phage display as a tool for protease ligand discovery. *Curr. Pharm. Biotechnol.* **2002**, *3*, 1–12. [CrossRef] [PubMed]

122. Fernandez-Gacio, A.; Uguen, M.; Fastrez, J. Phage display as a tool for the directed evolution of enzymes. *Trends Biotechnol.* **2003**, *21*, 408–414. [CrossRef]

123. Li, B.; Tom, J.Y.; Oare, D.; Yen, R.; Fairbrother, W.J.; Wells, J.A.; Cunningham, B.C. Minimization of a polypeptide hormone. *Science* **1995**, *270*, 1657–1660. [CrossRef] [PubMed]

124. Kronqvist, N.; Malm, M.; Gostring, L.; Gunneriusson, E.; Nilsson, M.; Hoiden Guthenberg, I.; Gedda, L.; Frejd, F.Y.; Stahl, S.; Lofblom, J. Combining phage and staphylococcal surface display for generation of ErbB3-specific Affibody molecules. *Protein Eng. Des. Sel.* **2011**, *24*, 385–396. [CrossRef]

125. Rader, C.; Cheresh, D.A.; Barbas, C.F., 3rd. A phage display approach for rapid antibody humanization: Designed combinatorial V gene libraries. *Proc. Natl. Acad. Sci. USA* **1998**, *95*, 8910–8915. [CrossRef]

126. Kohler, G.; Milstein, C. Continuous cultures of fused cells secreting antibody of predefined specificity. *Nature* **1975**, *256*, 495–497. [CrossRef]

127. Emmons, C.; Hunsicker, L.G. Muromonab-CD3 (Orthoclone OKT3): The first monoclonal antibody approved for therapeutic use. *IOWA Med.* **1987**, *77*, 78–82.

128. Hwang, W.Y.K.; Foote, J. Immunogenicity of engineered antibodies. *Methods* **2005**, *36*, 3–10. [CrossRef]

129. Ishida, I.; Tomizuka, K.; Yoshida, H.; Tahara, T.; Takahashi, N.; Ohguma, A.; Tanaka, S.; Umehashi, M.; Maeda, H.; Nozaki, C.; et al. Production of Human Monoclonal and Polyclonal Antibodies in TransChromo Animals. *Cloning Stem Cells* **2002**, *4*, 91–102. [CrossRef]

130. Ma, B.; Osborn, M.J.; Avis, S.; Ouisse, L.-H.; Ménoret, S.; Anegon, I.; Buelow, R.; Brüggemann, M. Human antibody expression in transgenic rats: Comparison of chimeric IgH loci with human VH, D and JH but bearing different rat C-gene regions. *J. Immunol. Methods* **2013**, *400–401*, 78–86. [CrossRef] [PubMed]

131. Nelson, A.L.; Dhimolea, E.; Reichert, J.M. Development trends for human monoclonal antibody therapeutics. *Nat. Rev. Drug Discov.* **2010**, *9*, 767–774. [CrossRef]

132. Tsuruta, L.R.; Lopes dos, M.; Ana Maria Moro, A.M. Display Technologies for the Selection of Monoclonal Antibodies for Clinical Use. In *Antibody Engineering*; Böldicke, T., Ed.; InTech Open: Rijeka, Croatia, 2017. [CrossRef]

133. Frenzel, A.; Schirrmann, T.; Hust, M. Phage display-derived human antibodies in clinical development and therapy. *MAbs* **2016**, *8*, 1177–1194. [CrossRef]

134. Kaplon, H.; Reichert, J.M. Antibodies to watch in 2019. *MAbs* **2018**. [CrossRef]

135. Hust, M.; Meyer, T.; Voedisch, B.; Rülker, T.; Thie, H.; El-Ghezal, A.; Kirsch, M.I.; Schütte, M.; Helmsing, S.; Meier, D.; et al. A human scFv antibody generation pipeline for proteome research. *J. Biotechnol.* **2011**, *152*, 159–170. [CrossRef]

136. Schofield, D.J.; Pope, A.R.; Clementel, V.; Buckell, J.; Chapple, S.D.; Clarke, K.F.; Conquer, J.S.; Crofts, A.M.; Crowther, S.R.E.; Dyson, M.R.; et al. Application of phage display to high throughput antibody generation and characterization. *Genome Biol.* **2007**, *8*, R254. [CrossRef]

137. De Haard, H.J.; van Neer, N.; Reurs, A.; Hufton, S.E.; Roovers, R.C.; Henderikx, P.; de Bruïne, A.P.; Arends, J.-W.; Hoogenboom, H.R. A Large Non-immunized Human Fab Fragment Phage Library That Permits Rapid Isolation and Kinetic Analysis of High Affinity Antibodies. *J. Biol. Chem.* **1999**, *274*, 18218–18230. [CrossRef] [PubMed]

138. Hoet, R.M.; Cohen, E.H.; Kent, R.B.; Rookey, K.; Schoonbroodt, S.; Hogan, S.; Rem, L.; Frans, N.; Daukandt, M.; Pieters, H.; et al. Generation of high-affinity human antibodies by combining donor-derived and synthetic complementarity-determining-region diversity. *Nat. Biotechnol.* **2005**, *23*, 344. [CrossRef] [PubMed]

139. Mazor, Y.; Van Blarcom, T.; Carroll, S.; Georgiou, G. Selection of full-length IgGs by tandem display on filamentous phage particles and *Escherichia coli* fluorescence-activated cell sorting screening. *FEBS J.* **2010**, *277*, 2291–2303. [CrossRef] [PubMed]

140. Hoogenboom, H.R. Selecting and screening recombinant antibody libraries. *Nat. Biotechnol.* **2005**, *23*, 1105. [CrossRef]

141. Kennedy, P.J.; Oliveira, C.; Granja, P.L.; Sarmento, B. Monoclonal antibodies: Technologies for early discovery and engineering. *Crit. Rev. Biotechnol.* **2018**, *38*, 394–408. [CrossRef]

142. Vaughan, T.J.; Williams, A.J.; Pritchard, K.; Osbourn, J.K.; Pope, A.R.; Earnshaw, J.C.; McCafferty, J.; Hodits, R.A.; Wilton, J.; Johnson, K.S. Human Antibodies with Sub-nanomolar Affinities Isolated from a Large Non-immunized Phage Display Library. *Nat. Biotechnol.* **1996**, *14*, 309–314. [CrossRef]

143. Lloyd, C.; Lowe, D.; Edwards, B.; Welsh, F.; Dilks, T.; Hardman, C.; Vaughan, T. Modelling the human immune response: Performance of a 1011 human antibody repertoire against a broad panel of therapeutically relevant antigens. *Protein Eng. Des. Sel.* **2008**, *22*, 159–168. [CrossRef]

144. Kügler, J.; Wilke, S.; Meier, D.; Tomszak, F.; Frenzel, A.; Schirrmann, T.; Dübel, S.; Garritsen, H.; Hock, B.; Toleikis, L.; et al. Generation and analysis of the improved human HAL9/10 antibody phage display libraries. *BMC Biotechnol.* **2015**, *15*, 10. [CrossRef] [PubMed]

145. Nelson, B.; Sidhu, S.S. Synthetic Antibody Libraries. In *Therapeutic Proteins: Methods and Protocols*; Voynov, V., Caravella, J.A., Eds.; Humana Press: Totowa, NJ, USA, 2012; pp. 27–41. [CrossRef]

146. Osbourn, J.; Groves, M.; Vaughan, T. From rodent reagents to human therapeutics using antibody guided selection. *Methods* **2005**, *36*, 61–68. [CrossRef] [PubMed]

147. Burmester, G.R.; Panaccione, R.; Gordon, K.B.; McIlraith, M.J.; Lacerda, A.P. Adalimumab: Long-term safety in 23 458 patients from global clinical trials in rheumatoid arthritis, juvenile idiopathic arthritis, ankylosing spondylitis, psoriatic arthritis, psoriasis and Crohn's disease. *Ann. Rheum. Dis.* **2013**, *72*, 517–524. [CrossRef]

148. Group, C.R.; Martin, D.F.; Maguire, M.G.; Ying, G.S.; Grunwald, J.E.; Fine, S.L.; Jaffe, G.J. Ranibizumab and bevacizumab for neovascular age-related macular degeneration. *N. Engl. J. Med.* **2011**, *364*, 1897–1908. [CrossRef]

149. Stohl, W.; Hilbert, D.M. The discovery and development of belimumab: The anti-BLyS-lupus connection. *Nat. Biotechnol.* **2012**, *30*, 69–77. [CrossRef] [PubMed]

150. Navarra, S.V.; Guzman, R.M.; Gallacher, A.E.; Hall, S.; Levy, R.A.; Jimenez, R.E.; Li, E.K.; Thomas, M.; Kim, H.Y.; Leon, M.G.; et al. Efficacy and safety of belimumab in patients with active systemic lupus erythematosus: A randomised, placebo-controlled, phase 3 trial. *Lancet* **2011**, *377*, 721–731. [CrossRef]

151. Mazumdar, S. Raxibacumab. *MAbs* **2009**, *1*, 531–538. [CrossRef] [PubMed]

152. Aprile, G.; Bonotto, M.; Ongaro, E.; Pozzo, C.; Giuliani, F. Critical appraisal of ramucirumab (IMC-1121B) for cancer treatment: From benchside to clinical use. *Drugs* **2013**, *73*, 2003–2015. [CrossRef]

153. Dienstmann, R.; Felip, E. Necitumumab in the treatment of advanced non-small cell lung cancer: Translation from preclinical to clinical development. *Expert Opin. Biol.* **2011**, *11*, 1223–1231. [CrossRef] [PubMed]

154. Chin, K.; Chand, V.K.; Nuyten, D.S.A. Avelumab: Clinical trial innovation and collaboration to advance anti-PD-L1 immunotherapy. *Ann. Oncol.* **2017**, *28*, 1658–1666. [CrossRef] [PubMed]

155. Machado, A.; Torres, T. Guselkumab for the Treatment of Psoriasis. *BioDrugs* **2018**, *32*, 119–128. [CrossRef]

156. Kenniston, J.A.; Faucette, R.R.; Martik, D.; Comeau, S.R.; Lindberg, A.P.; Kopacz, K.J.; Conley, G.P.; Chen, J.; Viswanathan, M.; Kastrapeli, N.; et al. Inhibition of plasma kallikrein by a highly specific active site blocking antibody. *J. Biol. Chem.* **2014**, *289*, 23596–23608. [CrossRef]

157. Banerji, A.; Busse, P.; Shennak, M.; Lumry, W.; Davis-Lorton, M.; Wedner, H.J.; Jacobs, J.; Baker, J.; Bernstein, J.A.; Lockey, R.; et al. Inhibiting Plasma Kallikrein for Hereditary Angioedema Prophylaxis. *N. Engl. J. Med.* **2017**, *376*, 717–728. [CrossRef] [PubMed]

158. Kreitman, R.J.; Pastan, I. Antibody fusion proteins: Anti-CD22 recombinant immunotoxin moxetumomab pasudotox. *Clin. Cancer Res.* **2011**, *17*, 6398–6405. [CrossRef]

159. Kreitman, R.J.; Dearden, C.; Zinzani, P.L.; Delgado, J.; Karlin, L.; Robak, T.; Gladstone, D.E.; le Coutre, P.; Dietrich, S.; Gotic, M.; et al. Moxetumomab pasudotox in relapsed/refractory hairy cell leukemia. *Leukemia* **2018**, *32*, 1768–1777. [CrossRef]

160. Markland, W.; Ley, A.C.; Ladner, R.C. Iterative optimization of high-affinity protease inhibitors using phage display. 2. Plasma kallikrein and thrombin. *Biochemistry* **1996**, *35*, 8058–8067. [CrossRef]

161. Farkas, H.; Varga, L. Ecallantide is a novel treatment for attacks of hereditary angioedema due to C1 inhibitor deficiency. *Clin. Cosmet. Investig. Derm.* **2011**, *4*, 61–68. [CrossRef]

162. Nixon, A.E.; Sexton, D.J.; Ladner, R.C. Drugs derived from phage display: From candidate identification to clinical practice. *MAbs* **2014**, *6*, 73–85. [CrossRef] [PubMed]

163. Molineux, G.; Newland, A. Development of romiplostim for the treatment of patients with chronic immune thrombocytopenia: From bench to bedside. *Br. J. Haematol.* **2010**, *150*, 9–20. [CrossRef]

164. Kelley, B.D.; Booth, J.; Tannatt, M.; Wub, Q.L.; Ladner, R.; Yuc, J.; Potter, D.; Ley, A. Isolation of a peptide ligand for affinity purification of factor VIII using phage display. *J. Chromatogr. A* **2004**, *1038*, 121–130. [CrossRef] [PubMed]

165. Kelley, B.; Jankowski, M.; Booth, J. An improved manufacturing process for Xyntha/ReFacto AF. *Haemophilia* **2010**, *16*, 717–725. [CrossRef] [PubMed]

166. Mimmi, S.; Maisano, D.; Quinto, I.; Iaccino, E. Phage Display: An Overview in Context to Drug Discovery. *Trends Pharmacol. Sci.* **2019**, *40*, 87–91. [CrossRef] [PubMed]

167. Fala, L. Tanzeum (Albiglutide): A Once-Weekly GLP-1 Receptor Agonist Subcutaneous Injection Approved for the Treatment of Patients with Type 2 Diabetes. *Am. Health Drug Benefits* **2015**, *8*, 126–130. [PubMed]

168. Bao, W.; Holt, L.J.; Prince, R.D.; Jones, G.X.; Aravindhan, K.; Szapacs, M.; Barbour, A.M.; Jolivette, L.J.; Lepore, J.J.; Willette, R.N.; et al. Novel fusion of GLP-1 with a domain antibody to serum albumin prolongs protection against myocardial ischemia/reperfusion injury in the rat. *Cardiovasc. Diabetol.* **2013**, *12*, 148. [CrossRef] [PubMed]

169. Wilson, P.C.; Andrews, S.F. Tools to therapeutically harness the human antibody response. *Nat. Rev. Immunol.* **2012**, *12*, 709. [CrossRef] [PubMed]

170. Brown, S. Metal-recognition by repeating polypeptides. *Nat. Biotechnol.* **1997**, *15*, 269–272. [CrossRef]

171. Brown, S. Engineered iron oxide-adhesion mutants of the Escherichia coli phage lambda receptor. *Proc. Natl. Acad. Sci. USA* **1992**, *89*, 8651–8655. [CrossRef]

172. Whaley, S.R.; English, D.S.; Hu, E.L.; Barbara, P.F.; Belcher, A.M. Selection of peptides with semiconductor binding specificity for directed nanocrystal assembly. *Nature* **2000**, *405*, 665–668. [CrossRef]

173. Seker, U.O.S.; Wilson, B.; Dincer, S.; Kim, I.W.; Oren, E.E.; Evans, J.S.; Tamerler, C.; Sarikaya, M. Adsorption Behavior of Linear and Cyclic Genetically Engineered Platinum Binding Peptides. *Langmuir* **2007**, *23*, 7895–7900. [CrossRef] [PubMed]

174. Sanghvi, A.B.; Miller, K.P.H.; Belcher, A.M.; Schmidt, C.E. Biomaterials functionalization using a novel peptide that selectively binds to a conducting polymer. *Nat. Mater.* **2005**, *4*, 496–502. [CrossRef] [PubMed]

175. So, C.R.; Hayamizu, Y.; Yazici, H.; Gresswell, C.; Khatayevich, D.; Tamerler, C.; Sarikaya, M. Controlling Self-Assembly of Engineered Peptides on Graphite by Rational Mutation. *ACS Nano* **2012**, *6*, 1648–1656. [CrossRef] [PubMed]

176. Peelle, B.R.; Krauland, E.M.; Wittrup, K.D.; Belcher, A.M. Design Criteria for Engineering Inorganic Material-Specific Peptides. *Langmuir* **2005**, *21*, 6929–6933. [CrossRef]

177. Hnilova, M.; Oren, E.E.; Seker, U.O.S.; Wilson, B.R.; Collino, S.; Evans, J.S.; Tamerler, C.; Sarikaya, M. Effect of Molecular Conformations on the Adsorption Behavior of Gold-Binding Peptides. *Langmuir* **2008**, *24*, 12440–12445. [CrossRef]

178. Korkmaz Zirpel, N.; Arslan, T.; Lee, H. Engineering filamentous bacteriophages for enhanced gold binding and metallization properties. *J. Colloid Interface Sci.* **2015**, *454*, 80–88. [CrossRef]

179. Tamerler, C.; Kacar, T.; Sahin, D.; Fong, H.; Sarikaya, M. Genetically engineered polypeptides for inorganics: A utility in biological materials science and engineering. *Mater. Sci. Eng. C* **2007**, *27*, 558–564. [CrossRef]

180. Gaskin, D.J.H.; Starck, K.; Vulfson, E.N. Identification of inorganic crystal-specific sequences using phage display combinatorial library of short peptides: A feasibility study. *Biotechnol. Lett.* **2000**, *22*, 1211–1216. [CrossRef]

181. Sarikaya, M.; Tamerler, C.; Schwartz, D.T.; Baneyx, F.O. Materials assembly and formation using engineered polypeptides. *Annu. Rev. Mater. Res.* **2004**, *34*, 373–408. [CrossRef]

182. Estephan, E.; Saab, M.-B.; Larroque, C.; Martin, M.; Olsson, F.; Lourdudoss, S.; Gergely, C. Peptides for functionalization of InP semiconductors. *J. Colloid Interface Sci.* **2009**, *337*, 358–363. [CrossRef]

183. Sano, K.-I.; Sasaki, H.; Shiba, K. Specificity and Biomineralization Activities of Ti-Binding Peptide-1 (TBP-1). *Langmuir* **2005**, *21*, 3090–3095. [CrossRef] [PubMed]

184. Wang, S.; Humphreys, E.S.; Chung, S.-Y.; Delduco, D.F.; Lustig, S.R.; Wang, H.; Parker, K.N.; Rizzo, N.W.; Subramoney, S.; Chiang, Y.-M.; et al. Peptides with selective affinity for carbon nanotubes. *Nat. Mater.* **2003**, *2*, 196–200. [CrossRef]

185. Chen, H.; Su, X.; Neoh, K.-G.; Choe, W.-S. QCM-D Analysis of Binding Mechanism of Phage Particles Displaying a Constrained Heptapeptide with Specific Affinity to SiO2 and TiO$_2$. *Anal. Chem.* **2006**, *78*, 4872–4879. [CrossRef] [PubMed]

186. Huang, Y.; Chiang, C.-Y.; Lee, S.K.; Gao, Y.; Hu, E.L.; Yoreo, J.D.; Belcher, A.M. Programmable Assembly of Nanoarchitectures Using Genetically Engineered Viruses. *Nano Lett.* **2005**, *5*, 1429–1434. [CrossRef]

187. Cui, Y.; Kim, S.N.; Jones, S.E.; Wissler, L.L.; Naik, R.R.; McAlpine, M.C. Chemical Functionalization of Graphene Enabled by Phage Displayed Peptides. *Nano Lett.* **2010**, *10*, 4559–4565. [CrossRef]

188. Galloway, J.M.; Staniland, S.S. Protein and peptide biotemplated metal and metal oxide nanoparticles and their patterning onto surfaces. *J. Mater. Chem.* **2012**, *22*, 12423–12434. [CrossRef]

189. Mao, C.; Solis, D.J.; Reiss, B.D.; Kottmann, S.T.; Sweeney, R.Y.; Hayhurst, A.; Georgiou, G.; Iverson, B.; Belcher, A.M. Virus-Based Toolkit for the Directed Synthesis of Magnetic and Semiconducting Nanowires. *Science* **2004**, *303*, 213–217. [CrossRef]

190. Nam, Y.S.; Park, H.; Magyar, A.P.; Yun, D.S.; Pollom, T.S.; Belcher, A.M. Virus-templated iridium oxide-gold hybrid nanowires for electrochromic application. *Nanoscale* **2012**, *4*, 3405–3409. [CrossRef] [PubMed]

191. Mao, C.; Flynn, C.E.; Hayhurst, A.; Sweeney, R.; Qi, J.; Georgiou, G.; Iverson, B.; Belcher, A.M. Viral assembly of oriented quantum dot nanowires. *Proc. Natl. Acad. Sci. USA* **2003**, *100*, 6946–6951. [CrossRef] [PubMed]

192. Flynn, C.E.; Mao, C.; Hayhurst, A.; Williams, J.L.; Georgiou, G.; Iverson, B.; Belcher, A.M. Synthesis and organization of nanoscale II-VI semiconductor materials using evolved peptide specificity and viral capsid assembly. *J. Mater. Chem.* **2003**, *13*, 2414–2421. [CrossRef]

193. Reiss, B.D.; Mao, C.; Solis, D.J.; Ryan, K.S.; Thomson, T.; Belcher, A.M. Biological Routes to Metal Alloy Ferromagnetic Nanostructures. *Nano Lett.* **2004**, *4*, 1127–1132. [CrossRef]

194. Loo, L.; Guenther, R.H.; Basnayake, V.R.; Lommel, S.A.; Franzen, S. Controlled Encapsidation of Gold Nanoparticles by a Viral Protein Shell. *J. Am. Chem. Soc.* **2006**, *128*, 4502–4503. [CrossRef] [PubMed]

195. Huang, X.; Bronstein, L.M.; Retrum, J.; Dufort, C.; Tsvetkova, I.; Aniagyei, S.; Stein, B.; Stucky, G.; McKenna, B.; Remmes, N.; et al. Self-Assembled Virus-like Particles with Magnetic Cores. *Nano Lett.* **2007**, *7*, 2407–2416. [CrossRef]

196. Minten, I.J.; Hendriks, L.J.A.; Nolte, R.J.M.; Cornelissen, J.J.L.M. Controlled Encapsulation of Multiple Proteins in Virus Capsids. *J. Am. Chem. Soc.* **2009**, *131*, 17771–17773. [CrossRef]

197. Wen, A.M.; Shukla, S.; Saxena, P.; Aljabali, A.A.A.; Yildiz, I.; Dey, S.; Mealy, J.E.; Yang, A.C.; Evans, D.J.; Lomonossoff, G.P.; et al. Interior Engineering of a Viral Nanoparticle and Its Tumor Homing Properties. *Biomacromolecules* **2012**, *13*, 3990–4001. [CrossRef] [PubMed]

198. Loo, L.; Guenther, R.H.; Lommel, S.A.; Franzen, S. Infusion of dye molecules into Red clover necrotic mosaic virus. *Chem. Commun.* **2008**, 88–90. [CrossRef]

199. Ashley, C.E.; Carnes, E.C.; Phillips, G.K.; Durfee, P.N.; Buley, M.D.; Lino, C.A.; Padilla, D.P.; Phillips, B.; Carter, M.B.; Willman, C.L.; et al. Cell-Specific Delivery of Diverse Cargos by Bacteriophage MS2 Virus-like Particles. *ACS Nano* **2011**, *5*, 5729–5745. [CrossRef]

200. Wu, M.; Sherwin, T.; Brown, W.L.; Stockley, P.G. Delivery of antisense oligonucleotides to leukemia cells by RNA bacteriophage capsids. *Nanomed. Nanotechnol. Biol. Med.* **2005**, *1*, 67–76. [CrossRef]

201. Wei, B.; Wei, Y.; Zhang, K.; Wang, J.; Xu, R.; Zhan, S.; Lin, G.; Wang, W.; Liu, M.; Wang, L.; et al. Development of an antisense RNA delivery system using conjugates of the MS2 bacteriophage capsids and HIV-1 TAT cell penetrating peptide. *Biomed. Pharmacother.* **2009**, *63*, 313–318. [CrossRef]

202. Glasgow, J.E.; Capehart, S.L.; Francis, M.B.; Tullman-Ercek, D. Osmolyte-Mediated Encapsulation of Proteins inside MS2 Viral Capsids. *ACS Nano* **2012**, *6*, 8658–8664. [CrossRef]

203. Capehart, S.L.; Coyle, M.P.; Glasgow, J.E.; Francis, M.B. Controlled Integration of Gold Nanoparticles and Organic Fluorophores Using Synthetically Modified MS2 Viral Capsids. *J. Am. Chem. Soc.* **2013**, *135*, 3011–3016. [CrossRef] [PubMed]

204. Liu, C.; Chung, S.-H.; Jin, Q.; Sutton, A.; Yan, F.; Hoffmann, A.; Kay, B.K.; Bader, S.D.; Makowski, L.; Chen, L. Magnetic viruses via nano-capsid templates. *J. Magn. Magn. Mater.* **2006**, *302*, 47–51. [CrossRef]

205. Edgar, R.; McKinstry, M.; Hwang, J.; Oppenheim, A.B.; Fekete, R.A.; Giulian, G.; Merril, C.; Nagashima, K.; Adhya, S. High-sensitivity bacterial detection using biotin-tagged phage and quantum-dot nanocomplexes. *Proc. Natl. Acad. Sci. USA* **2006**, *103*, 4841–4845. [CrossRef]

206. Oh, M.H.; Yu, J.H.; Kim, I.; Nam, Y.S. Genetically Programmed Clusters of Gold Nanoparticles for Cancer Cell-Targeted Photothermal Therapy. *ACS Appl. Mater. Interfaces* **2015**, *7*, 22578–22586. [CrossRef] [PubMed]

207. Aanei, I.L.; Huynh, T.; Seo, Y.; Francis, M.B. Vascular Cell Adhesion Molecule-Targeted MS2 Viral Capsids for the Detection of Early-Stage Atherosclerotic Plaques. *Bioconj. Chem.* **2018**, *29*, 2526–2530. [CrossRef]

208. Farkas, M.E.; Aanei, I.L.; Behrens, C.R.; Tong, G.J.; Murphy, S.T.; O'Neil, J.P.; Francis, M.B. PET Imaging and Biodistribution of Chemically Modified Bacteriophage MS2. *Mol. Pharm.* **2013**, *10*, 69–76. [CrossRef]

209. Anderson, E.A.; Isaacman, S.; Peabody, D.S.; Wang, E.Y.; Canary, J.W.; Kirshenbaum, K. Viral Nanoparticles Donning a Paramagnetic Coat: Conjugation of MRI Contrast Agents to the MS2 Capsid. *Nano Lett.* **2006**, *6*, 1160–1164. [CrossRef]

210. Ghosh, D.; Bagley, A.F.; Na, Y.J.; Birrer, M.J.; Bhatia, S.N.; Belcher, A.M. Deep, noninvasive imaging and surgical guidance of submillimeter tumors using targeted M13-stabilized single-walled carbon nanotubes. *Proc. Natl. Acad. Sci. USA* **2014**, *111*, 13948–13953. [CrossRef] [PubMed]

211. Ghosh, D.; Lee, Y.; Thomas, S.; Kohli, A.G.; Yun, D.S.; Belcher, A.M.; Kelly, K.A. M13-templated magnetic nanoparticles for targeted in vivo imaging of prostate cancer. *Nat. Nano* **2012**, *7*, 677–682. [CrossRef] [PubMed]

212. Xu, Z.; Sun, H.; Gao, F.; Hou, L.; Li, N. Synthesis and magnetic property of T4 virus-supported gold-coated iron ternary nanocomposite. *J. Nanopart. Res.* **2012**, *14*, 1267. [CrossRef]

213. Hou, L.; Gao, F.; Li, N. T4 Virus-Based Toolkit for the Direct Synthesis and 3D Organization of Metal Quantum Particles. *Chem. Eur. J.* **2010**, *16*, 14397–14403. [CrossRef]

214. Hou, L.; Tong, D.; Jiang, Y.; Gao, F. Synthesis and organization of platinum nanoparticles and nanoshells on a native virus bioscaffold. *Nano* **2014**, *9*, 1450058. [CrossRef]

215. Frenkel, D.; Solomon, B. Filamentous phage as vector-mediated antibody delivery to the brain. *Proc. Natl. Acad. Sci. USA* **2002**, *99*, 5675–5679. [CrossRef] [PubMed]

216. Bar, H.; Yacoby, I.; Benhar, I. Killing cancer cells by targeted drug-carrying phage nanomedicines. *BMC Biotechnol.* **2008**, *8*, 37. [CrossRef]
217. Suthiwangcharoen, N.; Li, T.; Li, K.; Thompson, P.; You, S.; Wang, Q. M13 bacteriophage-polymer nanoassemblies as drug delivery vehicles. *Nano Res.* **2011**, *4*, 483–493. [CrossRef]
218. Ren, Y.; Wong, S.M.; Lim, L.-Y. Folic Acid-Conjugated Protein Cages of a Plant Virus: A Novel Delivery Platform for Doxorubicin. *Bioconj. Chem.* **2007**, *18*, 836–843. [CrossRef]
219. Pitek, A.S.; Hu, H.; Shukla, S.; Steinmetz, N.F. Cancer Theranostic Applications of Albumin-Coated Tobacco Mosaic Virus Nanoparticles. *ACS Appl. Mater. Interfaces* **2018**, *10*, 39468–39477. [CrossRef]
220. Le, D.H.T.; Lee, K.L.; Shukla, S.; Commandeur, U.; Steinmetz, N.F. Potato virus X, a filamentous plant viral nanoparticle for doxorubicin delivery in cancer therapy. *Nanoscale* **2017**, *9*, 2348–2357. [CrossRef]
221. Alemzadeh, E.; Izadpanah, K.; Ahmadi, F. Generation of recombinant protein shells of Johnson grass chlorotic stripe mosaic virus in tobacco plants and their use as drug carrier. *J. Virol. Methods* **2017**, *248*, 148–153. [CrossRef]
222. Alemzadeh, E.; Dehshahri, A.; Dehghanian, A.R.; Afsharifar, A.; Behjatnia, A.A.; Izadpanah, K.; Ahmadi, F. Enhanced anti-tumor efficacy and reduced cardiotoxicity of doxorubicin delivered in a novel plant virus nanoparticle. *Colloids Surf. B Biointerfaces* **2019**, *174*, 80–86. [CrossRef]
223. Kumar, K.; Kumar Doddi, S.; Kalle Arunasree, M.; Paik, P. CPMV-induced synthesis of hollow mesoporous SiO2 nanocapsules with excellent performance in drug delivery. *Dalton Trans.* **2015**, *44*, 4308–4317. [CrossRef]
224. Yildiz, I.; Lee, K.L.; Chen, K.; Shukla, S.; Steinmetz, N.F. Infusion of imaging and therapeutic molecules into the plant virus-based carrier cowpea mosaic virus: Cargo-loading and delivery. *J. Control. Release* **2013**, *172*, 568–578. [CrossRef]
225. Liu, X.; Liu, B.; Gao, S.; Wang, Z.; Tian, Y.; Wu, M.; Jiang, S.; Niu, Z. Glyco-decorated tobacco mosaic virus as a vector for cisplatin delivery. *J. Mater. Chem. B* **2017**, *5*, 2078–2085. [CrossRef]
226. Franke, C.E.; Czapar, A.E.; Patel, R.B.; Steinmetz, N.F. Tobacco Mosaic Virus-Delivered Cisplatin Restores Efficacy in Platinum-Resistant Ovarian Cancer Cells. *Mol. Pharm.* **2018**, *15*, 2922–2931. [CrossRef]
227. DePorter, S.M.; McNaughton, B.R. Engineered M13 Bacteriophage Nanocarriers for Intracellular Delivery of Exogenous Proteins to Human Prostate Cancer Cells. *Bioconj. Chem.* **2014**, *25*, 1620–1625. [CrossRef]
228. Sánchez-Sánchez, L.; Cadena-Nava, R.D.; Palomares, L.A.; Ruiz-Garcia, J.; Koay, M.S.T.; Cornelissen, J.J.M.T.; Vazquez-Duhalt, R. Chemotherapy pro-drug activation by biocatalytic virus-like nanoparticles containing cytochrome P450. *Enzym. Microb. Technol.* **2014**, *60*, 24–31. [CrossRef]
229. Hoskins, J.M.; Carey, L.A.; McLeod, H.L. CYP2D6 and tamoxifen: DNA matters in breast cancer. *Nat. Rev. Cancer* **2009**, *9*, 576. [CrossRef]
230. Yacoby, I.; Shamis, M.; Bar, H.; Shabat, D.; Benhar, I. Targeting Antibacterial Agents by Using Drug-Carrying Filamentous Bacteriophages. *Antimicrob. Agents Chemother.* **2006**, *50*, 2087–2097. [CrossRef] [PubMed]
231. Brasch, M.; de la Escosura, A.; Ma, Y.; Uetrecht, C.; Heck, A.J.R.; Torres, T.; Cornelissen, J.J.L.M. Encapsulation of Phthalocyanine Supramolecular Stacks into Virus-like Particles. *J. Am. Chem. Soc.* **2011**, *133*, 6878–6881. [CrossRef]
232. Lee, K.L.; Carpenter, B.L.; Wen, A.M.; Ghiladi, R.A.; Steinmetz, N.F. High Aspect Ratio Nanotubes Formed by Tobacco Mosaic Virus for Delivery of Photodynamic Agents Targeting Melanoma. *ACS Biomater. Sci. Eng.* **2016**, *2*, 838–844. [CrossRef] [PubMed]
233. Chariou, P.L.; Wang, L.; Desai, C.; Park, J.; Robbins, L.K.; von Recum, H.A.; Ghiladi, R.A.; Steinmetz, N.F. Let There Be Light: Targeted Photodynamic Therapy Using High Aspect Ratio Plant Viral Nanoparticles. *Macromol. Biosci.* **2019**, e1800407. [CrossRef]
234. Ngweniform, P.; Abbineni, G.; Cao, B.; Mao, C. Self-Assembly of Drug-Loaded Liposomes on Genetically Engineered Target-Recognizing M13 Phage: A Novel Nanocarrier for Targeted Drug Delivery. *Small* **2009**, *5*, 1963–1969. [CrossRef]
235. Sawada, T.; Yanagimachi, M.; Serizawa, T. Controlled release of antibody proteins from liquid crystalline hydrogels composed of genetically engineered filamentous viruses. *Mater. Chem. Front.* **2017**, *1*, 146–151. [CrossRef]
236. Cao, J.; Liu, S.; Chen, Y.; Shi, L.; Zhang, Z. Synthesis of end-functionalized boronic acid containing copolymers and their bioconjugates with rod-like viruses for multiple responsive hydrogels. *Polym. Chem.* **2014**, *5*, 5029–5036. [CrossRef]

237. Zhi, X.; Zheng, C.; Xiong, J.; Li, J.; Zhao, C.; Shi, L.; Zhang, Z. Nanofilamentous Virus-Based Dynamic Hydrogels with Tunable Internal Structures, Injectability, Self-Healing, and Sugar Responsiveness at Physiological pH. *Langmuir* **2018**, *34*, 12914–12923. [CrossRef]

238. Honarbakhsh, S.; Guenther, R.H.; Willoughby, J.A.; Lommel, S.A.; Pourdeyhimi, B. Polymeric Systems Incorporating Plant Viral Nanoparticles for Tailored Release of Therapeutics. *Adv. Healthc. Mater.* **2013**, *2*, 1001–1007. [CrossRef]

239. Tian, Y.; Wu, M.; Liu, X.; Liu, Z.; Zhou, Q.; Niu, Z.; Huang, Y. Probing the Endocytic Pathways of the Filamentous Bacteriophage in Live Cells Using Ratiometric pH Fluorescent Indicator. *Adv. Healthc. Mater.* **2015**, *4*, 413–419. [CrossRef] [PubMed]

240. Koudelka, K.J.; Destito, G.; Plummer, E.M.; Trauger, S.A.; Siuzdak, G.; Manchester, M. Endothelial targeting of cowpea mosaic virus (CPMV) via surface vimentin. *PLoS Pathog.* **2009**, *5*, e1000417. [CrossRef] [PubMed]

241. Srivastava, A.S.; Kaido, T.; Carrier, E. Immunological factors that affect the in vivo fate of T7 phage in the mouse. *J. Virol. Methods* **2004**, *115*, 99–104. [CrossRef] [PubMed]

242. Aanei, I.L.; ElSohly, A.M.; Farkas, M.E.; Netirojjanakul, C.; Regan, M.; Taylor Murphy, S.; O'Neil, J.P.; Seo, Y.; Francis, M.B. Biodistribution of Antibody-MS2 Viral Capsid Conjugates in Breast Cancer Models. *Mol. Pharm.* **2016**, *13*, 3764–3772. [CrossRef] [PubMed]

243. Wu, M.; Shi, J.; Fan, D.; Zhou, Q.; Wang, F.; Niu, Z.; Huang, Y. Biobehavior in Normal and Tumor-Bearing Mice of Tobacco Mosaic Virus. *Biomacromolecules* **2013**, *14*, 4032–4037. [CrossRef] [PubMed]

244. Petrenko, V.A. Autonomous self-navigating drug-delivery vehicles: From science fiction to reality. *Ther. Deliv.* **2017**, *8*, 1063–1075. [CrossRef]

245. Le, D.H.T.; Méndez-López, E.; Wang, C.; Commandeur, U.; Aranda, M.A.; Steinmetz, N.F. Biodistribution of Filamentous Plant Virus Nanoparticles: Pepino Mosaic Virus versus Potato Virus X. *Biomacromolecules* **2019**, *20*, 469–477. [CrossRef]

246. Yip, Y.L.; Hawkins, N.J.; Smith, G.; Ward, R.L. Biodistribution of filamentous phage-Fab in nude mice. *J. Immunol. Methods* **1999**, *225*, 171–178. [CrossRef]

247. Wu, Z.; Chen, K.; Yildiz, I.; Dirksen, A.; Fischer, R.; Dawson, P.E.; Steinmetz, N.F. Development of viral nanoparticles for efficient intracellular delivery. *Nanoscale* **2012**, *4*, 3567–3576. [CrossRef] [PubMed]

248. Raja, K.S.; Wang, Q.; Gonzalez, M.J.; Manchester, M.; Johnson, J.E.; Finn, M.G. Hybrid Virus–Polymer Materials. 1. Synthesis and Properties of PEG-Decorated Cowpea Mosaic Virus. *Biomacromolecules* **2003**, *4*, 472–476. [CrossRef] [PubMed]

249. Steinmetz, N.F.; Manchester, M. PEGylated Viral Nanoparticles for Biomedicine: The Impact of PEG Chain Length on VNP Cell Interactions In Vitro and Ex Vivo. *Biomacromolecules* **2009**, *10*, 784–792. [CrossRef] [PubMed]

250. Bruckman, M.A.; Randolph, L.N.; VanMeter, A.; Hern, S.; Shoffstall, A.J.; Taurog, R.E.; Steinmetz, N.F. Biodistribution, pharmacokinetics, and blood compatibility of native and PEGylated tobacco mosaic virus nano-rods and -spheres in mice. *Virology* **2014**, *449*, 163–173. [CrossRef] [PubMed]

251. Bludau, H.; Czapar, A.E.; Pitek, A.S.; Shukla, S.; Jordan, R.; Steinmetz, N.F. POxylation as an alternative stealth coating for biomedical applications. *Eur. Polym. J.* **2017**, *88*, 679–688. [CrossRef]

252. Pitek, A.S.; Jameson, S.A.; Veliz, F.A.; Shukla, S.; Steinmetz, N.F. Serum albumin 'camouflage' of plant virus based nanoparticles prevents their antibody recognition and enhances pharmacokinetics. *Biomaterials* **2016**, *89*, 89–97. [CrossRef] [PubMed]

253. Krag, D.N.; Shukla, G.S.; Shen, G.-P.; Pero, S.; Ashikaga, T.; Fuller, S.; Weaver, D.L.; Burdette-Radoux, S.; Thomas, C. Selection of Tumor-binding Ligands in Cancer Patients with Phage Display Libraries. *Cancer Res.* **2006**, *66*, 7724–7733. [CrossRef] [PubMed]

254. Staquicini, F.I.; Cardó-Vila, M.; Kolonin, M.G.; Trepel, M.; Edwards, J.K.; Nunes, D.N.; Sergeeva, A.; Efstathiou, E.; Sun, J.; Almeida, N.F.; et al. Vascular ligand-receptor mapping by direct combinatorial selection in cancer patients. *Proc. Natl. Acad. Sci. USA* **2011**, *108*, 18637–18642. [CrossRef] [PubMed]

255. Schubbert, R.; Renz, D.; Schmitz, B.; Doerfler, W. Foreign (M13) DNA ingested by mice reaches peripheral leukocytes, spleen, and liver via the intestinal wall mucosa and can be covalently linked to mouse DNA. *Proc. Natl. Acad. Sci. USA* **1997**, *94*, 961–966. [CrossRef] [PubMed]

pharmaceutics

MDPI

Review

Challenges and Recent Progress in Oral Drug Delivery Systems for Biopharmaceuticals

Bahman Homayun, Xueting Lin and Hyo-Jick Choi *

Department of Chemical and Materials Engineering, University of Alberta, Edmonton, AB T6G 1H9, Canada; homayun@ualberta.ca (B.H.); xueting2@ualberta.ca (X.L.)
* Correspondence: hyojick@ualberta.ca; Tel.: +1-780-248-1666

Received: 6 February 2019; Accepted: 14 March 2019; Published: 19 March 2019

Abstract: Routes of drug administration and the corresponding physicochemical characteristics of a given route play significant roles in therapeutic efficacy and short term/long term biological effects. Each delivery method has favorable aspects and limitations, each requiring a specific delivery vehicles design. Among various routes, oral delivery has been recognized as the most attractive method, mainly due to its potential for solid formulations with long shelf life, sustained delivery, ease of administration and intensified immune response. At the same time, a few challenges exist in oral delivery, which have been the main research focus in the field in the past few years. The present work concisely reviews different administration routes as well as the advantages and disadvantages of each method, highlighting why oral delivery is currently the most promising approach. Subsequently, the present work discusses the main obstacles for oral systems and explains the most recent solutions proposed to deal with each issue.

Keywords: oral delivery; biological barriers; co-delivery; throughput; sustained delivery

1. Introduction

Intravenous (IV), intramuscular (IM), intranasal (IN), intradermal (ID)/transdermal and oral administration are the main drug delivery routes. Other routes, such as ocular delivery, have also been developed for localized, site-specific drug administration without unwanted systemic side effects [1]. Each administration method faces specific barriers against the delivery of the drugs. In addition, drugs can be incorporated into delivery devices, which considerably contribute to preservation of the drug, targeting and therapeutic efficacy. In this review, we first present an overview of the various administration routes, then focus on oral delivery systems as the most attractive route. We explain the main challenges associated with such methods and review the most recent solutions developed to address them.

The absorption mechanism as well as the nature of the drug are the fundamental factors that determine the appropriate delivery systems for achieving the highest bioavailability and effectivity. For instance, IM and ID administration are usually the preferred vaccination routes, depending upon the desired immune response mechanisms. On the other hand, researchers from both academia and industry have shown great interest in IN and oral vaccination systems, since these routes can induce both systemic and mucosal immune responses. In IV administration, the drug is rapidly injected into blood vessels through needles, and a high concentration of the drug is able to bypass the physiological barriers against drug absorption, providing the highest bioavailability and the fastest effect among all delivery routes. Therefore, such a parenteral administration is the preferred route for acute and emergency responses, while non-invasive methods are more suitable for sustained therapy and chronic delivery [2]. The abundance of blood vessels in muscles paves the way for the absorption of drugs injected via needles through IM administration. IM delivery bypasses the body's first defense barrier (skin) [3]. In comparison with oral administration, drugs administered through the IM route avoid the gastrointestinal (GI) environment. However, the injection can cause significant problems, including needle-associated phobia and pain, unsafe needle use and improper disposal, the need for trained healthcare personnel, muscle atrophy, and injuries to bones and nerves [4]. Additionally, there is concern over the direct injection of drugs into the bloodstream through IM administration, necessitating constant close observation to minimize adverse effects [5,6].

Biopharmaceuticals such as vaccines are of particular interest in drug delivery because of their specific challenges. The majority of the available vaccines are administered through IM injection [7]. This is mainly due to the poor permeability of macromolecular biopharmaceuticals across the mucosal layer in the non-parenteral route and the destructive effects of proteases in the GI tract [8]. Silica and polymer mesoporous structures can also be successfully used to preserve drugs in various biological surroundings and accurately control their release behavior in topical injections [9–11]. However, it should be noted that IM administration is not the ideal delivery route for peptides and proteins, compared to subcutaneous or IV injection, mainly because of the low immunogenicity and bioavailability achieved in IM administration [12]. Although IM vaccination is widely used commercially and the immune response in this system can be easily induced by the local depot at the injection site, this route is not the best choice for the delivery of peptides/proteins due to the possible aggregation of the drug [13].

The transdermal route concerns the delivery of drugs across skin layers to the blood circulatory system [14]. Drug absorption in this case mainly occurs through the intercellular, transcellular and transappendageal pathways. Intercellular and transcellular transports enable the permeation through the stratum corneum [15]. In the transappendageal pathway, the drug penetrates via the sweat ducts or the hair follicles with their associated sebaceous glands [14]. The transdermal route also avoids the challenges that the oral route faces, including the metabolism and the difficulties associated with the GI environment. Moreover, it can provide a sustained drug plasma level, and convenience of discontinuation of the drug in case adverse reactions occur. Recently, nanoparticles (NPs) were successfully employed on nano/micro-engineered needle patches to minimize the bacterial risks associated with transdermal delivery techniques [16]. Nanoimprint lithography has also been proposed as a fabrication technique for these structures to address the commercial requirements. Kim et al. also proposed a new deposition-etching protocol for the fabrication of flexible silicon nanoneedles to solve the structural mismatch and optical inconsistency between the Si needle wafers and soft tissues [17]. The patches fabricated by this technique exhibited successful injection of biomolecules into living tissues. Nonetheless, the most significant challenge associated with transdermal administration is the restriction over the size of the drug molecules that can be successfully delivered. Penetration of large molecules (>500 Da) through the stratum corneum is difficult. Additionally, the drugs are required to be soluble, so they can cross the outermost skin barrier [18]. Chemical enhancers (concerning the delivery of small-sized molecules) and physical approaches (for the delivery of macromolecules) have been developed to improve drug absorption levels in the transdermal delivery route. Physical skin permeability enhancement regards electrically-assisted methods such as electroporation, iontophoresis and sonophoresis. Chemical enhancers include fatty acids, surfactants, terpenes and solvents, which improve skin permeability by disrupting the highly ordered lipids and modifying the stratum corneum microstructure [19,20]. However, toxicity and skin irritation are the major concerns to consider when developing chemical enhancer formulations. Karande et al. used a library of more than 4000 binary formulations to systematically investigate the synergistic effects of different enhancers, and found some fundamental rules in terms of developing new formulations [21]. While these rules are well-known and widely accepted, the mechanisms behind such potential synergistic effects and the interactions each individual chemical enhancer may have with other enhancers or the stratum corneum still remain unclear.

Medications can sometimes be injected directly into veins. This is known as intravenous administration. In this scheme, the drug directly enters the circulatory system without facing any physical, chemical or biological barriers. Since the absorption of the drug is guaranteed and immediate, IV administration is the best route for emergency situations. Additionally, this method uniquely provides a very precise control over the dosage and speed of administration, making it the best approach for drugs requiring a stringent dosage [22]. On the other hand, the IV route entails risks of injury, infection by the needle at the site of injection, circulatory overload, phlebitis and thrombosis [23,24]. IV administration is commonly used for the delivery of biopharmaceuticals. The bioavailability of the drug injected through IV route is theoretically 100%, making this system outperform other delivery routes. However, it should be noted that IV administration is not the ideal route for the delivery of vaccines. This is due to the difficulties associated with inducing effective immune responses via IV administration, since the IV route does not provide an adequate local depot of antigens to stimulate/activate the innate immune response and induce the long-term secretion of antibodies [25]. In addition, it is not applicable to implement a mass administration (such as nation-wide vaccinations), due to the skills required for the practitioner, safety issues and patients' compliance.

Intranasal drug delivery entails the infusion of the drug into the highly vascularized mucosal layer of the nose to subsequently reach systemic circulation [26]. IN drug delivery is crucially significant for neurological diseases, where drugs are required to reach the central nervous system (CNS) by bypassing the blood-brain barrier (BBB). In general, the IN route is preferred for local diseases due to its limited systemic effects compared to the other methods. The IN route also has its own specific physiological and physicochemical barriers. The physiological barriers include capillary barriers, nasal mucus, mucus clearance and nasal metabolism. Other factors such as pH, possible drug-mucus interactions and the viscosity of the mucus may also influence drug diffusion and absorption in IN administration. Mucoadhesive microencapsulation systems have been developed to deal with the mucus-associated barriers of IN delivery and improve the bioavailability of nasally-administered drugs. For example, Nanaki et al. coated nasal microcapsules with thiolated chitosan to aid the mucoadhesion of the system, both physically (electrostatic attractions) and chemically (disulfide bonds) [27]. However, it should be noted that mucociliary clearance and discharge considerably limits the drug residence time, even if the carriers develop a strong bonding/binding with the mucus [28]. Physicochemical barriers concern the molecular weight of the drug, and its lipophilicity and degree of ionization, which also define the absorption mechanisms [29]. Despite multiple challenges, IN is an advantageous route for the delivery of a variety of the drugs. It avoids first-pass metabolism and GI complications. Due to its considerable absorption rate, IN is an applicable method for emergency cases and rapid drug action. Neurological drugs especially can be transported directly to the CNS through the IN route.

The nasal-associated lymphoid tissue (NALT) is the principal target for inducing mucosal immunity in nasal vaccination. The innate immunity is achieved by macrophages and dendritic cells, and the adaptive immunity at the mucosal layer is induced by IgA [30]. Hence, IN vaccination can induce both mucosal and systemic immunity. However, physiological barriers, especially mucociliary clearance, need to be adequately addressed for the effective delivery of antigens to the target site. Furthermore, due to the limitations of the nasal cavity, as well as the narrow passages beneath the thick mucus, nasal vaccination is only permitted for small dosage and low molecular weight compounds. In order to pass biosafety requirements, IN delivery devices need to be suitable for narrow nasal entrances and the complex geometry of the nasal passage. On top of that, lung exposure should be properly addressed in these systems. Among the various physical states of drugs (gels, droplets, powders, or aerosol sprays), aerosol sprays are usually preferred for IN administration. New technologies are constantly under development to improve the dispersion of the drugs and modify the deposition and clearance behavior through combining solid and liquid phases. However, human anatomy as well as the physiology of the nasal cavity and passage still limit clinical applications and delivery efficiency. Low bioavailability (<5%) causes another major challenge for IN administration systems [31].

2. Oral Route

Among the various drug delivery routes, the oral pathway has attracted the most attention due to its unique advantages, including sustained and controllable delivery, ease of administration, feasibility for solid formulations, patient compliance and an intensified immune response in the case of vaccines [32–36]. In addition, a large surface area (>300 m^2) lined with a viscous mucosal layer paves the way for drug attachment and subsequent absorption [37,38]. Furthermore, drug molecules trapped within mucus are protected against the shear stresses caused by flowing gastric juices [39]. The epithelium of the human intestine is very absorptive due to the abundance of enterocytes in different parts of the intestine, especially microfold cells (M cells) covering the Peyer's patches, the lymphoid segment of the small intestine [40–44]. However, in comparison with other routes, the absorption mechanism of oral drugs is more complex. Oral drugs need to be soluble in gastric fluid so they can be absorbed in the stomach, the small intestine or the colon (Table 1). Orally administered drugs can be absorbed in four types of pathways: Transcellular, paracellular, carrier-mediated transcellular and facilitated transport. Among these pathways, the transcellular pathway is the main mechanism. The challenges of drug absorption/efficacy do not limit the barriers met in the gut, but they include the hepatic barriers after they enter the vessels under the intestinal epithelium as well. In summary, oral drugs are not applicable for emergencies due to their slow absorption as well as the multiple levels of barriers they need to deal with.

Table 1. Characteristics of different segments of the human gastrointestinal (GI) tract.

Segment	pH	Length (cm)	Mean Diameter (cm)	Mucus Average Thickness (µm)	Mucus Turnover (hour)
Stomach	0.8–5 [45]	20 [45]	NA	245 ± 200 [38]	24–48 [46]
Duodenum	~7 [38]	17–56 [47]	4 [48]	15.5 [45]	
Jejunum	≥ 7 [38]	280–1000 [47]	2–2.5 [48]	15.5 [45]	24–48 [46]
Ileum	≥ 7 [38]		3 [48]	15.5 [45]	
Colon	7–8 [38]	80–313 [47]	4–4.8 [48]	135 ± 25 [38]	24–48 [46]

Although the oral route is the most desirable administration method for small therapeutic molecules, there are not so many oral vaccines on the market due to the harsh conditions along the GI tract (Table 1), which can degrade/denature active antigens.

However, the attraction of the mucosal immunity, which appears to be induced by oral and nasal routes, promotes the study of oral vaccines [49]. Besides, the convenience and other advantages of oral delivery make it a very promising strategy for mass vaccination programs. The inductive sites in the GI tract consist of Peyer's patches, lymphoid follicles in lymph nodes and antigen presenting cells (APCs). Intestine mucosal immunity is similar to that of nasal mucosal immunity. The main barrier for vaccine delivery is the change of pH in different sites in the GI tract and various enzymes, making it hard to permeate the mucus and reach the inductive site in gut-associated lymphoid tissue (GALT) [50]. Additionally, the mucosa may lead to the structural change of proteins and peptides due to various possible interactions [38]. Hence, delivery vehicles and formulations should be developed to gain a stronger immunogenicity to meet the required therapeutic efficacy. Currently, seven live oral vaccines have already been approved by FDA.

To meet the increasing demand for biopharmaceutical oral products, research has been focused on developing devices for oral delivery. While still at an early stage, recent devices include intestinal patch systems, microneedle capsules and particulate systems [51]. The intestinal patch systems are based on a unidirectional drug release depot, which is similar to a microdevice adhered to the intestinal wall [52]. The microneedle capsule increases the penetration rate of drug molecules by piercing the mucosa directly with microneedles. A recent study developed a method to inflate a microneedle into the mucosa by responding to the change in pH [53]. Particulate devices are the most common oral vehicles, which have been investigated for the encapsulation and targeting of a vast variety of therapeutics. In general, the current technologies are still at the preclinical stage. Therefore, more research efforts should be directed to solve the existing technical challenges of oral drug delivery systems and prove the feasibility in clinical use.

3. Challenges Associated with Oral Delivery

Oral drugs are transported and absorbed in the GI tract, which is in the shape of a conduit. Some drugs have local effects in the gut, while most of them are sent to the bloodstream in the systemic circulation to act in other parts of the body. The GI tract can be divided into upper and lower parts. The upper GI tract includes the oral cavity, pharynx, esophagus, stomach and the initial part of the small intestine, known as the duodenum. The lower GI tract includes the rest of the small intestine (jejunum and ileum), as well as the large intestine segments: The cecum, colon and rectum [54,55]. The structure of the GI tract is similar in all segments. The lumen is enveloped by smooth muscle cells, covered by mucus, submucosa and several muscle layers [56]. The mucosal layer which lines the inner part of the GI tract consists of a layer of epithelial cells, lamina propria and muscularis mucosae, which play significant roles in food/drug molecule transport and gastrointestinal immunity [54,55]. A large absorption area and long residence time provides greater opportunities for drug absorption, which is one of the reasons why drug absorption mostly occurs in the small intestine. Further, between the three main parts of the small intestine (duodenum, jejunum and ileum), the jejunum and ileum have a higher absorption capability compared to the duodenum [57,58].

The environmental factors that influence drug integrity and absorption include the average length of the segment, pH, thickness of the mucus, residence time of the drug and the bacterial diversity/population in different segments [38,59,60]. The obstacles against oral administration may be broadly classified into biological barriers and technical challenges. Biological barriers include any biological factors that denature the orally administered drugs or prevent their successful absorption in the target. On the other hand, technical challenges relate to any difficulty in the fabrication process of the oral delivery devices. The technical challenges may either be issues with creating specific properties for addressing the biological barriers or complications with scaling up and commercializing a system. The details of each category will be discussed in the following sections.

3.1. Biological Barriers

Any digested ingredient will be dealing with three main biological environments along the gastrointestinal (GI) tract, regardless of its absorption mechanism or target. These environments are the lumen (i.e., the interior space), mucus and tissue. Each of these three environments may have interactions with the drug molecules.

3.1.1. Lumen

The first biological barrier against any orally administered drug is the harsh acidic conditions inside the stomach (pH 1–2.5), denaturing/depurinating most of the administered molecules, drastically lowering their effectivity [61–64]. In addition to stomach acid, gastric enzymes such as pepsin and gelatinase can degrade biopharmaceuticals. pH-responsive hydrogels can encapsulate the drugs and protect them not only against the harsh acidic environment, but against gastric enzymes as well. These materials can remain intact in unfavorable surrounding conditions to protect the loaded drug (promptly reacting to environmental stimuli such as the pH in the target) and release the cargo. For instance, Yamagata et al. confirmed that pH-sensitive hydrogel microparticles (MPs) can efficiently preserve sensitive drugs such as insulin against gastric/intestinal enzyme fluids [65]. In another study, Cerchiara et al. also developed an oral pH-responsive microencapsulation system and demonstrated its capability for protection against both gastric acidic and gastric enzymatic environments [66].

Notably, in addition to the gastric enzymes, there are also pancreatic enzymes synthesized inside the pancreas and secreted into the intestinal lumen. These enzymes include lipase (degrading fats), trypsin (decomposing proteins), amylase (degrading starch) and peptidases (disintegrating peptides), and are especially abundant at the main entrance of the small intestine (duodenum). They can readily decompose nucleic acids and reduce the gastric residential stability of biomolecules [67,68]. Although pancreatic enzymes are also introduced as biological barriers against oral delivery, they are not considered as a major challenge due to three principal reasons. First, these enzymes are mainly abundant in the duodenum, and their concentrations considerably decrease in the jejunum and later parts [69]. In addition, Layer et al. attempted to deliberately deliver pancreatic enzymes to the intestine and noticed that even the concentration of the delivered enzymes substantially decreases in the midjejunum compared to that in the duodenum (they reported up to a 90% drop in enzymatic activities) [70]. The second aspect is the short transit time of the digested food inside the duodenum (Table 1), which is not enough for the enzymes to degrade the drugs. Fallingborg et al. reported that duodenal residence comprises only 10% of the whole small intestine average residence time [71]. Lastly, the pH of the duodenum is lower than that of the later parts of the lower small intestine (Table 1), meaning that unwanted release of the drugs inside the duodenum can be successfully avoided by controlling (increasing) the pKa of the delivery carriers. For example, Lozoya-Agullo et al. employed poly(lactic-*co*-glycolic)acid (PLGA) nanoparticles for colon delivery and confirmed that duodenal release may be significantly avoided due to the insufficient environmental pH [72].

In addition to the acidic and enzymatic degradation, the lumen can cause other damages to drug molecules. Osmotic stresses along the GI tract, peristalsis of the GI muscles, as well as the shear stresses by the flow rate of the gastric juice inside the lumen are other factors that decrease drug efficiency due to mechanical degradation inside the lumen [32,38]. Flowing gastric juice may also decrease the contact time between the drug molecules and the epithelial layer, thus impeding their absorption [67]. Enveloped biologics such as viruses, vaccines and cells are usually the main components sensitive to mechanical destruction. Valon et al. studied the possible effects of mechanical stresses on various types of cells, and reported that shear stresses and compaction may lead to apoptosis and cell death [73]. Choi et al. also found that hyperosmotic pressure can destroy virus integrity in an acidic environment [32]. Although mechanical stresses by the lumen can destroy biological agents, microencapsulation can properly address these problems as well. Indeed, MPs are hardly studied for mechanical testing, and mechanical strength is usually discussed in regard to hydrogel properties, which in turn is a function of the molecular weight of the monomers as well as the degree of crosslinking of the matrix. Nanoparticles may also be added to hydrogels as reinforcing components. For instance, France et al. recently employed cellulose nanocrystals (CNCs) to create physical crosslinking in a hydrogel matrix, proving their significant impact on mechanical behavior (up to a 35-fold increase in the shear storage modulus of the final composite) [74]. Yang et al. also reported up to a 3.5-fold improvement in the Young's modulus of poly(ethylene glycol) (PEG) hydrogels through the addition of 1.2 vol % CNCs [75]. As an instance of ceramic-polymer composite systems, Gaharwar et al. used hydroxyapatite as the reinforcing agent in a ceramic-polymer composite and reported up to a 1.7-fold improvement in the compressive strength of the matrix. It should be noted that the addition of secondary components to form a composite will not only affect the mechanical properties, but may also affect the swelling behavior, gelation rate and degradation rate of the hydrogels as well [74,75]. Furthermore, the distribution mechanisms of CNCs throughout the polymer matrix are poorly understood, and there are underwhelming (sometimes contradictory) reports about the improvements in mechanical properties of composites. However, the greatest and most effective improvement reported by the formation of a hydrogel composite to date concerns the addition of CNCs to hydrogels, without the need for any surface or other secondary modification of the particles.

3.1.2. Mucus

Mucus is the second compartment of the GI tract, which any digested moiety interacts with. The entire GI tract is lined by mucus (a sticky, elastic and viscous layer), responsible for the capture of foreign moieties, especially hydrophobic molecules, impeding their contact with the underlying epithelial layer. The mucus then discards such foreign moieties, acting as one of the main compartments of the immune system, known as mucosal immunity [76–79]. The mucus itself is mainly composed of water and mucin protein molecules coated with proteoglycans, giving the mucus a negative charge [80]. Carbohydrates, salts, bacteria, antibodies and cellular remnants are the other compounds found in mucus [81]. The thickness of this layer is reported to vary along the GI tract, as summarized in Table 1 [38].

Mucins are large macromolecules (0.5–40 MDa) made of monomers connected through disulfide bonding, and these macromolecules are subsequently crosslinked to build up the mucosal layer [82,83]. The mucosal layer itself is usually composed of two separate overlaying layers: An outer loosely adherent layer and an inner firmly adherent layer (Figure 1) [84]. The inner layer is composed of glycoproteins, glycolipids and cell-bound mucin [85,86]. It has been claimed that the thicker outer mucus acts as a barrier against the transition of released drugs to the submucosal tissue. For instance, Marie Boegh and Hanne Mørck Nielsen studied the diffusion of peptides and proteins through mucus and found that mucus is the main obstacle against the bioavailability of oral drugs [87]. On the other hand, the narrower inner mucus is known to help the absorption/enhance the uptake efficiency of drugs, justifying the dual role of mucus in the absorption/desorption of orally delivered drugs (Figure 1) [88,89]. There is an equilibrium between mucin secretion, degradation and clearance in each segment of the GI tract to protect the epithelium and control the nutrition absorption rate, which in turn defines the final thickness of the regional mucosal layer (Table 1) [38,90].

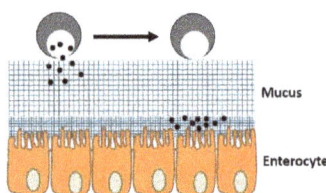

Figure 1. The structure and function of the mucus. The schematic shows the gastric mucus layer, the attachment of particles to the outer and inner mucus, and the drug delivery vehicle on the outer mucus.

The approaches taken regarding mucus in the field of oral delivery can be generally classified into two opposite mainstreams: Mucopenetration and mucoadhesion. Mucopenetrating oral vehicles may be made through controlling the hydrophobicity/hydrophilicity nature of the carriers' matrix. In other words, due to the hydrophobic nature of the outer mucus, it tends to have significantly less interactions with hydrophilic materials. Additionally, regulating the electrostatic interactions between the carriers and the mucus can dramatically facilitate the transition of particles through mucus. Li et al. incorporated Pluronic F127 (PF127) into the matrix of liposomes to induce hydrophilicity and nullify the electrostatic charge of the particles, demonstrating considerable improvements in the mucopenetration and uptake efficiency of the liposomes [91]. Cu et al. reported similar improvements in terms of the mucopenetration of poly(lactic-*co*-glycolic)acid (PLGA) nanoparticles by coating them using PEG, which neutralized the surface charge of the nanoparticles [92].

The other platform employed for penetrating through mucus is the application of mucolytic enzymes. Muller et al. fabricated nanoparticles made of complexes of poly(acrylic acid) (PAA) and papain (a mucolytic enzyme) through physical adsorption [93]. They showed that the enzyme-conjugated particles can reduce the viscosity of fresh mucus by up to 5 times, reflecting the potential of the system for improved oral bioavailability. Pereira de Sousa implemented a similar study comparing bromelain (another mucolytic enzyme) with papain for the same purpose [94]. They showed that the bromelain-modified particles exhibited a significantly enhanced mucopenetration behavior compared to the papain-conjugated particles (up to a 4-fold increase in mucopenetration compared to the papain-conjugated particles, and up to a 10-fold increase compared to the blank samples).

On the other hand, the protective and sticky properties of mucus can be exploited to protect the digested ingredients and extend their transition time along the GI tract. As a result, mucoadhesive drug carriers have attracted significant attention [95,96]. For this purpose, cationic hydrogels such as chitosan have been extensively investigated [97–99]. Recently, Kim et al. derived catechols from mussels and conjugated it with chitosan to develop a new complex with significantly better mucoadhesive properties and negligible cytotoxicity compared to chitosan alone [100]. Lectin functionalization has been another strategy employed for the same purpose. The study implemented by Ertl et al. is one of the first works with this approach, in which they conjugated wheat germ agglutinin with PLGA MPs and increased their mucoadhesion [101]. The advantage of this new system is that lectin-conjugated MPs exhibited improved adhesion not only to the mucus but also to the enterocyte (cell) surfaces, minimizing the problems associated with the short turnover period of mucus. Additionally, in specific parts of the GI tract where the mucus is not thick enough, this system can still show improved absorption efficiency.

In the case of anionic MPs, their surfaces have been functionalized with thiol functional groups to exhibit mucoadhesive properties by forming disulfide bonds with thiol groups existing in mucins [102–107]. Several studies have confirmed that such chemically-modified MPs show a significantly longer residence time in the target, acting as potential candidates for the sustained oral delivery of specific drugs, such as insulin and losartan [96,108]. For instance, Zhang et al. functionalized Eudragit L100 with thiol groups and confirmed improvement in the absorption and bioavailability of orally administered insulin compared to the unmodified polymer by monitoring the blood glucose concentration [108]. However, it should be noted that mucus is constantly secreted by goblet cells along the GI tract and is subsequently shed and cleared from tissues due to the rapid turnover of cells. This leads to a very short residence time for the attached agents to reach the epithelium for absorption (50–270 min) [109,110]. Mucus, with around a two day turnover period, plays the most significant role as a barrier [111,112]. Hence, sustained delivery using mucoadhesive particles may not represent the ideal strategy in this regard. In addition to this, it is difficult to model the real effects of mucus on delivery carriers in vitro due to its changing thickness along the GI tract and the constant effects of flowing gastric juice [38,67,112]. Since sustained oral delivery itself is a major challenge, it will be discussed under a separate subsection later in this review.

3.1.3. Tissue (Extracellular Barriers)

The characteristics of the drug molecule will determine its absorption sites and its pathway to cross through the intestinal epithelial cells in the GI tract. There are two main pathways for the absorption of oral MPs or drugs: Transcytosis by cells (transcellular route) and diffusion through the spaces between epithelial cells (paracellular route). Figure 2 schematically represents all the possible absorption scenarios that a digested molecule may encounter in the intestinal lumen. In the transcellular pathway, the drug molecules enter the enterocytes by crossing the membrane of the epithelial cells. The paracellular pathway permits only small hydrophilic molecules to be absorbed, playing a minor role in drug absorption as it is narrow and only occupies a small area fraction of the whole epithelium [87]. The principal extracellular biological barrier against oral delivery is known as tight junctions, which concern the paracellular absorption route for orally administered agents [113,114].

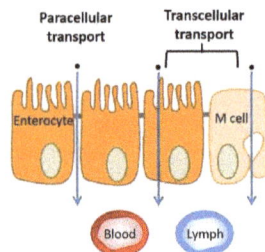

Figure 2. Absorption mechanisms through the mucosal layer. Paracellular route to lamina propia and transcellular route (enterocytes, M-cells, transfection of the epithelial cells, direct absorption through dendritic cells and active transport).

The transport of MPs or drug molecules through tissue depends on both their chemistry and their size. Generally, hydrophobic drug molecules, nanoparticles, vesicles and micelles prefer to be absorbed through transcellular routes due to their large size and chemistry, while hydrophilic small drug molecules prefer paracellular routes [67,115]. There are also barriers due to the structure of the cells. At the outer regions of the membrane bilayers, the high molecular density of the polar head groups of the lipid membranes makes it difficult for the drug molecules to pass through the cell membranes. Furthermore, after entering the cell membrane, cellular components such as enzymes may degrade/decompose the drug molecules in the cytosol, decreasing the therapeutic efficiency [89].

Passive diffusion of MPs through intercellular routes is only possible for agents with sizes of up to a few nanometers (0.5–3 nm), which is too small for the delivery of most drug molecules [116,117]. A notable example of this problem is the poor bioavailability of doxorubicin (DOX), which is attributed to its limited paracellular absorption in the intestine. Kim et al. developed a medium-sized chain glyceride-based water-in-oil (W/O) microemulsion system to overcome the paracellular barrier against DOX intestinal absorption [118]. The improvements in the absorption level of the drug were ascribed to the lipidic components of the carriers, inducing the paracellular enhancing effects. In addition to the chemistry of the carriers, there are few absorption enhancing agents that have been used in oral delivery systems. Sodium *N*-[8 (2-hydroxylbenzoyl) amino] caprylate (SNAC) is a paracellular permeability enhancer which has been recently used for clinical trials by Davies et al. [69]. The problem with this agent is that there is no distinct mechanism identified for its function. Recently, Taverner et al. identified other peptides (PIP peptides: 250 and 640), which could enhance the paracellular permeability of insulin through intestinal tissue [119]. They claimed that these peptides can dynamically adjust endogenous mechanisms, inducing myosin light chains (MLCs), opening the tight junctions and facilitating paracellular transition, especially for peptide therapeutics. Almansour et al. (in the same group) studied PIP 640 further and confirmed its stability in the intestinal lumen environment and explained its functioning mechanism: PIP 640 selectively enhances the MLC-pS^{19} levels of the cytoplasm of enterocytes in the epithelial layer [120]. Apart from the recent studies and the progress made, tight junctions are still one of the main challenges against the absorption of biopharmaceuticals. The lack of a fundamental understanding of the mechanisms controlling tissue permeability and the effects of various agents may be one of the principal reasons for this problem.

3.2. Technical Challenges

In addition to biological barriers, oral delivery systems face technical difficulties as well, in terms of deciding whether to induce new properties addressing biological barriers or to scale up existing systems for commercial purposes. In this section, most common oral delivery devices, sustained delivery strategies, solvent-free microencapsulation techniques, co-delivery systems and the challenges associated with the scaling-up of systems are analyzed.

3.2.1. Oral Delivery Devices and Materials

The devices developed for oral drug administration may be classified as intestinal patches, gastrointestinal microneedles and particulate carriers (including micro/nanoparticles, micelles and liposomes), as illustrated in Figure 3. Intestinal patches are millimeter-sized mucoadhesive blankets that attach to the inner walls of the GI tract, providing a drug reservoir at the target. These patches can protect the drug against the harsh environment and luminal loss, improving the bioavailability of the drug by providing a unidirectional diffusion regime towards the intestinal tissue. They have been especially attractive for improving the oral bioavailability of drugs and sustained delivery. Insulin, interferon-α and calcitonin are examples of drugs investigated for delivery using such devices [121]. As far as intestinal patches are concerned, the mucosal adhesion properties, loading capacity, release rate and release direction are the main factors to be considered. Mitragotri et al. have used mucoadhesive polymers, such as Eudragit copolymers or pectin, to prolong the gastric residence time of devices [122]. They also used impermeable ethyl cellulose sheets to create a unidirectional release pattern and seal the opposite side of the patches. Shen et al. also showed that incorporation of drug-loaded microspheres into the patches, instead of direct loading of the drugs, can provide significantly enhanced control over the release behavior of the drug [123]. Toorisaka et al. developed a lipophilic formulation (drug-in-oil formulation) to improve the compatibility of the system with intestinal cell lines and enhance the absorption of insulin [124]. However, their formulation lacked enough retention time at the target. They later solved this issue by designing a bilayer patch consisting of a drug-impermeable layer to guarantee a unidirectional drug release regime and a mucoadhesive layer to prolong the gastro-residence time [125]. The drug-in-oil formulation was also impregnated into the porous mucoadhesive layer. Although it has been more than two decades since the development of these oral devices, they have not attracted as much research attention as other designs, such as MPs. Oral patches are mainly applicable in the initial segments of the duodenum, since solid boluses of digested food can detach the patch from the lumen wall in later parts of the GI tract, significantly decreasing the transient time. Even in the duodenal part, the device needs strong bonding/binding with the mucus to avoid being washed away by gastric juice. Furthermore, the mucus turnover cycle limits the real application of these devices for sustained delivery.

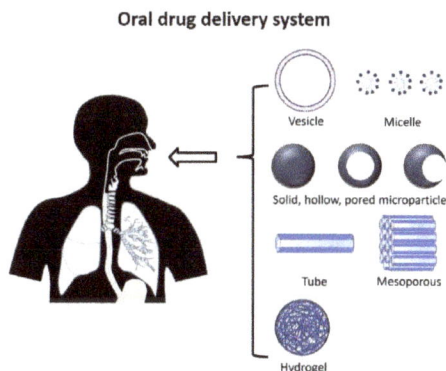

Figure 3. Schematic illustration of the principle oral vehicles designs.

Intestinal microneedles are the newest design developed as an oral delivery device. Microneedles were first developed for transdermal delivery (transdermal patches), and their application was subsequently extended to other administration routes including the vagina, anus and scalp [126]. In 2014, Ma et al. employed microneedles for oral vaccine delivery to the mouth cavity, which was the first time microneedles were used for oral administration [127]. They investigated the system for the delivery of two HIV antigens (DNA vaccines and virus-like particles), comparing the induced immune response with the response generated by the intramuscular injection of the drugs. They reported that only orally administered agents showed a stimulated antigen-specific IgA response in saliva. The limitation of this design is that it could only be employed for delivery to the oral cavity rather than the GI tract. Traverso et al. in 2015 tried to overcome the biological barriers of the GI tract using orally ingested microneedles and improved drug bioavailability [128]. They claimed that their device can be safely excreted from the GI tract. This design is especially promising for the delivery and successful absorption of large-size biomolecules. Their system demonstrated significant improvements in insulin bioavailability compared to the subcutaneous administration route. As a newly developed system, oral microneedles need to be studied considerably more, especially through clinical trials. There are currently not many investigations on this system.

Spherical carrier designs, including micelles, liposomes and MPs, are the most commonly studied oral delivery vehicles. Micelles are colloidal carriers (5–100 nm) developed to improve the aqueous solubility of hydrophobic pharmaceuticals and facilitate their oral delivery. They are made of amphiphilic molecules, enfolding the hydrophilic ingredients inside their hydrophobic core, with hydrophilic segments oriented on the outer walls. For instance, Dabholkar et al. developed a polymeric composition (polyethylene glycol-phosphatidylethanolamine conjugate) and increased the water solubility of paclitaxel (an anticancer drug) up to 5000 times [129]. Due to their very small average size (20–80 nm), micelles show spontaneous penetration through the interstitium of various tissues, improving their permeation efficiency and retention time. Yu et al. developed dual-responsive (pH and light) micelles to improve the passive tumor targeting of doxorubicin and address tumor resistance against the drug [130]. The micelles could trigger both deep tumor penetration and cytoplasm drug release, which in turn considerably improved treatment efficiency. Additionally, they found that the micelles could prolong the blood circulation cycle of the drug.

In addition to passive penetration, micelles can also help the active penetration of drugs. Suzuki et al. used cationic micelles (PLGA-*b*-bPEI-*b*-PLGA; Mw(PLGA): 36 kDa, Mw(bPEI): 25 kDa) for the encapsulation of doxorubicin and confirmed up to a 40-fold increase for in vitro drug penetration into multilayer cell cultures [131]. They claimed that iterative transcytosis via macropinocytosis and exocytosis are the main mechanisms for the penetration of the cationic micelles into cells. Although micelles are constantly proving their potential for improving therapeutic efficacy in bench-scale studies, they have hardly been commercialized. The clinical trials for these delivery devices are limited to a few cases for parenteral cancer therapies: Doxorubicin and its derivatives. A possible reason for this may be the safety of the materials, in terms of the physicochemical interactions between the carriers and mucus in real conditions [132]. On the other hand, liposomes are phospholipid vesicles (>200 nm) which can encapsulate both hydrophobic drugs in their hydrophobic compartment and hydrophilic drugs in their inner hydrophilic core. These carriers can be chemically modified through the immobilization of antibodies on their surfaces for improved target specificity. It should be noted that antibody-decorated liposomes may suffer from a short life cycle in blood circulation due to their accumulation in the liver, especially in the absence of sufficient target antigens [133].

MPs are the other common oral delivery architecture. The materials used in oral microencapsulation systems can be broadly classified into polymers and ceramics. Ceramics are usually safe materials for delivery applications due to their bio-inert nature. There are various ceramic materials used for delivery applications, including silica, alumina and calcium phosphate. For instance, cisplatin, methotrexate and hydrocortisone acetate have been successfully delivered using calcium phosphate carriers, and silica nanoparticles are dominantly used for chemotherapy [134]. Regarding the polymers employed for oral delivery, hydrogels are the most attractive structural materials, mainly due to their controllable chemical composition, tunable mechanical properties, water absorption, ability for internal material flow, and, above all, their capacity for stimuli-responsivity [135–139]. One of the primary features to control in hydrogels is porosity, since pore size can significantly affect the mechanical properties, uptake efficiency of extrinsic occupants, water/material flow through the polymer matrix and swelling ratio of the gel. The larger the mesh size, the easier the transport of materials through the structure is, and the higher the swelling ratio. For instance, Torres-Lugo et al. could regulate the release rate of salmon calcitonin from poly-(methacrylic acid) (PMAA) acid MPs by up to 50% by controlling the swelling ratio and mesh size of the hydrogel [140].

There are other strategies to increase the porosity of the hydrogel structure to a larger scale (micron scale) [141–143]. For instance, leaching out the template materials from the structure network is a common method for making porous materials [144]. Also, the number of crosslinking sites can be decreased through selective removal of one phase from the gel. Similarly, the aqueous/organic liquid absorbed inside the gel may be lyophilized, and the volume increase through the freezing process can create pores inside the polymer matrix [145–147]. The size of the pores created through this process depends upon the size of the ice crystals formed in the solid state and may range from nanoscale, by freezing the remaining molecular water contents inside the structure, to microscale macropores by concentrating the liquid at specific spots [148]. It should be noted that although removing crosslinking sites considerably affects the mechanical properties and the swelling ratio of the polymer, the porosity created by the frozen liquid does not change the hydrogel mesh size and consequently its properties/behavior [143].

In addition to achieving the optimum porosity for delivering small drug molecules, another major challenge is obtaining macropores for loading large biomolecules and cells into hydrogels [149,150]. If the pores are aligned in well-oriented geometries, the hydrogels can be used for directional cell growth and the creation of arrays [151]. Also, pore size and geometry define the nutrition transfer rate and cell migration pattern [152–154]. With all these aspects taken into account, developing methods for accurate control over the porosity of the hydrogel and size and morphology of pores in polymers has always been regarded as a main research topic for a range of various applications. France et al. recently reviewed the major methods developed for creating macroporous hydrogels [143].

3.2.2. Sustained Delivery

Mucoadhesive Carriers

The maintenance of a constant concentration of the drug in the blood stream is a favorable situation for the great majority of treatment cases [155]. On the other hand, severe fluctuations in drug concentration can cause serious problems such as toxicity or ineffective treatment, reflecting the possible negative effects of burst release and the need for sustained delivery approaches [156]. Although sustained delivery can be barely achieved in some administration routes, like transdermal approaches, oral delivery is known as one of the potential candidates for this goal. Materials developed for sustained delivery are either positively charged, in order to attach to mucus through electrostatic binding (e.g., cationic hydrogels made from chitosan), or thiol-functionalized hydrogels, to attach to the mucin glycoprotein through disulfide bonding [157,158]. Although these approaches have shown higher levels of attachment to mucus compared to the control samples, these particles can only slightly increase the delivery time (in the range of a few hours), due to the short turnover cycle of mucus in the intestine [67,80]. Furthermore, MPs employed for oral delivery are usually made of pH-responsive polymers, releasing their cargo only at pH values above the pKa of the hydrogel, which is usually >6 for oral applications. Although the pH in most of the spots of the intestine is usually above 6 inside the lumen (Table 1), which in turn may significantly affect the ionization/swelling/dissolution of the MPs embedded in the mucus [159]. As such, there are in vivo studies that report intact MPs made of Eudragit L100/S100 leave the body without releasing their cargo. Notably, mucoadhesion is sometimes described as increasing the friction between the drugs or the delivery device with the GI tract walls [160,161]. This approach may also extend the delivery time, however, only in the order of few hours, which may still not be long enough for many cases [160].

Recent Gastric-Resident Architectures

Apart from the two approaches already discussed, Traverso and Langer proposed the idea to make delivery architectures larger than specific gates along the GI tract, which would in turn substantially increase the retention time of the carrier inside the GI tract and achieve a prolonged delivery regime. They introduced two potential spots for this scheme: The cavity right before the anus and the pyloric sphincter (the end part of the stomach) [160]. Extended retention behind the pylorus (~1.5 cm in diameter) was first implemented by producing expandable structures, where the expanded size would reach up to more than 2 cm in diameter [162]. However, the original versions of this device were causing serious problems, such as damaging fractured pieces or causing lumen obstruction, due to the non-degradable/non-dissociable nature of the materials used, necessitating surgery for removal [163].

In a comprehensive study, Zhang et al. built an elastic foldable O-ring made of various hydrogels [164]. They controlled the degradation rate and the mechanical properties of the structure by modifying the chemical composition and adjusting the ratio of the hydrogels, finally encapsulating it in a degradable capsule. The capsule dissolved in the stomach, leaving the opened O-ring in the stomach cavity. The drug-loaded segments of the ring degraded gradually and released the drug over several days [164]. By controlling the degradation rate of the samples, one can tune the release pattern of the samples. There might be two concerns associated with this new design: (1) If the ring stays in the stomach, it may block/affect the passage of the digested food through the GI tract. (2) If such structures reside in the GI tract for a long time, they may cause or help spread adverse effects. As for the first issue, no negative effect on the stooling pattern was observed in the animals that were experimented on in the study. Regarding the second concern (the device staying in the stomach for a long time), this was compared with indigestible food masses trapped inside the GI tract, which do not cause serious problems unless they grow very large. Although there have been answers to these concerns, the issue of how to remove the device from the pylorus in the case of toxicity or adverse reaction by the drug being released needs to be addressed. Also, the possibility that the ring accidentally passes by the pylorus and enters the intestine lumen remains [160].

3.3. Solvent-Free Microencapsulation

3.3.1. Multiemulsion Systems

MPs have been the main candidates for oral microencapsulation systems [96,165–167]. MPs can be classified into two main categories: (1) Solid MPs, which are solid polymeric MPs with drug molecules dispersed in their matrix, and (2) hollow MPs, which are polymeric shells with hollow interior spaces that accommodate the delivered drug molecules [168,169]. In most of the protocols developed for the fabrication of MPs, drug molecules are directly involved in the fabrication process of the carrier, leading to direct contact between the drug molecules and the harsh organic solvents, raising concerns over drug denaturation, especially for biopharmaceuticals [170]. As such, the development of solvent-free microencapsulation technologies is of primary importance in drug delivery applications.

Multiemulsion systems (W/O/W and O/W/O) were originally developed to minimize contact between the drug and solvents, however, the traditional versions of these systems were difficult to make, control and stabilize, additionally being inefficient in terms of throughput [171–177]. While these systems minimized the contact between the drug molecules and organic solvents, they were not able to completely eliminate it due to partial contact at the O/W phase interface. For a wide range of applications of the emulsion technology, it is important to fully characterize and address the issues of drug denaturation along the fabrication process of the carriers, such as shear stresses caused by mechanical agitation or sonication and local temperature increases due to sonication.

Microfluidic devices were successful in addressing a number of fundamental problems associated with the emulsion technique: highly uniform MPs with very small polydispersity index and considerable stability in liquid state emulsion [178,179]. Also, making multi-channel devices with various geometries is more straightforward and significantly more efficient in terms of minimizing O/W phase contact. Furthermore, there is no need for sonication or mechanical agitation to stabilize the emulsion made by microfluidic devices, resolving various problems associated with traditional emulsification techniques [180]. Due to these advantages, this new approach has attracted considerable attention for cell encapsulation/delivery. For example, pancreatic islet cells, as one of the most significant treatments for type 1 diabetes, can be successfully encapsulated and delivered using microfluidic devices [181]. The size, morphology and loads of the carriers can be controlled by regulating the flow rate of the phases (flow rate of the continuous phase and infusion rate of the dispersed phase), the geometry of the channels and nozzles and the concentrations of the emulsifiers [182]. Although microfluidic devices contributed to advance the emulsion technology, there are still problems with these systems, including the clogging of channels, attachment of oil phase/polymer fragments to the channel walls and also O/W contact at the interface [183,184]. As the main challenge against the widespread application of these systems is their low throughput, this prevents them from being scaled up to commercial extents and entering the pharmaceutical industry for mass production.

3.3.2. Pored and Hollow Microencapsulation Systems

Another strategy for solvent-free microencapsulation is to employ separate processes for MP fabrication and drug encapsulation. That is, the drug can be loaded into the MPs in their favorable environmental conditions after the MP fabrication is complete, minimizing the possibility of drug denaturation due to the formation of MPs in the presence of the drug [185]. For this aim, several fabrication methods were previously developed, almost all of which were based on the same idea: coating solid spheres as templates with the desirable polymer material [186–188]. The templates (solid cores) were subsequently removed from the inside through either calcination or etching, leaving hollow polymeric spheres behind as the final product. These methods could hardly be successful due to the complications associated with diffusional material flow through the solid state polymeric shells. Since the drug can be loaded only by soaking the particles in a concentrated solution of drug, this method does not yield satisfactory levels of loading efficiency, limiting its general application [186,189,190].

Hyuk Im et al. proposed an idea to fabricate hollow polymeric microspheres with single surface pores using a solvent evaporation method [148]. Swollen solid polystyrene particles in organic solvents were plunged in liquid nitrogen, followed by slow evaporation of the solvent. The solvent would diffuse into the particles throughout the incubation step and would expand due to freezing, creating a hollow interior space inside the particles. The evaporation step would also generate surface pores, allowing for encapsulation of the cargo. It was demonstrated that the surface pores could then be closed through the thermal treatment of the particles above their glass transition temperature.

More recently, a simplified method of making single pored-MPs was reported by Kumar et al. based on ultrasonic O/W emulsification [185]. The obtained macropores could be used for the direct loading of the drugs in their favorable conditions by applying vacuum cycles, and then closed to protect the drug during transition inside the stomach (Figure 4) [185]. In this method, the surface pores were sealed through freeze drying, thus eliminating the concern over the possible thermal denaturation of the biopharmaceuticals during the heat treatment of the MPs. It has been proposed that the pore closure mechanism is due to polymer-polymer interactions, followed by the removal of water. This new design was later modified to make larger MPs with larger surface pores, suitable for the delivery of large biomolecules and cells.

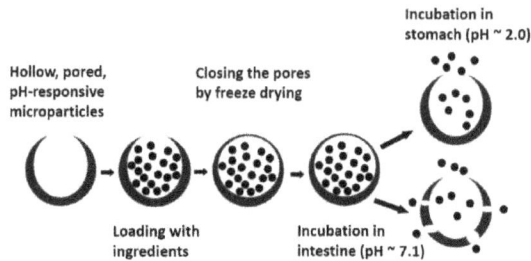

Figure 4. The schematic sequence of polymeric pored microencapsulation/release behavior.

3.4. Co-Delivery Systems

Although the co-delivery of drugs is not one of the core challenges associated with delivery systems, the administration of different drugs at the same time can play crucial roles in treatment efficacy in many cases. Delivering multiple drugs to target different sites at the same time can significantly reduce treatment time and the risk of failure [191]. Chemotherapy provides an example of co-delivery and its importance [191–194]. As another example, insulin delivery systems may cause a drop in the release of insulin due to frequent administration of the drug. The co-delivery of insulin and cyclic adenosine monophosphate (cAMP) was proposed to enhance the secretion of further insulin by activating the Ca^{2+} channels in beta cells of the pancreas [195]. It is thus of importance to develop co-delivery systems for the encapsulation and delivery of multi-target drugs.

A straightforward solution for co-delivery is the simultaneous administration of different drugs in separate delivery carriers [196]. Another solution is the concurrent encapsulation of different drugs in micelles, liposomes or MPs. The implementation of this idea dates back to 2011, when an anticancer drug (MEK inhibitor PD0325901) and a therapeutic gene (Mcl1-specific siRNA (siMcl1)) were concurrently loaded into N',N''-dioleylglutamide-containing cationic liposomes [197]. This new co-delivery strategy exhibited significantly enhanced anticancer activity both in vitro and in vivo. Cao et al. also simultaneously incorporated adenovirus encoding for murine interleukin-12 (Ad5) and paclitaxel (PTX) into anionic liposomes and performed in vitro/in vivo analyses, confirming that their co-delivery system (AL/Ad5/PTX) is an effective platform for treating melanoma [198].

Oral carriers based on SiO_2 are also one of the most commonly used delivery vehicles, especially for co-delivery purposes [192,194,199]. Mesoporous silica nanoparticles (MSNs) and halloysite nanotubes (HNTs) are the principal examples of SiO_2 structures investigated for such applications (Figure 5). HNTs are naturally forming structures, available in sufficient amounts in North America, China and New Zealand, making them a valuable candidate material in the pharmaceutical industry by meeting the requirements for scaling up and commercialization. As opposed to HNTs, MSNs require a time-consuming and expensive manufacturing process, although they have attracted significantly more attention as drug carriers compared to HNTs, mostly due to their larger interior space and higher loading capacity [200]. MSNs are nanoparticles made of silica with high chemical stability, a porous structure and the ability for surface modification/decoration [199,201]. Contrary to naturally forming HNTs, MSNs are artificially fabricated using a sol-gel method through the use of surfactants (Figure 5A). The idea of the addition of surfactants was first proposed by Kresge et al. and then modified by Inagaki et al. [202,203]. Concisely, hexagonal arrays of surfactant molecules are formed as micelles in a continuous aqueous phase, with hydrophobic segments in the center and hydrophilic fragments on the surface. Aluminosilicate is added afterwards to be adsorbed on the surface of the micelles and form the inorganic walls of the nanoparticles, and the organic core is finally extracted through thermal treatment. Co-delivery using MSNs has been investigated extensively by many researchers for the purpose of simultaneously delivering multiple drugs, and at the same time to minimize the multidrug resistance risk and lower the failure risk of the treatment. As an example,

Chen et al. designed a co-delivery system for the simultaneous administration of siRNA and DOX, considerably modified the toxic effects of DOX (by ~132 times) [191].

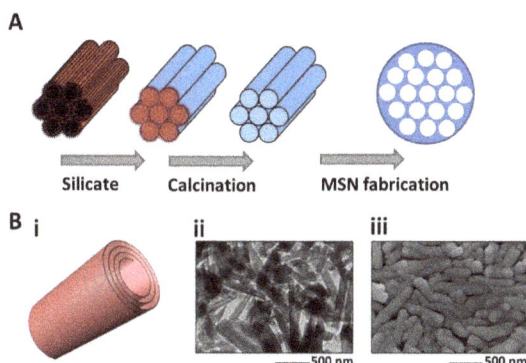

Figure 5. Oral delivery devices based on silica. (**A**) Fabrication of mesoporous silica nanoparticles (MSNs) (adapted with permission from [202]), (**B**) halloysite nanotube (HNT) microstructure ((**i**) schematic representation of a HNT, (**ii**) transmission electron microscopy (TEM) image and (**iii**) scanning electron microscopy (SEM) image).

HNTs are known as nonhazardous, non-degradable and biocompatible materials with no negative effects on different human cell lines, such as dermal fibroblasts and epithelial cells, making them favorable carriers for oral delivery, dermal treatments and cosmetics [191,204–206]. Notably, HNTs may not be suitable candidates for transdermal injections due to their size (0.4 to 1.5 µm long, which is slightly larger than the optimal length for biosafety, 1 µm). HNTs are spiral sheets of aluminosilicate kaolin, rolled 15–20 times over a single axis (15 nm inner diameter and 40–60 nm outer diameter) and have attracted attention as drug carriers, mainly for oral administration, however (Figure 5B) [200]. The inner and exterior sides of the lumen walls are made of alumina and silica, respectively. This double-layer structure enables the simultaneous loading of oppositely charged biomolecules. These are molecules with a negative zeta potential inside the lumen and those with a positive charge on the outer surface [207]. The outer negative charge can disperse particles in organic/aqueous environments and allow for further functionalization. The inner diameter of the lumen is also large enough to encapsulate both small drug molecules and larger macromolecules such as proteins [208].

To apply HNTs and MSNs to drug delivery systems, it is required to seal their pore gates using a third material, known as a cap or gate keeper. Stimuli-responsive hydrogels have been mainly investigated as sealing materials to confer target-specificity to MSNs (Figure 6) [209–213]. Different types of stimuli (i.e., magnetic, light, thermal and pH) have been employed to date for MSN systems. Stimuli-responsivity in this microencapsulation scheme can be implemented through two main approaches: (1) Cleavable covalent bonding/crosslinking between the carrier and the drug in response to stimuli, such as the cleavable bonding at pH values below the plasma pH, and (2) functionalization of the surface or coating of the channels, which can switch conformation based on the surrounding properties and stimuli-responsive caps [192]. As an example, doxorubicin was conjugated to the interior walls of the channels in MSNs through pH-sensitive hydrazine bonds, which can prevent any untimely release of the drug. That is, upon being taken up into the cell through endocytosis, the acidic conditions of the endosomal/lysosomal environment can trigger the release of drugs due to the protonation of the bonding [214]. Another example includes light-responsive MSNs, made by Mekaru et al. using photoactivated azobenzene, which triggered release upon excitation by an external source of light [215]. These light-responsive MSNs are among the most common carriers used for cancer therapies. Magnetic-responsive MSNs later replaced them for the same purpose, since light could not penetrate deep enough into the tissue, while magnetic fields could address any kind of tissue at any desirable depth. Iron oxide was one of the best options for inducing magnetic-responsivity in MSNs [216,217]. Although these systems were not specifically targeted for oral drug delivery, they demonstrate the versatile stimuli-responsive properties that can embedded into these materials, which would be advantageous to oral drug carriers.

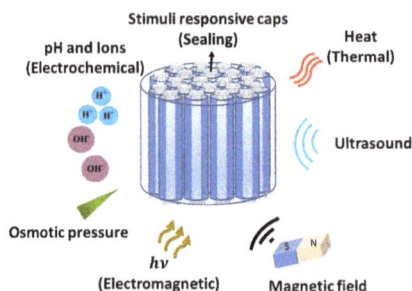

Figure 6. Different types of sealing strategies used for stimuli-responsive MSNs.

It should be noted that HNTs have attracted less attention than MSNs, although they are more readily available. HNTs would mainly be useful for applications requiring a small drug dosage due to their limited loading efficiency. The inner lumen usually shows about an 8–12% loading capacity, which can be increased to 30–40% by etching the HNTs in mild acidic environments [200]. Furthermore, MSNs structures can be modified to provide better control over the release pattern due to their tunable inner geometry.

3.5. Scaling up and Throughput

The technical challenges discussed so far mainly concern health-related issues. The throughput of the system, however, is the major challenge associated with the commercial implementation of the discussed technology. In general, almost all microencapsulation systems and protocols developed for the fabrication of microcarriers at a lab-scale suffer from low throughput and difficulty in scaling up.

Among the microencapsulation approaches under current investigation, spray drying generally shows the highest throughput, fastest production rate and greatest ease of operation [218]. However, the direct contact between the drug molecules (genes, proteins, vaccine, biopharmaceuticals) and the organic solvents can readily denature/deactivate the drugs. In addition, spraying/dispersing the drug molecules in hot chambers may easily lead to thermal denaturation, and not all kinds of drugs may have satisfactory solubility in volatile organic solvents [219]. As a result, there has been a constant need to develop a universal, scalable drug delivery system without modification of the drugs.

Despite the many advantages emulsion microencapsulation techniques provide, their general application to drug delivery systems faces major challenges to be overcome. These challenges relate to problems including yield, polydispersity and sonochemistry. As discussed previously, microfluidic systems could properly address most of the problems associated with traditional emulsification methods at the expense of yielding a lower output [220]. Regardless of its geometry, a single microfluidic channel can yield a throughput of emulsion droplets in the range of 0.1–10 mL/h with a very narrow size distribution (<5%), while the pharmaceutical industry is looking for much higher values (above 1 L/h) [221]. Traditional emulsification techniques may show a much higher yield (100–20,000 L/h) at the expense of the quality of the droplets and polydispersity range [222]. Ofner et al. proposed a multi-parallel channel microfluidic device to improve the throughput of microfluidic systems, reporting promising values that would satisfy commercial needs [219]. Tendulkar et al. implemented the same idea of the parallelization of several channels in a more complex design of microfluidic devices for the encapsulation of islet cells to treat type 1 diabetes [223]. They included an air jet supply to aid the detachment of the droplets from the T-junction nozzles and increase the throughput of the system, reporting a 8–64-fold increase in the production rate. Although multi-channel designs look to be a promising solution for increasing the throughput of MP fabrication, when comparing with the commercially required flow rates with bench-scale infusion rates from a single channel, about 1000 channels may be needed for attaining a satisfactory level of throughput. Nisisako et al. [224] and Conchouso et al. [225], in separate studies, proposed 3D circular arrays of channels to both minimize the flow rate distribution among the channels and increase the throughput of the system. These complex microfluidic devices can be made of metal, polymer or glass, and 3D printing, acid etching and computer numerical control CNC engraving are common practices for forming the channels [219,221,223–225]. It should be noted that systems' output in microfluidic devices is usually reported as the volume produced per minute [220,226], which may not be the ideal way for reporting the throughput of the system. In this regard, it would be necessary to better evaluate the performance of the system in terms of what fraction of the volume reported really represents the MPs produced, and what fraction is due to the contribution of other components of the system (e.g., continuous phase or random polymer fragments).

4. Conclusions

Although oral delivery is considered to be the most promising administration route due to its specific advantages, it faces substantial challenges that need to be addressed before oral systems can be commercially available for the delivery of biopharmaceuticals. Fabrication protocols of the carriers should adequately avoid any destructive effect (such as contact with organic solvents, shear stresses and local temperature increases) on the drug molecules, especially for biopharmaceutical encapsulation/delivery. Concurrently, the delivery material, design, size and polydispersity must be accurately controlled, due to their significant influence on treatment efficacy. Oral carriers deal with various biological barriers (the lumen, mucus and tissue of the GI tract) to successfully deliver drugs. Sustained delivery, solvent-free microencapsulation and co-delivery have also been major focuses in oral delivery studies. Nonetheless, the most significant issue against the commercialization of the oral systems is their low throughput. This review introduced the most promising solutions recently proposed for each barrier, which point to a positive progress in the field of oral drug delivery. In addition, the recent solutions proposed for scaling up the microencapsulation techniques, the parallelization of the microfluidic channels, are both significantly increasing the systems yield and addressing the other challenges as well. Due to the broadness, as well as the variety of the challenges against the oral drug delivery systems, we expect several different focuses for future research in this field. Considering the recent achievements in the case of technological challenges, we believe that future research in this field will mainly target biological barriers, especially the barriers associated with tissues. The main advantages of oral delivery systems, include sustained delivery, interaction with mucus and the capability for solid formulations that preserve pharmaceuticals, still making this the most attractive administration route for pharmaceuticals.

Author Contributions: Conceptualization: H.-J.C., B.H. Wring: B.H., X.L., H.-J.C. Editing: B.H., H.-J.C.

Funding: This research was funded by Alberta Innovates Technology Futures [AIF200900279] and University of Alberta. And the APC was funded by [AIF200900279].

Acknowledgments: The authors also thank Ilaria Rubino for helpful comments and suggestions on a previous draft.

Conflicts of Interest: The authors declare no conflict of interest.

References

1. Siafaka, P.I.; Titopoulou, A.; Koukaras, E.N.; Kostoglou, M.; Koutris, E.; Karavas, E.; Bikiaris, D.N. Chitosan derivatives as effective nanocarriers for ocular release of timolol drug. *Int. J. Pharm.* **2015**, *495*, 249–264. [CrossRef] [PubMed]

2. Ferraiolo, B.L.; Mohler, M.A.; Gloff, C.A. Volume 1: Protein pharmacokinetics and metabolism. In *Pharmaeutical Biotechnology.*; Borchard, R.T., Ed.; Springer Science+Business Media, LLC: New York, NY, USA, 1992; pp. 78–150.

3. Römgens, A.M.; Rem-bronneberg, D.; Kassies, R.; Hijlkema, M.; Bader, D.L.; Oomens, C.W.J.; Bruggen, M.P.B. Penetration and delivery characteristics of repetitive microjet injection into the skin. *J. Control. Release* **2016**, *234*, 98–103. [CrossRef] [PubMed]

4. Rodger, M.A.; King, L. Drawing up and administering intramuscular injections: A review of the literature. *J. Adv. Nurs.* **2000**, *31*, 574–582. [CrossRef] [PubMed]

5. Mishra, P.; Stringer, M.D. Sciatic nerve injury from intramuscular injection: A persistent and global problem. *Int. J. Clin. Pract.* **2010**, *64*, 1573–1579. [CrossRef] [PubMed]

6. Nicoll, L.H.; Hesby, A. Intramuscular injection: An integration research review and guideline for evidence-based practice. *Appl. Nurs. Res.* **2002**, *16*, 149–162. [CrossRef]

7. Liang, F.; Loré, K. Local innate immune responses in the vaccine adjuvant-injected muscle. *Clin. Transl. Immunol.* **2016**, *5*, 74–81. [CrossRef] [PubMed]

8. Herzog, R.W.; Hagstrom, J.N.; Kung, S.H.; Tai, S.J.; Wilson, J.M.; Fisher, K.J.; High, K.A. Stable gene transfer and expression of human blood coagulation factor IX after intramuscular injection of recombinant adeno-associated virus. *Proc. Natl. Acad. Sci. USA* **1997**, *94*, 5804–5809. [CrossRef] [PubMed]

9. Nanaki, S.; Siafaka, P.I.; Zachariadou, D.; Nerantzaki, M.; Giliopoulos, D.J.; Triantafyllidis, K.S.; Kostoglou, M.; Nikolakaki, E.; Bikiaris, D.N. PLGA/SBA-15 mesoporous silica composite microparticles loaded with paclitaxel for local chemotherapy. *Eur. J. Pharm. Sci.* **2017**, *99*, 32–44. [CrossRef]

10. Nanaki, S.; Tseklima, M.; Terzopoulou, Z.; Nerantzaki, M.; Giliopoulos, D.J.; Triantafyllidis, K.; Kostoglou, M.; Bikiaris, D.N. Use of mesoporous cellular foam (MCF) in preparation of polymeric microspheres for long acting injectable release formulations of paliperidone antipsychotic drug. *Eur. J. Pharm. Biopharm.* **2017**, *117*, 77–90. [CrossRef]

11. Fletcher, N.A.; Krebs, M.D. Sustained delivery of anti-VEGF from injectable hydrogel systems provides a prolonged decrease of endothelial cell proliferation and angiogenesis in vitro. *RSC Adv.* **2018**, *8*, 8999–9005. [CrossRef]

12. Moeller, E.H.; Jorgensen, L. Alternative routes of administration for systemic delivery of protein pharmaceuticals. *Drug Discov. Today Technol.* **2008**, *5*, 89–94. [CrossRef] [PubMed]

13. Kale, T.R. Needle free injection technology—An overview. *Inov. Pharm.* **2014**, *5*, 1–8. [CrossRef]

14. Brown, M.B.; Martin, G.P.; Jones, S.A.; Akomeah, F.K.; Brown, M.B.; Martin, G.P.; Jones, S.A.; Akomeah, F.K.; Brown, M.B.; Martin, G.P.; et al. Dermal and transdermal drug delivery systems: Current and future prospects. *Drug Deliv.* **2006**, *13*, 175–187. [CrossRef] [PubMed]

15. Ranade, V.V. Drug delivery systems. 6. Transdermal drug delivery. *J. Clin. Pharmacol.* **1991**, *31*, 401–418. [CrossRef]

16. Nerantzaki, M.; Kehagias, N.; Francone, A.; Ferna, A.; Torres, C.M.S.; Papi, R.; Choli-papadopoulou, T.; Bikiaris, D.N. Design of a multifunctional nanoengineered PLLA surface by maximizing the synergies between biochemical and surface design bactericidal effects. *ACS Omega* **2018**, *3*, 1509–1521. [CrossRef]

17. Kim, H.; Jang, H.; Kim, B.; Kim, M.K.; Wie, D.S.; Lee, H.S.; Kim, D.R.; Lee, C.H.F.; Jaganathan, K.S. Nasal vaccine delivery (Chapter fifteen). *Appl. Sci. Eng.* **2018**, *1*, 1–9.

18. Lee, S.; Mcauliffe, D.J.; Flotte, T.J.; Kollias, N.; Doukas, A.G. Photomechanical transcutaneous delivery of macromolecules. *J. Invest. Dermatol.* **1998**, *111*, 925–929. [CrossRef] [PubMed]

19. Prausnitz, M.R.; Langer, R. Transdermal drug delivery. *Nat. Biotechnol.* **2008**, *26*, 1261–1268. [CrossRef]

20. Chen, Y.; Quan, P.; Liu, X.; Wang, M.; Fang, L. Novel chemical permeation enhancers for transdermal drug delivery. *Asian J. Pharm. Sci.* **2014**, *9*, 51–64. [CrossRef]

21. Karande, P.; Jain, A.; Mitragotri, S. Insights into synergistic interactions in binary mixtures of chemical permeation enhancers for transdermal drug delivery. *J. Control. Release* **2006**, *115*, 85–93. [CrossRef] [PubMed]

22. Dougherty, L.; Lamb, J.; Elliott, T. Section 2. Practice. In *Intravenous Therapy in Nursing Practice*; Finlay, T., Lamb, J., Dougherty, L., Quinn, C., Eds.; Blackwell Publishing: Oxford, UK, 2008; pp. 143–225.

23. Maxwell, M.J.; Wilson, M.J.A. Complications of blood transfusion. *Contin. Educ. Anaesth. Crit. Care Pain* **2006**, *6*, 225–229. [CrossRef]

24. Korttila, K.; Aromaa, U. Venous complications after intravenous injection of diazepam, flunitrazepam, thiopentone and etomidate. *Acta Anaesthesiol. Scand.* **1980**, *24*, 227–230. [CrossRef] [PubMed]

25. Awate, S.; Babiuk, L.A.; Mutwiri, G. Mechanisms of action of adjuvants. *Front. Immunol.* **2013**, *4*, 1–10. [CrossRef]

26. Harshad, P.; Anand, B.; Dushyant, S. Recent techniques in nasal drug delivery: A review. *Int. J. Drug Dev. Res.* **2010**, *2*, 565–572.

27. Nanaki, S.; Tseklima, M.; Christodoulou, E.; Triantafyllidis, K.; Kostoglou, M.; Bikiaris, D.N. Thiolated chitosan masked polymeric microspheres with incorporated mesocellular silica foam (MCF) for intranasal delivery of paliperidone. *Polymers* **2017**, *9*, 617. [CrossRef]

28. Grassin-delyle, S.; Buenestado, A.; Naline, E.; Faisy, C.; Blouquit-laye, S.; Couderc, L.; Le, M.; Fischler, M.; Devillier, P. Intranasal drug delivery: An efficient and non-invasive route for systemic administration Focus on opioids. *Pharmacol. Ther.* **2012**, *134*, 366–379. [CrossRef]

29. Bhise, S.B.; Yadav, A.V.; Avachat, A.M.; Malayandi, R. Bioavailability of intranasal drug delivery system. *Asian J. Pharm.* **2008**, *2*, 201–215. [CrossRef]

30. Ramvikas, M.; Arumugam, M.; Chakrabarti, S.R.; Jaganathan, K.S. Nasal vaccine delivery (Chapter fifteen). In *Micro- and Nanotechnology in Vaccine Development*; Elsevier Inc.: Amsterdam, The Netherlands, 2017; pp. 279–301.

31. Bakri, W.; Donovan, M.D.; Cueto, M.; Wu, Y.; Orekie, C.; Yang, Z. Overview of intranasally delivered peptides: Key considerations for pharmaceutical development. *Expert Opin. Drug Deliv.* **2018**, *15*, 991–1005. [CrossRef]

32. Choi, H.J.; Kim, M.C.; Kang, S.M.; Montemagno, C.D. The osmotic stress response of split influenza vaccine particles in an acidic environment. *Arch. Pahrmacal Res.* **2014**, *37*, 1607–1616. [CrossRef]

33. Banerjee, A.; Qi, J.; Gogoi, R.; Wong, J.; Mitragotri, S. Role of nanoparticle size, shape and surface chemistry in oral drug delivery. *J. Control. Release* **2016**, *238*, 176–185. [CrossRef]

34. Araújo, F.; Pedro, J.; Granja, P.L.; Santos, H.A.; Sarmento, B. Functionalized materials for multistage platforms in the oral delivery of biopharmaceuticals. *Prog. Mater. Sceince* **2017**, *89*, 306–344. [CrossRef]

35. Hu, Q.; Luo, Y. Recent advances of polysaccharide-based nanoparticles for oral insulin delivery. *Int. J. Biol. Macromol.* **2018**, *120*, 775–782. [CrossRef]

36. Choi, H.-J.; Ebersbacher, C.F.; Kim, M.C.; Kang, S.M.; Montemagno, C.D. A mechanistic study on the destabilization of whole inactivated influenza virus vaccine in gastric environment. *PLoS ONE* **2013**, *8*, 1–14. [CrossRef]

37. Schenk, M.; Mueller, C. The mucosal immune system at the gastrointestinal barrier. *Best Pract. Res.* **2008**, *22*, 391–409. [CrossRef]

38. Ensign, L.M.; Cone, R.; Hanes, J. Oral drug delivery with polymeric nanoparticles: The gastrointestinal mucus barriers. *Adv. Drug Deliv. Rev.* **2012**, *64*, 557–570. [CrossRef]

39. Leal, J.; Smyth, H.D.C.; Ghosh, D. Physicochemical properties of mucus and their impact on transmucosal drug delivery. *Int. J. Pharm.* **2017**, *532*, 555–572. [CrossRef]

40. Fievez, V.; Garinot, M.; Schneider, Y.; Préat, V. Nanoparticles as potential oral delivery systems of proteins and vaccines: A mechanistic approach. *J. Control. Release* **2006**, *116*, 1–27.

41. Brayden, D.J.; Jepson, M.A.; Baird, A.W. Intestinal Peyer's patch M cells and oral vaccine targeting. *Drug Discov. Today* **2005**, *10*, 1145–1157. [CrossRef]

42. Kwon, K.; Daniell, H. Oral delivery of protein drugs bioencapsulated in plant cells. *Mol. Ther.* **2016**, *24*, 1342–1350. [CrossRef]

43. Ma, S.; Wang, L.; Huang, X.; Wang, X.; Chen, S.; Shi, W.; Qiao, X.; Jiang, Y. Oral recombinant Lactobacillus vaccine targeting the intestinal microfold cells and dendritic cells for delivering the core neutralizing epitope of porcine epidemic diarrhea virus. *Microb. Cell Fact.* **2018**, *17*, 1–12. [CrossRef]

44. Maharjan, S.; Singh, B.; Jiang, T.; Yoon, S.; Li, H.; Kim, G.; Jeong, M.; Ji, S.; Park, O.; Hyun, S.; et al. Systemic administration of RANKL overcomes the bottleneck of oral vaccine delivery through microfold cells in ileum. *Biomaterials* **2016**, *84*, 286–300. [CrossRef] [PubMed]

45. Varum, F.J.O.; Mcconnell, E.L.; Sousa, J.J.S.; Veiga, F.; Basit, A.W. Mucoadhesion and the gastrointestinal tract. *Crit. Rev. Ther. Drug Carrier Syst.* **2008**, *25*, 207–258. [CrossRef]
46. Dawson, M.; Krauland, E.; Wirtz, D.; Hanes, J. Transport of polymeric nanoparticle gene carriers in gastric mucus. *Biotechnol. Prog.* **2004**, *20*, 851–857. [CrossRef] [PubMed]
47. Hounnou, G.; Destrieux, C.; Desme, J.; Bertrand, P.; Velut, S. Anatomical study of the length of the human intestine. *Surg. Radiol. Anat.* **2002**, *24*, 290–294.
48. Helander, H.F.; Fändriks, L. Surface area of the digestive tract – revisited. *Scand. J. Gastroenterol.* **2014**, *49*, 681–689. [CrossRef]
49. Azizi, A.; Kumar, A.; Diaz-mitoma, F.; Mestecky, J. Enhancing oral vaccine potency by targeting intestinal M cells. *PLoS Pathog.* **2010**, *6*, 1001147–1001154. [CrossRef] [PubMed]
50. Mudie, D.M.; Amidon, G.L.; Amidon, G.E. Physiological parameters for oral delivery and in vitro testing. *Mol. Pharm.* **2010**, *7*, 1388–1405. [CrossRef]
51. Vllasaliu, D.; Thanou, M.; Stolnik, S.; Fowler, R. Recent advances in oral delivery of biologics: Nanomedicine and physical modes of delivery. *Expert Opin. Drug Deliv.* **2018**, *15*, 759–770. [CrossRef] [PubMed]
52. Tao, S.L.; Desai, T.A. Micromachined devices: The impact of controlled geometry from cell-targeting to bioavailability. *J. Control. Release* **2005**, *109*, 127–138. [CrossRef]
53. Rzhevskiy, A.S.; Raghu, T.; Singh, R.; Donnelly, R.F.; Anissimov, Y.G. Microneedles as the technique of drug delivery enhancement in diverse organs and tissues. *J. Control. Release* **2018**, *270*, 184–202. [CrossRef]
54. Dimmitt, R.A.; Sellers, Z.M.; Sibley, E. XIV-Gastrointestinal system-70 Gastrointestinal tract development. In *Avery's Diseases of the Newborn*; Elsevier Inc.: Amsterdam, The Netherlands, 2012; pp. 1032–1038.
55. Treuting, P.M.; Dintzis, S.M.; Montine, K. Upper gastrointestinal tract. In *Comparative Anatomy and Histology (Second Edition), A Mouse, Rat, and Human Atlas*; Academic Press, Elsevier: London, UK, 2018; pp. 190–211.
56. Cheng, H. Origin, differentiation and renewal of the four main epithelial cell types in the mouse small intestine. *Am. J. Anat.* **1974**, *141*, 481–502. [CrossRef]
57. Lennernas, H. Human intestinal permeability. *Int. J. Pharm. Sci.* **1998**, *87*, 403–410. [CrossRef] [PubMed]
58. Rubin, D.C.; Langer, J.C. Anatomy and development-small intestine: Anatomy and structural anomalies. In *Yamada's Atlas of Gastroenterology*; Podolsky, D.K., Camilleri, M., Shanahan, F., Fitz, J.G., Wang, T.C., Kalloo, A.N., Eds.; Wiley Blackwell: Oxford, UK, 2016; pp. 19–24.
59. Dressman, J.B.; Berardi, R.R.; Dermentzoglou, L.C.; Russell, T.L.; Schmaltz, S.P.; Barett, J.L.; Jarvenpaa, K.M. Upper gastrointestinal (GI) pH in young, healthy men and women. *Pharm. Res.* **1990**, *7*, 756–761. [CrossRef] [PubMed]
60. Rouge, N.; Buri, P.; Doelker, E. Drug absorption sites in the gastrointestinal tract and dosage forms for site-specific delivery. *Int. J. Pharm.* **1996**, *136*, 117–139. [CrossRef]
61. Moroz, E.; Matoori, S.; Leroux, J. Oral delivery of macromolecular drugs: Where we are after almost 100 years of attempts. *Adv. Drug Deliv. Rev.* **2016**, *101*, 108–121. [CrossRef] [PubMed]
62. Bar-zeev, M.; Assaraf, Y.G.; Livney, Y.D. β-casein nanovehicles for oral delivery of chemotherapeutic drug combinations overcoming P-glycoprotein-mediated multidrug resistance in human gastric cancer cells. *Oncotarget* **2016**, *7*, 23322–23335. [CrossRef] [PubMed]
63. Huang, J.; Shu, Q.; Wang, L.; Wu, H.; Wang, A.Y.; Mao, H. Layer-by-layer assembled milk protein coated magnetic nanoparticle enabled oral drug delivery with high stability in stomach and enzyme-responsive release in small intestine. *Biomaterials* **2015**, *39*, 105–113. [CrossRef] [PubMed]
64. Ruiz, G.A.; Opazo-Navarrete, M.; Meurs, M.; Minor, M.; Sala, G.; Van Boekel, M.; Stieger, M.; Janssen, A.E.M. Denaturation and in vitro gastric digestion of heat-treated quinoa protein isolates obtained at various extraction pH. *Food Biophys.* **2016**, *11*, 184–197. [CrossRef]
65. Yamagata, T.; Morishita, M.; Kavimandan, N.J.; Nakamura, K. Characterization of insulin protection properties of complexation hydrogels in gastric and intestinal enzyme fluids. *J. Control. Release* **2006**, *112*, 343–349. [CrossRef]
66. Cerchiara, T.; Abruzzo, A.; Parolin, C.; Vitali, B.; Bigucci, F.; Gallucci, M.C.; Nicoletta, F.P.; Luppi, B. Microparticles based on chitosan/carboxymethylcellulose polyelectrolyte complexes for colon delivery of vancomycin. *Carbohydr. Polym.* **2016**, *143*, 124–130. [CrossRef]
67. O'Neill, M.J.; Bourre, L.; Melgar, S.; O'Driscoll, C.M. Intestinal delivery of non-viral gene therapeutics: Physiological barriers and preclinical models. *Drug Discov. Today* **2011**, *16*, 203–218. [CrossRef]

68. Rawlings, N.D.; Barrett, A.J. Families of serine peptidases. In *Methods in Enzymology*; Academic Press, Elsevier, Inc.: Amsterdam, The Netherlands, 1994; Volume 244, pp. 19–61.
69. Davies, M.; Pieber, T.R.; Hartoft-Nielsen, M.L.; Hansen, O.K.H.; Jabbour, S.; Rosenstock, J. Effect of oral semaglutide compared with placebo and subcutaneous semaglutide on glycemic control in patients with type 2 diabetes a randomized clinical trial. *J. Am. Med. Assoc.* **2017**, *318*, 1460–1470. [CrossRef]
70. Layer, P.; Go, V.L.W.; Dimagno, E.P. Fate of pancreatic enzymes aboral transit in humans during small intestinal aboral transit in humans. *Am. J. Physiol.* **1986**, *251*, 475–480.
71. Fallingborg, J.; Christensen, L.A.; Ingeman-Nielsen, M.; Jacobsen, B.A.; Abildgaard, K.; Rasmussen, H.H. pH-profile and regional transit fimes of the normal gut measured by a radiotelemetry device. *Aliment. Pharmacol. Ther.* **1989**, *3*, 605–613. [CrossRef] [PubMed]
72. Lozoya-agullo, I.; Araújo, F.; González-álvarez, I.; Merino-sanjuán, M.; González-álvarez, M.; Bermejo, M.; Sarmento, B. PLGA nanoparticles are effective to control the colonic release and absorption on ibuprofen. *Eur. J. Pharm. Sci.* **2018**, *115*, 119–125. [CrossRef]
73. Valon, L.; Levayer, R. Dying under pressure: Cellular characterisation and in vivo functions of cell death induced by compaction. *Biol. Cell* **2019**, *111*, 1–16. [CrossRef] [PubMed]
74. De France, K.J.; Chan, K.J.W.; Cranston, E.D.; Hoare, T. Enhanced mechanical properties in cellulose nanocrystal−poly(oligoethylene glycol methacrylate) injectable nanocomposite hydrogels through control of physical and chemical cross-linking. *Biomacromolecules* **2016**, *17*, 649–660. [CrossRef] [PubMed]
75. Yang, J.; Zhao, J.; Xu, F.; Sun, R. Revealing strong nanocomposite hydrogels reinforced by cellulose nanocrystals: Insight into morphologies and interactions. *Appl. Mater. Interfaces* **2013**, *5*, 12960–12967. [CrossRef] [PubMed]
76. Mert, O.; Lai, S.K.; Ensign, L.; Yang, M.; Wang, Y.; Wood, J.; Hanes, J. A poly(ethylene glycol)-based surfactant for formulation of drug-loaded mucus penetrating particles. *J. Control. Release* **2012**, *157*, 455–460. [CrossRef] [PubMed]
77. Liu, Y.; Yang, T.; Wei, S.; Zhou, C.; Lan, Y.; Cao, A. Mucus adhesion- and penetration-enhanced liposomes for paclitaxel oral delivery. *Int. J. Pharm.* **2018**, *537*, 245–256. [CrossRef]
78. Shan, W.; Zhu, X.; Liu, M.; Li, L.; Zhong, J.; Sun, W.; Zhang, Z.; Huang, Y. Overcoming the diffusion barrier of mucus and absorption barrier of epithelium by self-assembled nanoparticles for oral delivery of insulin. *ACS Nano* **2015**, *9*, 2345–2356. [CrossRef] [PubMed]
79. Liu, M.; Zhang, J.; Zhu, X.; Shan, W.; Li, L.; Zhong, J.; Zhang, Z.; Huang, Y. Efficient mucus permeation and tight junction opening by dissociable "mucus-inert" agent coated trimethyl chitosan nanoparticles for oral insulin delivery. *J. Control. Release* **2016**, *222*, 67–77. [CrossRef] [PubMed]
80. Leal, J.; Dong, T.; Taylor, A.; Siegrist, E.; Gao, F.; Smyth, H.D.C. Mucus-penetrating phage-displayed peptides for improved transport across a mucus-like model. *Int. J. Pharm.* **2018**, *553*, 57–64. [CrossRef] [PubMed]
81. Zhang, X.; Cheng, H.; Dong, W.; Zhang, M.; Liu, Q.; Wang, X.; Guan, J. Design and intestinal mucus penetration mechanism of core-shell nanocomplex. *J. Control. Release* **2018**, *272*, 29–38. [CrossRef] [PubMed]
82. Navarro, L.A.; French, D.L.; Zauscher, S. Advances in mucin mimic synthesis and applications in surface science. *Curr. Opin. Colloid Interface Sci.* **2018**, *38*, 122–134. [CrossRef]
83. Kufe, D.W. Mucins in cancer: Function, prognosis and therapy Donald. *Nat. Rev. Cancer* **2009**, *9*, 874–885. [CrossRef] [PubMed]
84. Atuma, C.; Strugala, V.; Allen, A.; Holm, L. The adherent gastrointestinal mucus gel layer: Thickness and physical state in vivo. *Am. J. Physiol. Liver Physiol.* **2001**, *280*, 922–929. [CrossRef]
85. Chassaing, B.; Gewirtz, A.T. Identification of inner mucus-associated bacteria by laser capture microdissection. *Cell. Mol. Gastroenterol. Hepatol.* **2019**, *7*, 157–160. [CrossRef] [PubMed]
86. Bansil, R.; Turner, B.S. The biology of mucus: Composition, synthesis and organization. *Adv. Drug Deliv. Rev.* **2018**, *124*, 3–15. [CrossRef]
87. Boegh, M.; García-díaz, M.; Müllertz, A.; Nielsen, H.M. Steric and interactive barrier properties of intestinal mucus elucidated by particle diffusion and peptide permeation. *Eur. J. Pharm. Biopharm.* **2015**, *95*, 136–143. [CrossRef] [PubMed]
88. Hansson, G.C.; Johansson, M.E. V The inner of the two Muc2 mucin-dependent mucus layers in colon is devoid of bacteria. *Gut Microbes* **2010**, *1*, 51–54. [CrossRef] [PubMed]

89. Johansson, M.E.V.; Larsson, J.M.H.; Hansson, G.C. The two mucus layers of colon are organized by the MUC2 mucin, whereas the outer layer is a legislator of host–microbial interactions. *PNAS* **2011**, *108*, 4659–4665. [CrossRef]

90. Bajka, B.H.; Rigby, N.M.; Cross, K.L.; Macierzanka, A.; Mackie, A.R. The influence of small intestinal mucus structure on particle transport ex vivo. *Colloids Surf. B Biointerfaces* **2015**, *135*, 73–80. [CrossRef] [PubMed]

91. Li, X.; Chen, D.; Le, C.; Zhu, C.; Gan, Y.; Hovgaard, L.; Yang, M. Novel mucus-penetrating liposomes as a potential oral drug delivery system: Preparation, in vitro characterization, and enhanced cellular uptake. *Int. J. Nanomed.* **2011**, *6*, 3151–3162.

92. Cu, Y.; Saltzmanr, W.M. Controlled surface modification with poly(ethylene)glycol enhances diffusion of PLGA nanoparticles in human cervical mucus. *Mol. Pharm.* **2009**, *6*, 173–181. [CrossRef] [PubMed]

93. Muller, C.; Leithner, K.; Hauptstein, S.; Hintzen, F.; Salvenmoser, W.; Bernkop-Schnurch, A. Preparation and characterization of mucus-penetrating papain/poly(acrylic acid) nanoparticles for oral drug delivery applications. *J. Nanopart. Res.* **2013**, *15*, 1353–1366. [CrossRef]

94. DeSousa, I.P.; Cattoz, B.; Wilcox, M.D.; Griffiths, P.C.; Dalgliesh, R.; Rogers, S.; Bernkop-schnürch, A. Nanoparticles decorated with proteolytic enzymes, a promising strategy to overcome the mucus barrier. *Eur. J. Pharm. Biopharm.* **2015**, *97*, 257–264.

95. Moreno, J.A.S.; Mendes, A.C.; Stephansen, K.; Engwer, C. Development of electrosprayed mucoadhesive chitosan microparticles. *Carbohydr. Polym.* **2018**, *190*, 240–247. [CrossRef]

96. Park, C.G.; Huh, B.K.; Kim, S.; Lee, S.H.; Hong, H.R.; Choy, Y.B. Nanostructured mucoadhesive microparticles to enhance oral drug bioavailability. *J. Ind. Eng. Chem.* **2017**, *54*, 262–269. [CrossRef]

97. Krauland, A.H.; Guggi, D.; Bernkop-schnurch, A. Thiolated chitosan microparticles: A vehicle for nasal peptide drug delivery. *Int. J. Pharm.* **2006**, *307*, 270–277. [CrossRef] [PubMed]

98. Romero, G.B.; Keck, C.M.; Müller, R.H.; Bou-chacra, N.A. Development of cationic nanocrystals for ocular delivery. *Eur. J. Pharm. Biopharm.* **2016**, *107*, 215–222. [CrossRef]

99. De DeLima, J.A.; Paines, T.C.; Motta, M.H.; Weber, W.B.; Santos, S.S.; Cruz, L.; Silva, C.D.B. Novel Pemulen/Pullulan blended hydrogel containing clotrimazole-loaded cationic nanocapsules: Evaluation of mucoadhesion and vaginal permeation. *Mater. Sci. Eng. C* **2017**, *79*, 886–893. [CrossRef] [PubMed]

100. Kim, K.; Kim, K.; Hyun, J.; Lee, H. Chitosan-catechol: A polymer with long-lasting mucoadhesive properties. *Biomaterials* **2015**, *52*, 161–170. [CrossRef] [PubMed]

101. Ertl, B.; Heigl, F.; Wirth, M.; Gabor, F. Lectin-mediated bioadhesion: Preparation, stability and Caco-2 binding of wheat germ agglutinin-functionalized poly(D,L-lactic-co-glycolic acid)-microspheres. *J. Drug Target.* **2000**, *8*, 173–184. [CrossRef] [PubMed]

102. Anirudhan, T.S.; Parvathy, J. Novel thiolated chitosan-polyethyleneglycol blend/Montmorillonite composite formulations for the oral delivery of insulin. *Bioact. Carbohydr. Diet. Fibre* **2018**, *16*, 22–29. [CrossRef]

103. Bernkop-schnurch, A.; Hornof, M.; Guggi, D. Thiolated chitosans. *Eur. J. Pharm. Biopharm.* **2004**, *57*, 9–17. [CrossRef]

104. Deutel, B.; Laf, F.; Palmberger, T.; Saxer, A.; Thaler, M.; Bernkop-schnürch, A. In vitro characterization of insulin containing thiomeric microparticles as nasal drug delivery system. *Eur. J. Pharm. Sci.* **2016**, *81*, 157–161. [CrossRef] [PubMed]

105. Sajeesh, S.; Vauthier, C.; Gueutin, C.; Ponchel, G.; Sharma, C.P. Thiol functionalized polymethacrylic acid-based hydrogel microparticles for oral insulin delivery. *Acta Biomater.* **2010**, *6*, 3072–3080. [CrossRef] [PubMed]

106. Farris, E.; Heck, K.; Lampe, A.T.; Brown, D.M.; Ramer-tait, A.E.; Pannier, A.K. Oral non-viral gene delivery for applications in DNA vaccination and gene therapy. *Curr. Opin. Biomed. Eng.* **2018**, *7*, 51–57. [CrossRef]

107. Batista, P.; Castro, P.M.; Raquel, A.; Sarmento, B. Recent insights in the use of nanocarriers for the oral delivery of bioactive proteins and peptides. *Peptides* **2018**, *101*, 112–123. [CrossRef] [PubMed]

108. Zhang, Y.; Wu, X.; Meng, L.; Zhang, Y.; Ai, R.; Qi, N.; He, H.; Xu, H.; Tang, X. Thiolated Eudragit nanoparticles for oral insulin delivery: Preparation, characterization and in vivo evaluation. *Int. J. Pharm.* **2012**, *436*, 341–350. [CrossRef] [PubMed]

109. Cone, R.A. Barrier properties of mucus. *Adv. Drug Deliv. Rev.* **2009**, *61*, 75–85. [CrossRef] [PubMed]

110. Huckaby, J.T.; Lai, S.K. PEGylation for enhancing nanoparticle diffusion in mucus. *Adv. Drug Deliv. Rev.* **2018**, *124*, 125–139. [CrossRef] [PubMed]

111. Jung, T.; Kamm, W.; Breitenbach, A.; Kaiserling, E.; Xiao, J.X.; Kissel, T. Biodegradable nanoparticles for oral delivery of peptides: Is there a role for polymers to affect mucosal uptake? *Eur. J. Pharm. Biopharm.* **2000**, *50*, 147–160. [CrossRef]

112. Lai, S.K.; Wang, Y.Y.; Hanes, J. Mucus-penetrating nanoparticles for drug and gene delivery to mucosal tissues. *Adv. Drug Deliv. Rev.* **2009**, *61*, 158–171. [CrossRef] [PubMed]

113. Zhaeentana, S.; Amjadib, F.S.; Zandieb, Z.; Joghataei, M.T.; Bakhtiyari, M.; Aflatoonian, R. The effects of hydrocortisone on tight junction genes in an in vitro model of the human fallopian epithelial cells. *Eur. J. Obstet. Gynecol. Reprod. Biol.* **2018**, *229*, 127–131. [CrossRef] [PubMed]

114. Bein, A.; Eventov-friedman, S.; Arbell, D.; Schwartz, B. Intestinal tight junctions are severely altered in NEC preterm neonates. *Pediatr. Neonatol.* **2018**, *59*, 464–473. [CrossRef] [PubMed]

115. Gamboa, J.M.; Leong, K.W. In vitro and in vivo models for the study of oral delivery of nanoparticles. *Adv. Drug Deliv. Rev.* **2013**, *65*, 800–810. [CrossRef] [PubMed]

116. Linnankoski, J.; Makela, J.; Palmgren, J.; Mauriala, T.; Vedin, C.; Ungell, A.-L.; Artursson, P.; Urtti, A.; Yliperttula, M. Paracellular porosity and pore size of the human intestinal epithelium in tissue and cell culture models. *J. Pharm. Sci.* **2010**, *99*, 2166–2175. [CrossRef]

117. Salama, N.N.; Eddington, N.D.; Fasano, A. Tight junction modulation and its relationship to drug delivery. *Adv. Drug Deliv. Rev.* **2006**, *58*, 15–28. [CrossRef]

118. Kim, J.; Yoon, I.; Cho, H.; Kim, D.; Choi, Y.; Kim, D. Emulsion-based colloidal nanosystems for oral delivery of doxorubicin: Improved intestinal paracellular absorption and alleviated cardiotoxicity. *Int. J. Pharm.* **2014**, *464*, 117–126. [CrossRef]

119. Taverner, A.; Dondi, R.; Almansour, K.; Laurent, F.; Owens, S.; Eggleston, I.M.; Fotaki, N.; Mrsny, R.J. Enhanced paracellular transport of insulin can be achieved via transient induction of myosin light chain phosphorylation. *J. Control. Release* **2015**, *210*, 189–197. [CrossRef]

120. Almansour, K.; Taverner, A.; Eggleston, I.M.; Mrsny, R.J. Mechanistic studies of a cell-permeant peptide designed to enhance myosin light chain phosphorylation in polarized intestinal epithelia. *J. Control. Release* **2018**, *279*, 208–219. [CrossRef]

121. Banerjee, A.; Mitragotri, S. Intestinal patch systems for oral drug delivery. *Curr. Opin. Pharmacol.* **2017**, *36*, 58–65. [CrossRef]

122. Banerjee, A.; Lee, J.; Mitragotri, S. Intestinal mucoadhesive devices for oral delivery of insulin. *Bioeng. Transl. Med.* **2016**, *1*, 338–346. [CrossRef]

123. Shen, Z.; Mitragotri, S. Intestinal patches for oral drug delivery. *Pharm. Res.* **2002**, *19*, 391–395. [CrossRef]

124. Toorisaka, E.; Hashida, M.; Kamiya, N.; Ono, H. An enteric-coated dry emulsion formulation for oral insulin delivery. *J. Control. Release* **2005**, *107*, 91–96. [CrossRef]

125. Toorisaka, E.; Watanabe, K.; Ono, H.; Hirata, M.; Kamiya, N. Intestinal patches with an immobilized solid-in-oil formulation for oral protein delivery. *Acta Biomater.* **2012**, *8*, 653–658. [CrossRef]

126. Lee, J.W.; Prausnitz, M.R. Drug delivery using microneedle patches: Not just for skin. *Expert Opin. Drug Deliv.* **2018**, *15*, 541–543. [CrossRef]

127. Ma, Y.; Tao, W.; Krebs, S.J.; Sutton, W.F.; Haigwood, N.L.; Gill, H.S. Vaccine delivery to the oral cavity using coated microneedles induces systemic and mucosal immunity. *Pharm. Res.* **2014**, *31*, 2393–2403. [CrossRef]

128. Traverso, G.; Schoellhammer, C.M.; Schroeder, A.; Maa, R.; Lauwers, G.Y.; Polat, B.E.; Anderson, D.G.; Blankschtein, D.; Langer, R. Microneedles for Drug Delivery via the Gastrointestinal Tract. *J. Pharm. Sci.* **2015**, *104*, 362–367. [CrossRef]

129. Dabholkar, R.D.; Sawant, R.M.; Mongayt, D.A.; Devarajan, P.V.; Torchilin, V.P. Polyethylene glycol–phosphatidylethanolamine conjugate (PEG–PE)-based mixed micelles: Some properties, loading with paclitaxel, and modulation of P-glycoprotein-mediated efflux. *Int. J. Pharm.* **2006**, *315*, 148–157. [CrossRef] [PubMed]

130. Yu, H.; Cui, Z.; Yu, P.; Guo, C.; Feng, B.; Jiang, T. pH- and NIR light-responsive micelles with hyperthermia-triggered tumor penetration and cytoplasm drug release to reverse doxorubicin resistance in breast cancer. *Adv. Funct. Mater.* **2015**, *25*, 2489–2500. [CrossRef]

131. Suzuki, H.; Bae, Y.H. Evaluation of drug penetration with cationic micelles and their penetration mechanism using an in vitro tumor model. *Biomaterials* **2016**, *98*, 120–130. [CrossRef]

132. Sosnik, A.; Raskin, M.M. Polymeric micelles in mucosal drug delivery: Challenges towards clinical translation. *Biotechnol. Adv.* **2015**, *33*, 1380–1392. [CrossRef]

133. Torchilin, V.P. Fluorescence microscopy to follow the targeting of liposomes and micelles to cells and their intracellular fate. *Adv. Drug Deliv. Rev.* **2005**, *57*, 95–109. [CrossRef] [PubMed]

134. Byrne, R.S.; Deasy, P.B. Use of commercial porous ceramic particles for sustained drug delivery. *Int. J. Pharm.* **2002**, *246*, 61–73. [CrossRef]

135. Hoffman, A.S. Hydrogels for biomedical applications. *Adv. Drug Deliv. Rev.* **2012**, *64*, 18–23. [CrossRef]

136. Li, J.; Mooney, D.J. Designing hydrogels for controlled drug delivery. *Nat. Rev. Mater.* **2016**, *1*, 1–17. [CrossRef]

137. Chai, Q.; Jiao, Y.; Yu, X. Hydrogels for biomedical applications: Their characteristics and the mechanisms behind them. *Gels* **2017**, *3*, 6. [CrossRef]

138. Caló, E.; Khutoryanskiy, V. Biomedical applications of hydrogels: A review of patents and commercial products. *Eur. Polym. J.* **2015**, *65*, 252–267. [CrossRef]

139. Klouda, L. Thermoresponsive hydrogels in biomedical applications A seven-year update. *Eur. J. Pharm. Biopharm.* **2015**, *97*, 338–349. [CrossRef] [PubMed]

140. Torres-lugo, M.; Peppas, N.A. Molecular design and in vitro studies of novel pH-sensitive hydrogels for the oral delivery of calcitonin. *Macromolecules* **1999**, *32*, 6646–6651. [CrossRef]

141. Simpson, M.J.; Corbett, B.; Arezina, A.; Hoare, T. Narrowly dispersed, degradable, and scalable poly(oligoethylene glycol methacrylate)-based nanogels via thermal self-assembly. *Ind. Eng. Chem. Res.* **2018**, *57*, 7495–7506. [CrossRef]

142. Choi, J.; Moquin, A.; Bomal, E.; Na, L.; Maysinger, D.; Kakkar, A. Telodendrimers for physical encapsulation and covalent linking of individual or combined therapeutics. *Mol. Pharm.* **2017**, *14*, 2607–2615. [CrossRef]

143. DeFrance, K.J.; Xu, F.; Hoare, T. Structured macroporous hydrogels: Progress, challenges, and opportunities. *Adv. Healthc. Mater.* **2018**, *7*, 1–17.

144. Annabi, N.; Nichol, J.W.; Zhong, X.; Ji, C. Controlling the porosity and microarchitecture of hydrogels for tissue engineering. *Tissue Eng. Part B* **2010**, *16*, 371–385. [CrossRef] [PubMed]

145. Kim, U.; Park, J.; Li, C.; Jin, H.; Valluzzi, R.; Kaplan, D.L. Structure and properties of silk hydrogels. *Biomacromolecules* **2004**, *5*, 786–792. [CrossRef]

146. Yokoyama, E.; Masada, I.; Shimamura, K.; Ikawa, T.; Monobe, K. Morphology and structure of highly elastic poly(vinyl alcohol) hydrogel prepared by repeated freezing-and-melting. *Colloid Polym. Sci.* **1986**, *601*, 595–601. [CrossRef]

147. Hermansson, A.M.; Buchheim, W. Characterization of protein gels by scanning and transmission electron microscopy. *J. Colloid Interface Sci.* **1981**, *81*, 510–530. [CrossRef]

148. Hyuk Im, S.; Jeong, U.; Xia, Y. Polymer hollow particles with controllable holes in their surfaces. *Nat. Mater.* **2005**, *4*, 671–675. [CrossRef] [PubMed]

149. Staruch, R.M.T.; Glass, G.E.; Rickard, R.; Hettiaratchy, S.; Butler, P.E.M. Injectable pore-forming hydrogel scaffolds for complex wound tissue engineering: Designing and controlling their porosity and mechanical properties. *Tissue Eng. Part B* **2017**, *23*, 183–198. [CrossRef] [PubMed]

150. Loh, Q.L.; Choong, C. Three-dimensional scaffolds for tissue engineering applications: Role of porosity and pore size. *Tissue Eng. Part B* **2013**, *19*, 485–502. [CrossRef] [PubMed]

151. Li, Y.; Huang, G.; Zhang, X.; Wang, L.; Du, Y.; Jian, T.; Xu, F. Engineering cell alignment in vitro. *Biotechnol. Adv.* **2014**, *32*, 347–365. [CrossRef] [PubMed]

152. Vlierberghe, S.; Cnudde, V.; Dubruel, P.; Masschaele, B.; Cosijns, A.; De Paepe, I.; Jacobs, P.J.S.; Hoorebeke, L.; Remon, J.P.; Schacht, E. Porous gelatin hydrogels: 1. Cryogenic formation and structure analysis. *Biomacromolecules* **2007**, *8*, 331–337. [CrossRef] [PubMed]

153. Lai, J.; Li, Y. Functional assessment of cross-linked porous gelatin hydrogels for bioengineered cell sheet carriers. *Biomacromolecules* **2010**, *11*, 1387–1397. [CrossRef]

154. Lewus, R.K.; Carta, G. Protein transport in constrained anionic hydrogels: Diffusion and boundary-layer mass transfer. *Ind. Eng. Chem. Res.* **2001**, *40*, 1548–1558. [CrossRef]

155. Pundir, S.; Badola, A.; Sharma, D. Sustained release matrix technology and recent advance in matrix drug delivery system: A review. *Int. J. Drug Res. Technol.* **2013**, *3*, 12–20.

156. Thedrattanawong, C.; Manaspon, C.; Nasongkla, N. Controlling the burst release of doxorubicin from polymeric depots via adjusting hydrophobic/hydrophilic properties. *J. Drug Deliv. Sci. Technol.* **2018**, *46*, 446–451. [CrossRef]

157. Li, R.; Deng, L.; Cai, Z.; Zhang, S.; Wang, K.; Li, L.; Ding, S.; Zhou, C. Liposomes coated with thiolated chitosan as drug carriers of curcumin. *Mater. Sci. Eng. C* **2017**, *80*, 156–164. [CrossRef]

158. Martins, A.L.L.; Oliveira, A.C.; Nascimento, C.M.O.L.; Silva, D.; Gaeti, M.P.N.; Lima, E.M.; Taveira, S.F.; Fernandes, K.F.; Marreto, R.N. Mucoadhesive properties of thiolated pectin-based pellets prepared by extrusion-spheronization technique. *J. Pharm. Sci.* **2017**, *106*, 1363–1370. [CrossRef]

159. Lichtenberger, L.M. The hydrophobic barrier properties of gastrointestinal mucus. *Annu. Rev. Physiol.* **1995**, *57*, 565–583. [CrossRef] [PubMed]

160. Traverso, G.; Langer, R. Special delivery for the gut. *Nature* **2015**, *519*, S19. [CrossRef] [PubMed]

161. Lam, P.L.; Gambari, R. Advanced progress of microencapsulation technologies: In vivo and in vitro models for studying oral and transdermal drug deliveries. *J. Control. Release* **2014**, *178*, 25–45. [CrossRef] [PubMed]

162. Salessiotis, N. Measurement of the diameter of the Pylorus in man: Part I. Experimental project for clinical application. *Am. J. Surg.* **1972**, *124*, 331–333. [CrossRef]

163. Sultan, M.; Norton, R.A. Esophageal diameter and the treatment of achalasia. *Am. J. Dig. Dis.* **1969**, *14*, 611–618. [CrossRef] [PubMed]

164. Zhang, S.; Bellinger, A.M.; Glettig, D.L.; Barman, R.; Lee, Y.A.; Zhu, J.; Cleveland, C.; Montgomery, V.A.; Gu, L.; Nash, L.D.; et al. A pH-responsive supramolecular polymer gel as an enteric elastomer for use in gastric devices. *Nat. Mater.* **2015**, *14*, 1065–1071. [CrossRef]

165. Soudry-kochavi, L.; Naraykin, N.; DiPaola, R.; Gugliandolo, E.; Peritore, A.; Cuzzocrea, S.; Ziv, E.; Nassar, T.; Benita, S. Pharmacodynamical effects of orally administered exenatide nanoparticles embedded in gastro-resistant microparticles. *Eur. J. Pharm. Biopharm.* **2018**, *133*, 214–223. [CrossRef]

166. Agüero, L.; Zaldivar-silva, D.; Pena, L.; Dias, M.L. Alginate microparticles as oral colon drug delivery device: A review. *Carbohydr. Polym.* **2017**, *168*, 32–43. [CrossRef]

167. Chen, Q.; Gou, S.; Huang, Y.; Zhou, X.; Li, Q. Facile fabrication of bowl-shaped microparticles for oral curcumin delivery to ulcerative colitis tissue. *Colloids Surf. B Biointerfaces* **2018**, *169*, 92–98. [CrossRef]

168. Nadal, J.M.; Gomes, M.L.S.; Borsato, D.M.; Almeida, M.A.; Barboza, F.M.; Zawadzki, S.F.; Kanunfre, C.C.; Farago, P.V.; Zanin, S.M.W. Spray-dried Eudragit®L100 microparticles containing ferulic acid: Formulation, in vitro cytoprotection and in vivo anti-platelet effect. *Mater. Sci. Eng. C* **2016**, *64*, 318–328. [CrossRef]

169. Ratzinger, G.; Agrawal, P.; Korner, W.; Lonkai, J.; Sanders, H.M.H.F.; Terreno, E.; Wirth, M.; Strijkers, G.J.; Nicolay, K.; Gabor, F. Surface modification of PLGA nanospheres with Gd-DTPA and Gd-DOTA for high-relaxivity MRI contrast agents. *Biomaterials* **2010**, *31*, 8716–8723. [CrossRef] [PubMed]

170. Koch, B.; Rubino, I.; Quan, F.-S.; Yoo, B.; Choi, H.-J. Microfabrication for drug delivery. *Materials* **2016**, *9*, 646. [CrossRef] [PubMed]

171. Noguchi, H.; Takasu, M. Fusion pathways of vesicles: A Brownian dynamics simulation. *J. Chem. Phys.* **2001**, *115*, 9547–9551. [CrossRef]

172. Pekarek, K.J.; Jacob, J.S.; Mathiowitz, E. Double-walled polymer microspheres for controlled drug release. *Nature* **1994**, *367*, 258–260. [CrossRef]

173. Zolnik, B.S.; Burgess, D.J. Effect of acidic pH on PLGA microsphere degradation and release. *J. Control. Release* **2007**, *122*, 338–344. [CrossRef]

174. Wong, M.S.; Cha, J.N.; Choi, K.S.; Deming, T.J.; Stucky, G.D. Assembly of nanoparticles into hollow spheres using block copolypeptides. *Nano Lett.* **2002**, *2*, 583–587. [CrossRef]

175. Du, J.; O'Reilly, R.K. Advances and challenges in smart and functional polymer vesicles. *Soft Matter* **2009**, *5*, 3544–3561. [CrossRef]

176. Yokoyama, M.; Inoue, S.; Kataoka, K.; Yui, N.; Okano, T.; Sakurai, Y. Molecular design for missile drug: Synthesis of adriamycin conjugated with immunoglobulin G using poly(ethylene glycol)-block-poly(aspartic acid) as intermediate carrier. *Die Makromol. Chemie* **1989**, *190*, 2041–2054. [CrossRef]

177. Batycky, R.P.; Hanes, J.; Langer, R.; Edwards, D.A. A theoretical model of erosion and macromolecular drug release from biodegrading microspheres. *J. Pharm. Sci.* **1997**, *86*, 1464–1477. [CrossRef]

178. Steijn, V.; Korczyk, P.M.; Derzsi, L.; Abate, A.R.; Weitz, D.A.; Garstecki, P. Block-and-break generation of microdroplets with fixed volume. *Biomicrofluidics* **2013**, *7*, 24108. [CrossRef] [PubMed]

179. Abbaspourrad, A.; Carroll, N.J.; Kim, S.; Weitz, D.A. Polymer microcapsules with programmable active release. *J. Am. Chem. Soc.* **2013**, *135*, 7744–7750. [CrossRef] [PubMed]

180. Ostafe, R.; Prodanovic, R.; Ung, W.L.; Weitz, D.A.; Fischer, R. A high-throughput cellulase screening system based on droplet microfluidics. *Biomicrofluidics* **2014**, *8*, 041102. [CrossRef] [PubMed]

181. Bitar, C.; Markwick, K.E.; Hoesli, C.A. Encapsulation of Pancreatic Islet Cells for Type 1 Diabetes Treatment. In Proceedings of the XXV International Conference on Bioencapsulation, Nantes, France, 3–6 July 2017; Available online: http://bioencapsulation.net/221_newsletters/Bioencap_innov_2017_11/Bioencap_innov_2017_11.pdf (accessed on 18 March 2019).

182. Deng, N.-N.; Wang, W.; Ju, X.-J.; Xie, R.; Weitz, D.A.; Chu, L.-Y. Reply to the 'Comment on "Wetting-induced formation of controllable monodisperse multiple emulsions in microfluidics"' by J. Guzowski and P.; Garstecki, *Lab Chip*, 2014, 14, DOI:10.1039/ C3LC51229K. *Lab Chip* **2014**, *14*, 1479–1480. [CrossRef] [PubMed]

183. Massenburg, S.S.; Amstad, E.; Weitz, D.A. Clogging in parallelized tapered microfluidic channels. *Microfluid. Nanofluidics* **2016**, *20*, 1–5. [CrossRef]

184. Wyss, H.M.; Blair, D.L.; Morris, J.F.; Stone, H.A.; Weitz, D.A. Mechanism for clogging of microchannels. *Phys. Rev. E* **2006**, *74*, 1–4. [CrossRef] [PubMed]

185. Kumar, A.; Montemagno, C.; Choi, H.-j. Smart microparticles with a pH-responsive macropore for targeted oral drug delivery. *Sci. Rep.* **2017**, *7*, 061402. [CrossRef] [PubMed]

186. Caruso, F. Nanoengineering of particle surfaces. *Adv. Mater.* **2001**, *13*, 11–22. [CrossRef]

187. Cai, Y.; Chen, Y.; Hong, X.; Liu, Z.; Yuan, W. Porous microsphere and its applications. *Int. J. Nanomed.* **2013**, *8*, 1111–1120.

188. Kim, K.K.; Pack, D.W. Volume I: Biological and biomedical nanotechnology-Microspheres for drug delivery. In *BioMEMS and Biomedical Nanotechnology*; Ferrari, M., Lee, A., Lee, J., Eds.; Springer: Boston, MA, USA, 2006; pp. 19–50.

189. Homayun, B.; Sun, C.; Kumar, A.; Montemagno, C.; Choi, H.-J. Facile fabrication of microparticles with pH-responsive macropores for small intestine targeted drug formulation. *Eur. J. Pharm. Biopharm.* **2018**, *128*, 316–326. [CrossRef] [PubMed]

190. Homayun, B.; Kumar, A.; Nascimento, P.T.H.; Choi, H.-J. Macropored microparticles with a core–shell architecture for oral delivery of biopharmaceuticals. *Arch. Pharm. Res.* **2018**, *41*, 848–860. [CrossRef] [PubMed]

191. Chen, A.M.; Zhang, M.; Wei, D.; Stueber, D.; Taratula, O.; Minko, T.; He, H. Co-delivery of doxorubicin and Bcl-2 siRNA by mesoporous silica nanoparticles enhances the efficacy of chemotherapy in multidrug-resistant cancer cells. *Small* **2009**, *5*, 2673–2677. [CrossRef] [PubMed]

192. Wang, Y.; Zhao, Q.; Han, N.; Bai, L.; Li, J.; Liu, J.; Che, E.; Hu, L.; Zhang, Q.; Jiang, T.; et al. Mesoporous silica nanoparticles in drug delivery and biomedical applications. *Nanomed. Nanotechnol. Biol. Med.* **2015**, *11*, 313–327. [CrossRef]

193. Lai, C.Y.; Trewyn, B.G.; Jeftinija, D.M.; Jeftinija, K.; Xu, S.; Jeftinija, S.; Lin, V.S.Y. A mesoporous silica nanosphere-based carrier system with chemically removable CdS nanoparticle caps for stimuli-responsive controlled release of neurotransmitters and drug molecules. *J. Am. Chem. Soc.* **2003**, *125*, 4451–4459. [CrossRef] [PubMed]

194. Chen, Y.; Chen, H.; Ma, M.; Chen, F.; Guo, L.; Zhang, L.; Shi, J. Double mesoporous silica shelled spherical/ellipsoidal nanostructures: Synthesis and hydrophilic/hydrophobic anticancer drug delivery. *J. Mater. Chem.* **2011**, *21*, 5290–5298. [CrossRef]

195. Zhao, Y.; Trewyn, B.G.; Slowing, I.I.; Lin, S.-Y. Mesoporous silica nanoparticle-based double drug delivery system for glucose responsive controlled release of insulin and cyclic AMP. *J. Am. Chem. Soc.* **2009**, *131*, 8398–8400. [CrossRef] [PubMed]

196. Teo, P.Y.; Cheng, W.; Hedrick, J.L.; Yang, Y.Y. Co-delivery of drugs and plasmid DNA for cancer therapy. *Adv. Drug Deliv. Rev.* **2016**, *98*, 41–63. [CrossRef]

197. Kang, S.H.; Cho, H.; Shim, G.; Lee, S.; Kim, S.; Choi, H.; Kim, C.; Oh, Y.-K. Cationic liposomal co-delivery of small interfering RNA and a MEK inhibitor for enhanced anticancer efficacy. *Pharm. Res.* **2011**, *28*, 3069–3078. [CrossRef] [PubMed]

198. Cao, L.; Zeng, Q.; Xu, C.; Shi, S.; Zhang, Z.; Sun, X. Enhanced antitumor response mediated by the codelivery of paclitaxel and adenoviral vector expressing IL-12. *Mol. Pharm.* **2013**, *10*, 1804–1814. [CrossRef]

199. Mamaeva, V.; Sahlgren, C.; Lindén, M. Mesoporous silica nanoparticles in medicine-Recent advances. *Adv. Drug Deliv. Rev.* **2013**, *65*, 689–702. [CrossRef]

200. Santos, A.C.; Ferreira, C.; Veiga, F.; Ribeiro, A.J.; Panchal, A.; Lvov, Y.; Agarwal, A. Halloysite clay nanotubes for life sciences applications: From drug encapsulation to bioscaffold. *Adv. Colloid Interface Sci.* **2018**, *257*, 58–70. [CrossRef] [PubMed]

201. Slowing, I.I.; Vivero-escoto, J.L.; Wu, C.; Lin, V.S. Mesoporous silica nanoparticles as controlled release drug delivery and gene transfection carriers. *Adv. Drug Deliv. Rev.* **2008**, *60*, 1278–1288. [CrossRef] [PubMed]

202. Kresge, C.T.; Leonowicz, M.E.; Roth, W.J.; Vartuli, J.C.; Beck, J.S. Ordered mesoporous molecular sieves synthesized by a liquid crystal template mechanism. *Nature* **1992**, *359*, 710–712. [CrossRef]

203. Inagaki, S.; Fukushima, Y.; Kuroda, K. Synthesis of highly ordered mesoporous materials from a layered polysilicate. *J. Chem. Soc. Chem. Commun.* **1993**, *8*, 680–682. [CrossRef]

204. Gaaz, T.S.; Sulong, A.B.; Akhtar, M.N.; Kadhum, A.A.H.; Mohamad, A.B.; Al-Amiery, A.A. Properties and applications of polyvinyl alcohol, halloysite nanotubes and their nanocomposites. *Molecules* **2015**, *20*, 22833–22847. [CrossRef] [PubMed]

205. Lvov, Y.M.; Shchukin, D.G.; Mohwald, H.; Price, R.R. Halloysite clay nanotubes for controlled release of protective agents. *ACS Nano* **2008**, *2*, 814–820. [CrossRef]

206. Lvov, Y.M.; Devilliers, M.M.; Fakhrullin, R.F. The application of halloysite tubule nanoclay in drug delivery. *Expert Opin. Drug Deliv.* **2016**, *13*, 977–986. [CrossRef] [PubMed]

207. Fizir, M.; Dramou, P.; Zhang, K.; Sun, C.; Pham-Huy, C.; He, H. Polymer grafted-magnetic halloysite nanotube for controlled and sustained release of cationic drug. *J. Colloid Interface Sci.* **2017**, *505*, 476–488. [CrossRef]

208. Lvov, Y.; Wang, W.; Zhang, L.; Fakhrullin, R. Halloysite clay nanotubes for loading and sustained release of functional compounds. *Adv. Mater.* **2016**, *28*, 1227–1250. [CrossRef]

209. Baeza, A.; Colilla, M.; Vallet-regí, M. Advances in mesoporous silica nanoparticles for targeted stimuli-responsive drug delivery. *Expert Opin. Drug Deliv.* **2015**, *12*, 319–337. [CrossRef]

210. Croissant, J.G.; Zhang, D.; Alsaiari, S.; Lu, J.; Deng, L.; Tamanoi, F.; Almalik, A.M.; Zink, J.I.; Khashab, N.M. Protein-gold clusters-capped mesoporous silica nanoparticles for high drug loading, autonomous gemcitabine/doxorubicin co-delivery, and in-vivo tumor imaging. *J. Control. Release* **2016**, *229*, 183–191. [CrossRef] [PubMed]

211. Chen, Y.; Ai, K.; Liu, J.; Sun, G.; Yin, Q.; Lu, L. Multifunctional envelope-type mesoporous silica nanoparticles for pH-responsive drug delivery and magnetic resonance imaging. *Biomaterials* **2015**, *60*, 111–120. [CrossRef] [PubMed]

212. Niedermayer, S.; Weiss, V.; Herrmann, A.; Schmidt, A.; Datz, S.; Muller, K.; Wagner, E.; Bein, T.; Bräuchle, C. Multifunctional polymer-capped mesoporous silica nanoparticles for pH-responsive targeted drug delivery. *Nanoscale* **2015**, *7*, 7953–7964. [CrossRef] [PubMed]

213. Yao, X.; Niu, X.; Ma, K.; Huang, P.; Grothe, J.; Kaskel, S.; Zhu, Y. Graphene quantum dots-capped magnetic mesoporous silica nanoparticles as a multifunctional platform for controlled drug delivery, magnetic hyperthermia, and photothermal therapy. *Small* **2017**, *13*, 1–11. [CrossRef]

214. Lee, C.H.; Cheng, S.H.; Huang, I.P.; Souris, J.S.; Yang, C.S.; Mou, C.Y.; Lo, L.W. Intracellular pH-responsive mesoporous silica nanoparticles for the controlled release of anticancer chemotherapeutics. *Angew. Chem. Int. Ed.* **2010**, *49*, 8214–8219. [CrossRef] [PubMed]

215. Mekaru, H.; Lu, J.; Tamanoi, F. Development of mesoporous silica-based nanoparticles with controlled release capability for cancer therapy. *Adv. Drug Deliv. Rev.* **2015**, *95*, 40–49. [CrossRef]

216. Chen, P.J.; Hu, S.H.; Hsiao, C.S.; Chen, Y.Y.; Liu, D.M.; Chen, S.Y. Multifunctional magnetically removable nanogated lids of Fe3O4-capped mesoporous silica nanoparticles for intracellular controlled release and MR imaging. *Chem. J. Mater.* **2011**, *21*, 2535–2543. [CrossRef]

217. Saint-Cricq, P.; Deshayes, S.; Zink, J.I.; Kasko, A.M. Magnetic field activated drug delivery using thermodegradable azo-functionalised PEG-coated core–shell mesoporous silica nanoparticles. *Nanoscale* **2015**, *7*, 13168–13172. [CrossRef]

218. Gil, M.; Vicente, J.; Gaspar, F. Scale-up methodology for pharmaceutical spray drying. *Chem. Today* **2010**, *28*, 18–23.

219. Ofner, A.; Moore, D.G.; Rühs, P.A.; Schwendimann, P.; Eggersdorfer, M.; Amstad, E.; Weitz, D.A.; Studart, A.R. High-throughput step emulsification for the production of functional materials using a glass microfluidic device. *Macromol. Chem. Phys.* **2017**, *218*, 1–10. [CrossRef]

220. Stolovicki, E.; Ziblat, R.; Weitz, D.A. Throuput enhancement of parallel step emulsifier devices by shear-free and efficient nozzle clearance. *Lab Chip* **2018**, *18*, 132–138. [CrossRef]

221. Jeong, H.; Issadore, D.; Lee, D. Recent developments in scale-up of microfluidic emulsion generation via parallelization. *Korean J. Chem. Eng.* **2016**, *33*, 1757–1766. [CrossRef]

222. Holtze, C. Large-scale droplet production in microfluidic devices—An industrial perspective. *J. Phys. D. Appl. Phys.* **2013**, *46*, 1–10. [CrossRef]

223. Tendulkar, S.; Mirmalek-Sani, S.-H.; Childers, C.; Saul, J.; Opara, E.C.; Ramasubramanian, M.K. A three-dimensional microfluidic approach to scaling up microencapsulation of cells. *Biomed. Microdevices* **2012**, *14*, 9623–9626. [CrossRef]

224. Nisisako, T.; Ando, T.; Hatsuzawa, T. High-volume production of single and compound emulsions in a microfluidic parallelization arrangement coupled with coaxial annular world-to-chip interfaces. *Lab Chip* **2012**, *12*, 3426–3435. [CrossRef]

225. Conchouso, D.; Castro, D.; Khan, S.A.; Foulds, I.G. Three-dimensional parallelization of microfluidic droplet generators for a litre per hour volume production of single emulsions. *Lab Chip* **2014**, *14*, 3011–3020. [CrossRef]

226. Amstad, E.; Chemama, M.; Eggersdorfer, M.; Arriaga, L.R.; Brenner, M.P.; Weitz, D.A. Robust scalable high throughput production of monodisperse drops. *Lab Chip* **2016**, *16*, 4163–4172. [CrossRef]

pharmaceutics

MDPI

Review

Spatially Specific Liposomal Cancer Therapy Triggered by Clinical External Sources of Energy

Courtney van Ballegooie [1,2], Alice Man [3,†], Mi Win [4,†] and Donald T. Yapp [1,3,*]

1   Experimental Therapeutics, BC Cancer, Vancouver, BC V5Z 1L3, Canada; cballegooie@bccrc.ca
2   Faculty of Medicine, University of British Columbia, Vancouver, BC V6T 1Z3, Canada
3   Faculty of Pharmaceutical Sciences, University of British Columbia, Vancouver, BC V6T 1Z3, Canada;
    aliceman321@gmail.com
4   Department of Chemistry, Simon Fraser University, Burnaby, BC V5A 1S6, Canada; mi_mu_mu_win@sfu.ca
*   Correspondence: dyapp@bccrc.ca; Tel.: +1-604-675-8023
†   These authors contributed equally to this work.

Received: 26 February 2019; Accepted: 13 March 2019; Published: 16 March 2019

Abstract: This review explores the use of energy sources, including ultrasound, magnetic fields, and external beam radiation, to trigger the delivery of drugs from liposomes in a tumor in a spatially-specific manner. Each section explores the mechanism(s) of drug release that can be achieved using liposomes in conjunction with the external trigger. Subsequently, the treatment's formulation factors are discussed, highlighting the parameters of both the therapy and the medical device. Additionally, the pre-clinical and clinical trials of each triggered release method are explored. Lastly, the advantages and disadvantages, as well as the feasibility and future outlook of each triggered release method, are discussed.

Keywords: triggered drug release; liposomes; ultrasound; magnetic fields; radiation

1. Introduction

Spatially Specific Liposomal Cancer Therapy Utilizing Medical Devices as Triggering Mechanism

There are many factors that contribute to the successful treatment of cancer and maximize tumor control. Surgery, chemotherapy, and radiotherapy are used in combination depending on tumor stage and grade. Molecular interrogation of the tumor highlights therapeutic targets specific to the patient's tumor, and treatment options are optimized accordingly [1]. The importance of accurately imaging changes in the tumor volume and physiological functions has grown in tandem with the increasing use of targeted therapeutics [2]. Typically, the primary tumor is removed surgically, whenever possible, followed by adjuvant chemotherapy or radiation therapy. Chemotherapy is generally delivered systematically to kill cancer cells that may have migrated from the tumor, whereas radiation therapy is used locally to sterilize the surgical site. Unfortunately, each procedure carries its own adverse effects and risks; clinicians and patients must weigh the benefits against the risks before proceeding with a specific treatment regimen. In particular, using cytotoxic agents is associated with many unacceptable side-effects because these agents are potent cytotoxins that do not differentiate between normal and malignant cells. Unfortunately, despite efforts to mitigate the side effects, these negative effects can limit the drug dose that can be used with certain patients or reduce treatment compliance. Maximizing the anti-cancer activity of cytotoxic agents but minimizing their systemic toxicities, therefore, remains an important goal in optimizing chemotherapy treatments.

Liposomes have been particularly successful in modulating the biodistribution of cytotoxic drugs used in cancer treatment. In part, this is due to the versatility and classes of lipids that can be used to modify their distribution and release characteristics [3,4]. Liposomes are designed to encapsulate drugs, minimize drug release in circulation, accumulate at the tumor, and release drug locally when the bilayer is destabilized. This strategy enables drugs with proven activity to be preferentially delivered to the tumor site and reduce systemic side effects. However, the challenge is to balance drug encapsulation and release in a liposome so that the majority of the drug is released in the tumor, and not while in circulation.

More recently, there is growing interest in using external stimuli to trigger drug release from liposomes. The concept utilizes lipid carriers that are extremely stable and do not release significant amounts of drug at normal physiological conditions. However, the liposome would also be designed to be vulnerable to an external trigger that causes the liposome bilayer to become unstable and subsequently release its contents. Ideally, the stimuli would be focused on the tumor to ensure that only liposomes trapped there will release drug. Ultrasound (US) and magnetic fields (MF) used in magnetic resonance imaging (MRI) are primarily utilized for diagnostic purposes whereas radiation (RT) is used for imaging and treatment. These sources of external energies are of obvious interest for triggering drug release as they are already used clinically. The parameters for clinical imaging or therapy with US, MF, and RT are generally standardized whereas parameters used for drug release can vary greatly. Nonetheless, it is intriguing to speculate that clinically used parameters for US, MF, and RT could be modified to release drug from liposomes after imaging or concurrently during radiotherapy. Using imaging or therapeutic modalities to trigger intratumoral drug release on-command would help confine drug activity in the tumor and reduce systemic toxicities. However, clinical development of the strategies described could also open up novel treatments whereby, for example, imaging is used to confirm the presence of the drug carrier in the disease site before release. Furthermore, RT could be used to release drugs that potentiate the cell-killing effects of RT within the tumor. The latter strategy would facilitate drug-RT synergies that kill more cells than the sum of each separate approach. In this review, research work on the use US, MFs, and RT to trigger drug release from liposome drug carriers are summarized.

2. Ultrasound

2.1. Introduction

US triggered therapy relies on US waves compromising the integrity of a drug-loaded liposome to release its payload. By focusing the US transducer on the disease site, the emitted waves disrupt the bilayers of liposomes present in a defined area to release drug in a spatially-specific manner. The mechanisms causing drug release from the liposomes depends on the acoustic intensity, pulse frequency, pressure, duty cycle, and length of treatment. The effect of these parameters on the mechanism of release will also change depending on the tissue surrounding the drug carrier [5]. For example, tissues such as bone have a high absorption coefficient for acoustic waves and heat up more rapidly relative to other tissues with a low absorption coefficient [6,7]. Thus, a US treatment that is appropriate for soft tissue could potentially heat bone to lethal or damaging temperatures. In US-triggered drug release, acoustic parameters safe for the disease site must be matched to a liposome formulation that provides the desired mechanism of drug release. In the following section of the review, the mechanisms of drug release from liposomes using US and a summary of contributions from early pioneers in the field of US triggered release will be discussed.

2.2. Mechanism of Release

2.2.1. Thermally Induced Release

As US waves are propagated through tissue, the acoustic wave can be reflected, transmitted, scattered, or absorbed. When absorbed, the acoustic wave energy is transformed into heat; however, to create localized heat within the area of interest, absorption of the acoustic energy must be greater than its diffusion [8]. In most cases, local heat can be generated using moderate intensities (several

W/cm^2), high duty cycles (up to 100%), moderate pressures (hundreds of kPa to MPa range), high frequencies (>0.5 MHz), and long treatment times (minutes to hours) [5]. Focused ultrasound (FUS) transducers are preferred because they generate heat more specifically, and at deeper tissue depths than transducers which emit planar, less-focused wave patterns. This helps mitigate heat generation in non-disease areas and reduce damage to normal tissues. It should be noted that high intensity frequency ultrasound (HIFU), in the context of drug delivery, can be undesirable as it has the potential to cause tissue ablation [5,9,10]. Thus, there is interest in using US with temperature sensitive liposomes (TSLs) that have a Tm higher than physiological temperatures for controlled drug release.

TSLs release their payloads at temperatures near or above their Tm [11]. This is due to fluidity changes within the liposomal membrane as it transitions from its gel (solid-like) phase to its liquid-crystalline phase. Within the gel phase, the permeability of the lipid bilayer is orders of magnitude less than that of the liquid-crystalline phase. The drug, therefore, remains within the liposome while it circulates throughout the body, and the liposome only releases the drug when surrounding temperatures rise to, or above, its Tm [12,13].

Pioneers of TSLs, Yatvin et al., explored temperature-sensitive liposomes in the 1970s which were composed of 1,2-dipalmitoyl-*sn*-glycerophosphocholine (DPPC, Tm = 41.5 °C) and 1,2-distearoyl-*sn*-glycerophosphocholine (DSPC, Tm = 54.9 °C) for temperature-induced neomycin release in *E. coli*. Initially, work in temperature-sensitive liposomes displayed relatively slow drug release kinetics [11]. Despite this limitation, the DPPC DSPC TSL liposomes were tested in an in vivo subcutaneous Lewis lung tumor model where an implanted thermocouple maintained the temperature of the tumor at 42 °C. Although the authors were able to achieve a higher concentration of methotrexate (MTX) at the tumor site, confounding variables, such as increased blood flow and increased endothelial permeability, which occur due to continuous heating, made it unclear whether the increased MTX concentrations were due to the increased liposome accumulation at the site of the tumor or the increased liposomal drug release [14]. This study prompted further investigations of TSLs, such as improving the liposome's drug release kinetics as well as identifying a method that could heat the tumor rapidly and locally. Since then, the original thermo-sensitive liposome formulations by Yatvin have been modified to improve drug release. This was mainly achieved through the alteration of the liposome's lipid composition, such as including lysophospholipids, but the same has also been demonstrated through other methods, such as incorporating leucine-zippers to the membrane of the liposomes [12,15]. TSLs were later used with US induced heating to trigger drug release, as depicted in Figure 1, in the 1980s by Tracker and Anderson. It was then that the potential of using US to induce drug release in TSLs was first recognized due to the astounding 12-fold increase of drug accumulation at the site of interest under rapid, local US heating [16].

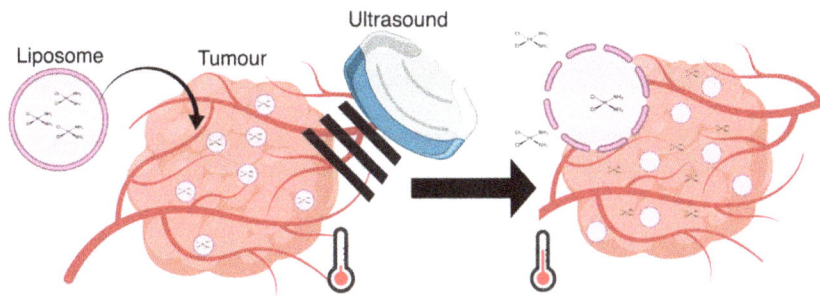

Figure 1. This figure depicts the accumulation of liposomes at the tumor site. Thermally induced release triggered by US then delivers the liposome's drug payload.

2.2.2. Mechanically Induced Release

Mechanical disruption of the liposome occurs via two mechanisms: stable cavitation coupled with radiation forces (RTFs) and inertial cavitation. Mechanical disruption of liposomes using US, however, requires the inclusion of compressible gaseous components, such as micelles, microbubbles (MBs), or liquid perfluorocarbon (PFC) droplets in the liposomal system. During cavitation, naturally nucleated or man-made gaseous materials contract and expand in response to the compression and refraction cycle of the acoustic wave. This, in turn, leads to a sustained oscillation of the bubble (stable cavitation) or to the rapid growth and ultimate collapse of the bubble (inertial cavitation). The type of cavitation that occurs depends highly upon the amplitude and frequency of the acoustic wave, as well as the size and material properties of the bubble [17–19]. The mechanical index (MI), the ratio of the in situ peak negative pressure (PnP) and the square root of the center frequency (F_c), is a predictor of which process will dominate. Typically, an MI less than 0.8 MPa·MHz$^{-1/2}$ results in stable cavitation while a higher MI leads to inertial cavitation [20]. Applications of cavitation not only include triggered release, as seen in Figure 2, but also include the modulation of blood perfusion, the permeabilization of the blood brain barrier (BBB), and even the breakdown of clots [21–23]. The modulation of perfusion and the permeabilization of the BBB allows for the modification of the nanoparticle's biodistribution so that it may accumulate preferentially at the tumor and cross the BBB, respectively [21,24].

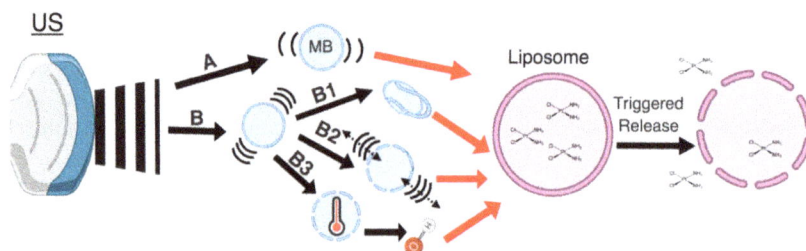

Figure 2. Schematic of the mechanisms associated with US that can be used to mechanically destabilize liposomes. The mechanisms are as follows: (**A**) US induced stable cavitation of a MB (**B**) US induced inertial cavitation of a MB; (**B1**) The production of a liquid microjets from a MB undergoing inertial cavitation; (**B2**) The production of shockwaves from a MB undergoing inertial cavitation; (**B3**) A MB collapsing and undergoing sonochemical changes.

Stable Cavitation and Radiation Forces: During stable cavitation, a bubble expands and contracts about an equilibrium value; the oscillation about the bubble's radius creates local swirling and fluid convection, termed micro-streaming, which induces shear stresses in the surrounding fluid [17]. This bubble-induced micro-streaming promotes the extravasation and delivery of circulating agents to target tissue [25]. Mechanistically, the shear stresses associated with US induced micro-streaming can rupture and deform liposomes or lyse the cell membrane (Figure 2B). These findings suggest that micro-streaming plays a pivotal role in drug delivery during instances of stable cavitation [26]. Micro-streaming caused by acoustic waves is a subset of forces, termed RTFs, that occur within the US field. These forces are able to displace particles and fluids not only via micro-streaming, but also through bulk streaming [25,27]. Bulk streaming occurs on a macro level where there is bulk, rather than local, fluid movement in the direction of the propagating acoustic wave. The RTFs generated via a stably cavitating bubble is highest at the driving frequencies near the microbubble's resonance frequency [20,25,26]. This concept has driven the design of the liposome carriers to include bubbles with a larger radius that will generate greater acoustic forces and require a lower resonance frequency. These bubbles commonly range in size from 0.8–3 μm [28,29].

The most prominent drug delivery method, capitalizing on the effects of stable cavitation and RTFs, is lipid coated microbubbles, also known as gas filled lipospheres. One of the rationales for this construct was to increase the loading capacity of hydrophobic drugs despite the large size of the bubbles needed. This, in turn, resulted in the inner membrane of the liposome being replaced with a layer of oil so that hydrophobic small molecule agents could be dissolved and encapsulated at the interface between the bubble and lipid layer. This configuration of monolayer lipids is outside of the scope of this paper and reviews on this subject can be found elsewhere [30]. It should be noted, however, that many similar applications involving liposomes, such as tumor drug delivery and gene therapy, and targeted imaging are being pursued in this field [31–36].

Inertial Cavitation: Unlike stable cavitation where bubbles oscillate around an equilibrium point, bubbles undergoing inertial cavitation oscillate with increasingly large amplitudes. The radius of the bubble increases until it exceeds a critical limiting value, the bubble resonant radius (BRR), whereupon it collapses [18]. Inertial cavitation is facilitated by rectified diffusion, a process where liquid vapors diffuse into the bubble faster than they diffuse out of the bubble. The differential in incoming and outgoing vapors arises when the bubble's radius and surface area increase due to the drop in internal pressure, which favors incoming over outgoing liquid vapors. Rectified diffusion is also affected by the change in concentration of the vapors in the bubble as the bubble oscillates due to the variation in the concentration gradient across the gas/liquid interface [18,19]. Factors that impact the resonant size of a bubble prior to collapse include the type of gas within the bubble, the surrounding medium, and the properties of the acoustic wave [18,37].

The three potential outcomes of inertial cavitation are (1) sonochemistry, (2) shockwaves, and (3) liquid microinjections [17,38] (Figure 2B). Sonochemistry is a sudden collapse of the bubble which generates momentarily high temperatures within the bubble's core. These temperatures have been shown to reach 5000 K but last only microseconds due to rapid cooling rates (10^{10} K/s) [39–41]. A secondary effect of these high temperatures is to generate reactive oxygen (ROS) species in the surrounding area [42,43]. Sonochemistry, and specifically sonodynamic therapy, was first introduced by Umemura et al. in 1989 where the synergistic effect of hematoporphyrin and US were observed during the treatment of both in-vivo and in-vitro tumor models [44]. Extensive work has since been performed within this field that has subsequently identified the types and levels of ROS produced, and the impact of the MB concentration, US irradiation time, amplitude, and pressure on the production of the ROS species [38,43,45–47]. For example, an in vitro study by He et al. used the change in the absorption and fluorescence spectra of bovine serum albumin (BSA) in the presence and absence of ROS scavengers to indicate the extent of the ROS induced damage of the protein. In this study, it was demonstrated that higher bubble concentrations and longer treatment times led to greater protein damage [45]. Although these effects were studied using proteins, both lipid and membrane damage has also been reported in the literature [48–50]. One of the most pre-clinically relevant studies, which gives insights to the parameters that generate the greatest radical production in vivo, was performed by Prieur et al. in 2015. The lipid-radical byproducts of internal cavitation, malondialdehyde (MDA) and hydroxyterephthalic acid (HTA), were quantified using varying US fields in freshly excised pig tissue. Briefly, they found that cavitation related oxidative stress increases with an increasing amount of bubbles present, treatment exposure time, and peak negative pressure [51]. Additional studies focusing on mechanically triggered drug release from liposomes will be discussed in the formulation factors section.

Inertial cavitation is also able to generate shockwaves that can exceed amplitudes of 10,000 atmospheres depending on the size of the bubbles. These shock waves increase the permeability of membranous structures (sonoporation) [52]. This effect is twofold when using liposomes as it not only encourages the release of the therapeutic agent from the liposome but also increases the cell's uptake of liposomes due to sonoporation [52,53]. Although the shock waves exist only for a short period of time, i.e., seconds to minutes, they have the ability to form large-spanning spatio-temporal pressure gradients [52]. Sonoporation has been demonstrated in vitro by Yudina et al. in 2010 where cell impermeable optical chromophores were added to a monolayer of C6 cells and were subjected to US. Only cells exposed to the US demonstrated fluorescent enhancement and this increased cell permeability phenomenon persisted even 24 h after exposure to US [54]. In vivo studies have further validated the presence of sonoporation, including delivering cell impermeable macromolecules (Bleomycin) to tumors, enhancing small molecule chemotherapy uptake in tumors, enhancing the blood brain barrier's permeability to previously poorly permeable chemotherapies, and delivering genetic material, such as DNA plasmids, siRNA, and pDNA [55–67].

Liquid microjets form in non-uniform environments where bubbles collapse near a surface and produce high-velocity projections. The velocity of these microjets can reach hundreds of meters per second and deposit significant energy densities at the site of impact. In doing so, it is thought that the microjet can penetrate the tissue or generate secondary stress waves in the tissue [37,38,68,69]. One of the first ever recorded evidence for microjets via US induced cavitation was demonstrated and characterized by Bowden and Brunton in 1958 and 1961. In this pioneering work, they demonstrated that (1) the jet velocities of bubbles above a few hundred m/sec acted like solid projectiles, (2) damage to the surface of impact contained two parts, an irreversible and often erosive deformation and a secondary shearing and tearing of the surface with subsequent fracturing, and (3) the pressure generated at impact could be approximated with the liquid density, velocity of the acoustic wave, and jet velocity [70]. Since then, extensive mathematical and experimental modeling has been performed [71–74]. In particular, a pivotal paper was released in 1998 by Kodama and Takayama which not only elucidated how the characteristics of the bubble impacted the microjet produced but also how the microjets interact and influence excised tissues. Briefly, they identified that the initial radius of the bubble had a logarithmic correlation to the penetration depth and pit size (i.e., the size of the damaged area) achieved. They also demonstrated that the liquid jet penetration into the liver induces a shear force between the hepatocytes, thereby leading to the elongation and splitting of the nuclei [75]. In the context of cancer, liquid microjets and their secondary shockwaves have been shown to (1) induce permeability of the cell to enhance chemotherapy uptake, and (2) cause cell death, membrane damage, and alterations in cellular metabolism [76,77].

Although the three outcomes of inertial cavitation have been discussed separately, identifying the primary mechanism of drug release is often difficult and it is thought that the therapeutic outcome results from two or more of the possible consequences of inertial cavitation. For the purposes of this review, inertial cavitation will be considered as a single entity regardless of the multiple mechanisms at play.

2.3. Formulation Factors

2.3.1. US Device Factors

As shown in Table 1, treatment parameters that can vary during treatment include (1) acoustic amplitude, (2) acoustic frequency, (3) duty factor, (4) pressure, and (5) treatment time. Often, varying one parameter will change the influence of another parameter on the type of triggered therapy achieved. An example of this was portrayed in Section 2.2 when describing the pressure and frequency parameters that would predict either stable or inertial cavitation. Table 1 also describes the common parameters used to achieve in vivo triggered release via the different mechanisms described in this review.

Table 1. Common parameters used to induce thermal, stable cavitation, or inertial cavitation in vivo using US.

Treatment Parameters	Thermal	Stable Cavitation	Inertial Cavitation
Acoustic Amplitude	High to Moderate (several W/cm^2) [5,78]	Low to Moderate (a few hundred mW/cm^2 or less) [78]	Low to Moderate (a few hundred mW/cm^2 or less) [78]
Acoustic Frequency	Moderate frequencies (0.5–1.5 MHz) [5]	Low to Moderate Frequencies (1 MHz or less) [51,78]	Low to Moderate Frequencies (1 MHz or less) [51,78]
Duty Factor	High duty cycles (up to 100%) [5]	Low duty cycles (as low as 1%) [51,78]	Low duty cycles (as low as 1%) [51,78]
Pressure	Moderate Pressure (100's of kPa to MPa) [5]	Low Pressure (below 500 kPa) [79]	Moderate Pressure (above 500 kPa) [79]
Treatment Time	Long treatment times (minutes to hours) [5]	Short treatment times (a few minutes or below) [51]	Short treatment times (a few minutes or below) [51]

Examples of detailed in vivo parameters can be seen in Table 2. Experimental, in-vitro tests and mathematical modeling have been performed to determine how US parameters impact drug release from liposomes [80]. Cavitation induced release will be discussed first, followed by thermal triggered release. Briefly, Schroeder et al. demonstrated that clinically approved liposomes, including Doxil®, Stealth™ Cisplatin, and methylpredinisolone hemisuccinate (MPS), when delivered under low frequency US (20 kHz), had a strong positive correlation of the % drug released with higher acoustic amplitudes (up to 7 W/cm^2) and irradiation time (up to 180 s). The impact of the increasing amplitudes continued with no maximal value achieved, while the irradiation time began to level off after 120 s of exposure. Additionally, the duty cycle, whether it be pulsed (<100%) or continuous (100%), had no impact on the % of drug released. It should be noted, however, that these experiments were performed in a glass scintillation vial with an immersed US probe [81]. Due to this experimental setup, the pressure was not varied or made to mimic conditions of the body, such as that found in the capillary vessels. A later study by Afadzi tested similar parameters in an insonication chamber on liposomes composed of 52 mol% DEPC, 5 mol% DSPC, 8 mol% DSPE-PEG, and 35 mol% cholesterol. Similar to Schroeder, they identified a positive correlation using low frequency US (300 kHz) between the % drug released and higher acoustic frequencies, with maximal release at 10 W/m^2, as well as a logarithmic correlation with exposure time. Contrary to Schroeder et al., however, they identified a positive correlation between the % drug released and the duty cycle (MI 2.4, 1.3 MPa, 180 s exposure, duty cycle ranged from 0–20%) [82]. This contrary finding could have been due to the % of duty cycles used. Further studies using a larger range of duty cycles should be performed.

Other US factors that can impact drug release include the (1) pulse duration (PD, i.e., the number of cycles multiplied by the inverse of the frequency), (2) pulse repetition frequency (PRF, i.e., the number of pulses per second), and (3) number of acoustic cycles (i.e., the number of acoustic oscillations per US pulse). Often, these parameters will not be specified, as they are related to the parameters in Table 1. For example, the duty factor and the PFR are directly related. Additionally, the PD is equivalent to the number of cycles multiplied by the inverse of the frequency. Therefore, because these factors are related to the initial five stated parameters, they will be covered only briefly in this review.

In the same study by Afadzi as described above, the % of drug release was also positively correlated with the PD and PRF [82]. A later study in 2016 by Lin et al. explored PD, PnP, and PRF in the context of the type and magnitude of cavitation induced. Here, they discovered that the onset of both stable and inertial cavitation exhibited a strong dependence on the PnP and PD and a relatively weak dependence on the PRF. Moreover, the amount of stable and inertial cavitation varied with the PRP. The amount of stable cavitation initially increased with increasing PnP until the pressure reached 0.5 MPa, where it rapidly decreased. By contrast, the amount of inertial cavitation recorded continuously increased with increasing PnP. Lastly, both PRF and PD positively correlated with both stable and inertial cavitation [83]. Another variable that was not previously discussed is the number of acoustic cycles applied. A paper by Mannaris and Averkiou identified the influence of the number of acoustic cycles applied on the microbubble by suspending the bubble in an enclosure that resembled capillaries. When applying the same acoustic parameters to a bubble (PRF = 100 Hz; f = 1 MHz) with an MI of 0.4, they found that increasing the number of cycles from 200 cycles to 1000 cycles had minimal effects on when the bubbles experienced inertial cavitation; this was likely caused by the high acoustic pressure used in the experiment. The authors speculate that the number of cycles could have a greater impact on the bubble's oscillation when exposed to nondestructive pressures [84]. However, it should also be noted that specific parameters of the experiment, such as the presence and size of the bubble used, will also impact the parameters utilized with the US device [28,29].

When considering heat induced release, many of the US parameters are limited by physiological factors. For example, although higher temperatures can be achieved with greater pressures, such as between 2–3 MPa at 1 MHz, kidney and lung hemorrhaging begin to appear at 3–5 MPa and 2 MPa respectively and can, therefore, not be achieved safely in vivo [85–87]. Additionally, the treatment time is dependent on the biology of the tumor and can range drastically based on the volume and location of the tumor tissue. Small, superficial tumors will take a fraction of the time to heat (approximately 1 h) relative to deep lying larger tumors (can be more than 6 h based on size and location) [88]. The influence of the acoustic amplitude, also known as intensity, with respect to the heating of tissue was mathematically derived by Pierce in 1981. Put simply, the power deposited per unit volume of tissue was found to equal two times the local acoustic intensity multiplied by the absorption coefficient of the tissue [7]. Therefore, higher intensities would lead to a greater energy deposition, and thus, a greater generation of heat. Reviews that detail the mathematical modeling of heat transfer and heat deposition using US can be found elsewhere [20]. The high intensities necessary to heat tissues is reflected in the first two preclinical trials listed in Table 2. Lastly, the duty cycle will influence how often the tissue is exposed to these high intensity US waves. The higher the duty cycle, the more energy deposition there is with the highest being a continuous wave (100% duty cycle) [87]. The duty cycle used will often reflect the amount of heat necessary at the site of the tumor. For instance, using a continuous exposure can result in the thermal ablation of tissue (>60 °C) while a pulsed exposure can achieve mild hyperthermia (37–45 °C) [89]. Controlling the energy deposition in vivo, whether it be through the intensity of the US wave and/or the duty cycle used, is critical as vascular damage is suggested to appear at a local energy density of 0.3 mJ/mm^2 [90].

2.3.2. Liposomal Factors

The major liposomal factors that contribute to the liposome's response to US include (1) the liposome's composition, and (2) the physical state of the liposome's bilayer. The liposome's composition can be further broken down into three categories: (a) the presence of thermo-sensitive lipids, such as 1,2-dimyristoyl-*sn*-glycero-3-phosphocholine (DMPC) or 1-myristoyl-2-palmitoyl-*sn*-glycero-3-phosphocholine (MPPC), (b) the presence of surface-active molecules, such as detergents, and (c) the presence of cholesterol and polyethylene glycol (PEG). In a paper studying the impact of thermosensitive lipids on TSLs, Needham et al. identified that MPPC and DMPC lipid-containing liposomes, which lower the phase Tm, enabled enhanced drug release by local hyperthermia. The enhanced drug release at the liposome's Tm was thought to occur due to the coexistence of the gel and liquid phase domains within the membrane. At the boundary regions between the two domains, a mismatch in molecular packing would occur, thereby facilitating the enhanced drug release. This phenomenon would be further enhanced by kinetically trapped MPPC lipids in the solid phase which, upon the gel-liquid crystalline phase transition, would leave the bilayer and enhance the permeability. In vitro findings by Needham et al. were later translated in vivo, and demonstrated significantly reduced tumor growth using Dox TSLs relative to the free Dox and a non-temperature sensitive Dox-containing liposome formulation [91]. Introducing structural irregularities within the membrane to disrupt the packing of the acyl chains is also the mechanism behind the increased drug release when introducing other unsaturated phospholipids. This hypothesis was tested by Huang and McDonald who showed that incorporating unsaturated diheptanoylphosphatidyl-choline (DHPC) into liposomes increased the release of encapsulated calcein upon US irradiation [92]. Surfactants, such as Triton and Tween, are also thought to destabilize the lipid bilayer. Indeed, a study involving two Triton and two Tween detergents showed a dramatically increased susceptibility of liposomes to US irradiation at concentrations that caused no observable increase in permeability in the absence of US [93]. Additionally, it was demonstrated that in cholesterol-free liposomes, Pluronic P105 sensitized liposomes to US irradiation when it was either in the presence of or directly incorporated with liposomes. The observed 10-fold increase in dye release, however, disappeared once cholesterol was incorporated into the lipid bilayer. This suggests that cholesterol has a protective effect between the interaction of Pluronic P105 and the lipid bilayer [94]. Interestingly, it was later demonstrated that increasing the cholesterol present in the lipid bilayer had a minimal but still statistically significant impact on dye release [95]. Lastly, the addition of PEG moieties will be discussed. While there are different methods of incorporating PEG into liposomal samples, such as the addition of PEG micelles or free PEG to the sample, PEGs covalently linked to phospholipids (PEG-lipids) and incorporated in the liposome's bilayer will solely be discussed due to their clinical relevance. The effect of the PEG length and molar ratios of PEG-lipids was studied using low frequency US (LFUS) by Lin and Thomas. Briefly, they identified that the length of the PEG (PEG350-DPPE and PEG2000-DPPE) had no impact on the amount of dye released when using concentrations below a mole ratio of 0.1 PEG-lipid to PC were utilized [93,94]. The influence of the PEG length on liposomal release only occurred when high molar ratios of PEG-lipid, above 0.1 PEG-lipid to PC, were used. These studies demonstrated that a shorter PEG length at high molar ratios yielded higher levels of dye release than a longer PEG length at high molar ratios [93]. It was speculated that this phenomenon occurred in part due to the acoustic absorption of PEG moieties as well as the potential of shorter PEGs hindering the resolution of deformities in the liposomal bilayers [93,96]. More research is required to confirm the mechanism behind this observation.

It should be noted that the above studies utilized liposomes primarily comprised of 1,2-Distearoyl-*sn*-glycero-3-phosphoethanolamine (DSPE), DPPC, or egg phosphocholine (egg PC). Recent studies have explored replacing the major lipid constituent with dioleoylphosphatidylethanolamine (DOPE) in order to create a sonosensitive liposomes. A study by Evjen et al. demonstrated a 30% increase in Dox release using DOPE liposomes compared to liposomes comprising DSPE, and a 9-fold improvement in release extent when compared to L-α-phosphatidylcholine (HSPC) pegylated liposome when irradiating with US for 6 min at 40 kHz [97]. When investigating the interaction of the physical state of the liposome's bilayer and US, Dunn and Tata and Maynard et al. identified enhanced US absorbance at the DMPC and DPPC liposome's Tm. Briefly, they subjected liposomes comprised of DMPC or DPPC to 1.42 and 2.11 MHz US, respectively, and recorded the ultrasonic absorption and velocity of the samples. Enhanced ultrasonic absorbance only occurred at the phase Tm; below the phase transition, it was observed that US was hardly absorbed by the membrane [98,99]. These findings suggest that, when working at temperatures below the phase transition of the liposome, the mechanism of release is independent of the liposome's absorbance of US but is dependent on the local cavitation and RTFs as well as heating of the surrounding tissue.

2.4. Future Perspectives

US is an emerging technology with the potential to be incorporated in the clinic for triggered delivery of liposomal drugs. As seen in Table 2, only the thermal release mechanism has proceeded to clinical trials at the time of writing of this review. This is in part due to the fact that US induced heating, such as HIFU, has already undergone multiple Phase II and III clinical trials and is currently in clinical practice in China [100]. Additionally, the safety of HIFU has been well documented experimentally in vivo and in patients. It was of initial concern that inertial cavitation and the shear forces produced using US would increase the cancer cell's ability to dissociate from the primary tumor and form a metastatic site at a distal location. This, however, was found not to be the case as HIFU treatment did not increase the number of metastatic sites nor the number of circulating tumor cells [89,101]. In fact, HIFU has demonstrated such promise that it may one day serve as an alternative to the surgical resection of tumors. It has been well documented that when primary malignant tumors are surgically resected, their distal metastases begin to rapidly progress. Although there are many proposed mechanisms, such as the secretion of growth factors in response to the surgery or a shift in the pro- and anti-angiogenic factors secreted from the tumor itself, the best understood mechanism for this phenomenon thus far is the suppression of the immune system. Recent studies have suggested that HIFU can enhance cancer-specific immunity after treatment. Specifically, HIFU is thought to enhance the T cell-mediated immune response [102,103]. Currently, the prevailing two mechanisms are (1) that the ablated tumor tissue acts as an antigen source for the generation of antitumor immunity and (2) that HIFU enhances the release of heat-shock proteins which can then stimulate cytotoxic T-cells [103–106]. Thus far the benefits of HIFU have been described but an important clinical consideration is the safety and side effect profile of the treatment. The most frequently occurring adverse events are moderate pain, with approximately <15% of patients experiencing this symptom, followed by transient fever and skin toxicities [107–110]. Interestingly, the Phase I clinical trials using mild hyperthermia for liposomal drug release (TARDOX and DIGNITY) reported either no adverse events or a low prevalence of grade 3–4 adverse events respectively. [111–113] This was likely due to the parameters used as the recorded level of tumor heating was found to be 40 °C rather than the 60 °C needed for tissue ablation [112,113]. Another advantage of US that was not discussed previously, but that is prevalent in these clinical trials, is the ability of the US treatments to be administered in a single treatment session rather than in multiple sessions as seen in radiotherapy. [111–113] While US does have obvious advantages as a treatment method, such as being minimally invasive and displaying a low adverse effect profile, there are some limitations to the technology. Specifically, there are three major disadvantages of US as a treatment method. The first is the inability of US to penetrate air-filled viscera. This will limit the ability of US to be utilized with tumors located in areas such as the lungs, intestines, or bladder. The second disadvantage also involves the tumor's location, particularly if there is no acoustic window for the US to reach the tissue of interest. For example, if there is a structure obscuring the tumor with a high absorption coefficient, such as bone, the acoustic wave may be unable to reach its intended target. The third major disadvantage to US is the long treatment times discussed in Section 2.3.1 [88]. Despite these limitations, the results of the clinical trials thus far and the benefits seen in HIFU treatment highlight the potential of combining liposomal triggered release, in conjunction with either mild or moderate hyperthermia, as a promising option for cancer treatment. While mechanically triggered release has shown promising results preclinically, more research is required to better evaluate this mechanism of release as an alternative option for cancer treatment.

Table 2. Summary of preclinical and clinical cancer treatments for US induced therapy release using liposomes. (Prf = pulse repetition frequency, TAT = total acoustic power, Statistically significant = *, cw = continuous wave, amp = amplitude, f = frequency, ns = not specified).

Delivery System	Release Type	Animal/Tumor Model	Dosing	Parameters: f, Duration, Amp, Pulse f	Outcome	Ref.
ThermoDox®	Thermal	Murine mammary adenocarcinoma; BALB/c	2 mg/kg single injection	Prf of 1 Hz for a total of 1 MHz; 15–20 min; 1300 W/cm², 120 pulses 10% duty cycle	* Tumor volume reduction	[114]
Prohance® & dox-loaded TTSL	Thermal	Rat subcutaneous 9 L gliosarcoma; 344	5 mg/kg single injection	1.4 MHz; 2 × 15 min; 117 W/cm²; cw	* Dox accumulation in the tumor	[115]
Prohance® & dox-loaded iLTSL	Thermal	Rabbit/VX2 tumor	5 mg/kg single injection	1.2 MHz; 4 × 10 min; ns; ns	ns	[116]
ThermoDox®	Thermal	Rabbit/VX2 tumor	5 mg/kg single injection	ns; 3 × 10 min; ns; ns MR-HIFU clinical system, parameters ns	* Dox accumulation in the tumor	[117]
Stealth™ cisplatin	Mechanical	Murine lymphoma (J6456); BALB/c	15 mg/kg single injection	20 kHz; 120 s; 5.9 W/cm²; cw	* Tumor volume reduction	[118]
DVDMS liposomes conjugated to MBs	Mechanical	Murine breast cancer (4T1); BALB/c	4.0, 2.0, or 0.4 µg/single injection	1.0 MHz; 3 min; TAT 3 W; 30% duty cycle	* Tumor volume reduction	[119]
Caelyx®	Mechanical	Murine prostate cancer (CWR22); BALB/c	3.5 mg/kg single injection	40 kHz; 4 min; 12 W/cm²; ns	Tumor volume reduction	[120]
DEPC-based Dox-loaded liposomes	Mechanical	Murine prostate adenocarcinoma (PC-3); BALB/c	Not specified	Prf of 250 Hz for a total of 300 kHz or 1 MHz; 10 min; ns; 5% duty cycle	* Dox accumulation in the tumor	[53]
Doxil®	Mechanical and permeabilization	Rat 9 L gliosarcoma; Sprague-Dawley	5.67 mg/kg single injection	Prf of 1 Hz for a total of 1.7 MHz; pressure 1.2 MPa, burst length: 10 ms, duration: 60–120 s	* Tumor regression and long-term survival	[21]
Doxil®	Mechanical and permeabilization	Rat 9 L gliosarcoma; Sprague-Dawley	5.67 mg/kg single injection	690 kHz; pressures amp 0.55–0.81 MPa; burst length: 10 ms; prf: 1 Hz; duration: 60	* Tumor regression and long-term survival	[121]
ThermoDox®	Thermal	Phase I DIGNITY Clinical Trial; Breast Cancer	20 mg/m²–50 mg/m², up to 6 doses, 21 days apart	ns; 1 h; ns; ns	Safe to move onto Phase II Clinical Trial	[111]
ThermoDox®	Thermal	Phase II DIGNITY Clinical Trial; Breast Cancer	40 mg/m²–50 mg/m², up to 6 doses, 21 days apart	ns; 1 h; ns; ns	Expansion of Phase II Clinical Trial	[122]
ThermoDox®	Thermal	Phase I TARDOX Clinical Trial; Liver Metastases	50 mg/m², up to 6 doses, 21 days apart	0.96 MHz; 33.2–80.0 min	Safe to move onto Phase II Clinical Trial	[112,113]

3. Magnetism for Triggered Drug Release

3.1. Introduction

MFs are an attractive way to release drugs from liposomes due to the technique's non-invasiveness, absence of ionizing radiation, and physiologically benign field frequencies and amplitudes [123]. In the clinical management of cancer, MFs are used primarily with contrast agents to diagnose and stage tumors. The technique is commonly used in cases where the tumor is made of softer tissue, as MRI scans provide images that enhance soft tissue contrast [124]. Aside from MRI, one of the best-known uses of magnetism in cancer is alternating magnetic field (AMF) induced hyperthermia. AMFs are characterized by rapid and regular changes in the MF's direction. Radiofrequency (low frequency AMFs) can penetrate deep into the body unhindered. However, high frequency, high amplitude AMFs will induce electric currents in tissue and can raise bulk tissue temperatures to lethal limits because of the tissue's resistance to electrical currents. If lethal temperatures are reached (>42 °C), the heat-induced damage to cancer cells cannot be repaired and the cells die [125]. Unfortunately, the effects of AMFs are not specific to malignant tissue and the surrounding normal tissue in the field may also be damaged, thereby limiting the maximum frequency and amplitude of AMFs that can be safely used in humans. Currently, there is no consensus on the safety limits for AMFs, but AMFs of 100 kHz and amplitude <10 kA/m have been used safely in clinical trials [126,127]. Additionally, the penetration depth and safety profile of low energy AMFs are advantageous when compared to other external triggers for activating nanomaterials, such as light or X-rays, which are limited by their shallow penetration depths and ionizing damage to normal tissue, respectively. [128].

The heat generated by magnetic nanoparticles (MNPs) upon application of AMFs has been used to raise the bulk temperatures of malignant tissue to lethal limits in a process called magnetic hyperthermia [129]. The bulk heating of tissue with magnetic hyperthermia has also been used to release drug from nano-composites, liposomes, polymers lipid structures, or cyclodextrin conjugated to MNPs [130–137]. In the following section of the review, the use of MFs to disrupt liposomes associated with MNPs, by either inducing heat or by mechanical motion to cause the release of their payload, will be discussed (Figure 3).

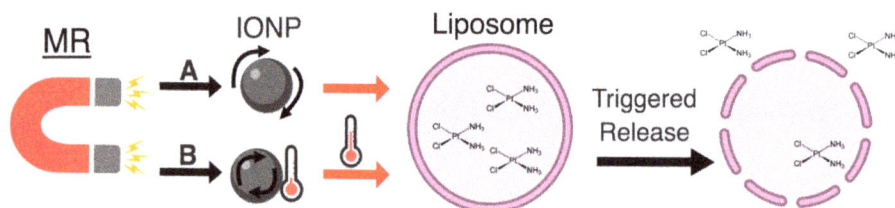

Figure 3. Schematic of the mechanisms associated with MR that can be used to destabilize liposomes. The mechanisms are: (**A**) MR induced mechanical disruption; (**B**) MR induced hyperthermia.

3.2. Delivery Using Heating and Mechanical Motion

In the field of temperature induced drug delivery, liposomes have been used to deliver drugs to tumors by taking advantage of the slightly elevated temperatures in malignant tissue. With a Tm only a few degrees above physiological temperatures, these liposomes were able to release their encapsulated drugs within the malignant tissue [138]. However, due to the liposome's Tm, these TSLs did not prevent drug loss as the liposomes circulated throughout the body. To circumvent the issue of nonspecific drug release, liposomes with Tms significantly higher than normal tissue temperatures were subsequently used in conjunction with tissue heating. Drug release would, therefore, not be triggered by normal body temperatures until they reached the disease site where temperatures were increased past the liposome's Tm with induced hyperthermia. Tissue hyperthermia can be induced in various ways, including the use of US (as seen previously in Section 2.2.1) and, in some cases, magnetism—particularly MFs interacting with liposomes and magnetic particles (MPs) [139–141].

Using MNPs in the tumor to potentiate the tissue heating effects of AMFs can raise tissue temperatures at lower magnetic frequencies and amplitudes that spare normal tissue from damage [127]. The incorporation of MNPs within the liposome itself has the potential to limit AMF-induced heating to within the drug carrier so that bulk heating of the tissue is unnecessary. A study demonstrated selective hypothermia in vivo and in vitro using magnetoliposomes under a low-frequency AMF to promote lipid membrane permeability from local heating. However, the relatively high heating of the composite particles led to concerns of overheating normal body tissues [142]. In order to overcome injury to healthy tissues caused by the overheating of the particles, other groups embedded the MNPs within the drug carrier to minimize heating of the tissue directly for their drug release studies. For example, Amstad et al. encapsulated iron oxide particles (IOPs) within liposomes and succeeded in controlling the timing of release by increasing the permeability of the liposomes without destroying them [143]. They concluded that AMF-induced heat was confined to the liposomes and subsequently spared normal tissue. Similarly, another study triggered the release of carboxyfluorescein from TSLs containing IOPs through AMF induced local heating [144].

Disrupting a liposome's lipid bilayer by local heating using encapsulated MNPs (magnetic fluid hyperthermia) to release drugs is common, but efforts to mechanically disrupt the bilayer with MFs have also been made. Most work investigating drug release by mechanical means have been done with MNPs located within the lipid bilayer. In an article studying the toxicity effects on cells, Kim et al. used microdiscs that oscillated when an alternating magnetic force is applied. Cell membrane integrity was compromised partly due to the oscillating microdiscs attached to the cell surface; therefore, it is not inconceivable that liposomal membranes containing MPs under the field would also be disrupted [145]. Drug release has also been observed from magnetoliposomes due to the mechanical vibrations of the IOPs [146]. Furthermore, the rotation of IOPs, induced through a dynamic magnetic field (DMF), can injure cell membranes. In contrast to using AMFs, DMF uses lower frequency parameters that cause unique rotations of individual particles around their own axes. This produces rotational shear forces that lower membrane integrity, without thermal effects [147]. Mechanical disruption of the membrane can also be achieved with pulses of an MF, as opposed to an alternating one. In this study, the authors showed drug release from liposomes after treatment with short magnetic pulses that disrupted the lipid bilayer. They further concluded that drug release from mechanical disruption of the liposomes was less harmful to the drug payload as an increase in temperature could potentially damage the drug [148].

3.3. Mechanisms of Release

The inherent magnetism of MNPs distinguishes them from other types of particles. MNPs behave as a single magnetic moment with an absolute value several orders of magnitude higher than that of single atoms, and can be remotely actuated or detected by an MF [149]. When MNPs are exposed to AMFs, their magnetic moments move to align with the field direction, but 'relax' or rotate back to their original alignment when the field is removed. The realignment or 'relaxation' of the magnetic moment represents a net energy loss that is released as heat. The MNP can physically rotate in the tissue and release heat into the surrounding tissue (Brownian relaxation), or the MNP remains stationary while its internal magnetic moment rotates with the field and releases heat at the surface of the MNP (Néel relaxation) [150]. Under the action of AMFs, this process happens many times, causing significant increases in temperature [151]. The amount of heat released by MNPs depends on the core magnetic material, the hydrodynamic diameter and shape of the particle, and the frequency and amplitude of the AMFs [152]. Brownian relaxation forms the basis for heating up bulk tissue in magnetic hyperthermia and the subsequent release of drug from TSLs. In contrast, Néel relaxation is thought to have less impact on the surrounding tissue; reports indicate that temperatures >40 °C have been estimated at MNP surfaces when subjected to AMFs, but that the temperature gradients between MNPs and their surroundings fall off very steeply [153,154]. The use of magnetic nanoparticles that undergo Néel relaxation for local heating of the lipid bilayer with AMFs benign to normal tissue would alleviate unintended heating of normal tissue.

MNPs can also behave as nanomagnets that align themselves to the plane of an MF, and this alignment can also be used to kill cancer cells. When the field is rotated or changes its direction, the particles move along with the field. Studies in which the particles are attached to cell membranes indicate that the particles can create mechanical forces strong enough to rupture the cell and cause cell death [155,156].

3.4. Formulation Factors

MNP induced liposomal drug release relies on both the properties of the MNP as well as the liposome. These components, therefore, allow for a range of conditions that can be tuned to change the release characteristics of the delivery system. The three main parameters that can be modified for drug release include (1) the MF's frequency and amplitude, (2) the composition of the lipid bilayer, and (3) the properties of the MNP such as the shape, size and composition. The tunability of these parameters are important and must be considered when developing delivery systems that are biocompatible and only release drug at the disease site in response to a magnetic field.

The properties of the MF are important when considering the thermal or mechanical disruption of the liposome for drug release [148,157]. The frequency of the MF is perhaps the factor most commonly changed, and a range of AMF frequencies have been used in studies for drug release, with higher frequencies leading to the higher motion of the nanoparticles as well as local heat production. This is also true of the strength of the field [158]. MFs can generate enough heat to irreversibly damage tumor cells yet may also damage healthy tissue at very high strengths. The suggested safe range for strength and frequency is up to 37 kA/m and 500 kHz [159]. Alternatively, pulses of strong MFs can be used to disrupt the lipid bilayer by using the motion of the nanoparticles, as opposed to generated heat [148].

Lipid bilayers used in drug delivery vary greatly in composition, due to the vast selection of fatty acids available for liposome production. To ensure that the liposomes destabilize at the proper temperature and release their drug payload, it is important to choose bilayer components carefully. Pradhan et al. used a liposomal formulation of DPPC:cholesterol:DSPE-PEG2000:DSPE-PEG2000-Folate at an 80:20:4.5:0.5 molar ratio in a delivery system containing MNPs. Their experiment demonstrated a significant release of the drug payload when exposed to an MF that increase the temperature a few degrees above ambient body temperature [160]. Peller et al. also used a DPPC, DSPC, and 1,2-dipalmitoyl-*sn*-glycero-3-phosphodiglycerol (DPPG2) liposome (Tm~43 °C) at a molar ratio of 50:20:30, respectively, for their TSLs in a magnetic hyperthermia study observing drug release using MRI markers. Here, they were able to successfully target drug delivery using temperature control [161].

MNPs can have an infinite number of formulations, with options including, but not limited to, their size, shape and composition. Generally, their size ranges from 1 to 100 nm in diameter [162]. These particles are commonly made from multiple elements including iron, cobalt, nickel and platinum. MPs are mainly classified based on their structure between magnetic alloy particles and magnetic metal oxide particles, the latter of which is used in drug delivery [163]. Metal oxides, particularly iron oxides, have already been seen as a promising candidate in magnetic hyperthermia, demonstrating abilities to kill cells locally through magnetically induced heating [129]. The properties of the IOPs impact the efficiency of these particles to confer heating to their immediate environments. For example, there is an increase in the specific absorption rate (SAR), the rate at which radiofrequency energy is absorbed, when nanorods are used compared to spherical and cubic forms due to their 1-dimensional nature. SAR is an important aspect to triggering the release of drugs as a higher absorption rate equates to particles heating up more as a certain amount of energy is applied. Therefore, lower energies and fewer particles are needed within the tissues for drug release. Das et al. found that the SAR can be changed by adjusting the aspect ratio of nanorods; higher aspect ratio of nanoparticles resulted in higher SAR values [164]. Furthermore, the ellipsoidal shape of magnetic nanorods can influence two effects when in an AMF; extra heat is released compared to nanospheres due to shape anisotropy and the nanorods dynamically reorient to the field [165]. These two properties are important as the former equates to extra heat release efficiency meaning fewer particles and lower field intensities are required during hyperthermia treatments, while the latter effect could be used to develop nanorobots in magnetic hyperthermia through controlled motion and orientation. The hyperthermic efficiency of nanorods, relative to their cubic and spherical counterparts of similar magnetic volumes, was further confirmed elsewhere. One study also compared the heating efficiency of nanospheres versus deformed cubes (orthopods) ranging from 17–47 nm. Throughout this size range, orthopods had a higher heating capacity and changing the size and shape of these particles changes the SAR [166]. Additonally, iron oxide nano-octopods were found to have better heating efficiency as compared to spheres [167]. Another factor that can impact the SAR of the NPs is their composition. A 2018 paper by Espinosa et al. compared the SAR values of maghemite-based IONPs (Fe_3O_4) and cobalt ferrite NPs ($CoFe_2O_4$) at clinically relevant settings (470 kHz) and found a small, but statistically significant, increase in SAR using $CoFe_2O_4$ NPs [168]. While other compositions have been studied, including $MnFe_2O_4$ and $NiFe_2O_4$, another factor that has been suggested to impact the IONP's SAR value is the iron oxide's oxidation state [169]. Overall, these experiments demonstrate that the heating efficiencies of MNPs can be modulated by the MNP's shape, size, and composition.

3.5. Future Perspectives

Using MFs as a release trigger is relatively new. The use of AMFs and IOPs encapsulated within liposomes has not yet been examined in clinical trials. Using MNPs in conjunction with thermosensitive liposomes for triggered release is a promising modality for cancer treatment. Some challenges, however, remain in the clinical application of these systems. One of the most pressing is determining the optimal MF parameters that maximize targeted heating of tissue or particles without damage to healthy cells. In this case, using MNPs to potentiate the heating effects of the MF are an advantage. Increasing MF strength may increase the SAR or heating potential of injected MPs, but too high of a field strength would lead to non-specific heating of the tissues [126]. Because of this, there has been much interest in producing MNPs with superior SAR values [164].

It is important to understand the toxicity, biocompatibility, and biodistribution of MNPs when using them in a triggering system. The biocompatibility of MNPs is linked to both the immune system response following its administration and to the intrinsic toxicity of the MNP and/or of its biodegradation metabolites. Factors that can influence the MNP's toxicity include their surface coating, size, and surface charge. Typically, toxicities can be avoided using compatible coatings (such as PEG and starches), small sizes (within the nanoparticle range) with appropriate doses, and nearly neutral charges ($+/-10$ mV). The chemical composition of MNPs also plays a role in their toxicity. IOPs, for example, have been found to be safe at high doses (100 s to 1000 s of mg/kg) via oral, intravenous, intraperitoneal, and subcutaneous administration. Once IOPs are injected, the IOPs are exposed to opsonization and accumulate in macrophages of the reticuloendothelial system. This includes organs such as the liver, spleen, and bone marrow. Despite this accumulation, major toxic side effects are rare as cells are able to incorporate the iron from the IOPs into their endogenous iron metabolism. If an iron overload does occur, however, the tissues may experience oxidative stress and injury to their cell membrane [149]. Cellular iron overload is rare but can be overcome by changing the biodistribution of IOPs through their magnetic properties. This allows the MNPs, via an external MF, to be guided to the site of interest [170–173]. Although MNPs may accumulate in tissues that are not of interest in the absence of guiding MFs, there seems to be promise in the sequestering capabilities of the particles in the context of cancer. One study done on fibrosarcoma tumor bearing mice looked at a novel formulation of co-encapsulated $La_{0.75}Sr_{0.25}MnO_3$ and IOPs that, under hyperthermia, resulted in tumor reduction by up to 3.6 fold, with little to no drainage of the particles to other organs in the body [140]. Future prospects of utilizing MNPs in conjunction with liposomes is promising as both components have been clinically approved as single agents (such as Doxil®, ThermoDox®, Caelyx®

Feraheme®, Feridex®, Gastromark®, etc.). There is potential for developing MPs as agents that react to hyperthermia and release liposome encapsulated drugs on-demand. At present, however, most studies are pre-clinical and much work remains to identify specific applications that take advantage of their unique, synergistic properties for clinical use.

4. X-ray Radiation

4.1. Introduction

Radiation therapy is one of the most effective modes of cancer treatment given the recent advancements in defining the spatial precision and depth penetration of ionizing radiation. More than 50% of all cancer patients receive radiotherapy over the course of their treatment with a curative or palliative intent [174]. By irradiating tumors with high energy photons or ion beams, cancer cells and the surrounding vasculature are irreparably damaged leading to tumor death [175,176]. Although radiotherapy is non-specific and can damage healthy tissue along the path of the photons, it remains the major course of treatment for primary non-metastasized solid tumors [174,177]. Concurrent RT and chemotherapy are also used in cancer treatment, particularly for unresectable tumors. Clinically, head and neck cancer patients with unresectable disease are treated with concurrent RT and cisplatin (CPT) to take advantage of drug-X-ray synergies for tumor control [178,179]. Unfortunately, in many patients, the systemic toxicity of CPT (hearing loss, kidney and nerve damage) is dose limiting [179,180]. Thus, using a delivery system where CPT is released locally at the irradiated site by X-rays would (1) minimize systemic chemotoxicity; (2) potentiate the efficacy of X-rays, and (3) potentially control the tumor with lower doses of radiation, chemotherapy, or both. Liposomes have demonstrated utility in chemotherapy as drug delivery vehicles by prolonging circulation time and increasing drug retention in tumors with several formulations used clinically [181]. Thus, it is not inconceivable that radiation-sensitive liposomes could be incorporated into pre-existing treatment plans for concurrent use with traditional radiotherapy. In comparison to other spatially-specific release systems, X-ray-triggered liposomal drug release is a relatively new concept. However, based on recent research focused on radiosensitization with gold nanoparticles, there is strong evidence suggesting that more efficient and effective systems can be designed to use radiation as a modality for triggered drug release [182].

The mechanism responsible for inducing the destabilization of the liposomal membrane is the radiosensitization effect [183,184], Figure 4. Radiosensitizers, such as gold nanoparticles, enhance the local radiation dose through the increased absorption of low and medium-energy X-rays and subsequent ejection of reactive secondary electrons [185]. A study by Sicard-Roselli et al. describes the direct and indirect mechanisms by which hydroxyl radicals are produced from gold nanoparticles irradiated in water. The direct mechanism produces hydroxyl radicals through the emission of electrons or lower energy photons from gold nanoparticles which interact with water, while the indirect mechanism involves the interaction of radiolysis products with gold nanoparticles which then eject electrons that interact with water [184]. The hydroxyl radicals produced through both pathways react with nucleic acids, proteins, and lipids located within their vicinity. In particular, reactive oxygen species are known to simultaneously: (i) initiate lipid peroxidation, a process which entails the abstraction of hydrogen atoms from lipid fatty acid chains, (ii) form peroxyl radicals, and (iii) convert fatty acid side chains into lipid hydroperoxides [183]. Although there have been no direct studies examining the mechanisms of radiosensitization in conjunction with liposomes, in theory, the local production of hydroxyl radicals and secondary electrons mediated by embedded radiosensitizers should cause lipid peroxidation and liposomal bilayer destabilization when irradiated, thereby triggering drug release [183,184].

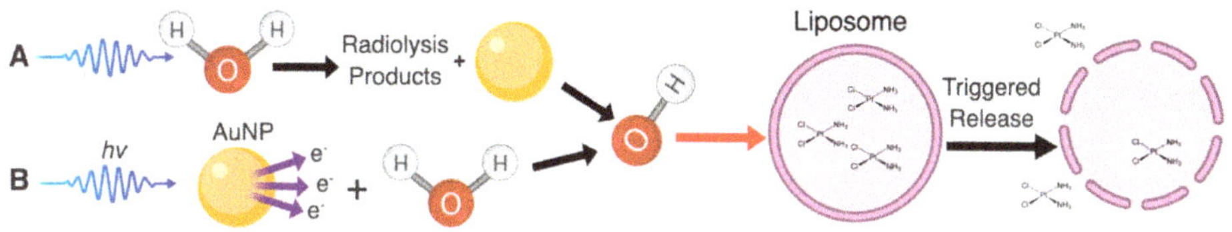

Figure 4. Schematic of the mechanisms associated with RT that can be used to destabilize liposomes. The mechanisms are as follows: (**A**) The interaction of radiation with water to produce radiolysis products that can interact with AuNPs to amplify hydroxy radical production. These radicals can then destabilize liposome bilayers for triggered drug release; (**B**) The interaction of radiation with AuNPs to produce secondary electrons (such as Compton scattering and Auger electrons) which can then interact with water to produce hydroxy radicals. These radicals can then destabilize liposome bilayers for triggered drug release.

To date, there have been several hundred studies exploring the potential therapeutic use of gold nanoparticles for radiosensitization [186]. The effect was initially shown by Hainfeld et al. who found a four-fold increase (86% versus 20%) in one-year survival rates for mice receiving both gold nanoparticles and X-ray therapy versus X-ray therapy alone. The mice were injected with 1.9 nm diameter gold nanoparticles (up to 2.7 g of Au/kg body mass) and irradiated with a 250 kVp X-ray beam [187]. In 2005, a Monte Carlo study, based on the aforementioned mouse study, was published by Cho, estimating a physical dose enhancement factor (DEF) of at least two-fold [188]. More than thirty reports have demonstrated radiosensitization effects in vitro with DEFs generally ranging from 1.1 to 1.9, while more than ten reports have shown radiosensitization effects in animal studies, showing that treatment with gold nanoparticles and X-ray cause tumor regression, or increased cell kill [189]. The initial understanding of the mechanism behind radiosensitization was attributed to physical factors, such as the high atomic number and photoelectric cross section of gold [190]. Monte Carlo simulations were used to discern the type and number of electrons emitted depending on variables such as the source, type, and energy of X-rays, as well as the size, concentration, and coating of gold nanoparticles. Generally, it was found that small gold nanoparticles at a high concentration, when irradiated with keV photon beams, generate the highest DEFs. However, as the in vivo and in vitro radiosensitization effects were often greater than predicted DEF values, it has now become well understood that complex chemical and biological interactions, such as the generation of hydroxyl radicals as mentioned above, are also involved radiosensitization, and therefore, require further investigation [189].

4.2. Formulation Factors

4.2.1. Radiation Type and Energy

Physical dose enhancement depends largely upon the energy and type of the incoming radiation. Kilovoltage (KV) photons with energies above the k-edge of gold (80.7 keV) have been shown to produce maximal dose enhancement due to the ability of these photons to excite the lowest-lying K-shell electrons. These electrons are ejected by the photoelectric effect and cause the subsequent emission of lower energy secondary electrons from the gold atom, known as the Auger cascade. Although a majority of the electrons are reabsorbed by other atoms in the gold nanoparticle, 1 to 7 electrons from each gold atom ultimately escape to interact with the environment [191]. Leung et al. reported that these electrons could travel 3 μm to 1 mm from the gold nanoparticle, while Jones et al. found that the dose enhancement effect was significant only a few microns away [192,193]. Photons with energies under the k-edge of gold (for example, 40–50 keV) have also been shown to produce radiosensitization effects through the ejection of higher shell (L, M, N) electrons and a localized Auger cascade, as the mass energy absorption coefficient of gold is over 100 times greater than soft tissue in the 40 to 50 keV energy range. However, due to the poor penetration of lower energy X-ray beams through soft tissue, this strategy would not be clinically feasible unless brachytherapy seeds were implanted in close proximity to the nanoparticles [190].

Although higher dose enhancement factors were observed using kilovoltage photon beams, megavoltage (MV) photons have also been shown to produce gold nanoparticle-mediated radiosensitization [194,195]. For 4 and 6 MV photon beams, dose enhancement factors ranging from 1.01 to 1.07 were predicted by Cho, and therefore, were not initially considered for dose enhancement [188]. However, in an in vivo mouse study, Chang et al. demonstrated that 25 Gy of 6 MeV radiation could produce significant tumor volume reduction in the presence of gold nanoparticles [196]. This could be explained by the increased absorption of secondary species produced by the ionization of water [177,190]. More recently, Yang et al. showed that the incorporation of gold nanoparticles into a chemoradiation regiment using cisplatin and 2 Gy of 6 MV radiation caused a 19% decrease in cell survival compared to cisplatin and radiation therapy alone [197]. Again, the observed dose enhancement was greater than predicted by Monte Carlo stimulations, indicating that physics dosimetry plays a smaller role in MV radiosensitization [191]. As MV photon beams are often used for radiotherapy and have a greater depth penetration in tissue, they are important for clinical applications of radiosensitization [198].

Gold nanoparticles have also been shown to interact with ion beams which is being explored for treatment because of its specificity in dose deposition attributed to its defined Bragg peak [199]. In a theoretical study, Verkhovtsev et al. showed that ion beams caused the collective electronic excitation of the surface plasmon in metal nanoparticles [200]. This effect was found to be particularly strong for noble metals due to the high excitability of the surface plasmon where the relaxation energy released causes the subsequent ejection of reactive electrons [177,200].

4.2.2. Gold Nanoparticle Size and Concentration

The size and concentration of the nanoparticles affect the degree of radiosensitization, since a greater number of gold atoms being irradiated generally causes greater dose enhancement [186]. Thus, higher concentrations of gold nanoparticles of the same size or larger gold nanoparticles at the same concentration produce greater dose enhancement effects [190,191]. However, when considering the optimal size of gold nanoparticles for a given mass of gold, smaller nanoparticle clusters produce greater dose enhancement effects. As the diameter of the gold nanoparticles increases, more of the secondary electrons and radiation are absorbed by the nanoparticle core, thereby reducing the energy available to interact with the surrounding environment [191]. In a Monte Carlo study, Lechtman et al. (2011) showed that a greater number of low energy Auger electrons were released by smaller nanoparticles while a greater number of high energy photoelectrons were released by larger nanoparticles [191]. This size effect has also been shown in a simulated cell study where 2 nm gold nanoparticles produced greater cell deaths than 50 nm gold nanoparticles when irradiated [201].

4.3. Future Perspectives

Currently, studies on X-ray triggered liposomal drug release are very limited. To our knowledge, only one study exists—that by Deng et al.—in which the authors used X-rays to trigger release from liposomes in vitro and in vivo. The system consisted of liposomes embedded with gold nanoparticles and verteporfin, a photosensitizer, where 19% of encapsulated calcein was released upon irradiation with 4 Gy of 6 MeV photons. Increased gene silencing and cell death were observed in vitro with the triggered release of antisense oligonucleotides and chemotherapy drugs, respectively. In a xenograft mouse model, X-ray triggered liposomes were shown to produce a 74% reduction in colorectal tumor volume compared to the control with phosphate buffered saline [182]. In a different study by Lukianova-Hleb et al., the authors relied on the increased endosomal uptake of AuNPs and liposomes in cancer cells to colocalize the NPs for triggered release. This paper demonstrated the synergy between liposomes, AuNPs, low-energy short laser pulses, and X-rays to induce plasmonic nanobubbles and

ROS formation for the destabilization of the liposomes. Despite the absence of AuNP conjugation or encapsulation within the liposome, this paper alludes to the possibility of relying on the tumor's biology to bring a carrier and triggering component in close enough proximity to each other to allow for triggered release [202]. Although these studies showed that X-rays are a promising modality for X-ray triggered liposomal drug release, many areas of possible development remain to increase the amount of drug released. For example, the amount of drug released could be improved using liposomal formulations containing higher concentrations of embedded gold nanoparticles [190]. Additionally, protein nanoparticles, such as Albumin, could possibly be used in place of liposomes since hydroxyl radicals interact strongly with proteins as well [203].

As X-ray-triggered drug release is still in its early stages of development, a better understanding of gold nanoparticle embedded liposomes, their toxicities, and the effects of radiotherapy fractionation is needed before clinical translation can be considered. Formulation factors affecting radiosensitization using gold nanoparticles have been well studied, but not in the context of liposomes. Since the incorporation of more gold nanoparticles increases local dose enhancement but destabilizes the membrane at high concentrations, an optimization of the two factors is needed [190,204]. To date, there has been one study outlining the in vitro pharmacokinetics of gold nanoparticle embedded liposomes. The intravenous injection of gold nanoparticle embedded liposomes (100 to 120 nm) into a fibrosarcoma mouse model resulted in the accumulation of gold nanoparticles in the liver, spleen, kidney, and intestines, but none in the tumor site [205]. This could possibly be due to the specific formulation of the liposome, as liposomes with diameters ranging from 100 to 300 nm have typically been found to accumulate near tumors due to the enhanced permeability and retention (EPR) effect [206,207]. Additionally, actively targeting the tumor using targeting ligands could increase the uptake efficiency of NPs in cancer cells. Significant progress has been made in this area of research both at the preclinical and at the clinical level and can be found in detail elsewhere [208]. A better understanding of the long-term toxicology of gold nanoparticles is also necessary, especially at the clinical level. Since surface chemistry, routes of administration, and dosages used vary extensively across pre-clinical toxicology studies, different results are found in the literature for nanoparticles of a given size [209]. For example, one study showed that gold nanoparticles ranging from 8–37 nm caused hepatocellular toxicities, while another showed that 13 nm PEG-capped gold nanoparticles caused no systemic toxicities [210,211]. Although the FDA has not approved any gold nanoparticle-based drugs for clinical use, a clinical trial of Aurimune®, which carries tumor necrosis factor into tumors, has successfully passed its first phase [212,213].

As radiotherapy can cause damage to normal tissue surrounding the tumor, careful consideration must be taken to limit the dose of radiation delivered during RT [177]. Therefore, the use of liposomes in conjunction with gold nanoparticles that promote radiosensitization is an attractive triggered therapy approach when taking into account the negative side effects that come from high dose RT. A feasible strategy for clinical translation would be to incorporate X-ray triggered drug release into existing treatment plans which use concurrent radiotherapy with chemotherapy. Examples include head and neck cancers, upper esophagus cancers, small cell lung cancer, and cervical cancer [179,214–217]. In particular, head and neck cancers and small cell lung cancer treatment plans recommend the use of over 60 Gy of radiation with concomitant chemotherapy where the total dose is fractionated to 2 Gy daily [217]. To maximize X-ray triggered drug release, irradiation should occur when the greatest concentration of drug loaded liposomes are found in the tumor site. Additionally, larger doses of radiation could be incorporated at this point to increase drug release if the treatment allows. Lastly, synergistically radiation-activated drugs, such as those used for photodynamic therapy, could be encapsulated and used for deep lying tumors that near infrared or visible light would be unable to reach [218]. Future prospects in this field are promising and experimental simulation of a clinical dosing schedule should be explored to better characterize the X- ray triggered liposomal release system as a whole.

5. Conclusions

This review explores the use of energy sources, including US, MFs, and external beam radiation, to trigger the delivery of drugs from liposomes in a tumor in a spatially-specific manner. The mechanism(s) of drug release that can be achieved using liposomes in conjunction with the external trigger were investigated for each of the energy sources. Figure 5 summarizes the mechanisms that can be achieved in each modality and the commonalities, such as hyperthermia and hydroxyl radical formation, found between some of the devices. While this paper identified the growing interest and advantages in using external stimuli to trigger drug release from liposomes, it also demonstrated the drawbacks associated with each method. Themes such as the range of penetration depth and off-target tissue damage, or lack thereof, were discussed in the advantages and disadvantages for each of the energy sources. Considerations such as these must be taken into account on a case by case basis and will impact the types of cancers that can be targeted with each modality. Furthermore, this review also detailed the treatment's formulation factors and explored the parameters of both the therapy and the energy source. Each energy source identified a correlation between the size and concentration of their corresponding mutually exclusive particle (such as the MB's, IONPs, or AuNPs) with the treatment's experimental impact. Understanding each method's formulation factors will aid in the development of future therapies which are susceptible to influence from external stimuli. Additionally, the pre-clinical and clinical trials of each triggered release method were explored. At the time of writing this review, only US used in conjunction with liposomes, specifically HIFU induced liposomal release, had clinical trials in motion as a method for cancer therapy. While these represent only three clinical trials, this only further highlights the feasibility and positive future outlook of utilizing the energy sources found in medical devices as external stimuli to induce liposomal release in the context of cancer therapy.

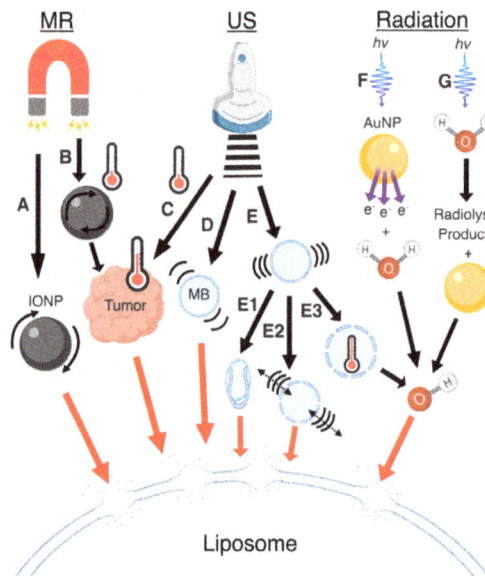

Figure 5. Schematic of the mechanisms associated with MR, US, and RT that can be used to destabilize liposomes. The mechanisms are as follows: (**A**) MR induced mechanical disruption; (**B**) MR induced hyperthermia; (**C**) US induced hyperthermia; (**D**) US induced stable cavitation of a MB; (**E**) US induced inertial cavitation of a MB; (**E1**) The production of a liquid microjets from a MB undergoing inertial cavitation; (**E2**) The production of shockwaves from a MB undergoing inertial cavitation; (**E3**) A MB collapsing and undergoing sonochemical changes; (**F,G**) The interaction of radiation with AuNPs and water to produce hydroxyl radicals that destabilize liposome bilayers.

Funding: This research received no external funding.

Conflicts of Interest: The authors declare no conflict of interest.

Abbreviations

1,2-dimyristoyl-*sn*-glycero-3-phosphocholine	(DMPC)
1,2-dipalmitoyl-*sn*-glycero-3-phosphodiglycerol	(DPPG2)
1,2-dipalmitoyl-*sn*-glycerophosphocholine	(DPPC)
1,2-distearoyl-*sn*-glycerophosphocholine	(DSPC)
1-myristoyl-2-palmitoyl-*sn*-glycero-3-phosphocholine	(MPPC)
Alternating magnetic field	(AMF)
Amplitude	(Amp)
Blood brain barrier	(BBB)
Bovine serum albumin	(BSA)
Bubble resonant radius	(BRR)
Center frequency	(F_c)
Cisplatin	(CPT)
Continuous wave	(CW)
Diheptanoylphosphatidyl-choline	(DHPC)
Dioleoylphosphatidylethanolamine	(DOPE)
Distearoyl-*sn*-glycero-3-phosphoethanolamine	(DSPE)
Dose enhancement factor	(DEF)
Dynamic magnetic field	(DMF)
Enhanced permeability and retention	(EPR)
Egg phosphocholine	(egg PC)
Focused ultrasound	(FUS)
Frequency	(*f*)
High Intensity Focused Ultrasound	(HIFU)
Hydroxyterephthalic acid	(HTA)
Iron oxide particles	(IOPs)
Kilovoltage	(KV)
L-α-phosphatidylcholine	(HSPC)
Low frequency US	(LFUS)
Magnetic fields	(MF)
Magnetic nanoparticles	(MNPs)
Magnetic particles	(MPs)
Magnetic resonance imaging	(MRI)
Malondialdehyde	(MDA)
Mechanical index	(MI)
Megavoltage	(MV)
Methotrexate	(MTX)
Methylpredinisolone hemisuccinate	(MPS)
Microbubbles	(MBs)
Not specified	(NS)
Peak negative pressure	(PnP)
Perfluorocarbon	(PFC)
Polyethylene glycol	(PEG)
Pulse duration	(PD)
Pulse repetition frequency	(PFR)
Radiation	(RT)
Radiation forces	(RTFs)
Reactive oxygen	(ROS)

Specific absorption rate	(SAR)
Temperature sensitive liposomes	(TSLs)
Total acoustic power	(TAT)
Ultrasound	(US)

References

1. Tsimberidou, A.-M. Targeted Therapy in Cancer. *Cancer Chemother. Pharmacol.* **2015**, *76*, 1113–1132. [CrossRef]
2. Gerwing, M.; Herrmann, K.; Helfen, A.; Schliemann, C.; Berdel, W.E.; Eisenblätter, M.; Wildgruber, M. The Beginning of the End for Conventional RECIST—Novel Therapies Require Novel Imaging Approaches. *Nat. Rev. Clin. Oncol.* **2019**. [CrossRef] [PubMed]
3. Zylberberg, C.; Matosevic, S. Pharmaceutical Liposomal Drug Delivery: A Review of New Delivery Systems and a Look at the Regulatory Landscape. *Drug Deliv.* **2016**, *23*, 3319–3329. [CrossRef] [PubMed]
4. Allen, T.M.; Cullis, P.R. Liposomal Drug Delivery Systems: From Concept to Clinical Applications. *Adv. Drug Deliv. Rev.* **2013**, *65*, 36–48. [CrossRef] [PubMed]
5. Mitragotri, S. Healing Sound the Use of Ultrasound in Drug Delivery and Other Therapeutic Applications. *Nat. Rev. Drug Discov.* **2005**, *4*, 5–10. [CrossRef] [PubMed]
6. Goss, S.A.; Johnston, R.L.; Dunn, F. Compilation of Empirical Ultrasonic Properties of Mammalian Tissues. II. *J. Acoust. Soc. Am.* **1980**, *68*, 93–108. [CrossRef]
7. Pierce, A.D. Acoustics: An Introduction to Its Physical Principles and Applications. *Phys. Today* **1981**, *34*, 56–57. [CrossRef]
8. Fry, W.J.; Wulff, V.J.; Tucker, D.; Fry, F.J. Physical Factors Involved in Ultrasonically Induced Changes in Living Systems: I. Identification of Non-Temperature Effects. *J. Acoust. Soc. Am.* **1950**, *22*, 867–876. [CrossRef]
9. Halliwell, M.A. Tutorial on Ultrasonic Physics and Imaging Techniques. *Proc. Inst. Mech. Eng. Part H J. Eng. Med.* **2010**, *224*, 127–142. [CrossRef]
10. Bailey, M.R.; Khokhlova, V.A.; Sapozhnikov, O.A.; Kargl, S.G.; Crum, L.A. Physical Mechanisms of the Therapeutic Effect of Ultrasound (a Review). *Acoust. Phys.* **2003**, *49*, 369–388. [CrossRef]
11. Yatvin, M.B.; Weinstein, J.N.; Dennis, W.H.; Blumenthal, R. Design of Liposomes for Enhanced Local Release of Drugs by Hyperthermia. *Science* **1978**, *202*, 1290–1293. [CrossRef] [PubMed]
12. Mabrey, S.; Sturtevant, J.M. Investigation of Phase Transitions of Lipids and Lipid Mixtures by Sensitivity Differential Scanning Calorimetry. *Proc. Natl. Acad. Sci. USA* **1976**, *73*, 3862–3866. [CrossRef] [PubMed]
13. Ruocco, M.J.; Siminovitch, D.J.; Griffin, R.G. Comparative Study of the Gel Phases of Ether- and Ester-Linked Phosphatidylcholines. *Biochemistry* **1985**, *24*, 2406–2411. [CrossRef] [PubMed]
14. Weinstein, J.N.; Magin, R.L.; Yatvin, M.B.; Zaharko, D.S. Liposomes and Local Hyperthermia: Selective Delivery. *Science* **1979**, *204*, 188–191. [CrossRef]
15. Al-Ahmady, Z.S.; Al-Jamal, W.T.; Bossche, J.V.; Bui, T.T.; Drake, A.F.; Mason, A.J.; Kostarelos, K. Lipid-Peptide Vesicle Nanoscale Hybrids for Triggered Drug Release by Mild Hyperthermia in Vitro and in Vivo. *ACS Nano* **2012**, *6*, 9335–9346. [CrossRef] [PubMed]
16. Tacker, J.R.; Anderson, R.U. Delivery of Antitumor Drug to Bladder Cancer by Use of Phase Transition Liposomes and Hyperthermia. *J. Urol.* **1982**, *127*, 1211–1214. [CrossRef]
17. Margulis, M.A. *Sonochemistry and Cavitation*; Gordon Breach Science Publishers: Langhorne, PA, USA, 1993.
18. Young, F.R. *Cavitation*; McGraw-Hill: Maidenhead, UK, 1989. [CrossRef]
19. Crum, L.A. Acoustic Cavitation Series: Part Five Rectified Diffusion. *Ultrasonics* **1984**, *22*, 215–223. [CrossRef]
20. Coussios, C.C.; Roy, R.A. Applications of Acoustics and Cavitation to Noninvasive Therapy and Drug Delivery. *Annu. Rev. Fluid Mech.* **2008**, *40*, 395–420. [CrossRef]
21. Treat, L.H.; McDannold, N.; Zhang, Y.; Vykhodtseva, N.; Hynynen, K. Improved Anti-Tumor Effect of Liposomal Doxorubicin after Targeted Blood-Brain Barrier Disruption by MRI-Guided Focused Ultrasound in Rat Glioma. *Ultrasound Med. Biol.* **2012**, *38*, 1716–1725. [CrossRef] [PubMed]
22. Ammi, A.Y.; Linder, J.R.; Zhao, Y.; Porter, T.; Siegel, R.; Kaul, S. Efficacy and spatial distribution of ultrasound-mediated clot lysis in the absence of thrombolytics. *Thromb. Haemost.* **2015**, *113*, 1357–1369. [CrossRef] [PubMed]

23. Belcik, J.T.; Davidson, B.P.; Xie, A.; Wu, M.D.; Yadava, M.; Qi, Y.; Liang, S.; Chon, C.R.; Ammi, A.Y.; Field, J.; et al. Augmentation of Muscle Blood Flow by Ultrasound Cavitation is Mediated by Atp and Purinergic Signaling. *Circulation* **2017**, *135*, 1240–1252. [CrossRef] [PubMed]

24. Burke, C.W.; Alexander, I.V., 4th; Timbie, K.; Kilbanov, A.L.; Price, R.J. Ultrasound-activated Agents Comprised of 5FU-bearing Nanoparticles Bonded to Microbubbles Inhibit Solid Tumor Growth and Improve Survival. *Mol. Ther.* **2014**, *22*, 321–328. [CrossRef] [PubMed]

25. Doinikov, A.A.; Bouakaz, A. Acoustic Microstreaming around an Encapsulated Particle. *J. Acoust. Soc. Am.* **2010**, *127*, 1218–1227. [CrossRef] [PubMed]

26. Marmottant, P.S. Controlled Vesicle Deformation and Lysis by Single Oscillating Bubbles. *Nature* **2003**, *423*, 153–156. [CrossRef]

27. Shi, X.; Martin, R.W.; Vaezy, S.; Crum, L.A. Quantitative Investigation of Acoustic Streaming in Blood. *J. Acoust. Soc. Am.* **2002**, *111*, 1110–1121. [CrossRef] [PubMed]

28. Zhao, S.; Borden, M.; Bloch, S.H.; Kruse, D.; Ferrara, K.W.; Dayton, P.A. Radiation-Force Assisted Targeting Facilitates Ultrasonic Molecular Imaging. *Mol. Imaging* **2004**, *3*, 135–148. [CrossRef] [PubMed]

29. Zhang, Y.; Guo, X.; Zhang, D.; Gong, X. Evaluation of the Effects of Secondary Radiation Force on Aggregation of Ultrasound Contrast Agents. In Proceedings of the 20th International Congress on Accoustics, Sydney, Australia, 23–27 August 2010.

30. Unger, E.C.; Porter, T.; Culp, W.; Labell, R.; Matsunaga, T.; Zutshi, R. Therapeutic Applications of Lipid-Coated Microbubbles. *Adv. Drug Deliv. Rev.* **2004**, *56*, 1291–1314. [CrossRef]

31. Unger, E.C.; McCreery, T.P.; Sweitzer, R.H.; Caldwell, V.E.; Wu, Y. Acoustically Active Liposheres Containing Paclitaxel: A New Therapeutic Ultrasound Contrast Agent. *Investig. Radiol.* **1998**, *33*, 886–892. [CrossRef]

32. Suzuki, R.; Namai, E.; Oda, Y.; Nishiie, N.; Otake, S.; Koshima, R.; Hirata, K.; Taira, Y.; Utoguchi, N.; Negishi, Y.; et al. Cancer Gene Therapy by IL-12 Gene Delivery Using Liposomal Bubbles and Tumoral Ultrasound Exposure. *J. Control. Release* **2010**, *142*, 245–250. [CrossRef]

33. Shortencarier, M.J.; Dayton, P.A.; Bloch, S.H.; Schumann, P.A.; Matsunaga, T.O.; Ferrara, K.W. A Method for Radiation-Force Localized Drug Delivery Using Gas-Filled Liposheres. *IEEE Trans. Ultrason. Ferroelectr. Freq. Control* **2004**, *51*, 822–831. [CrossRef]

34. Yang, C.Y.; Liu, H.W.; Tsai, Y.C.; Tseng, J.Y.; Liang, S.C.; Chen, C.Y.; Lian, W.N.; Wei, M.C.; Lu, M.; Lu, R.H.; et al. Interleukin-4 Receptor-Targeted Liposomal Doxorubicin as a Model for Enhancing Cellular Uptake and Antitumor Efficacy in Murine Colorectal Cancer. *Cancer Biol. Ther.* **2015**, *16*, 1641–1650. [CrossRef] [PubMed]

35. Wang, C.-H.; Huang, Y.-F.; Yeh, C.-K. Aptamer-Conjugated Nanobubbles for Targeted Ultrasound Molecular Imaging. *Langmuir* **2011**, *27*, 6971–6976. [CrossRef]

36. Unger, E.C.; Mccreery, T.P.; Shen, D.; Wu, G.; Sweitzer, R.; Wu, Q. Gas-Filled Liposomes as Ultrasound Contrast Agents for Blood Pool, Thrombus-Specific and Therapeutic Applications. *J. Acoust. Soc. Am.* **1998**, *103*, 3001. [CrossRef]

37. Suslick, K.S.; Nyborg, W.L. ULTRASOUND: Its Chemical, Physical and Biological Effects. *J. Acoust. Soc. Am.* **1990**, *87*, 919–920. [CrossRef]

38. Suslick, K.S. Ultrasound. In *Its Chemical, Physical, and Biological Effects*; VCH: New York, NY, USA, 1988. [CrossRef]

39. Suslick, K.S. Sonoluminescence and Sonochemistry. In *Encyclopedia of Physical Science and Technology*, 3rd ed.; Meyers, R.A., Ed.; Academic Press, Inc.: San Diego, CA, USA, 2001; Volume 1, pp. 523–532.

40. McNamara, W.B.; Didenko, Y.T.; Suslick, K.S. Sonoluminescence Temperatures during Multi-Bubble Cavitation. *Nature* **1999**, *401*, 772–775. [CrossRef]

41. Didenko, Y.T.; McNamara, W.B.; Suslick, K.S. Temperature of Multibubble Sonoluminescence in Water. *J. Phys. Chem. A* **1999**, *103*, 10783–10788. [CrossRef]

42. Weissler, A. Formation of Hydrogen Peroxide by Ultrasonic Waves: Free Radicals. *J. Am. Chem. Soc.* **1959**, *81*, 1077–1081. [CrossRef]

43. Riesz, P.; Christman, C.L. Sonochemical Free Radical Formation in Aqueous Solutions. *Fed. Proc.* **1986**, *45*, 2485–2492.

44. Umemura, S.; Yumita, N.; Nishigaki, R.; Umemura, K. Sonochemical Activation of Hematoporphyrin—A Potential Modality for Cancer Treatment. In Proceedings of the IEEE Ultrasonics Symposium, Montreal, QC, Canada, 3–6 October 1989. [CrossRef]

45. He, L.L.; Wang, X.; Wu, X.X.; Wang, Y.X.; Kong, Y.M.; Wang, X.; Liu, B.M.; Liu, B. Protein Damage and Reactive Oxygen Species Generation Induced by the Synergistic Effects of Ultrasound and Methylene Blue. *Spectrochim. Acta Part A Mol. Biomol. Spectrosc.* **2015**, *134*, 361–366. [CrossRef]

46. Kondo, T.; Mišík, V.; Riesz, P. Sonochemistry of Cytochrome C. Evidence for Superoxide Formation by Ultrasound in Argon-Saturated Aqueous Solution. *Ultrason. Sonochem.* **1996**, *3*, S193–S199. [CrossRef]

47. Miller, D.L.; Thomas, R.M.; Buschbom, R.L. Comet Assay Reveals DNA Strand Breaks Induced by Ultrasonic Cavitation in Vitro. *Ultrasound Med. Biol.* **1995**, *21*, 841–848. [CrossRef]

48. Leung, K.S.; Chen, X.; Zhong, W.; Yu, A.C.H.; Lee, C.-Y. Microbubble-Mediated Sonoporation Amplified Lipid Peroxidation of Jurkat Cells. *Chem. Phys. Lipids* **2014**, *180*, 53–60. [CrossRef]

49. Rahman, M.M.; Ninomiya, K.; Ogino, C.; Shimizu, N. Ultrasound-Induced Membrane Lipid Peroxidation and Cell Damage of Escherichia Coli in the Presence of Non-Woven TiO_2 Fabrics. *Ultrason. Sonochem.* **2010**, *17*, 738–743. [CrossRef] [PubMed]

50. Tsuru, H.; Shibaguchi, H.; Kuroki, M.; Yamashita, Y.; Kuroki, M. Tumor Growth Inhibition by Sonodynamic Therapy Using a Novel Sonosensitizer. *Free Radic. Biol. Med.* **2012**, *53*, 464–472. [CrossRef] [PubMed]

51. Prieur, F.; Pialoux, V.; Mestas, J.-L.; Mury, P.; Skinner, S.; Lafon, C. Evaluation of Inertial Cavitation Activity in Tissue through Measurement of Oxidative Stress. *Ultrason. Sonochem.* **2015**, *26*, 193–199. [CrossRef]

52. Pecha, R.; Gompf, B. Microimplosions: Cavitation Collapse and Shock Wave Emission on a Nanosecond Time Scale. *Phys. Rev. Lett.* **2000**, *84*, 1328–1330. [CrossRef] [PubMed]

53. Eggen, S.; Afadzi, M.; Nilssen, E.A.; Haugstad, S.B.; Angelsen, B.; Davies, C.d.L. Ultrasound Mediated Delivery of Liposomal Doxorubicin in Prostate Tumor Tissue. In Proceedings of the IEEE International Ultrasonics Symposium, Dresden, Germany, 7–10 October 2012. [CrossRef]

54. Yudina, A.; Lepetit-Coiffé, M.; Moonen, C.T.W. Evaluation of the Temporal Window for Drug Delivery Following Ultrasound-Mediated Membrane Permeability Enhancement. *Mol. Imaging Biol.* **2011**, *13*, 239–249. [CrossRef]

55. Larkin, J.O.; Casey, G.D.; Tangney, M.; Cashman, J.; Collins, C.G.; Soden, D.M.; O'Sullivan, G.C. Effective Tumor Treatment Using Optimized Ultrasound-Mediated Delivery of Bleomycin. *Ultrasound Med. Biol.* **2008**, *34*, 406–413. [CrossRef] [PubMed]

56. Nelson, J.L.; Roeder, B.L.; Carmen, J.C.; Roloff, F.; Pitt, W.G. Ultrasonically Activated Chemotherapeutic Drug Delivery in a Rat Model. *Cancer Res.* **2002**, *62*, 7280–7283.

57. Rapoport, N.Y.; Kennedy, A.M.; Shea, J.E.; Scaife, C.L.; Nam, K.-H. Controlled and Targeted Tumor Chemotherapy by Ultrasound-Activated Nanoemulsions/Microbubbles. *J. Control. Release* **2009**, *138*, 268–276. [CrossRef]

58. Park, E.-J.; Zhang, Y.-Z.; Vykhodtseva, N.; McDannold, N. Ultrasound-Mediated Blood-Brain/Blood-Tumor Barrier Disruption Improves Outcomes with Trastuzumab in a Breast Cancer Brain Metastasis Model. *J. Control. Release* **2012**, *163*, 277–284. [CrossRef] [PubMed]

59. Wei, K.-C.; Chu, P.-C.; Wang, H.-Y.; Huang, C.-Y.; Chen, P.-Y.; Tsai, H.-C.; Lu, Y.-J.; Lee, P.-Y.; Tseng, I.-C.; Feng, L.-Y.; et al. Focused Ultrasound-Induced Blood-Brain Barrier Opening to Enhance Temozolomide Delivery for Glioblastoma Treatment: A Preclinical Study. *PLoS ONE* **2013**, *8*, e58995. [CrossRef] [PubMed]

60. Kovacs, Z.; Werner, B.; Rassi, A.; Sass, J.O.; Martin-Fiori, E.; Bernasconi, M. Prolonged Survival upon Ultrasound-Enhanced Doxorubicin Delivery in Two Syngenic Glioblastoma Mouse Models. *J. Control. Release* **2014**, *187*, 74–82. [CrossRef] [PubMed]

61. McDannold, N.; Vykhodtseva, N.; Hynynen, K. Targeted Disruption of the Blood–Brain Barrier with Focused Ultrasound: Association with Cavitation Activity. *Phys. Med. Biol.* **2006**, *51*, 793–807. [CrossRef]

62. Hynynen, K.; McDannold, N.; Vykhodtseva, N.; Jolesz, F.A. Noninvasive MR Imaging–Guided Focal Opening of the Blood-Brain Barrier in Rabbits. *Radiology* **2001**, *220*, 640–646. [CrossRef] [PubMed]

63. Treat, L.H.; McDannold, N.; Vykhodtseva, N.; Zhang, Y.; Tam, K.; Hynynen, K. Targeted Delivery of Doxorubicin to the Rat Brain at Therapeutic Levels Using MRI-Guided Focused Ultrasound. *Int. J. Cancer* **2007**, *121*, 901–907. [CrossRef]

64. Guthkelch, A.N.; Carter, L.P.; Cassady, J.R.; Hynynen, K.H.; Iacono, R.P.; Johnson, P.C.; Obbens, E.A.M.T.; Roemer, R.B.; Seeger, J.F.; Shimm, D.S.; et al. Treatment of Malignant Brain Tumors with Focused Ultrasound Hyperthermia and Radiation: Results of a Phase I Trial. *J. Neurooncol.* **1991**, *10*, 271–284. [CrossRef] [PubMed]

65. Saito, M.; Mazda, O.; Takahashi, K.A.; Arai, Y.; Kishida, T.; Shin-Ya, M.; Inoue, A.; Tonomura, H.; Sakao, K.; Morihara, T.; et al. Sonoporation Mediated Transduction of PDNA/SiRNA into Joint Synovium in Vivo. *J. Orthop. Res.* **2007**, *25*, 1308–1316. [CrossRef]
66. Kinoshita, M.; Hynynen, K. A Novel Method for the Intracellular Delivery of SiRNA Using Microbubble-Enhanced Focused Ultrasound. *Biochem. Biophys. Res. Commun.* **2005**, *335*, 393–399. [CrossRef]
67. Taniyama, Y.; Tachibana, K.; Hiraoka, K.; Namba, T.; Yamasaki, K.; Hashiya, N.; Aoki, M.; Ogihara, T.; Yasufumi, K.; Morishita, R. Local Delivery of Plasmid DNA Into Rat Carotid Artery Using Ultrasound. *Circulation* **2002**, *105*, 1233–1239. [CrossRef]
68. Krasovitski, B.; Kimmel, E. Shear Stress Induced by a Gas Bubble Pulsating in an Ultrasonic Field near a Wall. *IEEE Trans. Ultrason. Ferroelectr. Freq. Control* **2004**, *51*, 973–979. [CrossRef] [PubMed]
69. Catania, A.E.; Ferrari, A.; Manno, M.; Spessa, E. A Comprehensive Thermodynamic Approach to Acoustic Cavitation Simulation in High-Pressure Injection Systems by a Conservative Homogeneous Two-Phase Barotropic Flow Model. *J. Eng. Gas Turbines Power* **2006**, *128*, 434–445. [CrossRef]
70. Bowden, F.P.; Brunton, J.H. The Deformation of Solids by Liquid Impact at Supersonic Speeds. *Proc. R. Soc. A Math. Phys. Eng. Sci.* **1961**, *263*, 433–450. [CrossRef]
71. Pecha, R.; Wang, Z.Q.; Gompf, B.; Eisenmenger, W. Single-bubble Sonoluminescence: Investigation of the Emitted Pressure Wave with a Streak Camera and a Fiber-optic Probe Hydrophone. *J. Acoust. Soc. Am.* **1999**, *105*, 960. [CrossRef]
72. Prosperetti, A. A New Mechanism for Sonoluminescence. *J. Acoust. Soc. Am.* **1997**, *101*, 2003–2007. [CrossRef]
73. Brujan, E.A.; Nahen, K.; Schmidt, P.; Vogel, A. Dynamics of Laser-Induced Cavitation Bubbles near an Elastic Boundary. *J. Fluid Mech.* **2001**, *433*, 251–281. [CrossRef]
74. Hajri, Z.; Boukadoum, M.; Hamam, H.; Fontaine, R. An Investigation of the Physical Forces Leading to Thrombosis Disruption by Cavitation. *J. Thromb. Thrombolysis* **2005**, *20*, 27–32. [CrossRef]
75. Kodama, T.; Takayama, K. Dynamic Behavior of Bubbles During Extracorporeal. *Ultrasound Med. Biol.* **1998**, *24*, 723–738. [CrossRef]
76. Wörle, K.; Steinbach, P.; Hofstädter, F. The Combined Effects of High-Energy Shock Waves and Cytostatic Drugs or Cytokines on Human Bladder Cancer Cells. *Br. J. Cancer* **1994**, *69*, 58–65. [CrossRef]
77. Steinbach, P.; Hofstädter, F.; Nicolai, H.; Rössler, W.; Wieland, W. In Vitro Investigations on Cellular Damage Induced by High Energy Shock Waves. *Ultrasound Med. Biol.* **1992**, *18*, 691–699. [CrossRef]
78. Couture, O.; Foley, J.; Kassell, N.; Larrat, B.; Aubry, J.-F. Review of Ultrasound Mediated Drug Delivery for Cancer Treatment: Updates from Pre-Clinical Studies. *Transl. Cancer Res.* **2014**, *3*, 494–511. [CrossRef]
79. Geers, B.; Dewitte, H.; De Smedt, S.C.; Lentacker, I. Crucial Factors and Emerging Concepts in Ultrasound-Triggered Drug Delivery. *J. Control. Release* **2012**, *164*, 248–255. [CrossRef]
80. Enden, G.; Schroeder, A. A Mathematical Model of Drug Release from Liposomes by Low Frequency Ultrasound. *Ann. Biomed. Eng.* **2009**, *37*, 2640–2645. [CrossRef] [PubMed]
81. Schroeder, A.; Avnir, Y.; Weisman, S.; Najajreh, Y.; Gabizon, A.; Talmon, Y.; Kost, J.; Barenholz, Y. Controlling Liposomal Drug Release with Low Frequency Ultrasound: Mechanism and Feasibility. *Langmuir* **2007**, *23*, 4019–4025. [CrossRef]
82. Afadzi, M.; Davies, C.d.L.; Hansen, Y.H.; Johansen, T.; Standal, Ø.K.; Hansen, R.; Måsøy, S.E.; Nilssen, E.A.; Angelsen, B. Effect of Ultrasound Parameters on the Release of Liposomal Calcein. *Ultrasound Med. Biol.* **2012**, *38*, 476–486. [CrossRef] [PubMed]
83. Lin, Y.; Lin, L.; Cheng, M.; Jin, L.; Du, L.; Han, T.; Xu, L.; Yu, A.C.H.; Qin, P. Effect of Acoustic Parameters on the Cavitation Behavior of SonoVue Microbubbles Induced by Pulsed Ultrasound. *Ultrason. Sonochem.* **2016**, *35*, 176–184. [CrossRef] [PubMed]
84. Mannaris, C.; Averkiou, M.A. Investigation of Microbubble Response to Long Pulses Used in Ultrasound-Enhanced Drug Delivery. *Ultrasound Med. Biol.* **2012**, *38*, 681–691. [CrossRef] [PubMed]
85. Mayer, R.; Schenk, E.; Child, S.; Norton, S.; Cox, C.; Hartman, C.; Cox, C.; Carstensen, E. Pressure Threshold for Shock Wave Induced Renal Hemorrhage. *J. Urol.* **1990**, *144*, 1505–1509. [CrossRef]
86. Child, S.Z.; Hartman, C.L.; Schery, L.A.; Carstensen, E.L. Lung Damage from Exposure to Pulsed Ultrasound. *Ultrasound Med. Biol.* **1990**, *16*, 817–825. [CrossRef]
87. Coussios, C.C.; Farny, C.H.; ter Haar, G.; Roy, R.A. Role of Acoustic Cavitation in the Delivery and Monitoring of Cancer Treatment by High-Intensity Focused Ultrasound (HIFU). *Int. J. Hyperth.* **2007**, *23*, 105–120. [CrossRef]

88. Kennedy, J.E. High-Intensity Focused Ultrasound in the Treatment of Solid Tumours. *Nat. Rev. Cancer* **2005**, *5*, 321–327. [CrossRef]
89. Hancock, H.; Dreher, M.R.; Crawford, N.; Pollock, C.B.; Shih, J.; Wood, B.J.; Hunter, K.; Frenkel, V. Evaluation of Pulsed High Intensity Focused Ultrasound Exposures on Metastasis in a Murine Model. *Clin. Exp. Metastasis* **2009**, *26*, 729–738. [CrossRef]
90. Steinbach, P.; Hofstaedter, F.; Nicolai, H.; Roessler, W.; Wieland, W. Determination of the Energy-Dependent Extent of Vascular Damage Caused by High-Energy Shock Waves in an Umbilical Cord Model. *Urol. Res.* **1993**, *21*, 279–282. [CrossRef]
91. Needham, D.; Anyarambhatla, G.; Kong, G.; Dewhirst, M.W. A New Temperature-Sensitive Liposome for Use with Mild Hyperthermia: Characterization and Testing in a Human Tumor Xenograft Model. *Cancer Res.* **2000**, *60*, 1197–1201.
92. Huang, S.L.; MacDonald, R.C. Acoustically Active Liposomes for Drug Encapsulation and Ultrasound-Triggered Release. *Biochim. Biophys. Acta Biomembr.* **2004**, *1665*, 134–141. [CrossRef]
93. Lin, H.Y.; Thomas, J.L. PEG-Lipids and Oligo (Ethylene Glycol) Surfactants Enhance the Ultrasonic Permeabilizability of Liposomes. *Langmuir* **2003**, *19*, 1098–1105. [CrossRef]
94. Lin, H.Y.; Thomas, J.L. Pluronic P105 Sensitizes Cholesterol-Free Liposomes to Ultrasound. In Proceedings of the the IEEE Annual Northeast Bioengineering Conference, Newark, NJ, USA, 22–23 March 2003. [CrossRef]
95. Lin, H.Y.; Thomas, J.L. Factors Affecting Responsivity of Unilamellar Liposomes to 20 KHz Ultrasound. *Langmuir* **2004**, *20*, 6100–6106. [CrossRef]
96. Tirosh, O.; Barenholz, Y.; Katzhendler, J.; Priev, A. Hydration of Polyethylene Glycol-Grafted Liposomes. *Biophys. J.* **1998**, *74*, 1371–1379. [CrossRef]
97. Evjen, T.J.; Nilssen, E.A.; Barnert, S.; Schubert, R.; Brandl, M.; Fossheim, S.L. Ultrasound-Mediated Destabilization and Drug Release from Liposomes Comprising Dioleoylphosphatidylethanolamine. *Eur. J. Pharm. Sci.* **2011**, *41*, 380–386. [CrossRef]
98. Tata, D.B.; Dunn, F. Ultrasound and Model Membrane Systems: Analyses and Predlctions. *J. Phys. Chem.* **1992**, *96*, 3548–3555. [CrossRef]
99. Maynard, V.M.; Magin, R.L.; Storm-Jensen, P.R.; Dunn, F. Ultrasonic Absorption by Liposomes. *Ultrason. Symp.* **1983**, 806–809. [CrossRef]
100. Wu, F.; Wang, Z.-B.; Chen, W.-Z.; Zou, J.-Z.; Bai, J.; Zhu, H.; Li, K.-Q.; Jin, C.-B.; Xie, F.-L.; Su, H.-B. Advanced Hepatocellular Carcinoma: Treatment with High-Intensity Focused Ultrasound Ablation Combined with Transcatheter Arterial Embolization. *Radiology* **2005**, *235*, 659–667. [CrossRef] [PubMed]
101. Wu, F.; Wang, Z.-B.; Jin, C.-B.; Zhang, J.-P.; Chen, W.-Z.; Bai, J.; Zou, J.-Z.; Zhu, H. Circulating Tumor Cells in Patients with Solid Malignancy Treated by High-Intensity Focused Ultrasound. *Ultrasound Med. Biol.* **2004**, *30*, 511–517. [CrossRef]
102. Mafune, K.; Tanaka, Y. Influence of Multimodality Therapy on the Cellular Immunity of Patients with Esophageal Cancer. *Ann. Surg. Oncol.* **2000**, *7*, 609–616. [CrossRef] [PubMed]
103. Wu, F.; Wang, Z.-B.; Lu, P.; Xu, Z.-L.; Chen, W.-Z.; Zhu, H.; Jin, C.-B. Activated Anti-Tumor Immunity in Cancer Patients after High Intensity Focused Ultrasound Ablation. *Ultrasound Med. Biol.* **2004**, *30*, 1217–1222. [CrossRef]
104. Kramer, G.; Steiner, G.E.; Gröbl, M.; Hrachowitz, K.; Reithmayr, F.; Paucz, L.; Newman, M.; Madersbacher, S.; Gruber, D.; Susani, M.; et al. Response to Sublethal Heat Treatment of Prostatic Tumor Cells and of Prostatic Tumor Infiltrating T-Cells. *Prostate* **2004**, *58*, 109–120. [CrossRef] [PubMed]
105. Schueller, G.; Kettenbach, J.; Sedivy, R.; Bergmeister, H.; Stift, A.; Fried, J.; Gnant, M.; Lammer, J. Expression of Heat Shock Proteins in Human Hepatocellular Carcinoma after Radiofrequency Ablation in an Animal Model. *Oncol. Rep.* **2004**, *12*, 495–499. [CrossRef]
106. Den Brok, M.H.M.G.M.; Sutmuller, R.P.M.; van der Voort, R.; Bennink, E.J.; Figdor, C.G.; Ruers, T.J.M.; Adema, G.J. In Situ Tumor Ablation Creates an Antigen Source for the Generation of Antitumor Immunity. *Cancer Res.* **2004**, *64*, 4024–4029. [CrossRef]
107. Leslie, T.; Ritchie, R.; Illing, R.; Ter Haar, G.; Phillips, R.; Middleton, M.; Bch, B.; Wu, F.; Cranston, D. High-Intensity Focused Ultrasound Treatment of Liver Tumours: Post-Treatment MRI Correlates Well with Intra-Operative Estimates of Treatment Volume. *Br. J. Radiol.* **2012**, *85*, 1363–1370. [CrossRef]

108. Wu, F.; Wang, Z.-B.; Chen, W.-Z.; Zou, J.-Z.; Bai, J.; Zhu, H.; Li, K.-Q.; Xie, F.-L.; Jin, C.-B.; Su, H.-B.; et al. Extracorporeal Focused Ultrasound Surgery for Treatment of Human Solid Carcinomas: Early Chinese Clinical Experience. *Ultrasound Med. Biol.* **2004**, *30*, 245–260. [CrossRef]
109. Visioli, A.; Rivens, I.; ter Haar, G.; Horwich, A.; Huddart, R.; Moskovic, E.; Padhani, A.; Glees, J. Preliminary Results of a Phase I Dose Escalation Clinical Trial Using Focused Ultrasound in the Treatment of Localised Tumours. *Eur. J. Ultrasound* **1999**, *9*, 11–18. [CrossRef]
110. Vallancien, G.; Harouni, M.; Guillonneau, B.; Veillon, B.; Bougaran, J. Ablation of Superficial Bladder Tumors with Focused Extracorporeal Pyrotherapy. *Urology* **1996**, *47*, 204–207. [CrossRef]
111. Zagar, T.M.; Vujaskovic, Z.; Formenti, S.; Rugo, H.; Muggia, F.; O'Connor, B.; Myerson, R.; Stauffer, P.; Hsu, I.C.; Diederich, C.; et al. Two Phase I Dose-Escalation/Pharmacokinetics Studies of Low Temperature Liposomal Doxorubicin (LTLD) and Mild Local Hyperthermia in Heavily Pretreated Patients with Local Regionally Recurrent Breast Cancer. *Int. J. Hyperth.* **2014**, *30*, 285–294. [CrossRef]
112. Gray, M.D.; Lyon, P.C.; Mannaris, D.C.; Folkes, L.K.; Stratford, M.; Campo, L.; Chung, D.D.Y.F.; Scott, F.S.; Gleeson, F.V.; Coussios, F.C.C. Focused Ultrasound Hyperthermia for Targeted Drug Release from Thermosensitive Liposomes: Results from a Phase I Trial. *Radiology* **2019**. [CrossRef] [PubMed]
113. Lyon, P.C.; Gray, M.D.; Mannaris, C.; Folkes, L.K.; Stratford, M.; Campo, L.; Chung, D.Y.F.; Scott, S.; Anderson, M.; Goldin, R.; et al. Safety and Feasibility of Ultrasound-Triggered Targeted Drug Delivery of Doxorubicin from Thermosensitive Liposomes in Liver Tumours (TARDOX): A Single-Centre, Open-Label, Phase 1 Trial. *Lancet Oncol.* **2018**, *19*, 1027–1039. [CrossRef]
114. Dromi, S.; Frenkel, V.; Luk, A.; Traughber, B.; Angstadt, M.; Bur, M.; Poff, J.; Xie, J.; Libutti, S.K.; Li, K.C.P.; et al. Pulsed-High Intensity Focused Ultrasound and Low Temperature—Sensitive Liposomes for Enhanced Targeted Drug Delivery and Antitumor Effect. *Clin. Cancer Res.* **2007**, *13*, 2722–2727. [CrossRef]
115. De Smet, M.; Heijman, E.; Langereis, S.; Hijnen, N.M.; Grüll, H. Magnetic Resonance Imaging of High Intensity Focused Ultrasound Mediated Drug Delivery from Temperature-Sensitive Liposomes: An in Vivo Proof-of-Concept Study. *J. Control. Release* **2011**, *150*, 102–110. [CrossRef]
116. Negussie, A.H.; Yarmolenko, P.S.; Partanen, A.; Ranjan, A.; Jacobs, G.; Woods, D.; Bryant, H.; Thomasson, D.; Dewhirst, M.W.; Wood, B.J.; et al. Formulation and Characterisation of Magnetic Resonance Imageable Thermally Sensitive Liposomes for Use with Magnetic Resonance-Guided High Intensity Focused Ultrasound. *Int. J. Hyperth.* **2011**, *27*, 140–155. [CrossRef]
117. Ranjan, A.; Jacobs, G.C.; Woods, D.L.; Negussie, A.H.; Partanen, A.; Yarmolenko, P.S.; Gacchina, C.E.; Sharma, K.V.; Frenkel, V.; Wood, B.J.; et al. Image-Guided Drug Delivery with Magnetic Resonance Guided High Intensity Focused Ultrasound and Temperature Sensitive Liposomes in a Rabbit Vx2 Tumor Model. *J. Control. Release* **2012**, *158*, 487–494. [CrossRef]
118. Schroeder, A.; Honen, R.; Turjeman, K.; Gabizon, A.; Kost, J.; Barenholz, Y. Ultrasound Triggered Release of Cisplatin from Liposomes in Murine Tumors. *J. Control. Release* **2009**, *137*, 63–68. [CrossRef]
119. Li, Y.; An, H.; Wang, X.; Wang, P.; Qu, F.; Jiao, Y.; Zhang, K.; Liu, Q. Ultrasound-Triggered Release of Sinoporphyrin Sodium from Liposome-Microbubble Complexes and Its Enhanced Sonodynamic Toxicity in Breast Cancer. *Nano Res.* **2018**, *11*, 1038–1056. [CrossRef]
120. Hagtvet, E.; Evjen, T.J.; Olsen, D.R.; Fossheim, S.L.; Nilssen, E.A. Ultrasound Enhanced Antitumor Activity of Liposomal Doxorubicin in Mice. *J. Drug Target.* **2011**, *19*, 701–708. [CrossRef] [PubMed]
121. Aryal, M.; Vykhodtseva, N.; Zhang, Y.-Z.; Park, J.; McDannold, N. Multiple Treatments with Liposomal Doxorubicin and Ultrasound-Induced Disruption of Blood-Tumor and Blood-Brain Barriers Improve Outcomes in a Rat Glioma Model. *J. Control. Release* **2013**, *169*, 103–111. [CrossRef] [PubMed]
122. Rugo, H.; Pabbathi, H.; Shrestha, S.; Aithal, S.; Borys, N.; Musso, L.; Zoberi, I. Lyso-Thermosensitive Liposomal Doxorubicin shows efficacy with minimal adverse events in patients with breast cancer recurrence at the chest wall. In Proceedings of the the San Antonio Breast Cancer Symposium, San Antonio, TX, USA, 8–12 December 2015.
123. Delfino, J.G.; Woods, T.O. New Developments in Standards for MRI Safety Testing of Medical Devices. *Curr. Radiol. Rep.* **2016**, *4*, 28. [CrossRef]
124. Noebauer-Huhmann, I.M.; Weber, M.A.; Lalam, R.K.; Trattnig, S.; Bohndorf, K.; Vanhoenacker, F.; Tagliafico, A.; Van Rijswijk, C.; Vilanova, J.C.; Afonso, P.D.; et al. Erratum: Soft Tissue Tumors in Adults: ESSR-Approved Guidelines for Diagnostic Imaging. *Semin. Musculoskelet. Radiol.* **2015**, *19*, 475. [CrossRef]

125. Sanz, B.; Calatayud, M.P.; Torres, T.E.; Fanarraga, M.L.; Ibarra, M.R.; Goya, G.F. Magnetic Hyperthermia Enhances Cell Toxicity with Respect to Exogenous Heating. *Biomaterials* **2017**, *114*, 62–70. [CrossRef]
126. Kozissnik, B.; Bohorquez, A.C.; Dobson, J.; Rinaldi, C. Magnetic Fluid Hyperthermia: Advances, Challenges, and Opportunity. *Int. J. Hyperth.* **2013**, *29*, 706–714. [CrossRef]
127. Maier-Hauff, K.; Ulrich, F.; Nestler, D.; Orawa, H.; Budach, V.; Jordan, A. Efficacy and Safety of Intratumoral Thermotherapy Using Magnetic Iron-Oxide Nanoparticles Combined with External Beam Radiotherapy on Patients with Recurrent Glioblastoma Multiforme. *J. Neurooncol.* **2011**, *103*, 317–324. [CrossRef]
128. Mura, S.; Nicolas, J.; Couvreur, P. Stimuli-Responsive Nanocarriers for Drug Delivery. *Nat. Mater.* **2013**, *12*, 991–1003. [CrossRef]
129. Hilger, I.; Kaiser, W.A. Iron Oxide-Based Nanostructures for MRI and Magnetic Hyperthermia. *Nanomedicine* **2012**, *7*, 1443–1459. [CrossRef]
130. Bonini, M.; Berti, D.; Baglioni, P. Current Opinion in Colloid & Interface Science Nanostructures for Magnetically Triggered Release of Drugs and Biomolecules. *Curr. Opin. Colloid Interface Sci.* **2013**, *18*, 459–467. [CrossRef]
131. Gautier, J.; Munnier, E.; Soucé, M.; Chourpa, I. Recent Advances in Theranostic Nanocarriers of Doxorubicin Based on Iron Oxide and Gold Nanoparticles. *J. Control. Release* **2013**, *169*, 48–61. [CrossRef]
132. Clares, B.; Biedma-ortiz, R.A.; Sáez-fernández, E.; Prados, J.C.; Melguizo, C.; Cabeza, L.; Ortiz, R.; Arias, J.L. Nano-Engineering of 5-Fluorouracil-Loaded Magnetoliposomes for Combined Hyperthermia and Chemotherapy against Colon Cancer. *Eur. J. Pharm. Biopharm.* **2013**, *85*, 329–338. [CrossRef] [PubMed]
133. Lee, J.H.; Chen, K.J.; Noh, S.H.; Garcia, M.A.; Wang, H.; Lin, W.Y.; Jeong, H.; Kong, B.J.; Stout, D.B.; Cheon, J.; et al. On-Demand Drug Release System for In Vivo Cancer Treatment through Self-Assembled Magnetic Nanoparticles. *Angew. Chem. Int. Ed. Engl.* **2013**, *52*, 4384–4388. [CrossRef]
134. Hayashi, K.; Nakamura, M.; Miki, H.; Ozaki, S.; Abe, M. Magnetically Responsive Smart Nanoparticles for Cancer Treatment with a Combination of Magnetic Hyperthermia and Remote-Control Drug Release. *Theranostics* **2014**, *4*, 834–844. [CrossRef] [PubMed]
135. Deok, S.; Sartor, M.; Hu, C.J.; Zhang, W.; Zhang, L.; Jin, S. Magnetic Field Activated Lipid–Polymer Hybrid Nanoparticles for Stimuli-Responsive Drug Release. *Acta Biomater.* **2013**, *9*, 5447–5452. [CrossRef] [PubMed]
136. Hayashi, K.; Ono, K.; Suzuki, H.; Sawada, M.; Moriya, M. Drug Release from Magnetic Nanoparticle/Organic Hybrid Based on Hyperthermic Effect. *ACS Appl. Mater. Interfaces* **2010**, *2*, 1903–1911. [CrossRef]
137. Chiang, W.; Ke, C.; Liao, Z.; Chen, S.; Chen, F. Pulsatile Drug Release from PLGA Hollow Microspheres by Controlling the Permeability of Their Walls with a Magnetic Field. *Small* **2012**, *8*, 3584–3588. [CrossRef]
138. Dou, Y.; Hynynen, K.; Allen, C. To Heat or Not to Heat: Challenges with Clinical Translation of Thermosensitive Liposomes. *J. Control. Release* **2017**, *249*, 63–73. [CrossRef]
139. Guo, Y.; Zhang, Y.; Ma, J.; Li, Q.; Li, Y.; Zhou, X.; Zhao, D.; Song, H.; Chen, Q.; Zhu, X. Light/Magnetic Hyperthermia Triggered Drug Released from Multi-Functional Thermo-Sensitive Magnetoliposomes for Precise Cancer Synergetic Theranostics. *J. Control. Release* **2018**, *272*, 145–158. [CrossRef]
140. Gogoi, M.; Jaiswal, M.K.; Sarma, H.D.; Bahadur, D.; Banerjee, R. Biocompatibility and Therapeutic Evaluation of Magnetic Liposomes Designed for Self-Controlled Cancer Hyperthermia and Chemotherapy. *Integr. Biol. (U. K.)* **2017**, *9*, 555–565. [CrossRef]
141. Wang, L.; Zhang, J.; An, Y.; Wang, Z.; Liu, J.; Li, Y.; Zhang, D. A Study on the Thermochemotherapy Effect of Nanosized As_2O_3/MZF Thermosensitive Magnetoliposomes on Experimental Hepatoma in Vitro and in Vivo. *Nanotechnology* **2011**, *22*, 315102. [CrossRef]
142. Nappini, S.; Fogli, S.; Castroflorio, B.; Bonini, M.; Baldelli Bombelli, F.; Baglioni, P. Magnetic Field Responsive Drug Release from Magnetoliposomes in Biological Fluids. *J. Mater. Chem. B* **2016**, *4*, 716–725. [CrossRef]
143. Amstad, E.; Kohlbrecher, J.; Müller, E.; Schweizer, T.; Textor, M.; Reimhult, E. Triggered Release from Liposomes through Magnetic Actuation of Iron Oxide Nanoparticle Containing Membranes. *Nano Lett.* **2011**, *11*, 1664–1670. [CrossRef]
144. Tai, L.A.; Tsai, P.J.; Wang, Y.C.; Wang, Y.J.; Lo, L.W.; Yang, C.S. Thermosensitive Liposomes Entrapping Iron Oxide Nanoparticles for Controllable Drug Release. *Nanotechnology* **2009**, *20*, 135101. [CrossRef]
145. Kim, D.H.; Rozhkova, E.A.; Ulasov, I.V.; Bader, S.D.; Rajh, T.; Lesniak, M.S.; Novosad, V. Biofunctionalized Magnetic-Vortex Microdiscs for Targeted Cancer-Cell Destruction. *Nat. Mater.* **2010**, *9*, 165–171. [CrossRef]
146. Joniec, A.; Sek, S.; Krysinski, P. Magnetoliposomes as Potential Carriers of Doxorubicin to Tumours. *Chem. A Eur. J.* **2016**, *22*, 17715–17724. [CrossRef]

147. Zhang, E.; Kircher, M.F.; Koch, M.; Eliasson, L.; Goldberg, S.N.; Renström, E. Dynamic Magnetic Fields Remote-Control Apoptosis via Nanoparticle Rotation. *ACS Nano* **2014**, *8*, 3192–3201. [CrossRef]

148. Podaru, G.; Ogden, S.; Baxter, A.; Shrestha, T.; Ren, S.; Thapa, P.; Dani, R.K.; Wang, H.; Basel, M.T.; Prakash, P.; et al. Pulsed Magnetic Field Induced Fast Drug Release from Magneto Liposomes via Ultrasound Generation. *J. Phys. Chem. B* **2014**, *118*, 11715–11722. [CrossRef]

149. Reddy, L.H.; Arias, J.L.; Nicolas, J.; Couvreur, P. Magnetic Nanoparticles: Design and Characterization, Toxicity and Biocompatibility, Pharmaceutical and Biomedical Applications. *Chem. Rev.* **2012**, *112*, 5818–5878. [CrossRef]

150. Deatsch, A.E.; Evans, B.A. Heating Efficiency in Magnetic Nanoparticle Hyperthermia. *J. Magn. Magn. Mater.* **2014**, *354*, 163–172. [CrossRef]

151. Rosensweig, R.E. Heating Magnetic Fluid with Alternating Magnetic Field. *J. Magn. Magn. Mater.* **2002**, *252*, 370–374. [CrossRef]

152. Carrey, J.; Mehdaoui, B.; Respaud, M. Simple Models for Dynamic Hysteresis Loop Calculations of Magnetic Single-Domain Nanoparticles: Application to Magnetic Hyperthermia Optimization. *J. Appl. Phys.* **2015**, *109*, 083921. [CrossRef]

153. Riedinger, A.; Guardia, P.; Curcio, A.; Garcia, M.A.; Cingolani, R.; Manna, L.; Pellegrino, T. Subnanometer Local Temperature Probing and Remotely Controlled Drug Release Based on Azo-Functionalized Iron Oxide Nanoparticles. *Nano Lett.* **2013**, *13*, 2399–2406. [CrossRef]

154. Dutz, S.; Hergt, R. Magnetic Nanoparticle Heating and Heat Transfer on a Microscale: Basic Principles, Realities and Physical Limitations of Hyperthermia for Tumour Therapy. *Int. J. Hyperth.* **2013**, *6736*, 790–800. [CrossRef] [PubMed]

155. Cheng, Y.; Muroski, M.E.; Petit, D.C.M.C.; Mansell, R.; Vemulkar, T.; Morshed, R.A.; Han, Y.; Balyasnikova, I.V.; Horbinski, C.M.; Huang, X.; et al. Rotating Magnetic Field Induced Oscillation of Magnetic Particles for in Vivo Mechanical Destruction of Malignant Glioma. *J. Control. Release* **2016**, *223*, 75–84. [CrossRef] [PubMed]

156. Zhang, L.; Zhao, Y.; Wang, X. Nanoparticle-Mediated Mechanical Destruction of Cell Membranes: A Coarse-Grained Molecular Dynamics Study. *ACS Appl. Mater. Interfaces* **2017**, *9*, 26665–26673. [CrossRef] [PubMed]

157. Cheng, D.; Li, X.; Zhang, G.; Shi, H. Morphological Effect of Oscillating Magnetic Nanoparticles in Killing Tumor Cells. *Nanoscale Res. Lett.* **2014**, *9*, 195. [CrossRef]

158. Rikken, R.S.M.; Nolte, R.J.M.; Maan, J.C.; Van Hest, J.C.M.; Wilson, D.A.; Christianen, P.C.M. Manipulation of Micro- and Nanostructure Motion with Magnetic Fields. *Soft Matter* **2014**, *10*, 1295–1308. [CrossRef]

159. Ortega, D.; Pankhurst, Q.A. Magnetic Hyperthermia. *Nanoscience* **2013**, *1*, 60–88. [CrossRef]

160. Pradhan, P.; Giri, J.; Rieken, F.; Koch, C.; Mykhaylyk, O.; Döblinger, M.; Banerjee, R.; Bahadur, D.; Plank, C. Targeted Temperature Sensitive Magnetic Liposomes for Thermo-Chemotherapy. *J. Control. Release* **2010**, *142*, 108–121. [CrossRef] [PubMed]

161. Peller, M.; Willerding, L.; Limmer, S.; Hossann, M.; Dietrich, O.; Ingrisch, M.; Sroka, R.; Lindner, L.H. Surrogate MRI Markers for Hyperthermia-Induced Release of Doxorubicin from Thermosensitive Liposomes in Tumors. *J. Control. Release* **2016**, *237*, 138–146. [CrossRef] [PubMed]

162. Lu, A.H.; Salabas, E.L.; Schüth, F. Magnetic Nanoparticles: Synthesis, Protection, Functionalization, and Application. *Angew. Chem. Int. Ed.* **2007**, *46*, 1222–1244. [CrossRef] [PubMed]

163. Hedayatnasab, Z.; Abnisa, F.; Daud, W.M.A.W. Review on Magnetic Nanoparticles for Magnetic Nanofluid Hyperthermia Application. *Mater. Des.* **2017**, *123*, 174–196. [CrossRef]

164. Das, R.; Alonso, J.; Nemati Porshokouh, Z.; Kalappattil, V.; Torres, D.; Phan, M.H.; Garaio, E.; García, J.Á.; Sanchez Llamazares, J.L.; Srikanth, H. Tunable High Aspect Ratio Iron Oxide Nanorods for Enhanced Hyperthermia. *J. Phys. Chem. C* **2016**, *120*, 10086–10093. [CrossRef]

165. Simeonidis, K.; Morales, M.P.; Marciello, M.; Angelakeris, M.; de la Presa, P.; Lazaro-Carrillo, A.; Tabero, A.; Villannueva, A.; Chubykalo, O.; Serantes, D. In-Situ Particles Reorientation during Magnetic Hyperthermia Application: Shape Matters Twice. *Sci. Rep.* **2016**, *6*, 38382. [CrossRef] [PubMed]

166. Abenojar, E.C.; Wickramasinghe, S.; Bas-Concepcion, J.; Samia, A.C.S. Structural Effects on the Magnetic Hyperthermia Properties of Iron Oxide Nanoparticles. *Prog. Nat. Sci. Mater. Int.* **2016**, *26*, 440–448. [CrossRef]

167. Nemati, Z.; Alonso, J.; Martinez, L.M.; Khurshid, H.; Garaio, E.; Garcia, J.A.; Phan, M.H.; Srikanth, H. Enhanced Magnetic Hyperthermia in Iron Oxide Nano-Octopods: Size and Anisotropy Effects. *J. Phys. Chem. C* **2016**, *120*, 8370–8379. [CrossRef]

168. Espinosa, A.; Kolosnjaj-Tabi, J.; Abou-Hassan, A.; Plan Sangnier, A.; Curcio, A.; Silva, A.K.A.; Di Corato, R.; Neveu, S.; Pellegrino, T.; Liz-Marzán, L.M.; et al. Magnetic (Hyper)Thermia or Photothermia? Progressive Comparison of Iron Oxide and Gold Nanoparticles Heating in Water, in Cells, and In Vivo. *Adv. Funct. Mater.* **2018**, *28*, 1–16. [CrossRef]
169. Walter, A.; Billotey, C.; Garofalo, A.; Ulhaq-Bouillet, C.; Lefevre, C.; Taleb, J.; Laurent, S.; Elst, L.V.; Uller, R.; Lartigue, L.; et al. Mastering the Shape and Composition of Dendronized Iron Oxide Nanoparticles to Tailor Magnetic Resonance Imaging and Hyperthermia. *Chem. Mater.* **2014**, *26*, 5252–5264. [CrossRef]
170. Shen, Y.; Wu, C.; Uyeda, T.Q.P.; Plaza, G.R.; Liu, B.; Han, Y.; Lesniak, M.S.; Cheng, Y. Elongated Nanoparticle Aggregates in Cancer Cells for Mechanical Destruction with Low Frequency Rotating Magnetic Field. *Theranostics* **2017**, *7*, 1735–1748. [CrossRef] [PubMed]
171. Tietze, R.; Zaloga, J.; Unterweger, H.; Lyer, S.; Friedrich, R.P.; Janko, C.; Pöttler, M.; Dürr, S.; Alexiou, C. Magnetic Nanoparticle-Based Drug Delivery for Cancer Therapy. *Biochem. Biophys. Res. Commun.* **2015**, *468*, 463–470. [CrossRef] [PubMed]
172. Dai, M.; Wu, C.; Wang, X.; Fang, H.-M.; Li, L.; Yan, J.B.; Zeng, D.L.; Zou, T. Thermo-Responsive Magnetic Liposomes for Hyperthermia-Triggered Local Drug Delivery. *J. Control. Release* **2017**, *259*, e93. [CrossRef]
173. García-Jimeno, S.; Escribano, E.; Queralt, J.; Estelrich, J. External Magnetic Field-Induced Selective Biodistribution of Magnetoliposomes in Mice. *Nanoscale Res. Lett.* **2012**, *7*, 452. [CrossRef] [PubMed]
174. Delaney, G.P.; Barton, M.B. Evidence-Based Estimates of the Demand for Radiotherapy. *Clin. Oncol.* **2015**, *27*, 70–76. [CrossRef] [PubMed]
175. Hainfeld, J.F.; Dilmanian, F.A.; Slatkin, D.N.; Smilowitz, H.M. Radiotherapy Enhancement with Gold Nanoparticles. *J. Pharm. Pharmacol.* **2008**, *60*, 977–985. [CrossRef] [PubMed]
176. Schulz-Ertner, D.; Tsujii, H. Particle Radiation Therapy Using Proton and Heavier Ion Beams. *J. Clin. Oncol.* **2007**, *25*, 953–964. [CrossRef] [PubMed]
177. Haume, K.; Rosa, S.; Grellet, S.; Śmiałek, M.A.; Butterworth, K.T.; Solov'yov, A.V.; Prise, K.M.; Golding, J.; Mason, N.J. Gold Nanoparticles for Cancer Radiotherapy: A Review. *Cancer Nanotechnol.* **2016**, *7*, 8. [CrossRef]
178. Lo Nigro, C.; Denaro, N.; Merlotti, A.; Merlano, M. Head and neck cancer: improving outcomes with a multidisciplinary approach. *Cancer Manag. Res.* **2017**, *9*, 363–731. [CrossRef]
179. Carlsson, L.; Bratman, S.V.; Siu, L.L.; Spreafico, A. The Cisplatin Total Dose and Concomitant Radiation in Locoregionally Advanced Head and Neck Cancer: Any Recent Evidence for Dose Efficacy? *Curr. Treat. Options Oncol.* **2017**, *18*, 39. [CrossRef]
180. Fayette, J.; Molin, Y.; Lavergne, E.; Montbarbon, X.; Racadot, S.; Poupart, M.; Ramade, A.; Zrounba, P.; Ceruse, P.; Pommier, P. Radiotherapy potentiation with weekly cisplatin compared to standard every 3 weeks cisplatin chemotherapy for locoregionally advanced head and neck squamous cell carcinoma. *Drug Des. Dev. Ther.* **2015**, *9*, 6203–6210. [CrossRef]
181. Chang, H.-I.; Yeh, M.-K. Clinical Development of Liposome-Based Drugs: Formulation, Characterization, and Therapeutic Efficacy. *Int. J. Nanomed.* **2012**, *7*, 49–60. [CrossRef]
182. Deng, W.; Chen, W.; Clement, S.; Guller, A.; Zhao, Z.; Engel, A.; Goldys, E.M. Controlled Gene and Drug Release from a Liposomal Delivery Platform Triggered by X-Ray Radiation. *Nat. Commun.* **2018**, *9*, 2713. [CrossRef]
183. Gutteridge, J.M.C.; Halliwell, B. The Measurement and Mechanism of Lipid Peroxidation in Biological Systems. *Trends Biochem. Sci.* **1990**, *15*, 129–135. [CrossRef]
184. Sicard-Roselli, C.; Brun, E.; Gilles, M.; Baldacchino, G.; Kelsey, C.; McQuaid, H.; Polin, C.; Wardlow, N.; Currell, F. A New Mechanism for Hydroxyl Radical Production in Irradiated Nanoparticle Solutions. *Small* **2014**, *10*, 3338–3346. [CrossRef]
185. Ashton, J.R.; Castle, K.D.; Qi, Y.; Kirsch, D.G.; West, J.L.; Badea, C.T. Dual-Energy CT Imaging of Tumor Liposome Delivery after Gold Nanoparticle-Augmented Radiation Therapy. *Theranostics* **2018**, *8*, 1782–1797. [CrossRef]
186. Schuemann, J.; Berbeco, R.; Chithrani, D.B.; Cho, S.H.; Kumar, R.; McMahon, S.J.; Sridhar, S.; Krishnan, S. Roadmap to Clinical Use of Gold Nanoparticles for Radiation Sensitization. *Int. J. Radiat. Oncol. Biol. Phys.* **2016**, *94*, 189–205. [CrossRef]
187. Hainfeld, J.F.; Slatkin, D.N.; Smilowitz, H.M. The Use of Gold Nanoparticles to Enhance Radiotherapy in Mice. *Phys. Med. Biol.* **2004**, *49*, N309. [CrossRef]

188. Cho, S.H. Estimation of Tumour Dose Enhancement Due to Gold Nanoparticles During Typical Radiation Treatments: A Preliminary Monte Carlo Study. *Phys. Med. Biol.* **2005**, *50*, N163–N173. [CrossRef]

189. Her, S.; Jaffray, D.A.; Allen, C. Gold Nanoparticles for Applications in Cancer Radiotherapy: Mechanisms and Recent Advancements. *Adv. Drug Deliv. Rev.* **2017**, *109*, 84–101. [CrossRef]

190. Brun, E.; Sanche, L.; Sicard-Roselli, C. Parameters Governing Gold Nanoparticle X-Ray Radiosensitization of DNA in Solution. *Colloids Surf. B Biointerfaces* **2009**, *72*, 128–134. [CrossRef]

191. Lechtman, E.; Chattopadhyay, N.; Cai, Z.; Mashouf, S.; Reilly, R.; Pignol, J.P. Implications on Clinical Scenario of Gold Nanoparticle Radiosensitization in Regards to Photon Energy, Nanoparticle Size, Concentration and Location. *Phys. Med. Biol.* **2011**, *56*, 4631–4647. [CrossRef]

192. Leung, M.K.K.; Chow, J.C.L.; Chithrani, B.D.; Lee, M.J.G.; Oms, B.; Jaffray, D.A. Irradiation of Gold Nanoparticles by X-Rays: Monte Carlo Simulation of Dose Enhancements and the Spatial Properties of the Secondary Electrons Production. *Med. Phys.* **2011**, *38*, 624–631. [CrossRef]

193. Jones, B.L.; Krishnan, S.; Cho, S.H. Estimation of Microscopic Dose Enhancement Factor around Gold Nanoparticles by Monte Carlo Calculations. *Med. Phys.* **2010**, *37*, 3809–3816. [CrossRef]

194. Hwang, C.; Kim, J.M.; Kim, J. Influence of Concentration, Nanoparticle Size, Beam Energy, and Material on Dose Enhancement in Radiation Therapy. *J. Radiat. Res.* **2017**, *58*, 405–411. [CrossRef]

195. Butterworth, K.T.; Mcmahon, S.J.; Taggart, L.E.; Prise, K.M. Radiosensitization by Gold Nanoparticles: Effective at Megavoltage Energies and Potential Role of Oxidative Stress. *Transl. Cancer Res.* **2013**, *2*, 269–279. [CrossRef]

196. Chang, M.Y.; Shiau, A.L.; Chen, Y.H.; Chang, C.J.; Chen, H.H.W.; Wu, C.L. Increased Apoptotic Potential and Dose-Enhancing Effect of Gold Nanoparticles in Combination with Single-Dose Clinical Electron Beams on Tumor-Bearing Mice. *Cancer Sci.* **2008**, *99*, 1479–1484. [CrossRef]

197. Yang, C.; Bromma, K.; Sung, W.; Schuemann, J.; Chithrani, D. Determining the Radiation Enhancement Effects of Gold Nanoparticles in Cells in a Combined Treatment with Cisplatin and Radiation at Therapeutic Megavoltage Energies. *Cancers* **2018**, *10*, 150. [CrossRef] [PubMed]

198. Rahman, W.N.; Bishara, N.; Ackerly, T.; He, C.F.; Jackson, P.; Wong, C.; Davidson, R.; Geso, M. Enhancement of Radiation Effects by Gold Nanoparticles for Superficial Radiation Therapy. *Nanomed. Nanotechnol. Biol. Med.* **2009**, *5*, 136–142. [CrossRef]

199. Schardt, D.; Elsässer, T.; Schulz-Ertner, D. Heavy-Ion Tumor Therapy: Physical and Radiobiological Benefits. *Rev. Mod. Phys.* **2010**, *82*, 383–425. [CrossRef]

200. Verkhovtsev, A.V.; Korol, A.V.; Solov'Yov, A.V. Revealing the Mechanism of the Low-Energy Electron Yield Enhancement from Sensitizing Nanoparticles. *Phys. Rev. Lett.* **2015**, *114*, 63401. [CrossRef] [PubMed]

201. Lin, Y.; McMahon, S.J.; Paganetti, H.; Schuemann, J. Biological Modeling of Gold Nanoparticle Enhanced Radiotherapy for Proton Therapy. *Phys. Med. Biol.* **2015**, *60*, 4149–4168. [CrossRef] [PubMed]

202. Lukianova-Helb, E.Y.; Ren, X.; Sawant, R.R.; Wu, X.; Torchilin, V.P.; Lapotko, D.O. On-demand intracellular amplification of chemoradiation with cancer-specific plasmonic nanobubbles. *Nat. Med.* **2014**, *20*, 778–784. [CrossRef]

203. Davies, M.J. Protein Oxidation and Peroxidation. *Biochem. J.* **2016**, *473*, 805–825. [CrossRef] [PubMed]

204. Park, S.H.; Oh, S.G.; Mun, J.Y.; Han, S.S. Loading of Gold Nanoparticles inside the DPPC Bilayers of Liposome and Their Effects on Membrane Fluidities. *Colloids Surf. B Biointerfaces* **2006**, *48*, 112–118. [CrossRef] [PubMed]

205. Rengan, A.K.; Bukhari, A.B.; Pradhan, A.; Malhotra, R.; Banerjee, R.; Srivastava, R.; De, A. In Vivo Analysis of Biodegradable Liposome Gold Nanoparticles as Efficient Agents for Photothermal Therapy of Cancer. *Nano Lett.* **2015**, *15*, 842–848. [CrossRef] [PubMed]

206. Hobbs, S.K.; Monsky, W.L.; Yuan, F.; Roberts, W.G.; Griffith, L.; Torchilin, V.P.; Jain, R.K. Regulation of Transport Pathways in Tumor Vessels: Role of Tumor Type and Microenvironment. *Proc. Natl. Acad. Sci. USA* **1998**, *95*, 4607–4612. [CrossRef]

207. Wicki, A.; Witzigmann, D.; Balasubramanian, V.; Huwyler, J. Nanomedicine in Cancer Therapy: Challenges, Opportunities, and Clinical Applications. *J. Control. Release* **2015**, *200*, 138–157. [CrossRef]

208. Lammers, T.; Kiessling, F.; Hennink, W.E.; Storm, G. Drug targeting to tumors: Principles, pitfalls and (pre-) clinical progress. *J. Control. Release* **2012**, *161*, 175–187. [CrossRef]

209. Dreaden, E.C.; Alkilany, A.M.; Huang, X.; Murphy, C.J.; El-Sayed, M.A. The Golden Age: Gold Nanoparticles for Biomedicine. *Chem. Soc. Rev.* **2012**, *41*, 2740–2779. [CrossRef]

210. Chen, Y.S.; Hung, Y.C.; Liau, I.; Huang, G.S. Assessment of the in Vivo Toxicity of Gold Nanoparticles. *Nanoscale Res. Lett.* **2009**, *4*, 858–864. [CrossRef] [PubMed]

211. Lasagna-Reeves, C.; Gonzalez-Romero, D.; Barria, M.A.; Olmedo, I.; Clos, A.; Sadagopa Ramanujam, V.M.; Urayama, A.; Vergara, L.; Kogan, M.J.; Soto, C. Bioaccumulation and Toxicity of Gold Nanoparticles after Repeated Administration in Mice. *Biochem. Biophys. Res. Commun.* **2010**, *393*, 649–655. [CrossRef] [PubMed]

212. Libutti, S.K.; Paciotti, G.F.; Byrnes, A.A.; Alexander, H.R., Jr.; Gannon, W.E.; Walker, M.; Seidel, G.D.; Yuldasheva, N.; Tamarkin, L. Phase I and Pharmacokinetic Studies of CYT-6091, a Novel PEGylated Colloidal Gold-RhTNF Nanomedicine. *Clin. Cancer Res.* **2010**, *16*, 6139–6149. [CrossRef] [PubMed]

213. Singh, P.; Pandit, S.; Mokkapati, V.R.S.S.; Garg, A.; Ravikumar, V.; Mijakovic, I. Gold Nanoparticles in Diagnostics and Therapeutics for Human Cancer. *Int. J. Mol. Sci.* **2018**, *19*, 1979. [CrossRef]

214. Jeene, P.M.; van Laarhoven, H.W.M.; Hulshof, M.C.C.M. The Role of Definitive Chemoradiation in Patients with Non-Metastatic Oesophageal Cancer. *Best Pract. Res. Clin. Gastroenterol.* **2018**, *36–37*, 53–59. [CrossRef]

215. Faivre-Finn, C.; Snee, M.; Ashcroft, L.; Appel, W.; Barlesi, F.; Bhatnagar, A.; Bezjak, A.; Cardenal, F.; Fournel, P.; Harden, S.; et al. Concurrent Once-Daily versus Twice-Daily Chemoradiotherapy in Patients with Limited-Stage Small-Cell Lung Cancer (CONVERT): An Open-Label, Phase 3, Randomised, Superiority Trial. *Lancet Oncol.* **2017**, *18*, 1116–1125. [CrossRef]

216. Gadducci, A.; Barsotti, C.; Laliscia, C.; Cosio, S.; Fanucchi, A.; Tana, R.; Fabrini, M.G. Dose-Dense Paclitaxel- and Carboplatin-Based Neoadjuvant Chemotherapy Followed by Surgery or Concurrent Chemoradiotherapy in Cervical Cancer: A Preliminary Analysis. *Anticancer Res.* **2017**, *37*, 1249–1256. [CrossRef]

217. Hoskin, P.; Bloomfield, D.; Dickson, J.; Jena, R.; Misra, V.; Prestwich, R. Radiotherapy Dose Fractionation. *Clin. Oncol.* **2016**, *2*, 1–136. [CrossRef]

218. Kamkaew, A.; Chen, F.; Zhan, Y.; Majewski, R.L.; Cai, W. Scintillating Nanoparticles as Energy Mediators for Enhanced Photodynamic Therapy. *ACS Nano* **2016**, *10*, 3918–3935. [CrossRef]

pharmaceutics

MDPI

Review

What Drives Innovation: The Canadian Touch on Liposomal Therapeutics

Ada W. Y. Leung [1,2,3,*], Carolyn Amador [3], Lin Chuan Wang [3], Urmi V. Mody [3] and Marcel B. Bally [1,3,4,5]

1   Cuprous Pharmaceuticals Inc., Vancouver, BC V6T 1Z4, Canada; mbally@bccrc.ca
2   Department of Chemistry, University of British Columbia, Vancouver, BC V6T 1Z1, Canada
3   Experimental Therapeutics, BC Cancer Research Centre, Vancouver, BC V5Z 1L3, Canada;
    carolyn.amador13@gmail.com (C.A.); lwang@bccrc.ca (L.C.W.); uvijay@bccrc.ca (U.V.M.)
4   Pathology and Laboratory Medicine, University of British Columbia, Vancouver, BC V6T 2B5, Canada
5   Pharmaceutical Sciences, University of British Columbia, Vancouver, BC V6T 1Z3, Canada
*   Correspondence: aleung@cuprous.ca

Received: 23 February 2019; Accepted: 12 March 2019; Published: 16 March 2019

Abstract: Liposomes are considered one of the most successful drug delivery systems (DDS) given their established utility and success in the clinic. In the past 40–50 years, Canadian scientists have made ground-breaking discoveries, many of which were successfully translated to the clinic, leading to the formation of biotech companies, the creation of research tools, such as the Lipex Extruder and the NanoAssemblr™, as well as contributing significantly to the development of pharmaceutical products, such as Abelcet®, MyoCet®, Marqibo®, Vyxeos®, and Onpattro™, which are making positive impacts on patients' health. This review highlights the Canadian contribution to the development of these and other important liposomal technologies that have touched patients. In this review, we try to address the question of what drives innovation: Is it the individual, the teams, the funding, and/or an entrepreneurial spirit that leads to success? From this perspective, it is possible to define how innovation will translate to meaningful commercial ventures and products with impact in the future. We begin with a brief history followed by descriptions of drug delivery technologies influenced by Canadian researchers. We will discuss recent advances in liposomal technologies, including the Metaplex technology from the author's lab. The latter exemplifies how a nanotechnology platform can be designed based on multidisciplinary groups with expertise in coordination chemistry, nanomedicines, disease, and business to create new therapeutics that can effect better outcomes in patient populations. We conclude that the team is central to the effort; arguing if the team is entrepreneurial and well positioned, the funds needed will be found, but likely not solely in Canada.

Keywords: liposomes; drug delivery systems; innovation; lipid nanoparticles; Metaplex

---

1. Perspective

Reviews are biased and this one is no exception. The senior author of this review completed his PhD training in the laboratory of Pieter Cullis, an innovator and serial entrepreneur. Although the PhD research was focused on gaining a better understanding of lipids in membranes, the outcome of this research changed the senior author's research directions and highlighted the importance of the team, solutions-oriented thinking, entrepreneurialism, and determination. In the past 30 years, the senior author has been pursuing his research under a common theme: There was never enough money. Yet the team that Cullis created chose to take on challenges that many told us were misguided. The underlying message—if one does not take on the task oneself, then it is very likely to never move forward and be successful. Of course, it is necessary to define success, which, from the perspective of the senior author (an academic by nature), is when the research efforts of trainees and collaborators touch a human. In this context, success can only be made in teams that were not intimidated by the initiation of companies that recognized innovative capabilities and captured intellectual property while continuing despite rejection. Money, always being an issue, is secondary. The question now is how success can be achieved faster and more frequently, noting that delays and innovation are mutually exclusive events.

2. A Brief History

The concept of liposomes was first described by Alec Bangham in the 1960s when he demonstrated the spontaneous assembly of egg lecithin into multilayer vesicular structures as phospholipids were introduced to aqueous solutions [1]. Liposomes first garnered scientific interest because of their structural similarity to cellular membranes [2]. This led to extensive studies exploring membrane structure, permeation, adhesion, and fusion as well as the roles of lipids within biological membranes [1,2]. Some of these works were pioneered by the Cullis group at the University of British Columbia (UBC) in collaboration with Ben de Kruijff (Utrecht University): Investigators who are internationally recognized for their discovery of lipid polymorphisms and the various behaviours of membrane phospholipids [3–5]. In the early 1970's, many compounds, such as lysozyme, chlorophyll a, and beta-fructofuranosidase, were investigated as candidates for liposomal encapsulation [6–8]. At this time, it became clear that certain compounds were not stored in the internal aqueous compartment; rather, they were associated with the lipid bilayer, suggesting that these compounds could become associated with liposomes by interacting with the hydrophobic regions of liposomes or by simple binding to the lipid membrane [6,7]. Additionally, it was found that encapsulation in liposomes resulted in localized cargo delivery [8]. This provided motivation for further investigation on compounds, such as actinomycin D and penicillin, by early pioneers, like Gregoriadis et al. from the United Kingdom [9]. His group also investigated the use of liposomes to carry other enzymes, as liposomes had the potential to protect enzymes from protease in the serum as well as the body's immune response [10,11]. Later, it was observed that packaging antigens into positively-charged liposomes lowered immune responses, suggesting that liposomal formulations could be key to preventing the development of severe allergic reactions [12,13]. Eventually, researchers became interested in examining the pharmacokinetic properties of liposomal drug formulations. Arakawa et al. used encapsulated 131I-insulin, 14C-sucrose, 14C-inulin, and 14C-cefazolin sodium as markers to evaluate the liposome elimination rate [14]. It was observed that drug-containing liposomes were eliminated more slowly than the unencapsulated "free" drug. However, drug absorption was also delayed, as the release of the drug from the liposome depended on the liposome's membrane composition and the loss of its structural integrity over time. To further complicate matters, it became apparent that the liposome composition, dose, size, and charge all affect the rate of elimination from the bloodstream [15,16]. These findings constitute the initial understanding of how liposomes interact with the body when given intravenously.

Anti-cancer drugs have been commonly selected for liposomal formulations, often in an attempt to reduce their toxic effects while maintaining or even enhancing antitumor activity. Initial attempts include the work by Steerenberg et al. demonstrating that the nephrotoxicity of cisplatin (CDDP) was decreased when the compound was encapsulated in liposomes. However, not only did antitumor activity decrease upon encapsulation, the tumors recurred and resistance to CDDP developed [17]. In contrast, the work by Sharma et al. showed a drastic increase in potency against models of ovarian cancer when N-(phosphonoacetyl)-L-aspartate was encapsulated in liposomes [18]. Aside from the typical preparation of liposomes for parenteral administrations [14,15,19], topical formulations were considered and these reduced the encapsulated drug's side effects due to the liposome's ability to increase the drug concentration at the target site while decreasing the drug exposure at off-target sites that often suffer from adverse effects [20,21]. One such example is the work completed by Harsani et al. from Michael Mezei's group in Dalhousie University, where they demonstrated that a liposomal formulation of radioactive triamcinolone acetonide palmitate (^{3}H-TRMAp) could be used as an effective topical treatment for oral ulcers [21]. Similarly, localization of the drug in the desired area improved the local anaesthetic effect of lidocaine when it was applied as a liposomal formulation instead of the conventional cream [22]. The preparation of lidocaine liposome for skin delivery continues to be of interest based on the recent development of penetration enhancer-containing vesicles by Fadda's group in Italy [23,24]. Liposomal antibiotics and antiviral drugs have also been successfully used in intravitreal applications to treat *Propionibacterium acnes* endophthalmitis and cytomegalovirus retinitis [25]. The years between 1980–2000 were fruitful in the context of liposomal pharmaceuticals as numerous products received regulatory approval for the treatment of cancer (Doxil®, DaunoXome®, Depocyt®, Myocet®), infectious diseases (Abelcet®, Ambisome®, Amphotec®), macular degeneration (Visudyne®), as well as for the prevention of viral infections in the form of virosomal vaccines (Epaxal® and Inflexal®) [26].

The development and advancement of liposomal technologies has proven successful in part due to the development of liposomal pharmaceuticals, but their real impact has been felt in the cosmetic industry, with products including Capture (C. Dior), Niosomes (L'Oréal), Revision (Revision SkinCare), and others. This review, however, will focus on Canada's contribution to the liposomal pharmaceutical field as a number of liposomal technologies have been developed in Canada, ranging from methods of liposome preparation to drug loading strategies, storage strategies, and targeted delivery. This focus is on how Canadian investigators and entrepreneurs impacted the field, but the success of this technology is global. We hope those reading this paper accept its focus and understand that we needed to neglect many other key individuals that made this field what it is today: Particularly teams in the USA, Japan, the Netherlands, and Israel. A timeline of the highlighted technologies is provided in Figure 1.

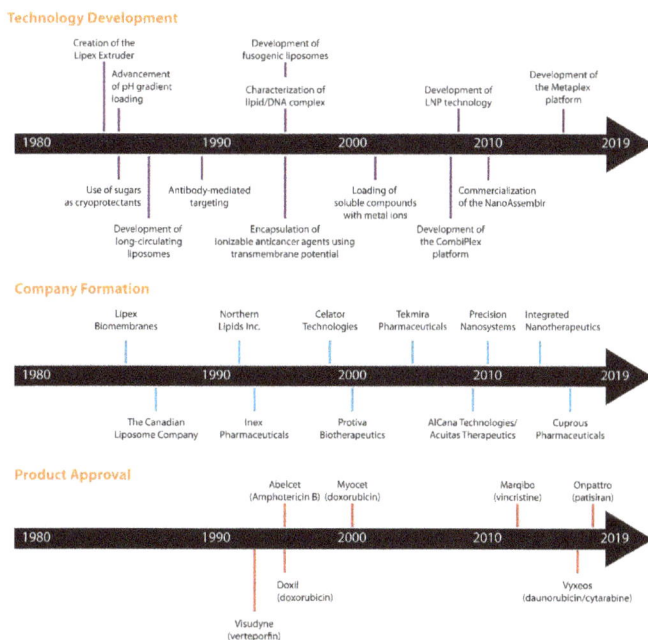

Figure 1. Canadian contribution to the development of liposomal technologies, formation of companies, and development of clinically approved formulations. Selected liposomal technologies are listed on the timeline based on the patent literature (top panel). These technologies led to the formation of companies, which are shown based on the year when they were established (middle panel). Regulatory approved liposomal formulations that were developed by Canadian researchers are shown on the timeline based on their year of approval (bottom panel).

3. Technologies for the Production of Liposomes

While liposomes are known to consist of phospholipids that self-assemble into multi-layer vesicles, uniformity of the liposomal structure is necessary for further pharmaceutical development. Liposomes for pharmaceutical applications are typically up to 200 nm in diameter, composed of a unilamellar or bilamellar bilayer and an aqueous core (Figure 2) [27,28]. Preparation of these homogeneous liposomal formulations was pioneered by Olson et al. from the laboratory of one of the pioneer liposomologists, Demetrios Papahadjopoulos (University of California San Francisco), where multilamellar vesicles were sequentially passed through polycarbonate membranes of 1.0, 0.8, 0.6, 0.4, and 0.2 μm pore sizes to yield a homogeneous preparation of liposomes with a mean diameter of about 270 nm [29]. Extrusion using this method could be completed using 10–12 mM lipid suspensions at low pressures (about 50 psi). While this method was sufficient to generate bench-scale formulations, it was challenging and time-consuming to prepare larger batches, which would be required for preclinical or clinical studies. Hope et al. from Pieter Cullis' group at UBC further advanced the extrusion technology: Lipid concentrations up to 300 mM could be used to extrude multilamellar vesicles through 100 nm pore size polycarbonate filters [30,31]. Using higher pressure (up to approximately 500 psi), unilamellar vesicles with a mean diameter of 60–100 nm could be produced within 10 min. This invention led to the creation of the Lipex® Extruder, which was first marketed by the spin-off company, Lipex Biomembranes, created in 1984. Lipex Biomembranes was bought by Northern Lipids, another biotechnology venture created in Vancouver, Canada. Northern Lipids was eventually purchased by Evonik Industries, a German multinational company. Evonik still retains the Northern Lipids enterprise in British Columbia and still markets the Lipex® Extruder today. This extrusion technology, ranging from simple devices for laboratory scale production to larger extrusion systems that can handle more than 100 L, continues to be the industrial standard for preparing pre-clinical and clinical batches of liposomal formulations.

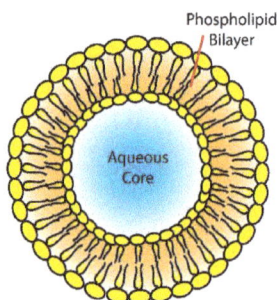

Figure 2. Structure of a liposome. A liposome consists of a phospholipid bilayer with an aqueous core.

As the technology of microfluidics emerged in the last two decades, novel methods of manufacturing liposomes have also been developed. Microfluidics systems manipulate and control the flow of fluids through networks of channels having cross-sectional dimensions of approximately 5 to 500 μm [32]. The first use of microfluidics mixing to generate liposomes was described by Jahn et al., where they demonstrated that hydrodynamically focusing of alcohol-dissolved lipids between two sheathed streams of aqueous buffer in a microfluidic channel could yield monodispersed liposomes ranging from 50 to 150 nm depending on the flow rate [33]. To better control the mixing process and to generate liposomes even more rapidly, Zhigaltsev et al. from the Cullis group here in Canada designed a microfluidic mixing device based on the concept of staggered herringbone mixing [34–36]. Using this method, liposomes of 20–50 nm in size could be prepared and loaded with small molecules, such as doxorubicin [35]. This work led to the creation of the NanoAssemblr™ device, which is commercialized by Precision NanoSystems (Vancouver, BC, Canada) for the preparation of liposomes encapsulating small molecules as well as macromolecules, such as nucleic acids and proteins, at bench and production scales [34,37]. This device is particularly well-suited for the self-assembly process used to prepare cationic lipid/anionic polymer (e.g., DNA, RNA, antisense oligonucleotide, siRNA, peptides, etc.) complexes, often referred to as lipid-nanoparticles (LNP). The ability of anionic polymers to bind cationic lipids to form complexes was first disclosed by Dr. Bally's team [38–40], and was later employed by Cullis and associates to define what is now a US Food and Drug Administration (FDA) approved siRNA formulation called Patisiran. It has been postulated that these structures are better defined as a particle, rather than a liposome because they likely do not comprise a lipid-bilayer structure.

4. Technologies for the Storage of Liposomes

Another practicality issue associated with the pharmaceutical development of liposomal formulations was the shelf-life of the products. Liposomal formulations typically require storage at 4 °C and are relatively unstable for long term (>2 years) storage compared to other pharmaceuticals that can be prepared as dried products. A discovery by J. Crowe, Louis Crowe, and Dennis Chapman overcame this problem. Crowe's team were trying to better understand membrane stability in the presence of carbohydrates, known to be secreted by anhydrobiotic organisms (e.g., nematode) that were able to survive drying/freezing conditions. They showed, in 1984, that carbohydrates, such as trehalose and sucrose, were able to stabilize the model membrane structure at low water contents [41]. This observation was applied by Madden et al. working in Vancouver, and it was shown that multiple types of sugars, including trehalose, sucrose, and lactose, could be effective at protecting liposomes during the dehydration-rehydration process when the sugar content was appropriately adjusted [42–45]. The first publication on this was released in July of 1985 by the Vancouver team, but was followed shortly thereafter by a publication from Crowe et al. in October of that year, showing that trehalose can be used to prevent liposomes from fusing during the freeze-drying process [41]. The use of carbohydrates to protect liposomes was disclosed in a patent with the Vancouver inventors and the technology was commercialized and functionalized with lactose being incorporated as a cryoprotectant in Amphotec®, Myocet®, and Visudyne®, and sucrose being added to AmBisome® to enable the preparation of lyophilized liposomal products. Perhaps most interesting, this approach worked very well in liposomes with no or low (<20 mol%) levels of cholesterol, where the ability to prevent aggregation and fusion during a freeze/thaw cycle was first demonstrated, even in the absence of carbohydrates, by Dos Santos et al. [46,47]. The first low-cholesterol product, Vyxeos®, is stored as a dehydrated powder using sucrose and has a shelf life of 2 years at 4 °C to 8 °C.

5. Optimization of Liposomes for Pharmaceutical Use

A summary of the main approaches discussed in this section is listed in Table 1.

Table 1. Strategies developed to optimize liposomal products for pharmaceutical use.

Method Developed	Utility	References
Dehydration-rehydration method	Improve passive encapsulation efficiency	[48]
Modulation of lipid fluidity	Improve passive encapsulation efficiency	[49]
pH gradient loading	Remote loading	[50–53]
Use of ionophore to load small molecules	Improve remote loading efficiency	[54]
Use of ethanol to load small molecules	Improve remote loading efficiency	[46]
Microencapsulation method	Improve loading efficiency of water soluble and insoluble compound	[55]
Layersomes	Improve liposome stability and oral delivery	[56]
Hyaluronan coating of liposomes	Enable topical applications	[57]
Use of metal ion gradient	Stabilize water-soluble compounds	[58,59]
Metaplex technology	Enable development of poorly soluble metal-binding compounds	[60,61]
Use of cationic lipids	Deliver nucleic acids	[39,40]
Lipid nanoparticle (LNP) technology	Optimize delivery of nucleic acids for clinical use	[62–64]

5.1. Improvement of Encapsulation Efficiency of Passive Loading

Following the initial forays into liposomal drug delivery, it became increasingly clear that more efficient ways to encapsulate drugs were needed. Typical drug entrapment efficiencies were at best 10% and this was due to a number of limitations [65,66]. In particular, given an optimal size of 100–200 nm for nanomedicines that attempt to leverage the enhanced permeability and retention (EPR) effect, a trapped volume of 1.5–2.5 µL/µmole, and a workable lipid concentration of 10–20 mM, it was just not possible to achieve trapping efficiencies above 10% [67]. Of course, methods that used increased lipid concentrations and/or association of the drug candidate with the membrane could be designed to achieve an improved trapping efficiency. In some cases, increasing the aqueous drug solubility of the compound of interest by changing the pH of the medium in which the compound is suspended or by increasing the temperature greatly enhanced the encapsulation efficiency. Regarding lipid concentration, the high pressure extrusion method allowed for the manufacturing certain liposomes at lipid concentrations of 300 mM and this could achieve trapping efficiencies as high as 80% [30]. Alternatively, in 1984, Kirby and Gregoriadis introduced the dehydration-rehydration method, where dehydrated liposomes were rehydrated in a small volume to increase the drug encapsulation efficiency to as high as 40–50% [48]. As suggested above, depending on the hydrophobicity of the compound being used, its association with the lipid membrane could be increased or the membrane's "fluidity" could be altered to enhance the association of these drugs [46,49,66]. However, these strategies came at a cost: Low drug-lipid ratio, which in one aspect meant a great deal of liposomal lipid was required to administer an effective dose of the therapeutic agent. Given the cost of lipids and drugs at the time, this made it unreasonable to pursue liposomal drugs as pharmaceuticals.

5.2. Development of Remote Loading Methods

In 1976, Nichols and Deamer demonstrated that catecholamines can accumulate within liposomes that have an established transmembrane pH gradient [68]. This concept was confirmed in 1985 by Bally et al., who discovered that lipophilic cations, like safranine O, could accumulate inside liposomes in response to an Na^+/K^+ electrochemical gradient where the liposome's interior was negative [50]. This resulted in an interior safranine concentration over 80 mM, many times greater than the solubility of the safranine. Research completed by Deamer's (USA) and Cullis' (Canada) groups set the foundation for remote or active loading, where it was possible to achieve a >98% encapsulation efficiency [46,50–53]. The pH gradient loading method, and varieties thereof, remains to be one of the most employed methods to encapsulate a drug or drug candidate in liposomes. Several methods for creating these gradients exist, such as using citrate buffer in the aqueous compartment, using an ionophore-mediated ion gradient to generate a pH gradient as originally described by the Vancouver team [54,69,70], or using transmembrane ammonia gradients as described by the Israel/US teams. Depending on the properties of the compound of interest, the Canadian team led by Marcel Bally showed that encapsulation efficiencies could be further improved by the addition of solvents, such as ethanol. While the use of ethanol could potentially double the encapsulation efficiency at lower temperatures, an excessive amount of ethanol could cause the liposomal structure to break down and collapse the pH gradient [46].

Although remote loading of small molecules in response to a transmembrane pH gradient has been widely applied, three issues still remain to be addressed: (1) Many small molecules do not have a protonizable amine function, which is required for efficient pH gradient loading; (2) some compounds, which are amenable to pH gradient loading, are associated with poor trapping efficiencies, which could be due to issues, such as proton leakage through the bilayer [53,71]; and (3) many therapeutically interesting compounds are poorly soluble in aqueous solution, but are not necessarily "hydrophobic". To increase drug encapsulation efficiency for various types of compounds, new encapsulation methods have been established. One was referred to as a microencapsulation method, in which a water/organic/water emulsion is agitated and used to prepare a liposomal suspension [55]. Other approaches to increasing the stability of liposomes include the development of "layersomes", where multiple layers of polyelectrolytes are added to conventional liposomes; this proved effective for the encapsulation of piroxicam [56]. Liposomes, functionalized with hyaluronic acid, have also been produced and were found to increase the bilayer packing order, reducing membrane flexibility and improving drug penetration in topical applications [57].

Another strategy that has been employed to improve the trapping efficiency of small molecules is the use of metal complexation. Initial studies were completed by the Cullis group in which doxorubicin was encapsulated into Sphingomyelin/Chol liposomes in response to a manganese (trapped $MnSO_4$) ion gradient [58]. Greater than 98% trapping efficiency was achieved, but the stability of the $MnSO_4$ solution required the use of a low pH. Although it was suggested that doxorubicin was capable of forming a metal complex with Mn^{2+}, the fact that the liposomes used also had a pH gradient confused the interpretation of the results. The observation was confirmed by Abraham et al. from the Bally group [59] in a manner that allowed that group to conclude that metal complexation could be the sole driver of encapsulation. Further, the technology appeared suitable for use with a number of other different metal ions, including copper and zinc [72], but the method always relied on the use of compounds that exhibited a solubility >1 mg/mL and the use of compounds that had a protonizable amine. The role of the pH gradient versus metal to encapsulate the drug was further confused by the Bally team, who discovered and patented that transition metals could be used in conjunction with a divalent metal ionophore (A23187) to generate a pH gradient. However, there was something about the use of copper that enhanced drug retention that was surprising and unexpected [73]. The product generated using this technology was focused on the camptothecins (irinotecan and topotecan) [54,74,75] and one of the resulting products, Irinophore C, was more active than any other previously described irinotecan formulations. It was disappointing that the resultant formulation never made it to the clinic. There are many reasons for this, including the development of another formulation of irinotecan (now approved and called Onivyde® [76]) that was clinically more advanced than the one created in Canada and the fact that the technology developed in Canada was licensed: The company that licensed the technology was not in a position to develop the product. The resulting delays and the early filing of intellectual property (which was granted in several countries) made it unlikely that further investments in funding clinical trials would result in meaningful returns to those that made that investment. The Bally's group was too slow to develop the technology.

It is worth noting that the incorporation of a transition metal into liposomes was critical to the creation of Vyxeos®: A copper-containing formulation wherein the formation of a copper-anthracycline complex is used to reduce the leakage rate of daunorubicin, while the use of low cholesterol liposomes was required to enhance the retention of cytrarabine. The resulting product was designed such that the two cytotoxic agents could be released from the liposomes at identical rates, ensuring the maintenance of a synergistic drug-to-drug ratio in vivo [77].

This transition metal-complexation technology has further evolved through works completed by the Bally's group, who is now working with other founders (Ada Leung and Thomas Redelmeier), the Vice President of research (Michael Abrams), and the entrepreneurial group at UBC (e@UBC and HATCH) to lead the development of what is referred to as Metaplex technology. The Metaplex technology is an active loading platform wherein a transition metal ion gradient is established across the membrane and used as the primary driving force to accumulate drugs inside liposomes; drugs that exhibit limited water solubility, may not contain a protonizable amine, but do contain a metal binding function [60,61]. In this technology, there is clear evidence that the selected drug has a metal binding function. By using lipid nanotechnology and metal coordination chemistry, this new formulation method created by the Bally group enables the development of drugs that are typically relegated to medicinal chemistry groups. Initial studies by Wehbe et al. explored formulation strategies for diethyldithiocarbamate (DDC), a metabolite generated after disulfiram is administered. Disulfiram is an anti-alcoholic agent known to inhibit aldehyde dehydrogenase [78]. DDC has long been known as a copper binding compound [79] and it has been shown to have copper-dependent anticancer activity [80–82]. Metaplex technology has been further expanded at the Vancouver-based Cuprous

Pharmaceuticals Inc. for two different classes of compounds: (1) Sparingly soluble small molecules that have metal-coordinating motifs, which could benefit from drug delivery technology; and (2) relatively inactive small molecules that become therapeutically active upon complexation with metal ions, such as Cu^{2+}. The preparation of nanoformulations for metal-dependent therapeutics was originally inspired by the increase in the number of publications in recent years demonstrating the therapeutic activity of copper complexes against a variety of disease indications, including cancer, inflammatory diseases, neurodegenerative diseases, and infectious diseases [83–88]. While metal complexes, or specifically copper complexes, hold promise as therapeutic agents based on in vitro data, there is a lack of preclinical evidence supporting their utility. The major reason for this is likely due to the fact that most of these therapeutically active metal complexes have poor solubility in aqueous solutions under physiological conditions, making it a challenge to test these agents in animals. The Metaplex technology addresses this problem by using liposomes as nano-scale reaction vessels in which metal coordination occurs [60]. Furthermore, these nanoparticles can be suspended in biocompatible aqueous buffers for parenteral administration into relevant preclinical models. The Bally group and Cuprous validated the concept of metal-dependent anticancer activity for copper-coordinating compounds through an in vitro screen on platinum-sensitive and platinum-resistant cell lines and prepared injectable copper-based formulations of DDC and clioquinol [80,89,90]. The Metaplex technology can also be used to reformulate sparingly soluble compounds that have metal-binding properties with the goal of either reducing toxicity or improving therapeutic activity. This was demonstrated by preparing liposomal formulations, CX-5461, an investigational compound that interacts with copper and when formulated using Metaplex technology, is more efficacious than the low pH clinical formulation currently used [91,92]. Cuprous is currently developing this technology for immuno-oncology treatments reliant on the use of small molecules rather than the more expensive antibody-based or cell-based therapeutics. Immunogenic cell death (ICD), a phenomenon wherein dying cancer cells emit specific molecular signals that ultimately lead to an anti-tumour adaptive immune response followed by long-term protection against recurrence, is a concept of immunotherapy that has recently garnered much attention due to its potential to treat metastatic disease and/or bring about long-term survival or cures [93,94]. Anthracyclines, like doxorubicin, are known to induce ICD as a secondary mechanism [93,95]. Most interestingly, metallic copper itself, is known to generate reactive oxygen species (ROS) and induce endoplasmic reticulum (ER) stress, which is required for ICD induction [96,97]. Cuprous is exploring exciting new opportunities for enhancing the delivery of such compounds using its proprietary platform technology [73,74].

Metaplex has the potential to work with a more diverse array of compounds than conventional pH gradient loading due to a larger chemical space that would satisfy the requirement for metal-ligand coordination to occur [86,87]. While the focus has thus far been on the use of Metaplex for oncology-based formulations, the potential application of this technology is much broader. Copper complexes of non-steroidal anti-inflammatory drugs (NSAIDs) have been shown to be associated with reduced toxicity and increased therapeutic activity [88,98]. Various copper complexes have exhibited hypoglycemic effects and may be suitable as diabetic treatments [99,100]. Other studies demonstrated that copper complexes could have potent antimicrobial activity and could be useful against infections by superbugs, which are becoming a global health concern [101,102]. Finally, it is known that metal imbalance is strongly associated with neurodegenerative diseases, such as amyotrophic lateral sclerosis, Alzheimer's, and Parkinson's diseases [83,103,104]. Strategies to adjust these imbalances using metal chelators are being evaluated [105,106]. All of these represent opportunities for the development of novel metal-based therapeutics, where the Metaplex platform could be used to design formulations for specific indications, further expanding the application of liposomal technologies non-oncology-based indications.

5.3. Development of Liposomes for Encapsulation of Nucleic Acids

The utility of liposomes has also been extended to nucleic acid delivery (DNA, mRNA, Antisense oligonucleotides, siRNA, etc.), typically for the purpose of genetic modification of target cells. Nucleic acids alone cannot pass through cellular bilayers, but liposomes have been designed to fuse with membranes and successfully deliver associated payloads [107]. The earliest examples of nucleic acid encapsulation were in the late 1970s: Dimitraidis et al. encapsulated mRNA, rRNA, and tRNA in large unilamellar liposomes, and Fraley et al. focused on the delivery of pBR322 bacterial plasmids [108,109]. Continued research demonstrated that factors, such as the presence of polyethylene glycol (PEG) or glycerol, liposome charge, and the number of lamellae, all affect nucleic acid infectivity or sequestration [110]. While early work supported the use of cationic liposomes as delivery agents of plasmid DNAs for transfection purposes, the physical characteristics of these liposome/DNA complexes were not well-defined. In 1995, Reimer et al. from the Bally group prepared and characterized, for the first time, hydrophobic complexes of cationic lipids and plasmid DNA that can be readily extracted in organic solvents [39,40]. They proposed these cationic lipid/DNA complexes as potential intermediates for the formation of particles suitable for gene delivery to cells [38–40]. This work was extended to the complexation of cationic lipids with antisense oligonucleotides designed for gene silencing and subsequently the addition of PEG to prevent aggregation of these lipid/nucleic acid complexes [111–113]. These works were further developed in the 1990s and 2000s through the Cullis group (UBC) and Vancouver-based companies, including Acuitas Therapeutics (formerly AlCana Technologies) and Inex Pharmaceuticals (now Arbutus Biopharma) [64,114]. These efforts led to the use of ionizable amino lipids for the delivery of nucleic acids [63] and the development of fusogenic liposomes: Liposomes that have an exchangeable PEG-lipid conjugate, which contributes to the in vivo stability of nanoparticles, particularly those encapsulated with antisense oligodeoxynucleotides [115,116]. These technologies seeded the evolution of nucleic acid drug delivery, leading to the creation of new lipids designed specifically for the encapsulation of RNA-based therapeutics and the development of the lipid nanoparticle (LNP) technology platform: The most advanced and currently the only clinically validated nucleic acid delivery system through the regulatory approval of the RNA interference (RNAi) therapeutic Onpattro® [5,62,64,114,117–120].

6. Other Key Canadian Discoveries that Impacted the Development of Therapeutically Interesting Drugs

Table 2 below highlights some of the Canadian discoveries that helped the development of liposomal pharmaceutical products evolve over time.

Table 2. Canadian discoveries that were involved in driving the advancement of liposomal pharmaceutical products.

Canadian Discoveries	References
Use of antibodies to mediate targeting with liposomes	[121,122]
Selective targeting of liposomes to the blood compartment	[123]
Use of GM_1 ganglioside in liposomes, leading to the development of "PEGylation"	[122]
Role of PEG in preventing liposome aggregation	[124]
Development of low-cholesterol liposomes with lipids that prevent aggregation	[124]
Maintenance of the drug-drug ratio for two drugs encapsulated in one liposome	[77]

6.1. Selective Drug Delivery with Liposomes

Shortly after liposomes were first described, their promise as selective delivery agents was considered. To this end, Gregoriadis et al. associated molecular probes to drug-containing liposomes and found that probes (Immunoglobulin G's (IgGs) raised against different types of cells) could mediate selective cellular uptake [125]. In the early 1980s, Leserman et al. coupled monoclonal antibodies to liposome surfaces, successfully demonstrating cell-specific liposome interaction [126]. A third example by Guru et al. demonstrated significant increases in the efficacy of sodium stibogluconate via encapsulation in tuftsin-bearing liposomes. Even liposomes carrying only tuftsin were found to make animals resistant to *Leishmania donovani* infections [127]. In addition to these early studies, Ryman's team explored the imaging potential of liposomes, highlighting the ability of liposomes to localize in lymph nodes by injecting technetium-99m labelled liposomes into rats and then studying tissue distribution via γ-camera imaging and radioassay [128], and Morgan et al. demonstrated that liposomes could be used to image staphylococcal infections [129]. Finally, Baldeschwieler's group highlighted the potential for liposomes to image tumours. All these studies were completed well before Matsumura and Maeda first published and described the EPR effect: Selective accumulation due to abnormally permeable vasculature found in tumours [130]. It is also worth noting that drug release mechanisms, such as endocytosis and fusion, were investigated early in the development of liposomal pharmaceuticals, where it was postulated that different uptake mechanisms could allow for selective delivery depending on how specific liposome formulations interacted with cells [65].

Clearly, an important rationale for developing more selective liposomes was based on strategies designed to increase interactions between the nanoparticles and target/disease cells while minimizing toxicities against healthy cells. One of the most commonly employed approaches still used today concerns the use of surface coatings. Liposomes with attached antibodies could bind to specific cell populations. Some of the earliest works by Papahadjopoulos' group demonstrated that liposomes coated with antibody fragments or immunoliposomes were able to bind human erythrocytes much more efficiently compared to non-targeting liposomes [131,132]. In Canada, some of the early studies investigating the use of antibody-mediated targeting of liposomes to treat cancer were conducted by Theresa (Terry) Allen's group at the University of Alberta [121]. For instance, Ahmad and Allen demonstrated that liposomes coated with antibodies targeting squamous carcinoma cells resulted in increased uptake and cytotoxic effects against KLN-205 lung squamous carcinoma cells relative to non-targeting liposomes [133]. Several studies have demonstrated that actively targeted liposomes may contribute to improved therapeutic activity in vivo [123,134,135]. While the initial focus was to alter the biodistribution of targeted liposomes for more efficient delivery to the target site (i.e., the tumour), it was discovered that delivery to the tumour for both immunoliposomes and conventional liposomes was primarily dependent on the EPR effect [136]. The differences in therapeutic activity reported were likely due to an increased uptake of immunoliposomes by cancer cells as a result of receptor-mediated endocytosis followed by the escape of the cytotoxic agent from endosomal/lysosomal degradation [136–139]. In recent years, researchers have also explored the use of peptide-mediated targeting [138,140,141]. Several excellent review articles are available describing the various functionalization strategies that have been employed in the development of active targeting nanomedicines [142–144]. Although first envisioned 40 years ago, there has yet to be a successful targeted formulation approved for clinical use. However, it is notable that the limitations for targeting solid tumours are clear. Allen's team was able to highlight the potential of targeting liposomes to cells within the vascular compartment [123]. It is hoped that this may prove to be of therapeutic value, particularly in light of some of the recent findings from the Bally group, which emphasize that therapeutic antibodies may exhibit improved therapeutic effects when attached to liposomes. These studies consider the potential of liposomes to deliver antibodies rather than antibodies to target liposomes [123].

6.2. The "PEGylation" Technology

The effects of various lipid compositions on the pharmacokinetics of liposomes were also explored with the goal of prolonging the presence of liposomes in the plasma compartment. The first attempt was made by Terry Allen's group through the addition of GM_1 ganglioside into liposomes, which reduced mononuclear phagocyte system (MPS) uptake, allowing the liposomes to remain in the blood stream for several hours [145–147]. Based on this pioneering work, scientists explored the incorporation of PEG into formulations as a steric stabilizer lipid (i.e., 1,2-Distearoyl-sn-glycero-3-phosphoethanolamine-polyethylene glycol (DSPE-PEG)) [147]. This "PEGylation" technology, otherwise known as stealth liposome technology, has been the most widely employed strategy since the 1990s [26,148–150]. Yet the role of surface coating of liposomes with PEG has likely been well overstated. It was the Canadian Bally group along with Christine Allen, Nancy Dos Santos, and others that first highlighted that the primary role of PEG on the surface of the liposomes was not to prevent protein association or even association with phagocytic cells, but to prevent surface–surface interactions that could lead to aggregation of liposomes [124]. This was elegantly proven by the work of Dos Santos, who demonstrated that selected liposomes could be prepared in the absence of cholesterol or low levels of cholesterol as long as they incorporated lipids that prevented their aggregation, such as PEG-modified lipids [124]. As suggested above, this technology was key to the development of Vyxeos®, a combination liposomal drug product now approved for treatment of acute myeloid leukemia.

6.3. Strategies to Encapsulate Multiple Agents

Related to the previous statement and existing evidence that cancer is a heterogeneous disease, which is most effectively treated with a combination of multiple therapeutic agents, there was significant interest in encapsulating multiple drugs in the same liposome. For example, daunorubicin and 6-mercaptopurine are a pair of chemotherapeutic compounds that were thought to act synergistically—one being a hydrogen acceptor and the other being a hydrogen donor. While this particular interaction was not observed, combining the two compounds in a dual drug liposome did appear to show synergistic cytotoxic effects [151]. Researchers in Vancouver (BC Cancer and Celator Pharmaceuticals) extensively studied the impact of drug-to-drug molar ratios on therapeutic outcomes in vitro. Drug combinations could result in synergistic or antagonistic treatment effects depending on the ratio used. It was logical to assume that if the effects of the anticancer drug in vitro were dependent on the drug-drug molar ratio in vitro then the same would hold true in vivo [152,153]. When encapsulating multiple drugs in the same liposome, or even different liposomes, the relative release rates of the two compounds must be considered, as the goal is to achieve an ideal ratio at which they could be administered (a fixed ratio product) and to maintain that ratio over time after administration to achieve optimal therapeutic effects [154]. Tardi et al. illustrated an example of a system in which cholesterol was used to control drug leakage rates and with this system, they were able to maintain a 1:1 synergistic ratio of irinotecan and floxuridine in vivo [72]. This observation proved to be a "patenting moment" that led to the development of liposomal combination products protected under the "Combiplex" patent [155]. This patent described the use of various drug delivery systems to be used to prepare products in a manner where the combination ratio could be maintained in vivo; a patent that first contemplated the use of daunorubicin and cytarabine as a liposomal combination product that eventually became Vyxeos [77].

7. The Canadian Impact on Regulatory Approved and Investigational Liposomal Formulations

Here, we provide a list of approved liposomal products and reiterate some of the information above to highlight how Canadian scientists influenced these products. An up-to-date list of all regulatory approved liposomal formulations is provided in Table 4.

Table 3. Regulatory approved liposomal formulations.

Approval Year	Trade Name	Active Agent	Lipid Composition	Approved Indication(s)	Current Ownership	References
1993	Epaxal (discontinued)	Inactivated hepatitis A virus (strain RGSB)	DOPC:DOPE (75:25 molar ratio)	Hepatitis A	Janssen Pharmaceuticals	[156,157]
1995	Doxil	Doxorubicin	HSPC:Cholesterol:PEG 2000-DSPE (56:39:5 molar ratio)	Ovarian, breast cancer, Kaposi's sarcoma	Janssen Pharmaceuticals	[158–160]
1995	Abelcet	Amphotericin B	DMPC:DMPG (7:3 molar ratio)	Invasive severe fungal infections	Leadiant Biosciences	[161–163]
1996	DaunoXome	Daunorubicin	DSPC:Cholesterol (2:1 molar ratio)	AIDS-related Kaposi's sarcoma	Galen Pharmaceuticals	[164,165]
1996	Amphotec	Amphotericin B	Cholesteryl sulphate:Amphotericin B (1:1 molar ratio)	Severe fungal infections	Kadmon Pharmaceuticals	[166]
1997	Ambisome	Amphotericin B	HSPC:DSPG:Cholesterol:Amphotericin B (2:0.8:1:0.4 molar ratio)	Presumed fungal infections	Astellas Pharma & Gilead Sciences	[167–169]
1997	Inflexal V (recalled)	Inactivated hemaglutinine of Influenza virus strains A and B	DOPC:DOPE (75:25 molar ratio)	Influenza	Crucell, Berna Biotech	[170]
1999	Depocyt (discontinued)	Cytarabine/Ara-C	Cholesterol:Triolein:DOPC:DPPG (11:1:7:1 molar ratio)	Neoplastic meningitis	Pacira Pharmaceuticals	[171,172]
2000	Myocet	Doxorubicin	EPC-Cholesterol (55:45 molar ratio)	Combination therapy with cyclophosphamide in metastatic breast cancer	Teva Pharmaceutical Industries	[173,174]
2000	Visudyne	Verteporfin	EPG:DMPC (3:5 molar ratio)	Choroidal neovascularisation	Cheplapharm Arzneimittel GmbH	[175,176]
2004	DepoDur (discontinued)	Morphine sulfate	Cholesterol:Triolein:DOPC:DPPG (11:1:7:1 molar ratio)	Pain management	Flynn Pharmaceuticals	[177]
2009	Mepact	Mifamurtide	DOPS:POPC (3:7 molar ratio)	High-grade, resectable, non-metastatic osteosarcoma	Takeda Pharmaceutical Ltd.	[178]
2011	Exparel	Bupivacaine	DEPC, DPPG, Cholesterol and Tricaprylin	Pain management	Pacira Pharmaceuticals, Inc.	[179,180]
2012	Marqibo	Vincristine	SM:Cholesterol (55:45 molar ratio)	Acute lymphoblastic leukemia	Spectrum Pharmaceuticals	[181,182]
2015	Onivyde	Irinotecan	DSPC:MPEG-2000:DSPE (3:2:0.015 molar ratio)	Combination therapy with fluorouracil and leucovorin in metastatic adenocarcinoma of the pancreas	Ipsen Biopharmaceuticals	[183]

Table 4. Regulatory approved liposomal formulations.

Approval Year	Trade Name	Active Agent	Lipid Composition	Approved Indication(s)	Current Ownership	References
2017	Vyxeos	Daunorubicin/Cytarabine	DSPC:DSPG:CHOL (7:2:1 molar ratio)	Therapy related acute myeloid leukemia (t-AML) or AML with myelodysplasia-related changes (AML-MRC)	Jazz Pharmaceuticals	[184,185]
2018	Onpattro	Patisiran	Dlin-MC3-DMA, PEG2000-C-DMG	Hereditary transthyretin-mediated amyloidosis	Alnylam Pharmaceuticals, Inc.	[185]

HSPC (hydrogenated soy phosphatidylcholine); PEG (polyethylene glycol); DSPE (distearoyl-sn-glycero-phosphoethanolamine); DSPC (distearoylphosphatidylcholine); DOPC (dioleoylphosphatidylcholine); DPPG (dipalmitoylphosphatidylglycerol); EPC (egg phosphatidylcholine); DOPS (dioleoylphosphatidylserine); POPC (palmitoylloleoylphosphatidylcholine);SM (sphingomyelin); MPEG (methoxy polyethylene glycol); DMPC (dimyristoyl phosphatidylcholine); DMPG (dimyristoyl phosphatidylglycerol); DSPG (distearoylphosphatidylglycerol); DEPC (dierucoylphosphatidylcholine); DOPE (dioleoly-sn-glycero-phophoethanolamine).

7.1. Liposomal Formulations of Amphotericin B: Abelcet® and iCo-019

Amphotericin B is an antifungal agent used to treat serious fungal infections and leishmaniasis. Multiple lipid-based formulations of amphotericin B (i.e., AmBisome®, Abelcet®, Amphotec®) are approved for use in various countries, of which Abelcet® presents a unique formulation wherein amphotericin B is complexed with dimyristoylphosphatidylcholine (DMPC) and dimyristoylphosphatidylglycerol (DMPG) at a 7:3 molar ratio, forming ribbon-like structures (hence known as amphotericin B lipid complexes), which are believed to have contributed to its favourable toxicity and therapeutic profiles [186]. The formation of Abelcet® was designed by Drs. Thomas Madden, Andrew Janoff, and Pieter Cullis at UBC. Abelcet® was the first drug from the Cullis group to reach the market. It was developed by The Liposome Company in association with the Canadian Liposome Company, a wholly owned subsidiary, and was approved in 1995 for the treatment of invasive fungal infections to which patients are non-responsive or cannot tolerate conventional amphotericin B treatments. Abelcet® is currently a Leadiant Biosciences product.

iCo-019 is an oral liposomal formulation developed by Kishor Wasan's group (from UBC and University of Saskatchewan). The formulation comprises Peceol and distearoylphosphatidylethanolamine (DSPE-PEG) and the resulting oral formulation was found to reduce the number of fungal colony formation units by more than 80% relative to untreated controls [187]. While the existing intravenous formulation of amphotericin B is effective at treating invasive fungal infections, safety issues associated with parenteral administrations, such as infection at the catheter, haemolysis, and renal toxicities, are concerning [187]. The oral formulation was designed to overcome these issues as well as to address the problem of these drugs being costly and difficult to administer in remote locations, where fungal infections are more problematic. The oral amphotericin B formulation is currently being developed by the Vancouver-based company, iCo Therapeutics, which has recently announced positive Phase I data on iCo-019.

7.2. Liposomal Formulations of Doxorubicin: Myocet® and Doxil®

Doxil® (Caelyx®) and Myocet® are perhaps the most well-known liposomal anticancer agents. Doxil® was first approved in 1995 for the treatment of acquired immunodeficiency syndrome (AIDS) related Kaposi's sarcoma [139]. It is now also being used to treat relapsed ovarian cancer, multiple myeloma, and locally advanced or metastatic breast cancer. Myocet® is another liposomal formulation of doxorubicin, which was approved in 2000 to be used in combination with cyclophosphamide to treat metastatic breast cancer in Europe. The main difference between Doxil® and Myocet® lies in their lipid composition, which ultimately affects their safety, drug release, and biodistribution profiles [188–190]. Doxil® is composed of hydrogenated soya phosphatidylcholine, cholesterol (Chol), and PEG-modified phosphatidylethanolamine (55:40:5 molar ratio) while Myocet® is a non-PEGylated liposomal formulation consisting of egg phosphatidylcholine (EPC) and Chol (55:45 molar ratio). Myocet® increases the circulation lifetime of doxorubicin by approximately three times relative to the free agent in mice [189,191]: An effect thought to be due to the toxicity of doxorubicin being delivered to phagocytic cells [192]. This is also known to occur for Doxil® because it is prepared with lipids that enhance drug retention, resulting in an increased blood residence time of doxorubicin and increased drug delivery to the skin, which made it an ideal formulation for the treatment of skin localized cancers, like Kaposi's sarcoma [189,193,194]. However, the increased skin delivery caused dose-limiting toxicities attributed to hand-and-foot syndrome [189]. Both liposomal formulations reduce cardiotoxicity, a major concern associated with free doxorubicin. Myocet was developed by the Vancouver group in association with the Canadian Liposome/The Liposome Company [173]. Doxil® originated from research completed by groups in California and Israel, but this product was greatly influenced by Terry Allen. The product was initially developed by Liposome Technology Inc. and is now owned by Johnson & Johnson. Myocet®, on the other hand, is now a product owned by Teva Pharmaceutical Industries [150,195].

7.3. Visudyne®

Aside from Abelcet®, another liposomal formulation that was not designed for oncology use is Visudyne® or liposomal verteporfin, which is a benzoporphyrin derivative that serves as a photosensitzer for photodynamic therapy in the treatment of age-related macular degeneration. The formulation consists of a mixture of DMPC and egg phosphatidyl glycerol (EPG) [196] and was designed by the Canadian scientist, Thomas Madden. Visudyne® is known for its selectivity against choroidal neovasculature arising from macular degeneration while minimizing the risk of severe visual acuity loss [197]. The product was developed by QLT Inc. (a spin-off company from UBC established in 1981) and is now a product owned by Bausch & Lomb Incorporated.

7.4. Marqibo®

Marqibo® is the liposomal formulation of vincristine developed to address dosing and pharmacokinetic limitations of the free agent. This formulation was designed by Bally, Mayer, and Cullis in the late 1980s and early 1990s. Although the product that arose from this research was originally owned by The Liposome Company through their agreements with UBC and the Canadian Liposome Company, the technology was eventually licensed back to a Vancouver-based start-up called Inex Pharmaceuticals. While the original product was prepared using DSPC/Chol liposomes, this product had an unacceptable storage life. This problem was overcome by using a new lipid composition of sphingomyelin and Chol in a 55:45 molar ratio [182,198]. This particular formulation exhibited a better storage shelf-life, and was associated with a surprising increase in drug retention and a profound improvement in therapeutic activity, exhibiting cures in multiple murine models of leukemia [181,199]. Marqibo® was originally developed by the Canadian company, Inex Pharmaceuticals Corporation, and is now a product of Spectrum Pharmaceuticals approved (2012) for the treatment of Philadelphia chromosome-negative acute lymphoblastic leukemia in adults.

7.5. Vyxeos®

Vyxeos® (formerly known as CPX-351) is the first dual-drug liposomal formulation to receive regulatory approval. It comprises cytarabine and daunorubicin packaged at a fixed 5:1 molar ratio inside 1,2-Distearoyl-sn-glycero-3-phosphocholine (DSPC)/1,2-distearoyl-sn-glycero-3-phospho-(1′-rac-glycerol) (DSPG)/Chol (70:20:10 molar ratio) liposomes [200]. Vyxeos® was developed based on the original concept of the CombiPlex® platform technology invented by Lawrence Mayer and Marcel Bally's group (Vancouver), where drug combinations exhibit synergistic anticancer activity when given at certain molar ratios and drug carriers could be used to maintain those ratios in vivo [152,201,202]. This technology demonstrated that (1) two drugs can be co-encapsulated into liposomes at a fixed molar ratio and (2) liposomes can be designed to optimize release kinetics such that the optimal therapeutic ratios can be achieved and maintained in vivo [203]. In 1999, Celator Pharmaceuticals Inc., was formed (Vancouver, BC, Canada) to develop the CPX product line. CPX-351(Vyxeos®) received regulatory approval for the treatment of treatment-related or secondary acute myeloid leukemia (AML) and AML with myelodysplasia-related changes in 2017. Just prior to this the company was acquired by Jazz Pharmaceuticals.

7.6. Onpattro®

The most recently approved liposomal formulation is Onpattro®, a nanomedicine that is revolutionary in multiple ways. Onpattro®, also known as patisiran, consists of an siRNA targeting the production of the transthyretin (TTR) protein, packaged inside LNPs, as described above. The lipid component contains Chol, DLin-MC3-DMA, DSPC, and PEG_{2000}-C-DMG at weight ratios of 6.2:13:3.3:1.6 per 2 mg of siRNA. By suppressing the production of wild-type and mutant TTR, patisiran reduces the accumulation of amyloid deposits in peripheral nerves, which would otherwise cause peripheral neuropathy [204]. Onpattro® is the first and only medication approved for the treatment of polyneuropathy caused by hereditary transthyretin-mediated amyloidosis. It is also the first and currently the only RNA interference therapeutic approved. Onpattro® is a product developed by Alnylam Pharmaceuticals using their proprietary siRNA and the LNP technology that originated from work completed by Jayaraman et al., including experienced Canadian scientists in the field: Pieter Cullis, Thomas Madden, Muthiah Manoharan, Steven Ansell, Jianxin Chen, and Michael Hope [64]. The development of Onpattro® resulted from collaborative efforts between Alnylam Pharmaceuticals and Acuitas Therapeutics (then AlCana Technologies). All commercialization work was conducted by Alnylam Pharmaceuticals.

8. Conclusions

Since the first description of liposomes in 1965, our knowledge about lipids and the role of lipids in membranes has expanded enormously. With this increase in understanding came innovations and discoveries that were impactful on patient treatment outcomes and quality of life, from reductions in adverse effects to controlled-release combinatory formulations. Nearly half of all regulatory approved liposomal formulations are Canadian inventions, highlighting the efforts of liposomologists from coast to coast. Researchers from Canada and around the world will endeavour to use liposomes to increase the therapeutic activity of promising compounds, making many more efficacious nanomedicines available to patients in the years to come.

There is a common theme to the success of the liposome technology developed by Canadians. First is the recognition of innovation and an aggressive patenting strategy that can protect the idea and its use. Next comes the "do it yourself" attitude: One that is most readily expressed in the context of new company formation. Finally, come partnerships, ones that include scientists, business development personnel, quality control staff, clinical trial specialists, clinicians, etc. However, the funding to develop this technology is great and, therefore, the efforts of venture capitalists and existing companies resourced to develop and commercialize therapeutic agents of value are also needed. Whether it is necessary to have a large company developed in Canada remains a question. In this context, success could be defined by partnerships with existing companies, even those in other countries. These partnerships should be highlighted as a Canadian success. Perhaps the only negative to all this is the fact that Canadian patients may be second, third or even fourth in line to have the opportunity to participate in clinical trials and access to the drug, if approved. It is more likely that successful products will be first marketed in larger economic markets, like the USA and Europe, before they reach Canadians. Further, in the absence of really compelling data, it is sometimes difficult to adopt new technology. Myocet®, for example, is approved in Europe and Canada, but it is rarely used in Canada, in part because it is not marketed there. This is despite evidence demonstrated in a large patient population that Myocet® does reduce the cardiotoxicity of doxorubicin and is a safer product. While Canadian access to new technology created in Canada may be limiting, the training opportunities here in Canada are fantastic, resulting in an international reputation of excellence and skills.

Finally, if one looks to the future, the strength of the liposome community as well as the drug delivery communities in general is very strong in Canada. The number of drug delivery/polymer/lipid technology based companies in operation are significant and these include the BC–based companies, Evonik, Precision Nanosystems, iCo Therapeutics, Acuitus, Sitka Biopharma, Cuprous Pharmaceuticals, Integrated Nanotherapeutics, Genevant Sciences, as well as Nanostics Precison Health (Alberta) and Nanovista (Ontario). The technology is creating jobs, training highly qualified personnel, and, most importantly, creating new products that are improving the health care of patients internationally.

Author Contributions: Conceptualization, M.B.B. and A.W.Y.L.; Writing—Original Draft preparation, A.W.Y.L., C.A., L.C.W. and U.V.M.; Writing—Review & Editing, M.B.B., A.W.Y.L., C.A., L.C.W. and U.V.M.; Figure and Table Preparation, A.W.Y.L. and L.C.W.

Funding: The research was funded by the Canadian Institutes of Health Research (PJT-153132) and the Canadian Cancer Society (705290).

Acknowledgments: A.W.Y. Leung would like to thank the Mitacs Elevate program for their financial support.

Conflicts of Interest: The authors declare no conflict of interest. The co-authors A.W.Y.L. and M.B.B. are Co-Founders of Cuprous Pharmaceuticals Inc. The company had no role in the design, writing, and decision to publish any of the content of the manuscript.

References

1. Shade, C.W. Liposomes as Advanced Delivery Systems for Nutraceuticals. *Integr. Med.* **2016**, *15*, 33–36.
2. Weissig, V. Liposomes Came First: The Early History of Liposomology. In *Liposomes: Methods and Protocols*; D'Souza, G.G.M., Ed.; Springer: New York, NY, USA, 2017; pp. 1–15.
3. Cullis, P.T.; Kruijff, B.D. Lipid polymorphism and the functional roles of lipids in biological membranes. *Biochim. Biophys. Acta (BBA)-Rev. Biomembr.* **1979**, *559*, 399–420. [CrossRef]
4. Cullis, P.; Kruijff, B.D.; Hope, M.; Nayar, R.; Schmid, S.J.C. Phospholipids and membrane transport. *Can. J. Biochem.* **1980**, *58*, 1091–1100. [CrossRef] [PubMed]
5. Tam, Y.K.; Madden, T.D.; Hope, M.J. Pieter Cullis' quest for a lipid-based, fusogenic delivery system for nucleic acid therapeutics: Success with siRNA so what about mRNA? *J. Drug Target.* **2016**, *24*, 774–779. [CrossRef] [PubMed]
6. Sessa, G.; Weissmann, G. Incorporation of lysozyme into liposomes. A model for structure-linked latency. *J. Biol. Chem.* **1970**, *245*, 3295–3301.
7. Trosper, T.; Raveed, D.; Ke, B. Chlorophyll a-containing liposomes. *Biochim. Biophys. Acta* **1970**, *223*, 463–465. [CrossRef]
8. Gregoriadis, G.; Ryman, B.E. Lysosomal localization of -fructofuranosidase-containing liposomes injected into rats. *Biochem. J.* **1972**, *129*, 123–133. [CrossRef] [PubMed]
9. Gregoriadis, G. Drug entrapment in liposomes. *FEBS Lett.* **1973**, *36*, 292–296. [CrossRef]
10. Gregoriadis, G.; Buckland, R.A. Enzyme-containing liposomes alleviate a model for storage disease. *Nature* **1973**, *244*, 170–172. [CrossRef]
11. Blomhoff, H.K.; Blomhoff, R.; Christensen, T.B. Enhanced stability of β-galactosidase in parenchymal and nonparenchymal liver cells by conjugation with dextran. *Biochim. Biophys. Acta (BBA) Gener. Subj.* **1983**, *757*, 202–208. [CrossRef]
12. Allison, A.G.; Gregoriadis, G. Liposomes as immunological adjuvants. *Nature* **1974**, *252*, 252. [CrossRef] [PubMed]
13. Gregoriadis, G.; Allison, A.C. Entrapment of proteins in liposomes prevents allergic reactions in pre-immunised mice. *FEBS Lett.* **1974**, *45*, 71–74. [CrossRef]
14. Arakawa, E.; Imai, Y.; Kobayashi, H.; Okumura, K.; Sezaki, H. Application of drug-containing liposomes to the duration of the intramuscular absorption of water-soluble drugs in rats. *Chem. Pharm. Bull.* **1975**, *23*, 2218–2222. [CrossRef] [PubMed]
15. Tanaka, T.; Taneda, K.; Kobayashi, H.; Okumura, K.; Muranishi, S.; Sezaki, H. Application of liposomes to the pharmaceutical modification of the distribution characteristics of drugs in the rat. *Chem. Pharm. Bull.* **1975**, *23*, 3069–3074. [CrossRef] [PubMed]
16. Juliano, R.L.; Stamp, D. The effect of particle size and charge on the clearance rates of liposomes and liposome encapsulated drugs. *Biochem. Biophys. Res. Commun.* **1975**, *63*, 651–658. [CrossRef]
17. Steerenberg, P.A.; Storm, G.; de Groot, G.; Claessen, A.; Bergers, J.J.; Franken, M.A.; van Hoesel, Q.G.; Wubs, K.L.; de Jong, W.H. Liposomes as drug carrier system for cis-diamminedichloroplatinum (II). II. Antitumor activity in vivo, induction of drug resistance, nephrotoxicity and Pt distribution. *Cancer Chemother. Pharmacol.* **1988**, *21*, 299–307. [CrossRef]
18. Sharma, A.; Straubinger, N.L.; Straubinger, R.M. Modulation of human ovarian tumor cell sensitivity to N-(phosphonacetyl)-L-aspartate (PALA) by liposome drug carriers. *Pharm. Res.* **1993**, *10*, 1434–1441. [CrossRef] [PubMed]
19. Desmukh, D.S.; Bear, W.D.; Wisniewski, H.M.; Brockerhoff, H. Long-living liposomes as potential drug carriers. *Biochem. Biophys. Res. Commun.* **1978**, *82*, 328–334. [CrossRef]
20. Mezei, M.; Gulasekharam, V. Liposomes—A selective drug delivery system for the topical route of administration. Lotion dosage form. *Life Sci.* **1980**, *26*, 1473–1477. [CrossRef]

21. Harsanyi, B.B.; Hilchie, J.C.; Mezei, M. Liposomes as drug carriers for oral ulcers. *J. Dent. Res.* **1986**, *65*, 1133–1141. [CrossRef]
22. Foldvari, M.; Gesztes, A.; Mezei, M. Dermal drug delivery by liposome encapsulation: Clinical and electron microscopic studies. *J. Microencapsul.* **1990**, *7*, 479–489. [CrossRef]
23. Caddeo, C.; Manconi, M.; Sinico, C.; Valenti, D.; Celia, C.; Monduzzi, M.; Fadda, A.M. Penetration Enhancer-Containing Vesicles: Does the Penetration Enhancer Structure Affect Topical Drug Delivery? *Curr. Drug Targets* **2015**, *16*, 1438–1447. [CrossRef] [PubMed]
24. Caddeo, C.; Valenti, D.; Nacher, A.; Manconi, M.; Fadda, A.M. Exploring the co-loading of lidocaine chemical forms in surfactant/phospholipid vesicles for improved skin delivery. *J. Pharm. Pharmacol.* **2015**, *67*, 909–917. [CrossRef]
25. Peyman, G.A.; Charles, H.C.; Liu, K.R.; Khoobehi, B.; Niesman, M. Intravitreal liposome-encapsulated drugs: A preliminary human report. *Int. Ophthalmol.* **1988**, *12*, 175–182. [CrossRef]
26. Bulbake, U.; Doppalapudi, S.; Kommineni, N.; Khan, W. Liposomal formulations in clinical use: An updated review. *Pharmaceutics* **2017**, *9*, 12. [CrossRef] [PubMed]
27. Kalra, J.; Bally, M.B. Liposomes. In *Fundamentals of Pharmaceutical Nanoscience*; Springer: Berlin/Heidelberg, Germany, 2013; pp. 27–63.
28. Leung, A.W.; Kalra, J.; Santos, N.D.; Bally, M.B.; Anglesio, M.S. Harnessing the potential of lipid-based nanomedicines for type-specific ovarian cancer treatments. *Nanomedicine* **2014**, *9*, 501–522. [CrossRef]
29. Olson, F.; Hunt, C.A.; Szoka, F.C.; Vail, W.J.; Papahadjopoulos, D. Preparation of liposomes of defined size distribution by extrusion through polycarbonate membranes. *Biochim. Biophys. Acta (BBA) Biomembr.* **1979**, *557*, 9–23. [CrossRef]
30. Mayer, L.D.; Hope, M.J.; Cullis, P.R. Vesicles of variable sizes produced by a rapid extrusion procedure. *Biochim. Biophys. Acta (BBA) Biomembr.* **1986**, *858*, 161–168. [CrossRef]
31. Cullis, P.R.; Hope, M.J.; Bally, M.B. Extrusion Technique for Producing Unilamellar Vesicles. U.S. Patent US 5,008,050, 16 April 1991.
32. Patil, Y.P.; Jadhav, S. Novel methods for liposome preparation. *Chem. Phys. Lipids* **2014**, *177*, 8–18. [CrossRef]
33. Jahn, A.; Vreeland, W.N.; Gaitan, M.; Locascio, L.E. Controlled Vesicle Self-Assembly in Microfluidic Channels with Hydrodynamic Focusing. *J. Am. Chem. Soc.* **2004**, *126*, 2674–2675. [CrossRef]
34. Evers, M.J.W.; Kulkarni, J.A.; van der Meel, R.; Cullis, P.R.; Vader, P.; Schiffelers, R.M. State-of-the-Art Design and Rapid-Mixing Production Techniques of Lipid Nanoparticles for Nucleic Acid Delivery. *Small Methods* **2018**, *2*, 1700375. [CrossRef]
35. Zhigaltsev, I.V.; Belliveau, N.; Hafez, I.; Leung, A.K.; Huft, J.; Hansen, C.; Cullis, P.R.J.L. Bottom-up design and synthesis of limit size lipid nanoparticle systems with aqueous and triglyceride cores using millisecond microfluidic mixing. *Langmuir* **2012**, *28*, 3633–3640. [CrossRef]
36. Belliveau, N.M.; Huft, J.; Lin, P.J.; Chen, S.; Leung, A.K.; Leaver, T.J.; Wild, A.W.; Lee, J.B.; Taylor, R.J.; Tam, Y.K. Microfluidic synthesis of highly potent limit-size lipid nanoparticles for in vivo delivery of siRNA. *Mol. Ther.-Nucleic Acids* **2012**, *1*, e37. [CrossRef] [PubMed]
37. Forbes, N.; Hussain, M.T.; Briuglia, M.L.; Edwards, D.P.; Horst, J.H.T.; Szita, N.; Perrie, Y. Rapid and scale-independent microfluidic manufacture of liposomes entrapping protein incorporating in-line purification and at-line size monitoring. *Int. J. Pharm.* **2018**, *556*, 68–81. [CrossRef] [PubMed]
38. Bally, M.B.; Zhang, Y.-P.; Reimer, D.L.; Wheeler, J.J. Lipid-Nucleic Acid Particles Prepared via a hydrophobic Lipid-Nucleic Acid Complex Intermediate and Use for Gene Transfer. U.S. Patent US 5,705,385, 6 January 1998.
39. Wong, F.M.; Reimer, D.L.; Bally, M.B. Cationic lipid binding to DNA: Characterization of complex formation. *Biochemistry* **1996**, *35*, 5756–5763. [CrossRef]
40. Reimer, D.L.; Zhang, Y.P.; Kong, S.; Wheeler, J.J.; Graham, R.W.; Bally, M.B. Formation of Novel Hydrophobic Complexes between Cationic Lipids and Plasmid DNA. *Biochemistry* **1995**, *34*, 12877–12883. [CrossRef] [PubMed]
41. Crowe, L.M.; Crowe, J.H.; Rudolph, A.; Womersley, C.; Appel, L. Preservation of freeze-dried liposomes by trehalose. *Arch. Biochem. Biophys.* **1985**, *242*, 240–247. [CrossRef]
42. Janoff, A.S.; Cullis, P.R.; Bally, M.B.; Fountain, M.W.; Ginsberg, R.S.; Hope, M.J.; Madden, T.D.; Schieren, H.P.; Jablonski, R.L. Methods of Dehydrating, Storing and Rehydrating Liposomes. U.S. Patent US 5,922,350, 13 July 1999.

43. Janoff, A.S.; Cullis, P.R.; Bally, M.B.; Fountain, M.W.; Ginsberg, R.S.; Hope, M.J.; Madden, T.D.; Schieren, H.P.; Jablonski, R.L. Dehydrated Liposomes. U.S. Patent US 4,880,635, 14 November 1989.

44. Janoff, A.S.; Cullis, P.R.; Bally, M.B.; Fountain, M.W.; Ginsberg, R.S.; Hope, M.J.; Madden, T.D.; Schieren, H.P.; Jablonski, R.L. Method of Dehydrating Liposomes Using Protective Sugars. U.S. Patent US 5,578,320, 26 November 1996.

45. Madden, T.D.; Bally, M.B.; Hope, M.J.; Cullis, P.R.; Schieren, H.P.; Janoff, A.S. Protection of large unilamellar vesicles by trehalose during dehydration: Retention of vesicle contents. *Biochim. Biophys. Acta (BBA) Biomembr.* **1985**, *817*, 67–74. [CrossRef]

46. Dos Santos, N.; Cox, K.A.; McKenzie, C.A.; van Baarda, F.; Gallagher, R.C.; Karlsson, G.; Edwards, K.; Mayer, L.D.; Allen, C.; Bally, M.B. pH gradient loading of anthracyclines into cholesterol-free liposomes: Enhancing drug loading rates through use of ethanol. *Biochim. Biophys. Acta (BBA) Biomembr.* **2004**, *1661*, 47–60. [CrossRef]

47. Cabral-lilly, D.; Mayer, L.; Tardi, P.; Watkins, D.; Zeng, Y. Method of Lyophilizing Liposomes. U.S. Patent 14/352,662, 24 July 2018.

48. Kirby, C.; Gregoriadis, G. Dehydration-Rehydration Vesicles: A Simple Method for High Yield Drug Entrapment in Liposomes. *Bio/Technology* **1984**, *2*, 979. [CrossRef]

49. Arora, A.; Byrem, T.M.; Nair, M.G.; Strasburg, G.M. Modulation of Liposomal Membrane Fluidity by Flavonoids and Isoflavonoids. *Arch. Biochem. Biophys.* **2000**, *373*, 102–109. [CrossRef] [PubMed]

50. Bally, M.B.; Hope, M.J.; Van Echteld, C.J.A.; Cullis, P.R. Uptake of safranine and other lipophilic cations into model membrane systems in response to a membrane potential. *Biochim. Biophys. Acta (BBA) Biomembr.* **1985**, *812*, 66–76. [CrossRef]

51. Fritze, A.; Hens, F.; Kimpfler, A.; Schubert, R.; Peschka-Süss, R. Remote loading of doxorubicin into liposomes driven by a transmembrane phosphate gradient. *Biochim. Biophys. Acta (BBA) Biomembr.* **2006**, *1758*, 1633–1640. [CrossRef] [PubMed]

52. Harrigan, P.R.; Wong, K.F.; Redelmeier, T.E.; Wheeler, J.J.; Cullis, P.R. Accumulation of doxorubicin and other lipophilic amines into large unilamellar vesicles in response to transmembrane pH gradients. *Biochim. Biophys. Acta (BBA) Biomembr.* **1993**, *1149*, 329–338. [CrossRef]

53. Madden, T.D.; Harrigan, P.R.; Tai, L.C.; Bally, M.B.; Mayer, L.D.; Redelmeier, T.E.; Loughrey, H.C.; Tilcock, C.P.; Reinish, L.W.; Cullis, P.R. The accumulation of drugs within large unilamellar vesicles exhibiting a proton gradient: A survey. *Chem. Phys. Lipids* **1990**, *53*, 37–46. [CrossRef]

54. Fenske, D.B.; Wong, K.F.; Maurer, E.; Maurer, N.; Leenhouts, J.M.; Boman, N.; Amankwa, L.; Cullis, P.R. Ionophore-mediated uptake of ciprofloxacin and vincristine into large unilamellar vesicles exhibiting transmembrane ion gradients. *Biochim. Biophys. Acta (BBA) Biomembr.* **1998**, *1414*, 188–204. [CrossRef]

55. Nii, T.; Ishii, F. Encapsulation efficiency of water-soluble and insoluble drugs in liposomes prepared by the microencapsulation vesicle method. *Int. J. Pharm.* **2005**, *298*, 198–205. [CrossRef]

56. Ciobanu, M.; Heurtault, B.; Schultz, P.; Ruhlmann, C.; Muller, C.D.; Frisch, B. Layersome: Development and optimization of stable liposomes as drug delivery system. *Int. J. Pharm.* **2007**, *344*, 154–157. [CrossRef]

57. Franze, S.; Marengo, A.; Stella, B.; Minghetti, P.; Arpicco, S.; Cilurzo, F. Hyaluronan-decorated liposomes as drug delivery systems for cutaneous administration. *Int. J. Pharm.* **2018**, *535*, 333–339. [CrossRef]

58. Cheung, B.C.L.; Sun, T.H.T.; Leenhouts, J.M.; Cullis, P.R. Loading of doxorubicin into liposomes by forming Mn^{2+}-drug complexes. *Biochim. Biophys. Acta (BBA) Biomembr.* **1998**, *1414*, 205–216. [CrossRef]

59. Abraham, S.A.; Edwards, K.; Karlsson, G.; MacIntosh, S.; Mayer, L.D.; McKenzie, C.; Bally, M.B. Formation of transition metal–doxorubicin complexes inside liposomes. *Biochim. Biophys. Acta (BBA) Biomembr.* **2002**, *1565*, 41–54. [CrossRef]

60. Wehbe, M.; Anantha, M.; Backstrom, I.; Leung, A.; Chen, K.; Malhotra, A.; Edwards, K.; Bally, M.B. Nanoscale Reaction Vessels Designed for Synthesis of Copper-Drug Complexes Suitable for Preclinical Development. *PLoS ONE* **2016**, *11*, e0153416. [CrossRef] [PubMed]

61. Wehbe, M.; Chernov, L.; Chen, K.; Bally, M.B. PRCosomes: Pretty reactive complexes formed in liposomes. *J. Drug Target.* **2016**, *24*, 787–796. [CrossRef]

62. Hope, M.J.; Madden, T.D.; Cullis, P.R.; Maier, M.; Jayaraman, M.; Rajeev, K.G.; Akinc, A.; Manoharan, M. Methods and Compositions for Delivery of Nucleic Acids. U.S. Patent Application US 15/152,216, 3 November 2016.

63. Hope, M.J.; Semple, S.C.; Chen, J.; Madden, T.D.; Cullis, P.R.; Ciufolini, M.A.; Mui, B.L.S. Amino Lipids and Methods for the Delivery of Nucleic Acids. U.S. Patent US 9,139,554, 22 September 2015.

64. Jayaraman, M.; Ansell, S.M.; Mui, B.L.; Tam, Y.K.; Chen, J.; Du, X.; Butler, D.; Eltepu, L.; Matsuda, S.; Narayanannair, J.K.; et al. Maximizing the Potency of siRNA Lipid Nanoparticles for Hepatic Gene Silencing In Vivo. *Angew. Chem. Int. Ed.* **2012**, *51*, 8529–8533. [CrossRef]

65. Fendler, J.H.; Romero, A. Liposomes as drug carriers. *Life Sci.* **1977**, *20*, 1109–1120. [CrossRef]

66. Stamp, D.; Juliano, R.L. Factors affecting the encapsulation of drugs within liposomes. *Can. J. Physiol. Pharmacol.* **1979**, *57*, 535–539. [CrossRef]

67. Bozzuto, G.; Molinari, A. Liposomes as nanomedical devices. *Int. J. Nanomed.* **2015**, *10*, 975–999. [CrossRef] [PubMed]

68. Nichols, J.W.; Deamer, D.W. Catecholamine uptake and concentration by liposomes maintaining pH gradients. *Biochim. Biophys. Acta (BBA) Biomembr.* **1976**, *455*, 269–271. [CrossRef]

69. Cullis, P.R.; Hope, M.J.; Bally, M.B.; Madden, T.D.; Mayer, L.D.; Fenske, D.B. Influence of pH gradients on the transbilayer transport of drugs, lipids, peptides and metal ions into large unilamellar vesicles. *Biochim. Biophys. Acta (BBA)-Rev. Biomembr.* **1997**, *1331*, 187–211. [CrossRef]

70. Wheeler, J.J.; Veiro, J.A.; Cullis, P.R. Ionophore-mediated loading of Ca^{2+} into large unilamellar vesicles in response to transmembrane pH gradients. *Mol. Membr. Biol.* **1994**, *11*, 151–157. [CrossRef]

71. Immordino, M.L.; Dosio, F.; Cattel, L. Stealth liposomes: Review of the basic science, rationale, and clinical applications, existing and potential. *Int. J. Nanomed.* **2006**, *1*, 297–315.

72. Tardi, P.; Johnstone, S.; Webb, M.; Bally, M.; Abraham, S. Liposome Loading with Metal Ions. U.S. Patent US 7,238,367, 3 July 2007.

73. Taggar, A.S.; Alnajim, J.; Anantha, M.; Thomas, A.; Webb, M.; Ramsay, E.; Bally, M.B. Copper–topotecan complexation mediates drug accumulation into liposomes. *J. Control. Release* **2006**, *114*, 78–88. [CrossRef] [PubMed]

74. Abraham, S.A.; Edwards, K.; Karlsson, G.; Hudon, N.; Mayer, L.D.; Bally, M.B. An evaluation of transmembrane ion gradient-mediated encapsulation of topotecan within liposomes. *J. Control. Release* **2004**, *96*, 449–461. [CrossRef]

75. Ramsay, E.; Alnajim, J.; Anantha, M.; Taggar, A.; Thomas, A.; Edwards, K.; Karlsson, G.; Webb, M.; Bally, M. Transition Metal-Mediated Liposomal Encapsulation of Irinotecan (CPT-11) Stabilizes the Drug in the Therapeutically Active Lactone Conformation. *Pharm. Res.* **2006**, *23*, 2799–2808. [CrossRef]

76. DiGiulio, S. FDA approves onivyde combo regimen for advanced pancreatic cancer. *Oncol. Times* **2015**. [CrossRef]

77. Tardi, P.; Johnstone, S.; Harasym, N.; Xie, S.; Harasym, T.; Zisman, N.; Harvie, P.; Bermudes, D.; Mayer, L. In vivo maintenance of synergistic cytarabine:daunorubicin ratios greatly enhances therapeutic efficacy. *Leukemia Res.* **2009**, *33*, 129–139. [CrossRef]

78. Banys, P. The clinical use of disulfiram (Antabuse®): A review. *J. Psychoact. Drugs* **1988**, *20*, 243–261. [CrossRef] [PubMed]

79. San Andres, M.P.; Marina, M.L.; Vera, S. Spectrophotometric determination of copper(II), nickel(II) and cobalt(II) as complexes with sodium diethyldithiocarbamate in cationic micellar medium of hexadecyltrimethylammonium salts. *Talanta* **1994**, *41*, 179–185. [CrossRef]

80. Wehbe, M.; Anantha, M.; Shi, M.; Leung, A.W.; Dragowska, W.H.; Sanche, L.; Bally, M.B. Development and optimization of an injectable formulation of copper diethyldithiocarbamate, an active anticancer agent. *Int. J. Nanomed.* **2017**, *12*, 4129–4146. [CrossRef]

81. Chen, D.; Cui, Q.C.; Yang, H.; Dou, Q.P. Disulfiram, a clinically used anti-alcoholism drug and copper-binding agent, induces apoptotic cell death in breast cancer cultures and xenografts via inhibition of the proteasome activity. *Cancer Res.* **2006**, *66*, 10425–10433. [CrossRef]

82. Liu, P.; Brown, S.; Goktug, T.; Channathodiyil, P.; Kannappan, V.; Hugnot, J.; Guichet, P.; Bian, X.; Armesilla, A.; Darling, J.L. Cytotoxic effect of disulfiram/copper on human glioblastoma cell lines and ALDH-positive cancer-stem-like cells. *Br. J. Cancer* **2012**, *107*, 1488. [CrossRef] [PubMed]

83. Duncan, C.; White, A.R. Copper complexes as therapeutic agents. *Metallomics* **2012**, *4*, 127–138. [CrossRef] [PubMed]

84. Santini, C.; Pellei, M.; Gandin, V.; Porchia, M.; Tisato, F.; Marzano, C.J. Advances in copper complexes as anticancer agents. *Chem. Rev.* **2013**, *114*, 815–862. [CrossRef] [PubMed]

85. Marzano, C.; Pellei, M.; Tisato, F.; Santini, C. Copper complexes as anticancer agents. *Anti-Cancer Agents Med. Chem.* **2009**, *9*, 185–211. [CrossRef]

86. Rafique, S.; Idrees, M.; Nasim, A.; Akbar, H.; Athar, A.J.B.; Reviews, M.B. Transition metal complexes as potential therapeutic agents. *Biotechnol. Mol. Biol. Rev.* **2010**, *5*, 38–45.

87. Wehbe, M.; Leung, A.W.; Abrams, M.J.; Orvig, C.; Bally, M.B. A Perspective–can copper complexes be developed as a novel class of therapeutics? *Dalton Trans.* **2017**, *46*, 10758–10773. [CrossRef] [PubMed]

88. Weder, J.E.; Dillon, C.T.; Hambley, T.W.; Kennedy, B.J.; Lay, P.A.; Biffin, J.R.; Regtop, H.L.; Davies, N.M. Copper complexes of non-steroidal anti-inflammatory drugs: An opportunity yet to be realized. *Coord. Chem. Rev.* **2002**, *232*, 95–126. [CrossRef]

89. Wehbe, M.; Lo, C.; Leung, A.W.Y.; Dragowska, W.H.; Ryan, G.M.; Bally, M.B. Copper (II) complexes of bidentate ligands exhibit potent anti-cancer activity regardless of platinum sensitivity status. *Investig. New Drugs* **2017**, *35*, 682–690. [CrossRef]

90. Wehbe, M.; Malhotra, A.K.; Anantha, M.; Lo, C.; Dragowska, W.H.; Dos Santos, N.; Bally, M.B. Development of a copper-clioquinol formulation suitable for intravenous use. *Drug Deliv. Transl. Res.* **2018**, *8*, 239–251. [CrossRef]

91. Prosser, K.E.; Leung, A.W.; Harrypersad, S.; Lewis, A.R.; Bally, M.B.; Walsby, C.J. Transition Metal Ions Promote the Bioavailability of Hydrophobic Therapeutics: Cu and Zn Interactions with RNA Polymerase I Inhibitor CX5461. *Chemistry* **2018**, *24*, 6334–6338. [CrossRef]

92. Leung, A.; Anantha, M.; Dragowska, W.; Wehbe, M.; Bally, M.J. Copper-CX-5461: A novel liposomal formulation for a small molecule rRNA synthesis inhibitor. *J. Control. Release* **2018**, *286*, 1–9. [CrossRef] [PubMed]

93. Kepp, O.; Senovilla, L.; Vitale, I.; Vacchelli, E.; Adjemian, S.; Agostinis, P.; Apetoh, L.; Aranda, F.; Barnaba, V.; Bloy, N. Consensus guidelines for the detection of immunogenic cell death. *Oncoimmunology* **2014**, *3*, e955691. [CrossRef]

94. Kroemer, G.; Galluzzi, L.; Kepp, O.; Zitvogel, L. Immunogenic cell death in cancer therapy. *Ann. Rev. Immunol.* **2013**, *31*, 51–72. [CrossRef]

95. Obeid, M.; Tesniere, A.; Ghiringhelli, F.; Fimia, G.M.; Apetoh, L.; Perfettini, J.-L.; Castedo, M.; Mignot, G.; Panaretakis, T.; Casares, N. Calreticulin exposure dictates the immunogenicity of cancer cell death. *Nat. Med.* **2007**, *13*, 54. [CrossRef] [PubMed]

96. Terenzi, A.; Pirker, C.; Keppler, B.K.; Berger, W. Anticancer metal drugs and immunogenic cell death. *J. Inorg. Biochem.* **2016**, *165*, 71–79. [CrossRef]

97. Mookerjee, A.; Basu, J.M.; Majumder, S.; Chatterjee, S.; Panda, G.S.; Dutta, P.; Pal, S.; Mukherjee, P.; Efferth, T.; Roy, S.J. A novel copper complex induces ROS generation in doxorubicin resistant Ehrlich ascitis carcinoma cells and increases activity of antioxidant enzymes in vital organs in vivo. *BMC Cancer* **2006**, *6*, 267. [CrossRef]

98. Boyle, E.; Freeman, P.; Goudie, A.; Mangan, F.; Thomson, M.J. The role of copper in preventing gastrointestinal damage by acidic anti-inflammatory drugs. *J. Pharm. Pharmacol.* **1976**, *28*, 865–868. [CrossRef] [PubMed]

99. Abdul-Ghani, A.-S.; Abu-Hijleh, A.-L.; Nahas, N.; Amin, R. Hypoglycemic effect of copper(II) acetate imidazole complexes. *Biol. Trace Elem. Res.* **1996**, *54*, 143–151. [CrossRef]

100. Yasumatsu, N.; Yoshikawa, Y.; Adachi, Y.; Sakurai, H. Antidiabetic copper(II)-picolinate: Impact of the first transition metal in the metallopicolinate complexes. *Bioorg. Med. Chem.* **2007**, *15*, 4917–4922. [CrossRef]

101. Djoko, K.Y.; Achard, M.E.S.; Phan, M.-D.; Lo, A.W.; Miraula, M.; Prombhul, S.; Hancock, S.J.; Peters, K.M.; Sidjabat, H.E.; Harris, P.N.; et al. Copper Ions and Coordination Complexes as Novel Carbapenem Adjuvants. *J. Antimicrob. Agents Chemother.* **2018**, *62*, e02280-17. [CrossRef]

102. Haeili, M.; Moore, C.; Davis, C.J.; Cochran, J.B.; Shah, S.; Shrestha, T.B.; Zhang, Y.; Bossmann, S.H.; Benjamin, W.H.; Kutsch, O.J.; et al. Copper complexation screen reveals compounds with potent antibiotic properties against methicillin-resistant Staphylococcus aureus. *Antimicrob. Agents Chemother.* **2014**, *58*, 3727–3736. [CrossRef] [PubMed]

103. Gaggelli, E.; Kozlowski, H.; Valensin, D.; Valensin, G.J. Copper homeostasis and neurodegenerative disorders (Alzheimer's, prion, and Parkinson's diseases and amyotrophic lateral sclerosis). *Chem. Rev.* **2006**, *106*, 1995–2044. [CrossRef]

104. Kozlowski, H.; Luczkowski, M.; Remelli, M.; Valensin, D.J. Copper, zinc and iron in neurodegenerative diseases (Alzheimer's, Parkinson's and prion diseases). *Coord. Chem. Rev.* **2012**, *256*, 2129–2141. [CrossRef]

105. Tisato, F.; Marzano, C.; Porchia, M.; Pellei, M.; Santini, C. Copper in diseases and treatments, and copper-based anticancer strategies. *Med. Res. Rev.* **2010**, *30*, 708–749. [CrossRef] [PubMed]
106. Giampietro, R.; Spinelli, F.; Contino, M.; Colabufo, N.A. The Pivotal Role of Copper in Neurodegeneration: A New Strategy for the Therapy of Neurodegenerative Disorders. *Mol. Pharm.* **2018**, *15*, 808–820. [CrossRef]
107. Gershon, H.; Ghirlando, R.; Guttman, S.B.; Minsky, A. Mode of formation and structural features of DNA-cationic liposome complexes used for transfection. *Biochemistry* **1993**, *32*, 7143–7151. [CrossRef]
108. Dimitraidis, G.J. Introduction of ribonucleic acids into cells by means of liposomes. *Nucleic Acids Res.* **1978**, *5*, 1381–1386. [CrossRef]
109. Fraley, R.T.; Fornari, C.S.; Kaplan, S. Entrapment of a bacterial plasmid in phospholipid vesicles: Potential for gene transfer. *Proc. Natl. Acad. Sci. USA* **1979**, *76*, 3348–3352. [CrossRef]
110. Fraley, R.; Subramani, S.; Berg, P.; Papahadjopoulos, D. Introduction of liposome-encapsulated SV40 DNA into cells. *J. Biol. Chem.* **1980**, *255*, 10431–10435.
111. Wong, F.M.P.; MacAdam, S.A.; Kim, A.; Oja, C.; Ramsay, E.C.; Bally, M.B. A Lipid-based Delivery System for Antisense Oligonucleotides Derived from a Hydrophobic Complex. *J. Drug Target.* **2002**, *10*, 615–623. [CrossRef]
112. Wheeler, J.; Bally, M.B.; Zhang, Y.-P.; Reimer, D.L.; Hope, M. Method of Preventing Aggregation of a Lipid: Nucleic Acid Complex. U.S. Patent US 6,858,224, 22 Febuary 2005.
113. Wheeler, J.J.; Palmer, L.; Ossanlou, M.; MacLachlan, I.; Graham, R.W.; Zhang, Y.P.; Hope, M.J.; Scherrer, P.; Cullis, P.R. Stabilized plasmid-lipid particles: Construction and characterization. *Gene Ther.* **1999**, *6*, 271–281. [CrossRef] [PubMed]
114. Semple, S.C.; Akinc, A.; Chen, J.; Sandhu, A.P.; Mui, B.L.; Cho, C.K.; Sah, D.W.Y.; Stebbing, D.; Crosley, E.J.; Yaworski, E.; et al. Rational design of cationic lipids for siRNA delivery. *Nat. Biotechnol.* **2010**, *28*, 172. [CrossRef]
115. Hu, Q.; Shew, C.R.; Bally, M.B.; Madden, T.D. Programmable fusogenic vesicles for intracellular delivery of antisense oligodeoxynucleotides: Enhanced cellular uptake and biological effects. *Biochim. Biophys. Acta (BBA) Biomembr.* **2001**. [CrossRef]
116. Holland, J.W.; Madden, T.D.; Cullis, P.R. Liposome Having an Exchangeable Component. U.S. Patent US 6,673,364, 6 January 2004.
117. Tam, Y.Y.; Chen, S.; Cullis, P.R. Advances in Lipid Nanoparticles for siRNA Delivery. *Pharmaceutics* **2013**, *5*, 498–507. [CrossRef]
118. Zhao, Y.; Huang, L. Chapter Two—Lipid Nanoparticles for Gene Delivery. In *Advances in Genetics*; Huang, L., Liu, D., Wagner, E., Eds.; Academic Press: Cambridge, MA, USA, 2014; Volume 88, pp. 13–36.
119. Heyes, J.; Palmer, L.; Bremner, K.; MacLachlan, I. Cationic lipid saturation influences intracellular delivery of encapsulated nucleic acids. *J. Control. Release* **2005**, *107*, 276–287. [CrossRef]
120. Maier, M.A.; Jayaraman, M.; Matsuda, S.; Liu, J.; Barros, S.; Querbes, W.; Tam, Y.K.; Ansell, S.M.; Kumar, V.; Qin, J.; et al. Biodegradable lipids enabling rapidly eliminated lipid nanoparticles for systemic delivery of RNAi therapeutics. *Mol. Ther.* **2013**, *21*, 1570–1578. [CrossRef] [PubMed]
121. Allen, T.M. Long-circulating (sterically stabilized) liposomes for targeted drug delivery. *Trends Pharmacol. Sci.* **1994**, *15*, 215–220. [CrossRef]
122. Allen, T.M.; Hansen, C.; Rutledge, J.J. Liposomes with prolonged circulation times: Factors affecting uptake by reticuloendothelial and other tissues. *Biochim. Biophys. Acta (BBA)-Biomembr.* **1989**, *981*, 27–35. [CrossRef]
123. de Menezes, D.E.L.; Pilarski, L.M.; Allen, T.M.J. In vitro and in vivo targeting of immunoliposomal doxorubicin to human B-cell lymphoma. *Cancer Res.* **1998**, *58*, 3320–3330.
124. Dos Santos, N. *Characterization of Cholesterol-Free Liposomes for Use in Delivery of Anti-Cancer Drugs*; University of British Columbia: Vancouver, BC, Canada, 2004.
125. Gregoriadis, G.; Neerunjun, E.D. Homing of liposomes to target cells. *Biochem. Biophys. Res. Commun.* **1975**, *65*, 537–544. [CrossRef]
126. Leserman, L.D.; Machy, P.; Barbet, J. Cell-specific drug transfer from liposomes bearing monoclonal antibodies. *Nature* **1981**, *293*, 226–228. [CrossRef] [PubMed]
127. Guru, P.Y.; Agrawal, A.K.; Singha, U.K.; Singhal, A.; Gupta, C.M. Drug targeting in Leishmania donovani infections using tuftsin-bearing liposomes as drug vehicles. *FEBS Lett.* **1989**, *245*, 204–208. [CrossRef]
128. Osborne, M.P.; Richardson, V.J.; Jeyasing, K.; Ryman, B.E. Radionuclide-labelled liposomes—A new lymph node imaging agent. *Int. J. Nuclear Med. Biol.* **1979**, *6*, 75–83. [CrossRef]

129. Morgan, J.R.; Williams, K.E.; Davies, R.L.; Leach, K.; Thomson, M.; Williams, L.A. Localisation of experimental staphylococcal abscesses by 99MTC-technetium-labelled liposomes. *J. Med. Microbiol.* **1981**, *14*, 213–217. [CrossRef]

130. Matsumura, Y.; Maeda, H. A New Concept for Macromolecular Therapeutics in Cancer Chemotherapy: Mechanism of Tumoritropic Accumulation of Proteins and the Antitumor Agent Smancs. *Cancer Res.* **1986**, *46*, 6387–6392. [PubMed]

131. Heath, T.D.; Fraley, R.T.; Papahdjopoulos, D. Antibody targeting of liposomes: Cell specificity obtained by conjugation of F(ab')2 to vesicle surface. *Science* **1980**, *210*, 539. [CrossRef] [PubMed]

132. Martin, F.J.; Hubbell, W.L.; Papahadjopoulos, D. Immunospecific targeting of liposomes to cells: A novel and efficient method for covalent attachment of Fab' fragments via disulfide bonds. *Biochemistry* **1981**, *20*, 4229–4238. [CrossRef]

133. Ahmad, I.; Allen, T.M. Antibody-mediated specific binding and cytotoxicity of liposome-entrapped doxorubicin to lung cancer cells in vitro. *Cancer Res* **1992**, *52*, 4817–4820.

134. Park, J.W.; Hong, K.; Kirpotin, D.B.; Colbern, G.; Shalaby, R.; Baselga, J.; Shao, Y.; Nielsen, U.B.; Marks, J.D.; Moore, D.; et al. Anti-HER2 immunoliposomes: Enhanced efficacy attributable to targeted delivery. *Clin Cancer Res.* **2002**, *8*, 1172–1181.

135. ElBayoumi, T.A.; Torchilin, V.P. Tumor-targeted nanomedicines: enhanced antitumor efficacy in vivo of doxorubicin-loaded, long-circulating liposomes modified with cancer-specific monoclonal antibody. *Clin. Cancer Res.* **2009**, *15*, 1973. [CrossRef]

136. Kirpotin, D.B.; Drummond, D.C.; Shao, Y.; Shalaby, M.R.; Hong, K.; Nielsen, U.B.; Marks, J.D.; Benz, C.C.; Park, J.W. Antibody Targeting of Long-Circulating Lipidic Nanoparticles Does Not Increase Tumor Localization but Does Increase Internalization in Animal Models. *Cancer Res.* **2006**, *66*, 6732. [CrossRef]

137. Sapra, P.; Allen, T.M. Internalizing antibodies are necessary for improved therapeutic efficacy of antibody-targeted liposomal drugs. *Cancer Res.* **2002**, *62*, 7190–7194. [PubMed]

138. Song, S.; Liu, D.; Peng, J.; Sun, Y.; Li, Z.; Gu, J.-R.; Xu, Y. Peptide ligand-mediated liposome distribution and targeting to EGFR expressing tumor in vivo. *Int. J. Pharm.* **2008**, *363*, 155–161. [CrossRef]

139. Allen, T.M.; Cullis, P.R. Liposomal drug delivery systems: From concept to clinical applications. *Adv. Drug Deliv. Rev.* **2013**, *65*, 36–48. [CrossRef]

140. Lee, T.-Y.; Lin, C.-T.; Kuo, S.-Y.; Chang, D.-K.; Wu, H.-C. Peptide-Mediated Targeting to Tumor Blood Vessels of Lung Cancer for Drug Delivery. *Cancer Res.* **2007**, *67*, 10958. [CrossRef]

141. Mai, J.; Song, S.; Rui, M.; Liu, D.; Ding, Q.; Peng, J.; Xu, Y. A synthetic peptide mediated active targeting of cisplatin liposomes to Tie2 expressing cells. *J. Control. Release* **2009**, *139*, 174–181. [CrossRef] [PubMed]

142. Lammers, T.; Hennink, W.E.; Storm, G. Tumour-targeted nanomedicines: Principles and practice. *Br. J. Cancer* **2008**, *99*, 392. [CrossRef]

143. Bertrand, N.; Wu, J.; Xu, X.; Kamaly, N.; Farokhzad, O.C. Cancer nanotechnology: The impact of passive and active targeting in the era of modern cancer biology. *Adv. Drug Deliv. Rev.* **2014**, *66*, 2–25. [CrossRef] [PubMed]

144. Andresen, T.L.; Jensen, S.S.; Jørgensen, K. Advanced strategies in liposomal cancer therapy: Problems and prospects of active and tumor specific drug release. *Prog. Lipid Res.* **2005**, *44*, 68–97. [CrossRef] [PubMed]

145. Allen, T.M.; Chonn, A. Large Unilamellar Liposomes with Low Uptake into the Reticuloendothelial System. *FEBS Lett.* **1987**, *223*, 42–46. [CrossRef]

146. Bakkerwoudenberg, I.A.J.M.; Lokerse, A.F.; Tenkate, M.T.; Storm, G. Enhanced Localization of Liposomes with Prolonged Blood-Circulation Time in Infected Lung-Tissue. *Biochim. Biophys. Acta* **1992**, *1138*, 318–326. [CrossRef]

147. Shen, Z.; Fisher, A.; Liu, W.K.; Li, Y. 1—PEGylated "stealth" nanoparticles and liposomes. In *Engineering of Biomaterials for Drug Delivery Systems*; Parambath, A., Ed.; Woodhead Publishing: Sawston, UK, 2018; pp. 1–26.

148. Blume, G.; Cevc, G. Liposomes for the sustained drug release in vivo. *Biochim. Biophys. Acta* **1990**, *1029*, 91–97. [CrossRef]

149. Kamps, J.A.; Swart, P.J.; Morselt, H.W.; Pauwels, R.; De Bethune, M.P.; De Clercq, E.; Meijer, D.K.; Scherphof, G.L. Preparation and characterization of conjugates of (modified) human serum albumin and liposomes: Drug carriers with an intrinsic anti-HIV activity. *Biochim. Biophys. Acta* **1996**, *1278*, 183–190. [CrossRef]

150. Allen, T.M.; Papahadjopoulos, D. STERICALLY STABILIZED ("STEALTH"). *Liposome Technol.* **1992**, *3*, 59.

151. Agrawal, V.; Paul, M.K.; Mukhopadhyay, A.K. 6-mercaptopurine and daunorubicin double drug liposomes-preparation, drug-drug interaction and characterization. *J. Liposome Res.* **2005**, *15*, 141–155. [CrossRef] [PubMed]

152. Ramsay, E.C.; Santos, N.D.; Dragowska, W.H.; Laskin, J.J.; Bally, M.B. The Formulation of Lipid-Based Nanotechnologies for the Delivery of Fixed Dose Anticancer Drug Combinations. *Curr. Drug Deliv.* **2005**, *2*, 341–351. [CrossRef] [PubMed]

153. Mayer, L.D.; Harasym, T.O.; Tardi, P.G.; Harasym, N.L.; Shew, C.R.; Johnstone, S.A.; Ramsay, E.C.; Bally, M.B.; Janoff, A.S. Ratiometric dosing of anticancer drug combinations: Controlling drug ratios after systemic administration regulates therapeutic activity in tumor-bearing mice. *Mol. Cancer Ther.* **2006**, *5*, 1854–1863. [CrossRef] [PubMed]

154. Mayer, L.D.; Janoff, A.S. Optimizing combination chemotherapy by controlling drug ratios. *Mol. Int.* **2007**, *7*, 216–223. [CrossRef] [PubMed]

155. Janoff, A.; Mayer, L.; Redman, J.; Swenson, C. Fixed Ratio Drug Combination Treatments for Solid Tumors. U.S. Patent US 7,842,676, 30 November 2010.

156. Bovier, P.A. Epaxal®: A virosomal vaccine to prevent hepatitis A infection. *Expert Rev. Vaccin.* **2008**, *7*, 1141–1150. [CrossRef]

157. Usonis, V.; Bakasenas, V.; Valentelis, R.; Katiliene, G.; Vidzeniene, D.; Herzog, C. Antibody titres after primary and booster vaccination of infants and young children with a virosomal hepatitis A vaccine (Epaxal®). *Vaccine* **2003**, *21*, 4588–4592. [CrossRef]

158. Gabizon, A.; Goren, D.; Fuks, Z.; Barenholz, Y.; Dagan, A.; Meshorer, A. Enhancement of adriamycin delivery to liver metastatic cells with increased tumoricidal effect using liposomes as drug carriers. *Cancer Res.* **1983**, *43*, 4730–4735.

159. Gabizon, A.; Barenholz, Y. *Adriamycin-Containing Liposomes in Cancer Chemotherapy*; Wiley: New York, NY, USA, 1988; pp. 365–379.

160. James, N.; Coker, R.; Tomlinson, D.; Harris, J.; Gompels, M.; Pinching, A.; Stewart, J. Liposomal doxorubicin (Doxil): An effective new treatment for Kaposi's sarcoma in AIDS. *Clin. Oncol.* **1994**, *6*, 294–296. [CrossRef]

161. Clark, J.M.; Whitney, R.R.; Olsen, S.J.; George, R.J.; Swerdel, M.R.; Kunselman, L.; Bonner, D.P. Amphotericin B lipid complex therapy of experimental fungal infections in mice. *Antimicrob. Agents Chemother.* **1991**, *35*, 615–621. [CrossRef] [PubMed]

162. Janoff, A.; Perkins, W.; Saletan, S.; Swenson, C. Amphotericin B lipid complex (ABLC™): A molecular rationale for the attenuation of amphotericin B related toxicities. *J. Liposome Res.* **1993**, *3*, 451–471. [CrossRef]

163. Adedoyin, A.; Bernardo, J.F.; Swenson, C.E.; Bolsack, L.E.; Horwith, G.; DeWit, S.; Kelly, E.; Klastersky, J.; Sculier, J.-P.; DeValeriola, D.; et al. Pharmacokinetic profile of ABELCET (amphotericin B lipid complex injection): Combined experience from phase I and phase II studies. *Antimicrob. Agents Chemother.* **1997**, *41*, 2201–2208. [CrossRef] [PubMed]

164. Gill, P.S.; Wernz, J.; Scadden, D.T.; Cohen, P.; Mukwaya, G.M.; von Roenn, J.H.; Jacobs, M.; Kempin, S.; Silverberg, I.; Gonzales, G.J. Randomized phase III trial of liposomal daunorubicin versus doxorubicin, bleomycin, and vincristine in AIDS-related Kaposi's sarcoma. *J. Clin. Oncol.* **1996**, *14*, 2353–2364. [CrossRef]

165. Forssen, E.A.; Ross, M.E. Daunoxome® Treatment of Solid Tumors: Preclinical and Clinical Investigations. *J. Liposome Res.* **1994**, *4*, 481–512. [CrossRef]

166. Noskin, G.; Pietrelli, L.; Gurwith, M.; Bowden, R.J.B.m.t. Treatment of invasive fungal infections with amphotericin B colloidal dispersion in bone marrow transplant recipients. *Bone Marrow Transplant.* **1999**, *23*, 697. [CrossRef]

167. Adler-Moore, J. AmBisome targeting to fungal infections. *Bone Marrow Transplant.* **1994**, *14*, S3–S7.

168. Adler-Moore, J.; Proffitt, R.T. Ambisome: Lipsomal formulation, structure, mechanism of action and pre-clinical experience. *J. Antimicrob. Chemother.* **2002**, *49*, 21–30. [CrossRef] [PubMed]

169. Proffitt, R.T.; Satorius, A.; Chiang, S.-M.; Sullivan, L.; Adler-Moore, J.P. Pharmacology and toxicology of a liposomal formulation of amphotericin B (AmBisome) in rodents. *J. Antimicrob. Chemother.* **1991**, *28*, 49–61. [CrossRef]

170. Herzog, C.; Hartmann, K.; Künzi, V.; Kürsteiner, O.; Mischler, R.; Lazar, H.; Glück, R. Eleven years of Inflexal® V—A virosomal adjuvanted influenza vaccine. *Vaccine* **2009**, *27*, 4381–4387. [CrossRef]

171. Glantz, M.J.; Jaeckle, K.A.; Chamberlain, M.C.; Phuphanich, S.; Recht, L.; Swinnen, L.J.; Maria, B.; LaFollette, S.; Schumann, G.B.; Cole, B.F. A randomized controlled trial comparing intrathecal sustained-release cytarabine (DepoCyt) to intrathecal methotrexate in patients with neoplastic meningitis from solid tumors. *Clin. Cancer Res.* **1999**, *5*, 3394–3402. [PubMed]
172. Murry, D.J.; Blaney, S.M. Clinical pharmacology of encapsulated sustained-release cytarabine. *Ann. Pharmacother.* **2000**, *34*, 1173–1178. [CrossRef] [PubMed]
173. Balazsovits, J.A.E.; Mayer, L.D.; Bally, M.B.; Cullis, P.R.; McDonell, M.; Ginsberg, R.S.; Falk, R.E. Analysis of the effect of liposome encapsulation on the vesicant properties, acute and cardiac toxicities, and antitumor efficacy of doxorubicin. *Cancer Chemother. Pharmacol.* **1989**, *23*, 81–86. [CrossRef] [PubMed]
174. Swenson, C.; Perkins, W.; Roberts, P.; Janoff, A.S. Liposome technology and the development of Myocet™(liposomal doxorubicin citrate). *Breast* **2001**, *10*, 1–7. [CrossRef]
175. Keam, S.J.; Scott, L.J.; Curran, M.P. Verteporfin. *Drugs* **2003**, *63*, 2521–2554. [CrossRef] [PubMed]
176. Frennesson, C.I.; Nilsson, S.E. Encouraging results of photodynamic therapy with Visudyne in a clinical patient material of age-related macular degeneration. *Acta Ophthalmol. Scand.* **2004**, *82*, 645–650. [CrossRef] [PubMed]
177. Kim, T.; Kim, J.; Kim, S. Extended-release formulation of morphine for subcutaneous administration. *Cancer Chemother. Pharmacol.* **1993**, *33*, 187–190. [CrossRef]
178. Biteau, K.; Guiho, R.; Chatelais, M.; Taurelle, J.; Chesneau, J.; Corradini, N.; Heymann, D.; Redini, F. L-MTP-PE and zoledronic acid combination in osteosarcoma: Preclinical evidence of positive therapeutic combination for clinical transfer. *Am. J. Cancer Res.* **2016**, *6*, 677.
179. Richard, B.M.; Newton, P.; Ott, L.R.; Haan, D.; Brubaker, A.N.; Cole, P.I.; Ross, P.E.; Rebelatto, M.C.; Nelson, K.G. The safety of EXPAREL®(bupivacaine liposome injectable suspension) administered by peripheral nerve block in rabbits and dogs. *J. Drug Deliv.* **2012**, *2012*, 962101. [CrossRef] [PubMed]
180. Mantripragada, S. A lipid based depot (DepoFoam® technology) for sustained release drug delivery. *Prog. Lipid Res.* **2002**, *41*, 392–406. [CrossRef]
181. Boman, N.L.; Masin, D.; Mayer, L.D.; Cullis, P.R.; Bally, M.B. Liposomal vincristine which exhibits increased drug retention and increased circulation longevity cures mice bearing P388 tumors. *Cancer Res.* **1994**, *54*, 2830–2833. [PubMed]
182. Webb, M.S.; Logan, P.; Kanter, P.M.; Onge, G.S.; Gelmon, K.; Harasym, T.; Mayer, L.D.; Bally, M.B. Preclinical pharmacology, toxicology and efficacy of sphingomyelin/cholesterol liposomal vincristine for therapeutic treatment of cancer. *Cancer Chemother. Pharmacol.* **1998**, *42*, 461–470. [CrossRef] [PubMed]
183. Kalra, A.V.; Kim, J.; Klinz, S.G.; Paz, N.; Cain, J.; Drummond, D.C.; Nielsen, U.B.; Fitzgerald, J.B. Preclinical activity of nanoliposomal irinotecan is governed by tumor deposition and intratumor prodrug conversion. *Cancer Res.* **2014**, *74*, 7003–7013. [CrossRef] [PubMed]
184. Lim, W.-S.; Tardi, P.G.; Dos Santos, N.; Xie, X.; Fan, M.; Liboiron, B.D.; Huang, X.; Harasym, T.O.; Bermudes, D.; Mayer, L.D. Leukemia-selective uptake and cytotoxicity of CPX-351, a synergistic fixed-ratio cytarabine: Daunorubicin formulation, in bone marrow xenografts. *Leuk. Res.* **2010**, *34*, 1214–1223. [CrossRef] [PubMed]
185. Bayne, W.F.; Mayer, L.D.; Swenson, C.E. Pharmacokinetics of CPX-351 (cytarabine/daunorubicin HCl) liposome injection in the mouse. *J. Pharm. Sci.* **2009**, *98*, 2540–2548. [CrossRef] [PubMed]
186. Janoff, A.S.; Boni, L.T.; Popescu, M.C.; Minchey, S.R.; Cullis, P.R.; Madden, T.D.; Taraschi, T.; Gruner, S.M.; Shyamsunder, E.; Tate, M.W. Unusual lipid structures selectively reduce the toxicity of amphotericin B. *Proc. Natl. Acad. Sci. USA* **1988**, *85*, 6122. [CrossRef] [PubMed]
187. Wasan, E.K.; Bartlett, K.; Gershkovich, P.; Sivak, O.; Banno, B.; Wong, Z.; Gagnon, J.; Gates, B.; Leon, C.G.; Wasan, K.M. Development and characterization of oral lipid-based Amphotericin B formulations with enhanced drug solubility, stability and antifungal activity in rats infected with Aspergillus fumigatus or Candida albicans. *Int. J. Pharm.* **2009**, *372*, 76–84. [CrossRef]
188. Waterhouse, D.N.; Tardi, P.G.; Mayer, L.D.; Bally, M.B. A comparison of liposomal formulations of doxorubicin with drug administered in free form: Changing toxicity profiles. *Drug Saf.* **2001**, *24*, 903–920. [CrossRef]
189. Abraham, S.A.; Waterhouse, D.N.; Mayer, L.D.; Cullis, P.R.; Madden, T.D.; Bally, M.B. The liposomal formulation of doxorubicin. In *Methods in Enzymology*; Elsevier: Amsterdam, The Netherlands, 2005; Volume 391, pp. 71–97.

190. Charrois, G.J.R.; Allen, T.M. Drug release rate influences the pharmacokinetics, biodistribution, therapeutic activity, and toxicity of pegylated liposomal doxorubicin formulations in murine breast cancer. *Biochim. Biophys. Acta (BBA) Biomembr.* **2004**, *1663*, 167–177. [CrossRef] [PubMed]

191. Bally, M.B.; Nayar, R.; Masin, D.; Hope, M.J.; Cullis, P.R.; Mayer, L.D. Liposomes with entrapped doxorubicin exhibit extended blood residence times. *Biochim. Biophys. Acta (BBA) Biomembr.* **1990**, *1023*, 133–139. [CrossRef]

192. Leonard, R.C.F.; Williams, S.; Tulpule, A.; Levine, A.M.; Oliveros, S. Improving the therapeutic index of anthracycline chemotherapy: Focus on liposomal doxorubicin (Myocet™). *Breast* **2009**, *18*, 218–224. [CrossRef] [PubMed]

193. Allen, C.; Dos Santos, N.; Gallagher, R.; Chiu, G.N.C.; Shu, Y.; Li, W.M.; Johnstone, S.A.; Janoff, A.S.; Mayer, L.D.; Webb, M.S.; et al. Controlling the Physical Behavior and Biological Performance of Liposome Formulations Through Use of Surface Grafted Poly(ethylene Glycol). *Biosci. Rep.* **2002**, *22*, 225. [CrossRef]

194. Parr, M.J.; Masin, D.; Cullis, P.R.; Bally, M.B. Accumulation of liposomal lipid and encapsulated doxorubicin in murine Lewis lung carcinoma: The lack of beneficial effects by coating liposomes with poly (ethylene glycol). *J. Pharmacol. Exp. Ther.* **1997**, *280*, 1319–1327. [PubMed]

195. Mayer, L.D.; Bally, M.B.; Cullis, P.R. Uptake of adriamycin into large unilamellar vesicles in response to a pH gradient. *Biochim. Biophys. Acta* **1986**, *857*, 123–126. [CrossRef]

196. Chowdhary, R.K.; Shariff, I.; Dolphin, D. Drug release characteristics of lipid based benzoporphyrin derivative. *J. Pharm. Pharm. Sci.* **2003**, *6*, 13–19.

197. Schmidt-Erfurth, U.; Hasan, T. Mechanisms of Action of Photodynamic Therapy with Verteporfin for the Treatment of Age-Related Macular Degeneration. *Surv. Ophthalmol.* **2000**, *45*, 195–214. [CrossRef]

198. Semple, S.C.; Leone, R.; Wang, J.; Leng, E.C.; Klimuk, S.K.; Eisenhardt, M.L.; Yuan, Z.-N.; Edwards, K.; Maurer, N.; Hope, M.J.; et al. Optimization and Characterization of a Sphingomyelin/Cholesterol Liposome Formulation of Vinorelbine with Promising Antitumor Activity. *J. Pharm. Sci.* **2005**, *94*, 1024–1038. [CrossRef]

199. Webb, M.S.; Harasym, T.O.; Masin, D.; Bally, M.B.; Mayer, L.D. Sphingomyelin-cholesterol liposomes significantly enhance the pharmacokinetic and therapeutic properties of vincristine in murine and human tumour models. *Br. J. Cancer* **1995**, *72*, 896. [CrossRef]

200. Dicko, A.; Kwak, S.; Frazier, A.A.; Mayer, L.D.; Liboiron, B.D. Biophysical characterization of a liposomal formulation of cytarabine and daunorubicin. *Int. J. Pharm.* **2010**, *391*, 248–259. [CrossRef] [PubMed]

201. Tardi, P.G.; Dos Santos, N.; Harasym, T.O.; Johnstone, S.A.; Zisman, N.; Tsang, A.W.; Bermudes, D.G.; Mayer, L.D. Drug ratio–dependent antitumor activity of irinotecan and cisplatin combinations in vitro and in vivo. *Mol. Cancer Ther.* **2009**. [CrossRef] [PubMed]

202. Tolcher, A.W.; Mayer, L.D. Improving combination cancer therapy: The CombiPlex® development platform. *Future Oncol.* **2018**, *14*, 1317–1332. [CrossRef] [PubMed]

203. Johnstone, S.; Harvie, P.; Shew, C.; Kadhim, S.; Harasym, T.; Tardi, P.; Harasym, N.; Bally, M.; Mayer, L.D. Synergistic antitumor activity observed for a fixed ratio liposome formulation of Cytarabine (Cyt):Daunorubicin (Daun) against preclinical leukemia models. *Cancer Res.* **2005**, *65*, 329.

204. Hoy, S.M. Patisiran: First Global Approval. *Drugs* **2018**, *78*, 1625–1631. [CrossRef] [PubMed]

pharmaceutics

MDPI

Perspective

The Development of Oral Amphotericin B to Treat Systemic Fungal and Parasitic Infections: Has the Myth Been Finally Realized?

Grace Cuddihy [1], Ellen K. Wasan [1], Yunyun Di [1] and Kishor M. Wasan [1,2,*]

[1] College of Pharmacy and Nutrition, University of Saskatchewan, Saskatoon, SK S7N 2Z4, Canada; cac579@mail.usask.ca (G.C.); ellen.wasan@usask.ca (E.K.W.); yud083@mail.usask.ca (Y.D.)

[2] Faculty of Pharmaceutical Sciences, University of British Columbia, Vancouver, BC V6T 1Z3, Canada

* Correspondence: kishor.wasan@usask.ca

Received: 30 January 2019; Accepted: 19 February 2019; Published: 26 February 2019

Abstract: Parenteral amphotericin B has been considered as first-line therapy in the treatment of systemic fungal and parasitic infections, however its use has been associated with a number of limitations including affordability, accessibility, and an array of systemic toxicities. Until very recently, it has been very challenging to develop a bioavailable formulation of amphotericin B due to its physical chemical properties, limited water and lipid solubility, and poor absorption. This perspective reviews several novel oral Amphotericin B formulations under development that are attempting to overcome these limitations.

Keywords: oral formulation; amphotericin B; fungal infections; parasitic infections; developing world; drug delivery

1. Preamble

One of the authors (KMW) was presenting grand rounds to the infectious disease group at Vancouver General Hospital, discussing combination therapies to treat systemic fungal infections, particularly those patients who were about to go through organ transplantation. An infectious disease physician asked if it was possible to develop an oral formulation of amphotericin B to treat patients. This physician commented that if an oral formulation could be developed, then it would be widely used, because it would have the potential to overcome many of the limitations of intravenous administration. These limitations include affordability, accessibility, and the well-known systemic toxicities associated with amphotericin B. At the time, KMW considered it extremely challenging to develop a bioavailable formulation of amphotericin B that would achieve the tissue concentrations required to have a pharmacological effect and ameliorating the dose-dependent nephrotoxicity associated with the drug. Factors include the large molecular weight of amphotericin B, its amphoteric physical chemical nature, very poor water and lipid solubility, as well as acid lability.

However, as KMW thought about it, it became clear that with a growing understanding of dietary and excipient lipid processing in the gastrointestinal tract (GIT) as well as associated new drug delivery technologies, it could in fact be possible to develop an efficacious oral formulation.

2. Purpose

The aim of this perspective is firstly to provide sufficient background information on both amphotericin B (AmB) and the target disease leishmaniasis, as well as to explain the need for an oral formulation of this life-saving medication. Secondly, our purpose is to describe pharmaceutical advances that have led to several novel AmB formulations which have emerged over the last decade.

Equally important is to discuss the role of formulation in reducing specific barriers to treatment in highly endemic regions of visceral leishmaniasis, such as cost and storage considerations.

3. Chemistry of AmB

3.1. Structure Overview

AmB has a large, highly complex structure (Figure 1). It is classified as a polyene macrolide antibiotic; specifically, it is known as a macrolide because it contains a polyketide that is linked to a mycosamine sugar. Furthermore, it is classified as a polyene macrolide due to the presence of the hydrophobic polyene subunit, which is attached to the hydrophilic polyol portion of the molecule [1]. Overall, it consists of a 38-membered macrolactone ring, which is β-glycosylated with mycosamine at the C-19 hydroxyl position [1]. Seven conjugated double bonds comprise the polyene subunit, while an ester and a ketone separated by 12 carbons and substituted with six hydroxyl groups comprise the polyol subunit of the molecule.

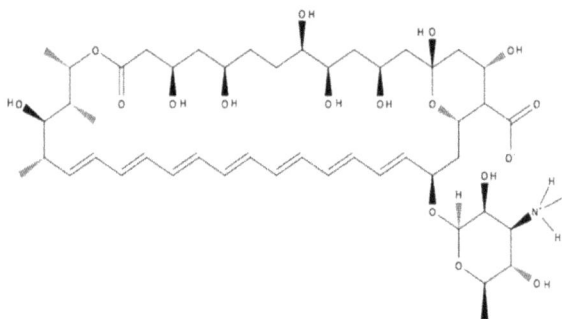

Figure 1. Structure of amphotericin B in zwitterionic form.

3.2. Structure-Activity Relationship

The structure-activity relationship of AmB has been the focus of numerous studies over the last three decades, which are briefly outlined below. Generalizations can be made with regard to the pharmacophore of this molecule: The positively charged amino group is required for activity; the polyene subunit is important for activity; if the carboxyl group is negatively charged, it leads to decreased selectivity for ergosterol over cholesterol; and conversely, N-aminoacylation leads to improved selectivity [1].

3.3. Mechanism of Action of AmB

Polyene antifungals such as AmB act by binding with ergosterol of the fungal cell walls and forming pores which permit leakage of cell contents, which eventually results in apoptosis [2,3]. This binding of AmB to ergosterol occurs through hydrophobic interactions, disrupting the lipid membrane integrity and resulting in the formation of pores [4,5]. These channels in the cell membrane allow efflux of small ions and other macromolecules, such as potassium and magnesium [6,7]. Simulation of the AmB–ergosterol structure finds that the formed pores promote water transport across the cell membrane, which might further disrupt the intracellular environment [8]. Recent evidence indicates that ergosterol binding and pore formation may not be the only mechanism leading to fungal cell death. It has been reported that AmB could kill yeasts by extracting ergosterols from cell membrane lipid layers [9,10]. Furthermore, it has been proposed that AmB causes accumulation of intracellular reactive oxygen species (ROS), which also contributes to the antifungal effect of this drug [11]. Several studies have described elevated ROS in fungal cells treated with AmB [12,13]. However, it is still not clear how AmB induces ROS production.

3.4. Bioavailability of AmB

The complexity of the AmB molecule is partly due to the individual functional groups present but also due to the asymmetry of the important subunits that make up the molecule. For instance, the polyol subunit is highly hydrophilic with many hydrogen bond donors and acceptors available to interact with molecules of water. By contrast, the polyene subunit is highly hydrophobic, as it consists of seven conjugated double bonds in a hydrocarbon chain 14 carbons in length. In addition to the amphiphilic nature of AmB, it also has a zwitterionic character on one portion of the molecule with the carboxylic acid and primary amine functional groups, which can be negatively and positively charged, respectively [14]. Therefore, overall, this asymmetric, amphiphilic molecule with zwitterionic character demonstrates a poor aqueous solubility of less than 1 mg/L at physiological pH, leading to its precipitation in aqueous media [14]. Lipinski's rule of five, which describes drug features that increase the probability of oral bioavailability based on passive diffusion though cellular membranes, can be applied to AmB with predictable results. AmB violates three out of four rules: AmB has more than 5 H-bond donors, more than 10 H-bond acceptors, and a molecular weight greater than 500 Da [14]. Thus, AmB will not easily be absorbed through the gastrointestinal mucosal membranes by passive transport following oral administration, which is confirmed by AmB's known low oral bioavailability of 0.2–0.9% [14]. Together with the aforementioned chemical complexities of AmB, there are significant barriers that must be overcome for an oral formulation of AmB to be developed.

4. Treating Visceral Leishmaniasis (VL)

Leishmaniasis is one of 20 conditions listed in the World Health Organization (WHO)'s list of "Neglected Tropical Diseases" [15]. Despite having effective treatments for the various presentations of the infection since the late 1950s, leishmaniasis is still a major concern in the 74 endemic countries identified by the WHO's Global Health Observatory data repository in 2016 [16]. Although the distribution of the disease is quite widespread, the large majority of new cases are limited to the following hyperendemic regions: Brazil, Ethiopia, India, Kenya, Somalia, South Sudan, and Sudan [17,18]. In 2016 alone, the number of reported cases of VL (or kala-azar), the most severe form of the disease, was: 6249 in India; 4285 in South Sudan; 3810 in Sudan; 3200 in Brazil; 1593 in Ethiopia; and 911 in Somalia [19].

Leishmaniasis is a vector-borne parasitic protozoan infection caused by more than 20 species of the *Leishmania* genus [20]. The known vectors of these parasites are the female sand flies of the genus *Phlebotomus*, which have a broad geographical distribution ranging from areas of the tropics, subtropics, and even temperate regions [21]. Additionally, domestic dogs are known reservoir hosts in the Mediterranean and New World regions [17].

Leishmania has a digenetic life cycle, switching between sand fly stages and human stages transmitted by sand flies biting humans. During a blood meal, the metacyclic promastigotes inoculated into the human skin immediately invade into macrophages, dendritic cells, fibroblasts, and keratinocytes and subsequently deactivate the host's complement system, suppressing the production of microbicidal molecules, such as superoxide and nitric oxide [22–24]. Although the parasites are found in these various cell types, macrophages are the main host cells where the metacyclic promastigotes differentiate into amastigotes [25]. The amastigotes continue to proliferate and disseminate into other tissues and organs, including the liver, spleen, and bone marrow. The existence of cutaneous or visceral leishmaniasis symptoms in humans depends on the parasite species, host conditions, and other factors [20]. At this point, if the sand fly bites the infected host again, the circulating amastigote-infected macrophages are likely to transmit to the new vector. In the gut of sand flies, amastigotes transformed into extracellular promastigotes, which takes approximately 7–14 days for transmissible infection to develop in the vector [26]. The promastigotes then migrate anteriorly to the stomodeal valve of the sand fly and undergo a series of developmental transitions to form infectious metacyclic promastigotes. Finally, during a blood meal on an appropriate host, a new digenetic cycle of leishmaniasis will begin.

4.1. Amphotericin B Parenteral Formulations

Amphotericin B (AmB) is a polyene macrolide antibiotic administered parenterally in the treatment of a variety of systemic fungal infections including candidiasis, aspergillosis, fusariosis, and zygomycosis [27]. In addition, AmB has exhibited antiparasitic activity for certain protozoan infections, including leishmaniasis as well as primary amoebic meningoencephalitis [28]. Prior to the development of lipid based formulations, the commercially available formulation used in the clinic was Fungizone®, a conventional micellar form of AmB in a complex with deoxycholate [29]. Unfortunately, the conventional form is associated with renal toxicity, which led to the development of other nonconventional formulations [30]. Nonconventional or lipid-based formulations have been developed to overcome some of the toxicity problems associated with the conventional formulation. There are several lipid-based parenteral formulations which have been marketed to treat fungal infections, which include the liposomal formulation AmBisome®, the lipid complex formulation Abelcet®, and a colloidal dispersion formulation Amphocil® (Amphotec) [31–33]. More recently, an emulsion form of AmB (Amphomul®) was developed and completed its Phase III clinical trial in 2014 [34]. The aim of this trial was to assess the safety and efficacy of the parenteral lipid emulsion formulation compared to AmBisome® as a single infusion treatment for VL [34]. Overall, the drawbacks of the conventional parenteral formulation are the administration route, treatment duration, infusion time, and most importantly, the toxicities associated with treatment. It is, however, still widely used in developing nations where patients do not have access to the safer yet more expensive nonconventional formulations [27].

4.2. Visceral Leishmaniasis Treatment Options and Limitations

Over the past few decades, treatment for VL is limited to pentavalent antimonials, AmB deoxycholate and pentamidine, and more recently, liposomal AmB, mitefosine, and paromomycin [35]. At present, in developed countries, the first-line therapy for VL in both immunocompetent and immunocompromised patients is short-course intravenous liposomal AmB, which has been demonstrated to have improved efficacy with reduced nephrotoxicity compared to conventional formulations [36]. However, more than 90% of global VL cases occur in developing countries, where conventional AmB is still considered first-line therapy for VL because it is the most affordable option [37,38]. An oral formulation of AmB would improve access to safe and effective treatment for VL in these affected regions worldwide by removing the barriers of high costs, the need for hospitalization, and a requirement for cold chain transport and storage conditions.

Cost of treatment is an important consideration for most patients; since liposomal AmB is 30 times more expensive than the conventional formulation, it is a huge limitation for patients in developing countries [17]. In 2010, the WHO released the "Costs of medicines in current use for the treatment of leishmaniasis" that included drug prices per unit and their estimated prices per VL treatment [39]. This document has the price per unit provided by the manufacturer, or the WHO-negotiated prices where applicable. They stated that the median cost per 50 mg of AmB deoxycholate to be $7.5 USD in comparison to the WHO negotiated price of $18 USD per 50 mg vial of AmBisome®. The estimated price per VL treatment was $252 USD per 2–4 day treatment with 20 mg/kg AmBisome® in comparison to $20 USD for a 30 day treatment (alternating days) with 1 mg/kg AmB deoxycholate [39]. However, these estimates were done for a patient weighing 35 kg (or 77 lbs); therefore, many patients' treatment would be appreciably more expensive. Moreover, the treatment regimen used for the estimation includes a shorter treatment duration than what is recommended, as previously described. It remains unclear which guidelines WHO used to determine the treatment regimen as it was not disclosed, and standard treatment regimens will vary by country. If the manufacturer recommendations for treatment duration were used, the estimated cost of treatment would undoubtedly increase. A reduction in the cost of treatment, in the form of an oral formulation of AmB, would greatly improve access to treatment for those where the financial burden of treatment is simply unreasonable.

Additionally, parenteral formulations must be administered in a hospital setting under the supervision of health care professionals. Beyond the direct costs of in-patient treatment, including admission, medical supplies, and charges for physician and laboratory services, there are numerous indirect costs which make this form of treatment impossible for many low-income populations [17,40]. Indirect costs may include: Travel to the healthcare facility, food for the patient and caregiver while in hospital, loss of income of the patient and/or their family members which accompany them, as well as any other unforeseen miscellaneous costs [17]. Oral AmB would permit out-patient treatment where patients could stay in their rural communities for the duration of their treatment, reducing the economic burden of in-patient health care costs and the detrimental indirect costs of treatment for patients and their families.

Although nonconventional formulations have improved the safety profile of AmB, there are some inherent drawbacks to using a parenteral formulation of any kind in the developing nations which are most affected by VL. The storage and transportation of liposomal AmB is a limitation, as the intact vials must be stored $\leq$25 °C and reconstituted vials are only stable for 24 h in 2–8 °C [41]. This is an important limitation if one considers that all of the hyperendemic regions occur in tropical or subtropical climates where proper refrigeration may not be feasible. In general, compared with parenteral formulations, oral dosage forms are more flexible with their required storage conditions in terms of temperatures and sterility, making them an attractive alternative. Another important limitation is the different aggregation states of AmB. This amphipathic molecule has the ability to self-aggregate in aqueous solution, which affects the safety profile of the different formulations of this drug [42]. For instance, the monomeric form of AmB remains the safest due to its ability to target ergosterol; thus, many formulations attempt to deliver AmB to target tissues in this form [42–44]. Conversely, the dimeric form of AmB, which is the most common state of reconstituted Fungizone®, is associated with the worst toxicity of AmB [42,45]. Furthermore, the poly-aggregated state is safer than the dimeric form [46,47].

5. Oral Formulations of AmB Currently in Development

5.1. Solid Lipid Nanoparticles

Chaudhari et al. (2015) developed solid lipid nanoparticles (SLNs) loaded with AmB (AmbiOnp) to overcome the poor oral bioavailability and kidney toxicity issues with AmB [47] (Table 1). The authors argued that producing a formulation that keeps AmB in its monomeric and/or super-aggregated form will keep the drug in a form which preferentially targets ergosterol, as opposed to its dimeric or oligomeric form, with its high affinity for cholesterol, which is responsible for the toxicity associated with conventional AmB. The AmbiOnp formulation was prepared by a probe sonication-assisted nanoprecipitation technique which produced a greater proportion of super-aggregated AmB that accumulated to a lesser extent in the kidneys, as reported in in biodistribution studies. This oral formulation was found to have a greatly improved safety profile compared to conventional IV-administered Fungizone®, with kidney tissue concentrations of approximately 84.5 $\pm$ 22.9 ng/g and 518.6 $\pm$ 31.5 ng/g, respectively, eight hours following administration [47]. Furthermore, the authors did not report any adverse reactions with the new formulation. In vivo pharmacokinetic studies demonstrated that orally administered AmbiOnp had a 1.05 relative bioavailability compared to intravenous Fungizone®, the long-standing gold standard of therapy for systemic fungal infections and VL. AmbiOnp demonstrated an optimal sustained release of AmB from the SLN delivery system, with 60% encapsulated in simulated intestinal fluid (SIF) over a period of 6 h. This formulation had the added benefit of an improved stability profile for storage conditions compared to conventional AmB:AmbiOnp was shown to be stable in 2–8 °C for 3 months or around 15 days when stored at 25 °C and 40 °C [47].

5.2. PLGA–PEG Nanoparticles

In contrast to the super-aggregate form of AmbiOnp, Radwan et al. (2017) formulated a nanoparticle formulation in the hopes that it would release AmB solely in its monomeric form [43] (Table 1). This formulation consisted of a poly(lactide-*co*-glycolide)–poly(ethylene glycol) (PLGA–PEG) copolymer loaded with AmB and glycyrrhizic acid as an absorption enhancer. This delivery system was formulated in hopes of an increase in the solubility of AmB, lessened toxicity, and the delivery of monomeric AmB to ensure the efficacy of the formulation. In vivo efficacy was not investigated; however, in vitro investigations found that the PLGA–PEG formulation had a greater antifungal activity with a minimum inhibitory concentration (MIC) reduction of fourfold or greater than that of Fungizone® 24 and 48 h after inoculation with *Candida albicans* in rats [43]. Pharmacokinetic studies found that the formulation had a 1.3 relative bioavailability compared to Fungizone® [43]. Kumar et al. (2015) also developed a PLGA–PEG encapsulated AmB formulation and tested the efficacy against *Leishmania donovani* in hamsters [48]. According to the report, this formulation was able to inhibit the parasite load in the liver by 93.2% compared to the free Amb group (74.6%) [48].

5.3. Chitosan-Coated Nanostructured Lipid Carriers

Ling Tan et al. (2018) designed a formulation consisting of a mixture of solid lipids and lipid oils, which they called nanostructured lipid carriers (NLC), with added chitosan coating for mucoadhesion [42] (Table 1). The authors' aim was to maximize lymphatic transport of their formulation to improve the oral bioavailability of AmB. Additionally, this formulation aimed to deliver AmB in its less toxic monomeric form. By one of the preparation methods tested, both the uncoated and chitosan coated NLC formulations were found to be stable in a predominantly monomeric form and, to a lesser extent, a poly-aggregate form for a 120-day period [42]. Encapsulation efficiency of the AmB-NLC formulation was 83.4 ± 0.72% and with a drug loading of 12.3 ± 0.11%, with the encapsulation efficiency significantly increasing with the chitosan coated form [42]. Both coated and uncoated forms demonstrated a biphasic release profile: An initial burst release phase followed by sustained release. The authors concluded that their formulation addressed the concerns of toxicity by keeping AmB in its monomeric and polyaggregated forms and that it has the potential to improve AmB's oral bioavailability due to the mucoadhesive properties of the NLCs which permit uptake in the small intestine. They plan to follow up with in vivo pharmacokinetic studies and safety studies to confirm their findings [42].

5.4. Lecithin-Based Mixed Polymeric Micelles

Chen et al. (2015) prepared a self-assembling lecithin-based mixed polymeric micellar formulation as an oral delivery system of AmB [49] (Table 1). This micellar formulation uses lecithin as the lipid component with a number of polymers (including but not limited to Pluronic® and 1,2-distearoyl-*sn*-glycero-3-phosphoethanolamine-*N*-methoxy(poly(ethylene glycol)-2000 (DSPE–PEG2K), which are loaded with AmB using a thin film method. Specifically, the authors' optimal formulation, which they named Ambicelles, consisted of AmB:lecithin:DSPE–PEG2K in a 1:1:10 mass ratio. Ambicelles were shown to increase the solubility of AmB from 0.001 to 5 mg/mL in addition to an improved relative oral bioavailability of 1.50 compared to that of Fungizone® in rats, which the authors attributed to the optimal sustained delivery of monomeric AmB. In vitro cytotoxicity studies showed that Ambicelles were less cytotoxic than Fungizone® and free AmB in a human colon adenocarcinoma cell line (HT29) [49].

5.5. O/W Microemulsion

Another approach to the oral delivery of AmB is in the form of an oil-in-water microemulsion (O/W ME) [50] (Table 1). Silva et al. (2013) prepared an O/W ME using a surfactant mixture of Tween 80® and Span 80® with a hydrophilic–lipophilic balance of approximately 13 and an oil phase consisting of Capryol® 90 or Capryol® PGMC. This ME formulation was able to increase the solubility of AmB 1000-fold when compared to the aqueous solubility of AmB as well as providing favorable rheological behavior for an oral delivery system. Time-dependent cytotoxicity results found the ME formulation to be slightly less toxic than AmB in DMSO at concentrations up to 25 μg/mL in a murine macrophage cell line [50]. The authors attributed this time-dependent toxicity to the discovery of the formation of AmB aggregates which must be addressed before the development of this formulation progresses [50].

5.6. Pickering Emulsion

In contrast to the traditional emulsion, a Pickering emulsion uses solid particles instead of surfactants or other emulsifiers in order to stabilize its internal phase [51]. Richter et al. (2018) formulated an AmB-loaded Pickering emulsion stabilized by self-assembled cashew tree gum grafted with polylactide nanoparticles [51] (Table 1). The results demonstrated a novel formulation which permitted the incorporation of this poorly water-soluble drug into their emulsions with a process efficiency of up to approximately 47% and without suboptimal aggregation of the drug, as seen in some commercial preparations. The authors plan to continue the development of this formulation with subsequent in vitro release and toxicity studies [51].

5.7. Tragacanth/Acrylic Acid Copolymer

Mohamed et al. (2017) prepared a hydrogel drug carrier consisting of tragacanth and acrylic acid (Aac) using gamma-irradiation [52] (Table 1). This pH-sensitive copolymer formulation was shown to protect the AmB in an aggregated form in simulated gastric fluid (pH = 1) while drug was released as the formulation dissociated in SIF (pH = 7). The authors suggested that the release rate and total amount of drug released was dependent on pH and the Aac content of the copolymer with the aforementioned variables increasing with Aac content. In vivo antifungal efficacy investigations against candidiasis in mice showed that the oral (Trag/Aac)–AmB formulation (dose equivalent to 1 mg/kg) resulted in 0% mortality compared to the 10% mortality eight days post intravenous inoculation when administered intravenously with free AmB (1 mg/kg). Oral administration of (Trag/Aac)–AmB had similar efficacy to that of free AmB as shown by the measured reduction of colony forming units (CFU) found in kidney and liver tissues; free AmB reduced CFUs by 93% in the kidneys and 95% in the liver, comparatively (Trag/Aac)–AmB reduced CFUs by 97% and 93%, respectively [52]. Moreover, assessment of serum antibodies against *C. albicans* found no significant difference between the formulation of interest and free AmB, thus providing further evidence of the comparable efficacy of the (Trag/Aac)–AmB formulation. Furthermore, the authors did not find that that their formulation produced significant levels of the cytokines: Tumor necrosis factor-αβ, interleukin-1β, and nitric oxide in the kidney and the liver when compared to the free AmB-treated animals, which they interpreted as evidence supporting the superior safety of their formulation. In vivo toxicity investigations found (Trag/Aac)–AmB to be relatively safe, with negligible reported nephrotoxicity as demonstrated by no significant increase in creatinine or blood urea nitrogen (BUN) levels when compared with the AmB-treated control. Similar results were reported for liver toxicity as measured by serum aspartate aminotransferase and alanine aminotransferase enzymes. Further histopathological examinations were completed by the authors, which demonstrated that their oral formulation caused minimal renal damage and notable reduction of injury of hepatocytes when compared with the degenerative effects following treatment with free AmB on the renal glomerular tuft and hepatocyte necrosis [52].

5.8. Chitosan (CS) and Porphyrin (POR) Polymeric Nanocarrier

Bhatia et al. (2014) suggested that loading AmB as a polyelectrolyte complex into a biodegradable polymeric nanocarrier is an optimal solution to the delivery of this problematic drug [53] (Table 1). Specifically, they chose to use chitosan and porphyrin as two oppositely charged polymers with AmB associated with them. Stability studies showed that their polyelectrolyte complex formulation (i.e., with or without tripolyphosphate as a crosslinking agent) showed less degradation in simulated gastric fluid and a superior release profile for up to 12 h, when compared to plain AmB and chitosan-only nanoparticles. Moreover, an in vitro antifungal activity study found the formulated nanoformulations to yield significantly higher antifungal activity, as measured by their IC_{50}, than the marketed formulations (AmB, Fungizone®, and AmBisome®) in the chosen fungal strains: *Aspergillus. fumigatus*, *Aspergillus. niger*, *Aspergillus. flavus*, and *Candida. albicans*. The most effective formulation was found to be the CS–POR–AmB formulation, with 23-fold greater activity than Ambisome® in *A. fumigatus* and 12- to 15- fold greater activity in *A. niger*, *A. flavus*, and *C. albicans*. An in vitro hemolytic study found the authors' nanoformulations to have less hemolytic toxicity than plain AmB and the chosen marketed formulations, proving the polyelectrolyte complexation (PEC) ormulation to be nontoxic up to concentrations of 55.5 µg/mL and with only approximately 4.1% hemolysis in the CS–POR–AmB formulation (compared to ~39.9% for plain AmB). However, the investigation into the in vivo toxicity of their POR formulations discovered an unexpected increase in platelet count and minimal decrease in red blood cell count, white blood cell count, hemoglobin, and hematocrit values when compared to the control. The authors suggested the platelet activation response may be due to the high sulfur content or due to the high anhydrogalactose (AGR) per mole concentration in their samples [53]. In vivo toxicity studies based on serum creatinine and blood urea nitrogen levels found that the renal toxicity at maximum dose was worst for Fungizone® followed by CS–AmB, CS–POR–AmB, CS, and lastly POR. The authors proposed this result may be due to the associated release rate of each formulation as the AGR and sulfur present in POR produce a gelling effect which may be better suited for a gastroretentive release of AmB [53].

5.9. Chitosan–Ethylenediamine Tetraacetic Acid (EDTA) Microparticles

Singh et al. (2013) characterized a novel solid self-nanoemulsifying drug delivery system (S-SNEDDS) formulation of AmB using spray dried covalently crosslinked EDTA–chitosan (COECH) microparticles for oral administration [54] (Table 1). They synthesized and characterized this formulation in hopes of developing an adequate delivery system for poorly water soluble and thermolabile drugs, such as AmB [54]. The authors reported that their formulation was indeed able to self-nanoemulsify into a thermodynamically stable delivery system once in contact with an aqueous environment. This formulation demonstrated a 12-fold improvement in in vitro dissolution relative to pure AmB. Overall, the authors concluded that their COECH–S-SNEDDS formulation prepared by spray drying technology was a reasonable approach which provided a solid substrate for the development of an AmB nanoemulsion for oral administration [54].

5.10. Carbon Nanotubes

Prajapati et al. (2012) used carbon nanotubes (CNTs), which they covalently attached AmB in order to create a potential formulation for oral administration [55] (Table 1). In this study addressing the in vivo antileishmanicidal efficacy of their oral formulation, the authors found that their nanovector delivery system, known as f-CNT–AmB, was able to inhibit the parasite load within the spleen in a dose-dependent manner with 90.2%, 96.5%, and 98.2% inhibition for 5 mg/kg, 10 mg/kg, and 15 mg/kg doses, respectively [55]. In addition, in this small study using a hamster model of infection at the highest oral dose of f-CNT–AmB at 15 mg/kg, it demonstrated comparable efficacy to a 5 mg/kg dose of interperitoneally administered liposomal AmB. Furthermore, the lowest administered dose of their oral formulation (5 mg/kg) had greater efficacy than the same dose of a currently marketed oral treatment for VL, namely miltefosine [55]. Previous work published by these researchers reported on the characterization of their formulation as well as the in vitro cytotoxicity (IC_{50} 0.00234 µg/mL compared to 0.03263 µg/mL for AmB, in a macrophage model), and in vivo safety and efficacy of their formulation following intraperitoneal administration in mice and hamster models (no evidence of toxicity and percent suppression of 89.8% for f-CNT–AmB compared with 68.9% for AmB) [56].

5.11. Cubosomes

Yang et al. (2012) formulated cubosomes as a lipid-based delivery vehicle for AmB, as they believed that their formulation would be able to overcome the molecule's major inherent drawback, i.e., poor bioavailability [57] (Table 1). In a small animal model, the authors found no indication of nephrotoxicity following a single dose of the oral AmB-loaded cubosomes at doses of 10–20 mg/kg, as measured by plasma BUN and plasma creatinine concentrations. Pharmacokinetic results determined that the cubosome formulation (10 mg/kg) had increased the oral bioavailability of AmB 285% compared with the oral administration of Fungizone®. In contrast with the nephrotoxicity results, a biodistribution study showed that the high dose of AmB-loaded cubosomes (20 mg/kg) demonstrated higher uptake of the drug in the kidneys in comparison with the liver and spleen. The liver and the spleen had the highest uptake of the lower dose of AmB-loaded cubosomes. The authors hypothesized that these results may indicate that their formulation would not be able to reduce the kidney toxicity associated with AmB. However, as previously mentioned, this is contradictory to their findings that neither dose caused an abnormal increase plasma BUN and creatinine concentrations 24 h following oral administration [57].

5.12. GCPQ Nanoparticles (Quaternary Ammonium Palmitoyl Glycol Chitosan)

Serrano et al. (2015) encapsulated AmB in quaternary ammonium palmitoyl glycol chitosan (GCPQ) nanoparticles in hopes that this self-assembling nanoparticle forming polymer would improve the oral bioavailability of AmB by exhibiting drug delivery to target organs while bypassing toxicity in nontarget organs [58] (Table 1). In order to test their hypothesis, the authors undertook detailed investigations in murine and canine animal models evaluating the efficacy of their formulation in systemic fungal infections, i.e., candidiasis and aspergillosis, in addition to VL. For all tested disease states, AmB–GCPQ had similar efficacy to the marketed parenteral lipid-based formulation AmBisome®. Serrano et al. (2015) found that their formulation improved the dissolution of AmB in simulated gastrointestinal fluid compared to conventional AmB. Pharmacokinetic studies showed that AmB–GCPQ delivered more drug to the target organs of pathology, namely, the liver, lung, and spleen, with relatively less delivered to the kidneys. Moreover, the formulation also delivered AmB to the bone marrow and the brain, which the authors argued would be beneficial for the clearance of the *Leishmania* parasite and the treatment of systemic infections, respectively. The reported relative oral bioavailability of the formulation was 24.7% [58].

The aforementioned studies have developed promising formulations which offer a wide range of diverse approaches to overcome the limitations in the development of a viable oral AmB formulation. However, to the best of our knowledge, these formulations have not progressed into the clinical trial stage of development. Conversely, the following two formulations to be discussed are the furthest in the advancement towards achieving the ultimate goal of developing an oral formulation of AmB and bringing it to the market, as they have successfully commenced clinical trials.

5.13. Cochleates

Zarif et al. (2008) published results from multiple investigations into the in vitro and in vivo safety and efficacy of their formulation which utilized lipid-based cochleates as a delivery system for AmB for use in *Candida* infections [59] (Table 1). The cochleates consist of solid lipid bilayers arranged in rolled-up sheets that are composed of phospholipid-cation precipitates, specifically phosphatidylserine and calcium, respectively [59]. The AmB encapsulated in the cochleates is thus protected from degradation in the GIT, permitting their use as an oral delivery system. The amphotericin B cochleates (CAMB) formulation was prepared using a hydrogel method and was found to be stable; no drug was lost from the delivery system for four months when stored at 4 °C. In murine models, biodistribution studies provided evidence that absorption through the GI mucosa had occurred, permitting adequate amounts of AmB to reach the target organs affected by systemic fungal infections (i.e., lungs, liver, spleen, and kidneys) following a 10-day oral administration of CAMB (10 mg/kg). The authors believe this absorption occurred due to the involvement of the gut associated lymphatic tissue (GALT), as a large concentration of AmB was found in the liver and spleen. The in vivo studies performed in murine models of *Candida albicans* infection demonstrated that at 0.5 mg/kg/day (up to 2.5 mg/kg/day), CAMB resulted in 100% survival 16 days post-infection compared to 30% mortality in mice treated with 1 mg/kg/day of parenteral Fungizone® (however, 2 mg/kg/day resulted in 100% survival), and 10% mortality resulted from 10 mg/kg/day AmBisome®. In addition, CAMB appeared to have comparable efficacy at 2.5 mg/kg/day with that of parenteral Fungizone® at 2 mg/kg/day, resulting in 3.5 log CFU count reduction in the kidneys and no detectable CFUs present in the lungs [59]. In vitro safety studies found no hemolytic effect of CAMB at concentrations up to 500 μg/mL AmB on RBCs. In vivo safety investigations found no abnormal changes in BUN levels and histopathology following 14 day treatment of 50 mg/kg doses of CAMB [59]. Further investigations include the in vivo efficacy in a murine model of *Aspergillus* infection [60]; the in vitro activity in *Leishmania chagasi*; and toxicokinetic studies in vivo in both rat and dog models [61]. All these studies had promising results, which resulted in approval for human trials on CAMB, now known as MAT2203, which is being investigated for the prevention of invasive fungal infections in patients with acute lymphoblastic leukemia [62]. Preliminary results from the Phase I study evaluating the safety, efficacy, tolerability and pharmacokinetics (PK) of CAMB in healthy volunteers has been released [63], demonstrating the potential use of the formulation in single doses of 200 and 400 mg. This study found these doses to be well tolerated with no serious adverse events or laboratory abnormalities and predictable plasma levels comparable to previous animal studies, thus providing evidence that will progress this formulation to the Phase II efficacy trials [63].

5.14. SEDDS (iCo-010/019)

Wasan et al. have also worked to solve the seemingly impossible task of developing an oral AmB formulation for many years [44]. Their approach was to develop a lipid-based self-emulsifying drug delivery system (SEDDS) for AmB to permit oral administration of this poorly bioavailable drug with an additional aim of lessening its nephrotoxicity while maintaining optimal antileishmanial activity [44,64–66]. The authors employed mono- and di-glycerides in addition to D-alpha-tocopheryl poly(ethylene glycol) succinate (vitamin E–TPGS). An additional goal was to provide stability for AmB in their delivery system in order to withstand tropical temperatures, considering the clinical target [66]. Before deciding on the iCo-010 formulation, which has recently completed Phase I clinical trials, many versions of the formulation were developed and tested for stability, safety, and efficacy [44,66]. iCo-010 was determined to be the most promising formulation with optimal stability (>75% over 60 days in 30 °C; >95% after 4 h in SIF) antileishmanial activity was observed in a murine model of VL, where <99% reduction in parasitic infection was achieved following 5 days of treatment with 10 mg/kg po bid and 95% inhibition following treatment with 20 mg/kg po qd for 5 days, relative to the control. This formulation also exhibited more desirable self-emulsifying properties compared to other versions of the formulation, namely, iCo-011, -012, -013 [66] (Table 1). The authors hypothesized that the desirable efficacy of their oral AmB formulation was likely a result of improved solubility, stability in the gastrointestinal tract, membrane permeability, and its ability to target the lymphatic transport system. The latter improvement may permit this formulation to target the greatest sites of infection in VL-infected organisms [66]. iCo-10 was also found to maintain AmB in monomeric form upon emulsification in simulated gastric fluid (Wasan lab, unpublished data). Further investigation into the safety of the iCo-010 formulation found no evidence of GI toxicity, hepatotoxicity, or nephrotoxicity following the oral administration of multiple doses in a murine model [65]. The biodistribution of the formulation in a mouse model showed uptake in the organs of the reticuloendothelial system at levels above the IC_{50} for the *leishmania* organism [65], which propelled iCo-010 into Phase I clinical trials. Furthermore, the potential use of iCo-010 for indications other than VL was explored, e.g., systemic candidiasis, which was found to be an effective once daily 5 day treatment for this indication in a rat model [67]. On June 27th, 2018, iCo Therapeutics announced a positive clinical outcome, as the primary safety and tolerability endpoint was met in the Phase I clinical trials of this oral AmB formulation, now known as iCo-019, further supporting the potential of iCo-019 to make it to the market and become accessible to those most affected by VL in endemic regions to have a safe and effective treatment with an oral form of AmB [68].

Table 1. AmB oral formulation summary.

AmB oral formulation	Efficacy	Stability
Solid lipid nanoparticle [47]	Lower kidney tissue concentration, 105% Fo of Fungizone®	2–8 °C for 3 months, 15 days ≥ 25 °C
PLGA–PEG nanoparticle [43,48]	Increase antifungal activity 4-fold in vitro Inhibit parasite load by 93.2% compared with free AmB group (74.6%) 130% Fo of Fungizone®	N/A
Chitosan-coated nanostructured lipid carriers [42]	N/A	63.9% AmB retained encapsulated after 30 min incubation in SIF
Lecithin-based mixed polymeric micelles [49]	Less toxic in HT29 cells 150% Fo of Fungizone®	Increase solubility
O/W microemulsion [50]	Slightly less toxic than free DMSO	Increase the solubility by 1000 folds
Pickering emulsion [51]	N/A	Stable one month under refrigeration
Tragacanth/acrylic acid copolymer [52]	No mortality observed in mice comparing with free AmB Improve oral bioavailability comparing with free AmB	N/A
Chitosan and porphyrin polymeric nanocarrier [53]	23-fold antifungal activity than Ambisome® Slightly less toxic than Fungizone®	Less degradation in SIF and a superior release profile for up to 12 h
Chitosan–EDTA microparticles [54]	N/A	12-fold improvement in in vitro dissolution relative to pure AmB
Carbon Nanotubes [55,56]	Inhibit the parasite load in a dose-dependent manner No evidence of toxicity in mice and hamster models	N/A
Cubosomes (cubic liquid crystal nanoparticles) [57]	low dose of AmB-loaded cubosomes shows low kidney concentration than Fungizone® 285% bioavailability of Fungizone®	74% detectable AmB after 3h in SIF
GCPQ nanoparticles [58]	Absolute Fo is 24.7% Higher concentration in liver, lung and spleen	Stable for a year on storage
Cochleate–CAMB/MAT2203 [59,60]	100% survival comparing with Fungizone® and AmBisome® No serious adverse event in Phase I study	Stable for 4 months at 4 °C
SEDDS (iCo-010/019) [66]	<99% reduction in parasitic infection in a murine model 95% inhibition when compared to control	>75% over 60 days in 30 °C; >95% after 4 h in SIF

Abbreviations: SIF, simulated intestinal fluid; Fo, oral bioavailability.

6. Discussion and Concluding Remarks

The abundance of data published on the topic of developing an oral form of AmB for the treatment of systemic infections such as VL alone supports the urgent need for a formulation to make it to the market. A number of researchers felt inclined to find a solution to overcome the barriers imposed by physicochemical properties of AmB. However, the majority of these formulations were unsuccessful, which demonstrates the difficulty of this task. Nevertheless, the two formulations which have made it to clinical trials with positive preliminary results provide us with evidence that a solution may finally be found which absolves the myth that an oral AmB formulation could not be developed.

Funding: The Wasan Lab was funded by the Canadian Institutes of Health Research (CIHR), iCo Therapeutics Inc. and the Consortium for Parasitic Drug Development via the Bill and Melinda Gates Foundation for the development of iCo 010/019.

Conflicts of Interest: The authors declare no conflict of interest in writing of this manuscript. The funders of the iCo 010/019 work that was reported in this manuscript had no role in the writing of the manuscript, or in the decision to publish this perspective.

Abbreviations

Acrylic acid	AAc	Anhydrogalactose	AGR
Amphotericin B	AmB	Blood urea nitrogen	BUN
Amphotericin B cochleates	CAMB	Colony forming units	CFU
Chitosan	CS	1,2-distearoyl-*sn*-glycero-3-phosphoethanolamine-N-methoxy(poly(ethylene glycol)-2000	DSPE-PEG 2K
Carbon nanotubes	CNTs	Ethylenediaminetet-raacetic acid	EDTA
Gut associated lymphatic tissue	GALT	Oral bioavailability	Fo
Hydrophilic–lipophilic balances	HLBs	Gastrointestinal tract	GIT
Oil-in-water microemulsion	O/W ME	Minimum inhibitory concentration	MIC
Polyelectrolyte complexation	PEC	Poly(lactide-*co*-glycolide)–poly(ethylene glycol)	PLGA–PEG
Phamacokinetics	PK	Porphyrin	POR
Quaternary ammonium palmitoyl glycol chitosan	GCPQ	Reactive oxygen species	ROS
Self-emulsifying drug delivery system	SEDDS	Simulated intestinal fluid	SIF
Solid lipid nanoparticles	SLNs	Visceral leishmaniasis	VL

References

1. Cereghetti, D.M.; Carreira, E.M. Amphotericin B: 50 years of chemistry and biochemistry. *Synthesis* **2006**, *6*, 0914–0942. [CrossRef]
2. Fernandez-Garcia, R.; de Pablo, E.; Ballesteros, M.P.; Serrano, D.R. Unmet clinical needs in the treatment of systemic fungal infections: The role of amphotericin B and drug targeting. *Int. J. Pharm.* **2017**, *525*, 139–148. [CrossRef] [PubMed]
3. Mesa-Arango, A.C.; Scorzoni, L.; Zaragoza, O. It only takes one to do many jobs: Amphotericin B as antifungal and immunomodulatory drug. *Front. Microbiol.* **2012**, *3*, 286. [CrossRef] [PubMed]
4. Gray, K.C.; Palacios, D.S.; Dailey, I.; Endo, M.M.; Uno, B.E.; Wilcock, B.C.; Burke, M.D. Amphotericin primarily kills yeast by simply binding ergosterol. *Proc. Natl. Acad. Sci. USA* **2012**, *109*, 2234–2239. [CrossRef] [PubMed]
5. Kaminski, D.M. Recent progress in the study of the interactions of amphotericin B with cholesterol and ergosterol in lipid environments. *Eur. Biophys. J.* **2014**, *43*, 453–467. [CrossRef] [PubMed]
6. Kinsky, S.C. Antibiotic interaction with model membranes. *Annu. Rev. Pharmacol.* **1970**, *10*, 119–142. [CrossRef] [PubMed]
7. Palacios, D.S.; Dailey, I.; Siebert, D.M.; Wilcock, B.C.; Burke, M.D. Synthesis-enabled functional group deletions reveal key underpinnings of amphotericin B ion channel and antifungal activities. *Proc. Natl. Acad. Sci. USA* **2011**, *108*, 6733–6738. [CrossRef] [PubMed]
8. Wu, H.-C.; Yoshioka, T.; Nakagawa, K.; Shintani, T.; Tsuru, T.; Saeki, D.; Shaikh, A.R.; Matsuyama, H. Preparation of Amphotericin B-Ergosterol structures and molecular simulation of water adsorption and diffusion. *J. Membrane Sci.* **2018**, *545*, 229–239. [CrossRef]
9. Grela, E.; Wieczor, M.; Luchowski, R.; Zielinska, J.; Barzycka, A.; Grudzinski, W.; Nowak, K.; Tarkowski, P.; Czub, J.; Gruszecki, W.I. Mechanism of Binding of Antifungal Antibiotic Amphotericin B to Lipid Membranes:

An Insight from Combined Single-Membrane Imaging, Microspectroscopy, and Molecular Dynamics. *Mol. Pharm.* **2018**, *15*, 4202–4213. [CrossRef] [PubMed]

10. Anderson, T.M.; Clay, M.C.; Cioffi, A.G.; Diaz, K.A.; Hisao, G.S.; Tuttle, M.D.; Nieuwkoop, A.J.; Comellas, G.; Maryum, N.; Wang, S.; et al. Amphotericin forms an extramembranous and fungicidal sterol sponge. *Nat. Chem. Biol.* **2014**, *10*, 400–406. [CrossRef] [PubMed]

11. Mesa-Arango, A.C.; Trevijano-Contador, N.; Román, E.; Sánchez-Fresneda, R.; Casas, C.; Herrero, E.; Argüelles, J.C.; Pla, J.; Cuenca-Estrella, M.; Zaragoza, O. The production of reactive oxygen species is an universal action mechanism of Amphotericin B against pathogenic yeasts and contributes to the fungicidal effect of this drug: AMPHORES study. *Antimicrob. Agents Chemother.* **2014**, *58*, 6627–6638. [CrossRef] [PubMed]

12. Guirao-Abad, J.P.; Sánchez-Fresneda, R.; Alburquerque, B.; Hernández, J.A.; Argüelles, J.-C. ROS formation is a differential contributory factor to the fungicidal action of amphotericin B and micafungin in Candida albicans. *Int. J. Med. Microbiol.* **2017**, *307*, 241–248. [CrossRef] [PubMed]

13. Phillips, A.J.; Sudbery, I.; Ramsdale, M. Apoptosis induced by environmental stresses and amphotericin B in Candida albicans. *Proc. Natl. Acad. Sci. USA* **2003**, *100*, 14327–14332. [CrossRef] [PubMed]

14. Serrano, D.R.; Lalatsa, A. Oral amphotericin B: The journey from bench to market. *J. Drug Deliv. Sci. Tech.* **2017**, *42*, 75–83. [CrossRef]

15. World Health Organization. Neglected Tropical Diseases. Available online: https://www.who.int/neglected_diseases/diseases/en/ (accessed on 15 September 2018).

16. Global Health Observatory Data Repository. Status of Endemicity of Visceral Leishmaniasis Data by Country. Available online: http://apps.who.int/gho/data/view.main.NTDLEISHVENDv (accessed on 15 September 2018).

17. Thornton, S.J.; Wasan, K.M.; Piecuch, A.; Lynd, L.L.D.; Wasan, E.K. Barriers to treatment for visceral leishmaniasis in hyperendemic areas: India, Bangladesh, Nepal, Brazil and Sudan. *Drug Dev. Ind. Pharm.* **2010**, *36*, 1312–1319. [CrossRef] [PubMed]

18. World Health Organization. Leishmaniasis: Epidemiological Situation. Available online: https://www.who.int/leishmaniasis/burden/en/ (accessed on 15 September 2018).

19. Global Health Observatory Data Repository. Number of Cases of Visceral Leishmaniasis Reported Data by Country. Available online: http://apps.who.int/gho/data/node.main.NTDLEISHVNUM?lang=en (accessed on 15 September 2018).

20. Torres-Guerrero, E.; Quintanilla-Cedillo, M.R.; Ruiz-Esmenjaud, J.; Arenas, R. Leishmaniasis: A review. *F1000Research* **2017**, *6*, 750. [CrossRef] [PubMed]

21. Maroli, M.; Feliciangeli, M.D.; Bichaud, L.; Charrel, R.N.; Gradoni, L. Phlebotomine sandflies and the spreading of leishmaniases and other diseases of public health concern. *Med. Vet. Entomol.* **2013**, *27*, 123–147. [CrossRef] [PubMed]

22. Walker, D.M.; Oghumu, S.; Gupta, G.; McGwire, B.S.; Drew, M.E.; Satoskar, A.R. Mechanisms of cellular invasion by intracellular parasites. *Cell Mol. Life Sci.* **2014**, *71*, 1245–1263. [CrossRef] [PubMed]

23. Mosser, D.M.; Edelson, P.J. The third component of complement (C3) is responsible for the intracellular survival of Leishmania major. *Nature* **1987**, *327*, 329–331. [CrossRef] [PubMed]

24. Kima, P.E. The amastigote forms of Leishmania are experts at exploiting host cell processes to establish infection and persist. *Int. J. Parasitol.* **2007**, *37*, 1087–1096. [CrossRef] [PubMed]

25. Beattie, L.; Kaye, P.M. Leishmania–host interactions: What has imaging taught us? *Cell Microbiol.* **2011**, *13*, 1659–1667. [CrossRef] [PubMed]

26. Bates, P.A. Revising Leishmania's life cycle. *Nat. Microbiol.* **2018**, *3*, 529–530. [CrossRef] [PubMed]

27. Thornton, S.J.; Wasan, K.M. The reformulation of amphotericin B for oral administration to treat systemic fungal infections and visceral leishmaniasis. *Expert Opin. Drug Deliv.* **2009**, *6*, 271–284. [CrossRef] [PubMed]

28. Grace, E.; Asbill, S.; Virga, K. Naegleria fowleri: Pathogenesis, diagnosis, and treatment options. *Antimicrob. Agents Chemother.* **2015**, *59*, 6677–6681. [CrossRef] [PubMed]

29. Belkherroubi-Sari, L.; Adida, H.; Seghir, A.; Boucherit, Z.; Boucherit, K. New strategy for enhancing the therapeutic index of Fungizone((R)). *J. Mycol. Med.* **2013**, *23*, 3–7. [CrossRef] [PubMed]

30. Pham, T.T.; Loiseau, P.M.; Barratt, G. Strategies for the design of orally bioavailable antileishmanial treatments. *Int. J. Pharm.* **2013**, *454*, 539–552. [CrossRef] [PubMed]

31. Stone, N.R.; Bicanic, T.; Salim, R.; Hope, W. Liposomal amphotericin B (AmBisome®): A review of the pharmacokinetics, pharmacodynamics, clinical experience and future directions. *Drugs* **2016**, *76*, 485–500. [CrossRef] [PubMed]

32. Lister, J. Amphotericin B Lipid Complex (Abelcet) in the treatment of invasive mycoses: The North American experience. *Eur. J. Haematol. Suppl.* **1996**, *57*, 18–23. [CrossRef] [PubMed]

33. Clemons, K.V.; Stevens, D.A. Comparative efficacies of four amphotericin B formulations—Fungizone, Amphotec (Amphocil), AmBisome, and Abelcet—against systemic murine aspergillosis. *Antimicrob. Agents Chemother.* **2004**, *48*, 1047–1050. [CrossRef] [PubMed]

34. Sundar, S.; Pandey, K.; Thakur, C.P.; Jha, T.K.; Das, V.N.; Verma, N.; Lal, C.S.; Verma, D.; Alam, S.; Das, P. Efficacy and safety of amphotericin B emulsion versus liposomal formulation in Indian patients with visceral leishmaniasis: A randomized, open-label study. *PLoS Negl. Trop. Dis.* **2014**, *8*, e3169. [CrossRef] [PubMed]

35. World Health Organization. Control of the leishmaniases. *World Health Organ. Tech. Rep. Ser.* **2010**, *949*, 57–59.

36. Sundar, S.; Chakravarty, J. Liposomal amphotericin B and leishmaniasis: Dose and response. *J. Glob. Infect. Dis.* **2010**, *2*, 159–166. [CrossRef] [PubMed]

37. Sundar, S.; Chakravarty, J. An update on pharmacotherapy for leishmaniasis. *Expert Opin. Pharmacother.* **2015**, *16*, 237–252. [CrossRef] [PubMed]

38. Alvar, J.; Velez, I.D.; Bern, C.; Herrero, M.; Desjeux, P.; Cano, J.; Jannin, J.; den Boer, M.; WHO Leishmaniasis Control Team. Leishmaniasis worldwide and global estimates of its incidence. *PLoS ONE* **2012**, *7*, e35671. [CrossRef] [PubMed]

39. World Health Organization. Costs of Medicines in Current Use for the Treatment of Leishmaniasis. 2010. Available online: https://www.who.int/leishmaniasis/research/978_92_4_12_949_6_Annex6.pdf?ua=1 (accessed on 15 September 2018).

40. de Assis, T.S.; Rosa, D.C.; de Morais Teixeira, E.; Cota, G.; Azeredo-da-Silva, A.L.; Werneck, G.; Rabello, A. The direct costs of treating human visceral Leishmaniasis in Brazil. *Rev. Soc. Bras. Med. Trop.* **2017**, *4*, 478–482.

41. Jensen, G.M. The care and feeding of a commercial liposomal product: Liposomal amphotericin B (AmBisome((R))). *J. Liposome Res.* **2017**, *27*, 173–179. [CrossRef] [PubMed]

42. Ling Tan, J.S.; Roberts, C.J.; Billa, N. Mucoadhesive chitosan-coated nanostructured lipid carriers for oral delivery of amphotericin B. *Pharm. Dev. Technol.* **2018**, *26*, 1–9. [CrossRef] [PubMed]

43. Radwan, M.A.; AlQuadeib, B.T.; Siller, L.; Wright, M.C.; Horrocks, B. Oral administration of amphotericin B nanoparticles: Antifungal activity, bioavailability and toxicity in rats. *Drug Deliv.* **2017**, *24*, 40–50. [CrossRef] [PubMed]

44. Wasan, E.K.; Bartlett, K.; Gershkovich, P.; Sivak, O.; Banno, B.; Wong, Z.; Gagnon, J.; Gates, B.; Leon, C.G.; Wasan, K.M. Development and characterization of oral lipid-based amphotericin B formulations with enhanced drug solubility, stability and antifungal activity in rats infected with Aspergillus fumigatus or Candida albicans. *Int. J. Pharm.* **2009**, *372*, 76–84. [CrossRef] [PubMed]

45. Barwicz, J.; Tancrede, P. The effect of aggregation state of amphotericin-B on its interactions with cholesterol- or ergosterol-containing phosphatidylcholine monolayers. *Chem. Phys. Lipids* **1997**, *85*, 145–155. [CrossRef]

46. Espada, R.; Valdespina, S.; Alfonso, C.; Rivas, G.; Ballesteros, M.P.; Torrado, J.J. Effect of aggregation state on the toxicity of different amphotericin B preparations. *Int. J. Pharm.* **2008**, *361*, 64–69. [CrossRef] [PubMed]

47. Chaudhari, M.B.; Desai, P.P.; Patel, P.A.; Patravale, V.B. Solid lipid nanoparticles of amphotericin B (AmbiOnp): In vitro and in vivo assessment towards safe and effective oral treatment module. *Drug Deliv. Transl. Res.* **2016**, *6*, 354–364. [CrossRef] [PubMed]

48. Kumar, R.; Sahoo, G.C.; Pandey, K.; Das, V.; Das, P. Study the effects of PLGA-PEG encapsulated amphotericin B nanoparticle drug delivery system against Leishmania donovani. *Drug Deliv.* **2015**, *22*, 383–388. [CrossRef] [PubMed]

49. Chen, Y.C.; Su, C.Y.; Jhan, H.J.; Ho, H.O.; Sheu, M.T. Physical characterization and in vivo pharmacokinetic study of self-assembling amphotericin B-loaded lecithin-based mixed polymeric micelles. *Int. J. Nanomed.* **2015**, *10*, 7265–7274. [CrossRef]

50. Silva, A.E.; Barratt, G.; Cheron, M.; Egito, E.S. Development of oil-in-water microemulsions for the oral delivery of amphotericin B. *Int. J. Pharm.* **2013**, *454*, 641–648. [CrossRef] [PubMed]

51. Richter, A.R.; Feitosa, J.P.A.; Paula, H.C.B.; Goycoolea, F.M.; de Paula, R.C.M. Pickering emulsion stabilized by cashew gum- poly-l-lactide copolymer nanoparticles: Synthesis, characterization and amphotericin B encapsulation. *Colloids Surf. B Biointerfaces* **2018**, *164*, 201–209. [CrossRef] [PubMed]

52. Mohamed, H.A.; Radwan, R.R.; Raafat, A.I.; Ali, A.E. Antifungal activity of oral (Tragacanth/acrylic acid) Amphotericin B carrier for systemic candidiasis: In vitro and in vivo study. *Drug Deliv. Transl. Res.* **2018**, *8*, 191–203. [CrossRef] [PubMed]

53. Bhatia, S.; Kumar, V.; Sharma, K.; Nagpal, K.; Bera, T. Significance of algal polymer in designing amphotericin B nanoparticles. *Sci. World J.* **2014**, *2014*, 564573. [CrossRef] [PubMed]

54. Singh, K.; Tiwary, A.; Rana, V. Spray dried chitosan–EDTA superior microparticles as solid substrate for the oral delivery of amphotericin B. *Int. J. Biol. Macromol.* **2013**, *58*, 310–319. [CrossRef] [PubMed]

55. Prajapati, V.K.; Awasthi, K.; Yadav, T.P.; Rai, M.; Srivastava, O.N.; Sundar, S. An oral formulation of amphotericin B attached to functionalized carbon nanotubes is an effective treatment for experimental visceral leishmaniasis. *J. Infect. Dis.* **2011**, *205*, 333–336. [CrossRef] [PubMed]

56. Prajapati, V.K.; Awasthi, K.; Gautam, S.; Yadav, T.P.; Rai, M.; Srivastava, O.N.; Sundar, S. Targeted killing of Leishmania donovani in vivo and in vitro with amphotericin B attached to functionalized carbon nanotubes. *J. Antimicrob. Chemother.* **2011**, *66*, 874–879. [CrossRef] [PubMed]

57. Yang, Z.; Tan, Y.; Chen, M.; Dian, L.; Shan, Z.; Peng, X.; Wu, C. Development of amphotericin B-loaded cubosomes through the SolEmuls technology for enhancing the oral bioavailability. *AAPS PharmSciTech* **2012**, *13*, 1483–1491. [CrossRef] [PubMed]

58. Serrano, D.R.; Lalatsa, A.; Dea-Ayuela, M.A.; Bilbao-Ramos, P.E.; Garrett, N.L.; Moger, J.; Guarro, J.; Capilla, J.; Ballesteros, M.P.; Schatzlein, A.G.; et al. Oral particle uptake and organ targeting drives the activity of amphotericin B nanoparticles. *Mol. Pharm.* **2015**, *12*, 420–431. [CrossRef] [PubMed]

59. Zarif, L.; Graybill, J.R.; Perlin, D.; Mannino, R.J. Cochleates: New lipid-based drug delivery system. *J. Liposome Res.* **2000**, *10*, 523–538. [CrossRef]

60. Delmas, G.; Park, S.; Chen, Z.W.; Tan, F.; Kashiwazaki, R.; Zarif, L.; Perlin, D.S. Efficacy of orally delivered cochleates containing amphotericin B in a murine model of aspergillosis. *Antimicrob. Agents Chemother.* **2002**, *46*, 2704–2707. [CrossRef] [PubMed]

61. Kalbag, S.; Lu, R.; Ngoje, J.; Mannino, R.J. Oral Administration of Amphotericin B: Toxicokinetic Studies in Animal Models. *Antimicrob. Agents Chemother.* **1992**, *12*, 2681–2685.

62. Matinas Biopharma. MAT2203: LNC Formulation of Amphotericin B. Available online: https://www.matinasbiopharma.com/pipeline/mat2203-lnc-formulation-of-amphotericin-b (accessed on 15 September 2018).

63. Mannino, R.; De, B.; Teae, A. Oral Administration of Amphotericin B (CAmB) in Humans: A Phase I Study of Tolerability and Pharmacokinetics Preliminary Pharmacokinetics. Available online: https://content.equisolve.net/_db6027646f523d19fe795801a0b7aff1/matinasbiopharma/db/128/510/pdf/Oral_Dosing_of_Encochleated_Amphotericin_B_%28CAmB%29__Rapid_Drug_Targeting_to_Infected_Tissues_in_Mice_with_Invasive_Candidiasis.pdf (accessed on 21 February 2019).

64. Wasan, K.M.; Wasan, E.K.; Gershkovich, P.; Zhu, X.; Tidwell, R.R.; Werbovetz, K.A.; Clement, J.G.; Thornton, S.J. Highly effective oral amphotericin B formulation against murine visceral leishmaniasis. *J. Infect. Dis.* **2009**, *200*, 357–360. [CrossRef] [PubMed]

65. Sivak, O.; Gershkovich, P.; Lin, M.; Wasan, E.K.; Zhao, J.; Owen, D.; Clement, J.G.; Wasan, K.M. Tropically stable novel oral lipid formulation of amphotericin B (iCo-010): Biodistribution and toxicity in a mouse model. *Lipids Health Dis.* **2011**, *10*, 135. [CrossRef] [PubMed]

66. Wasan, E.K.; Gershkovich, P.; Zhao, J.; Zhu, X.; Werbovetz, K.; Tidwell, R.R.; Clement, J.G.; Thornton, S.J.; Wasan, K.M. A novel tropically stable oral amphotericin B formulation (iCo-010) exhibits efficacy against visceral Leishmaniasis in a murine model. *PLoS Negl. Trop. Dis.* **2010**, *12*, e913. [CrossRef] [PubMed]

67. Ibrahim, F.; Sivak, O.; Wasan, E.K.; Bartlett, K.; Wasan, K.M. Efficacy of an oral and tropically stable lipid-based formulation of Amphotericin B (iCo-010) in an experimental mouse model of systemic candidiasis. *Lipids Health Dis.* **2013**, *12*, 158. [CrossRef] [PubMed]

68. Rae, A. iCo Therapeutics Announces Positive Oral Amphotericin Study. 2018. Available online: https://ceo.ca/@newsfile/ico-therapeutics-announces-positive-clinical-outcome (accessed on 15 September 2018).

MDPI

St. Alban-Anlage 66

4052 Basel

Switzerland

Tel. +41 61 683 77 34

Fax +41 61 302 89 18

www.mdpi.com

Pharmaceutics Editorial Office

E-mail: pharmaceutics@mdpi.com

www.mdpi.com/journal/pharmaceutics

www.ingramcontent.com/pod-product-compliance
Lightning Source LLC
Chambersburg PA
CBHW051711210326
41597CB00032B/5441